ILLUSTRATED PLANT GLOSSARY

ENID MAYFIELD

For my family.

For the families of plants and animals and the families of the people who have custody of this Earth.

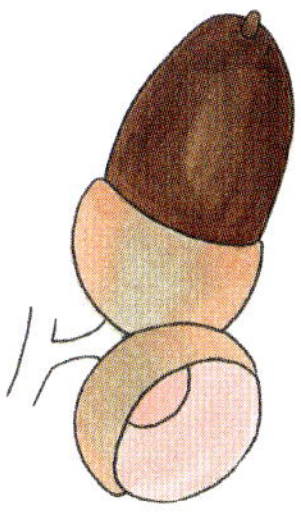

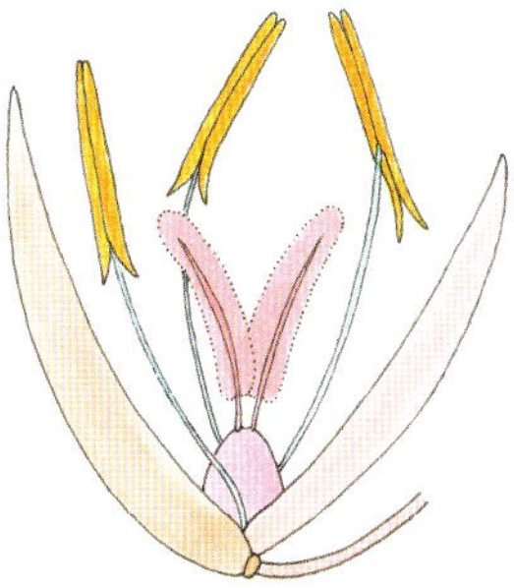

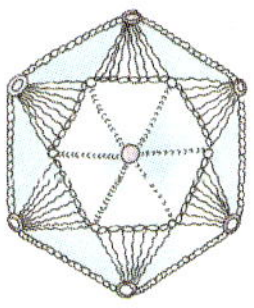

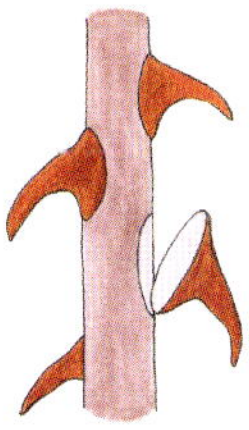

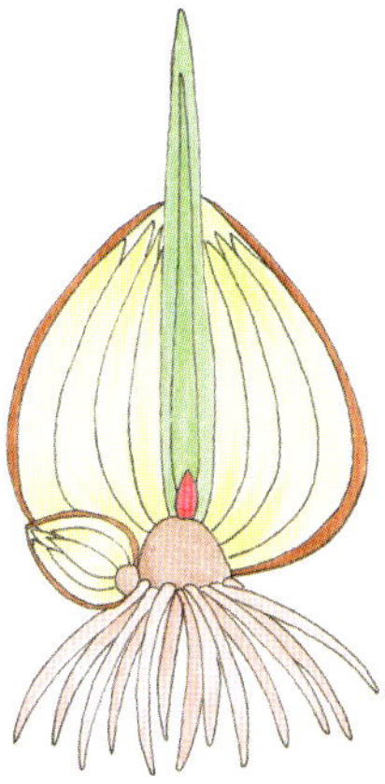

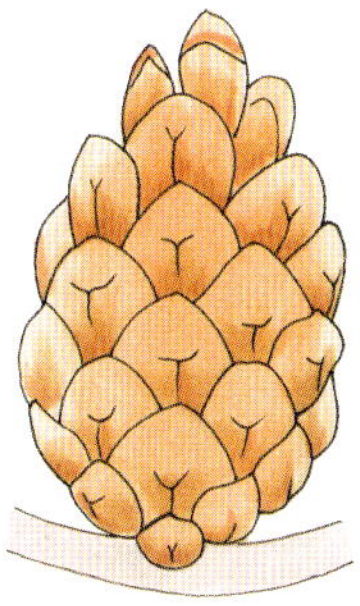

ILLUSTRATED PLANT GLOSSARY

ENID MAYFIELD

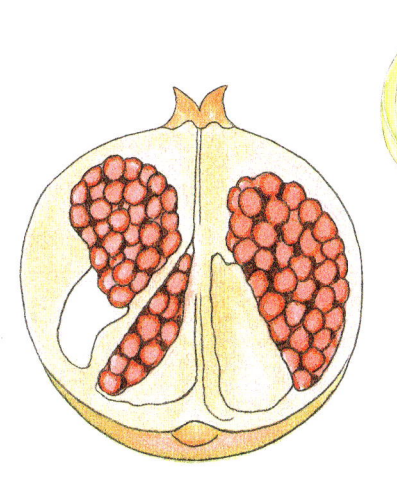

CSIRO PUBLISHING

CABI

A catalogue record for this book is available from the National Library of Australia.

ISBN: 9781486303533 (pbk)
ISBN: 9781486303540 (epdf)
ISBN: 9781486303557 (epub)

Published in print in Australia and New Zealand, and in all other formats throughout the world, by CSIRO Publishing.

CSIRO Publishing
Locked Bag 10
Clayton South VIC 3169
Australia

Telephone: +61 3 9545 8400
Email: publishing.sales@csiro.au
Website: www.publish.csiro.au
Sign up to our email alerts: publish.csiro.au/earlyalert

A catalogue record for this book is available from the British Library, London, UK.

Published in print only, throughout the world (except in Australia and New Zealand), by CABI.

ISBN 9781800620674

CABI
Nosworthy Way
Wallingford
Oxfordshire OX10 8DE
UK

CABI
We Work
One Lincoln Street, 24th Floor
Boston, MA 02111
USA

Tel: +44 (0)1491 832111
Fax: +44 (0)1491 833508
Email: info@cabi.org
Website: www.cabi.org

Tel: +1 (617)682-9015
E-mail: cabi-nao@cabi.org

All illustrations are by the author.

Cover design by Cath Pirret
Printed in China by Leo Paper Products Ltd

CSIRO Publishing publishes and distributes scientific, technical and health science books, magazines and journals from Australia to a worldwide audience and conducts these activities autonomously from the research activities of the Commonwealth Scientific and Industrial Research Organisation (CSIRO). The views expressed in this publication are those of the author(s) and do not necessarily represent those of, and should not be attributed to, the publisher or CSIRO. The copyright owner shall not be liable for technical or other errors or omissions contained herein. The reader/user accepts all risks and responsibility for losses, damages, costs and other consequences resulting directly or indirectly from using this information.

Acknowledgement
CSIRO acknowledges the Traditional Owners of the lands that we live and work on across Australia and pays its respect to Elders past and present. CSIRO recognises that Aboriginal and Torres Strait Islander peoples have made and will continue to make extraordinary contributions to all aspects of Australian life including culture, economy and science. CSIRO is committed to reconciliation and demonstrating respect for Indigenous knowledge and science. The use of Western science in this publication should not be interpreted as diminishing the knowledge of plants, animals and environment from Indigenous ecological knowledge systems.

The paper this book is printed on is in accordance with the standards of the Forest Stewardship Council® and other controlled material. The FSC® promotes environmentally responsible, socially beneficial and economically viable management of the world's forests.

MIX
Paper from
responsible sources
FSC® C020056

May21_01

CONTENTS

FOREWORD

Enid Mayfield's new plant book provides the reader with a remarkably comprehensive glossary including more than 4000 scientific and botanical terms, beautifully and accurately illustrated by her alone. This book will be invaluable to anyone with an interest in understanding plants, who may be frustrated when using other glossaries that can be incomplete, contradictory or lack 'instantly informative' illustrations (when a picture is worth a thousand words).

This particular glossary stands out in the way the material was researched and assembled under themes and sub-topics, finally arranged alphabetically with very useful cross-referencing. The themes include: anatomy, angiosperms, bryophytes, chemistry, cytology, ferns and fern allies, family specific terms, flowers, fruit, genetics, grasses, rushes and sedges, gymnosperms, habit and growth, habitat and ecology, indumentum, inflorescence, leaves, orchids, reproduction, roots, seeds and systematics. Enid researched scientific areas beyond those found in many traditional plant glossaries, including DNA-based terminology, palynology and modern systematics.

Also unique is Enid's own illustrations of terms using familiar examples – often diagrammatic but often easily identifiable plants including many Australian species. The use of drawings of plants and plant structures, rather than photographs, allows the author to emphasise the salient features to convey clearly the meaning of a botanical term. Enid's diagrammatic illustrations of the various (and at times bewildering) terms used to describe the various inflorescences of flowering plants (used in keys to identify species) are a good example of where she has achieved accurate, understandable definitions with common plant examples.

Enid's background is in education, discovering a talent for illustrating plants later in life, spending time in the National Herbarium of Victoria, and contributing artwork to the *Flora of Victoria* publications. She became interested in the botany of her local area: researching and publishing two illustrated books on the flora of the Otway Plain and Ranges when she lived in Geelong. She now resides in Queensland and is interested in plants of the Sunshine Coast and hinterland near her Noosa home.

Enid has shown quite extraordinary discipline and commitment over seven years, working closely with her scientific editor, Neville Walsh, Senior Botanist, Royal Botanic Gardens Victoria, and CSIRO Publishing. The result is this excellent and widely accessible *Illustrated Plant Glossary* that will be used by botanists, horticulturalists, ecologists, teachers, students and plant enthusiasts alike.

Professor Emeritus Pauline Ladiges AO FAA
The University of Melbourne

ACKNOWLEDGEMENTS

Neville Walsh is Senior Conservation Botanist at the Royal Botanic Gardens Victoria and editor of the four volume *Flora of Victoria*. He has an exceptional mind, a vast knowledge of plant sciences and a wonderful wit. It has been my privilege to have him as Scientific Editor for this publication.

Both Susan Howells and Jenny Stein gave many hours of invaluable work to this project in its early days.

I would also like to acknowledge the wonderful relationships I have had with staff in the Science Division of Royal Botanic Gardens Victoria during my time as an Honorary Associate.

The journey over the last seven years, however, has been largely a solitary one. Each theme was researched thoroughly, the terms defined and decisions made as to how best elucidate them with thousands of illustrations. All of this was then inserted into the text, ready to pass on to the publisher as a print-ready document. It has been a many-faceted project, always engaging and endlessly enjoyable.

If this work expands the knowledge of plant sciences and brings its many aspects into finer focus for my readers, then I am very happy.

ABOUT THE AUTHOR

Enid Mayfield's career as a teacher was varied. She worked part time as she had a young family. At secondary level her interest was innovative curriculum design and development for which she was recognised with awards. She also tutored at Deakin University in the Institute of Koorie Education.

When she discovered she could draw she studied Botanical Illustration at Burnley Horticultural College. She then illustrated for the *Flora of Victoria* at the Royal Botanic Gardens Victoria, in the Department of Plant Sciences (now Science Division). She contributed to the *Flora of Australia* project and many other publications. Her association with the Gardens was ongoing and she became an Honorary Associate there.

Her base was a studio and office space in the Geelong Botanic Gardens where she had a close relationship with the horticultural staff. For 12 years she went into the field to collect plants to illustrate for her two-volume *Flora of the Otway Plain and Ranges*.

Her skill is in being able to research complex scientific topics and write and illustrate them so that they are accessible to a wide audience.

She is married to Rob and they have two children, Anne who is married to Mark and David who is married to Rebecca. They have six grandsons, Darcy, Samuel, Julian, Jules, Raiph and Lachlan. Plants and planet are the forces that have guided her journey.

2n   The number of chromosomes in a somatic cell as opposed to a sex cell.
A diploid species, such as corn, has two sets of 10 chromosomes, so that $n = 10$ and $2n = 2x = 20$.
A hexaploid species, such as wheat, has six sets of seven chromosomes so that $n = 7$ and $2n = 6x = 42$.
see **chromosome set, ploidy**

a-, an-   A prefix meaning absent.

ab-   A prefix meaning away from.
cf. **ad-**

abaxial   The side of an organ that is facing away from the axis.
The undersurface of an erect leaf with respect to the vertical stem.
see also **resupinate**
cf. **adaxial**

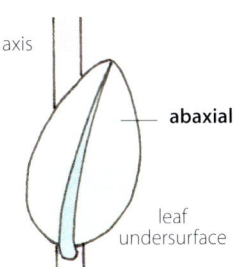

abbr.   Abbreviation.

abbreviation, *abbr.* **abbr.**   A shortened form of a word, phrase or name of an authority after a taxon name. The scientific name of the smooth blackberry is *Rubus canadensis* L., with the authority L. referring to Carl Linnaeus who named the species.

aberrant   Unusual or atypical. Different from the usual form. Abnormal, as an aberrant chromosome or aberrant plant growth.

abiotic   Non-living, including wind, water, sunlight, soil and minerals.
In an ecosystem, relating to or resulting from the non-living components.
cf. **biotic**

abortion   Failure to develop properly or to develop to completion, as the mericarps of some woodruffs (*Asperula*).
abort   To be checked in normal development so as to partially develop or completely fail to develop.

abortive   Of an arrest or failure of development. Imperfectly formed or developed.
cf. **rudimentary, vestigial**

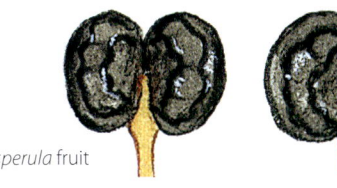
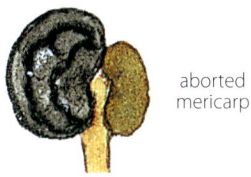

Abortion

Asperula fruit

aborted mericarp

abrupt   Appearing cut off at the base or the apex. Terminating suddenly.
= **truncate**

leaf tip **abrupt**

abruptly-pinnate
Of a pinnate leaf with leaflets arranged in pairs and terminating with a pair of leaflets.
= **even-pinnate, paripinnate**

abruptly-pinnate

abscise   Separate by abscission, fall off.

abscisic acid   A plant hormone associated with seed and bud dormancy, functioning of the stomata and responses to environmental stresses like drought. It is associated with the fall of leaves in evergreen and deciduous plants.
see **phytohormone**

abscission   The natural shedding of plant parts, typically leaves and fruit, caused by the breakdown of cells at the base of the structure. Protective scar tissue (periderm) develops at the point of abscission.

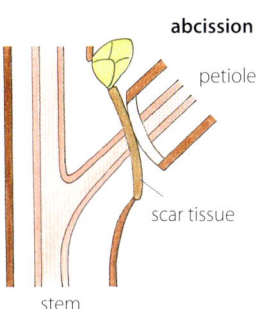

abcission

petiole

scar tissue

stem

abscission layer
A layer of cells that disintegrates to facilitate the fall of a plant part, as a leaf or fruit.
It forms in the abscission zone of some plants.
= **separation layer**

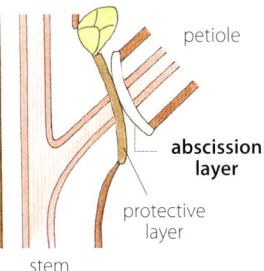

petiole

abscission layer

protective layer

stem

abscission zone

The zone, as at the base of a leaf, where shedding occurs. It includes the protection layer and the abscission layer.

= **separation zone**

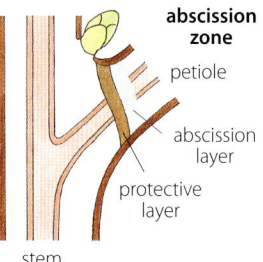

absorption   The process by which one substance takes in another substance, as roots that absorb dissolved nutrients from the soil.

acantha   A thorn, spine or prickle.
acanthaceous   Thorny, spiny or prickly.
acanthoid   Shaped like a spine, thorn or prickle.

acanthophyll

A spine derived from a leaflet, as those at the tip of the leaf in rattans and at the base of the frond of the Canary Island date palm (*Phoenix canariensis*).

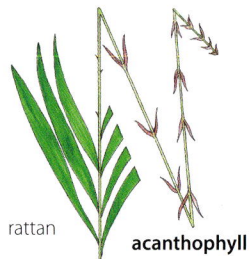

acarpic, acarpous   Producing no fruit, sterile.

acaulescent   Lacking an above-ground stem except for the inflorescence axis and bearing most leaves at ground level, as some sundews (*Drosera*).
= **acaulous**
cf. **caulescent**

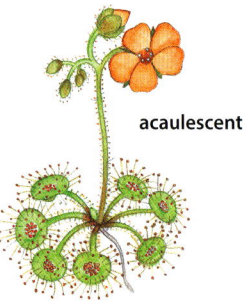

acaulous   Another name for acaulescent.

accepted   The published and correct scientific name of a plant.

accessory   Additional to the usual. Having a secondary or supplementary function to another organ.

accessory bud   An additional bud that occurs in the leaf axil beside, above or below the usually solitary bud.

accessory fruit   A fruit derived from a simple ovary or compound ovary and some additional non-ovarian tissue like the receptacle. A strawberry has the true fruits (achenes derived from the ovaries) embedded in the fleshy receptacle. Other accessory fruits include hips, pomes and pineapples.

= **false fruit, pseudocarp**

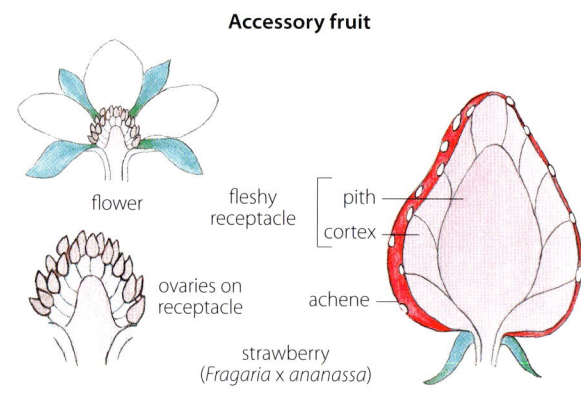

Accessory fruit

flower

fleshy receptacle

pith

cortex

ovaries on receptacle

achene

strawberry (*Fragaria* x *ananassa*)

accessory organs

Organs that assist the functioning of another organ, as the calyx and corolla of a flower that do not take part in the actual process of reproduction but assist by providing protection or attracting pollinators.
cf. **essential organs**

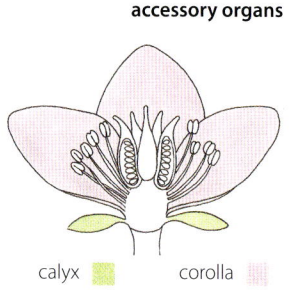

accessory organs

calyx

corolla

accrescent   Increasing in size with age and continuing to grow beyond what is normal, as the calyx of the Chinese lantern (*Physalis alkekengi*) that continues to grow after the corolla has fallen.

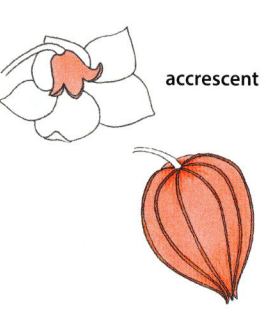

accrescent

accumbent   Of a plant part lying or folded against another part. Of cotyledons in a seed that are folded so that both lie against the hypocotyl-shoot axis, as bitter cress (*Barbarea sisymbrium*).

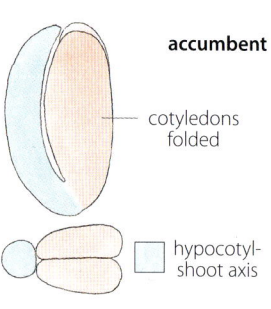

accumbent

cotyledons folded

hypocotyl-shoot axis

-aceous   A suffix denoting resemblance, belonging to or of the nature of.

aceriform   Having leaves similar to those of maple (*Acer*) trees.

acerose   Needle-shaped. Slender, stiff and pointed.
= **acicular**

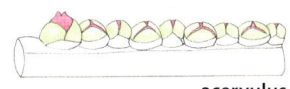

acerose

acervulus, *pl.* **acervuli**
Of palms (Arecaceae), a group of flowers borne in a line.

acervulus

acetolysis   Decomposition of an organic molecule through the action of acetic acid or acetic anhydride.
The technique is used to isolate pollen for study.

achene   A dry indehiscent fruit with one seed attached to the fruit wall (pericarp) at one point only.
Derived from a one-carpelled superior ovary, as buttercups (*Ranunculus*).
cf. **caryopsis, cypsela, diclesium**
achenoid   Of or like an achene.

Achene

buttercup
(*Ranunculus*)

pericarp
seed coat
seed
attachment

achenecetum
An aggregate fruit composed of a cluster of achenes, as the buttercup (*Ranunculus*).

achenes

achenecetum

achenocarp   A general name for a dry indehiscent fruit.

achlamydeous   Having no perianth (petals or sepals).
Naked, as the female flower of willows (*Salix*).
cf. **chlamydeous**

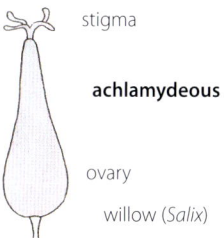

stigma

achlamydeous

ovary

willow (*Salix*)

achlorophyllous   Having no chlorophyll.

acicula, *pl.* **aciculae,**
acicle   A slender, stiff needle-like projection.
A needle-like prickle.
acicular   Needle-shaped. Slender, stiff and pointed.
= **acerose**
aciculate   Marked as if with pin pricks or needle scratches.

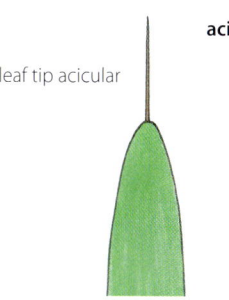

acicle

leaf tip acicular

acid   Any of a class of substances with a sour taste. An acid has the ability to turn blue litmus paper red (indicating a pH of 0 to 6), and to react with bases and certain metals to form salts.
acidic   Having a high concentration of acid.
Of a substance with a pH of 0 to 6.
see **litmus test**

acidity   Refers to the concentration of acids in a substance.

acidophile   A plant that has a preference for or grows exclusively on acidic soils that have a pH of less than 7.
acidophilous   Thriving in an acidic environment, as bogs and marshes.
cf. **basiphilous**

acidophyte   A plant growing only on acid soils with a pH of less than 5.0.
acidophytic   Growing on acid soils.

acinaceous
Having many small seeds like a grape (*Vitis vinifera*).

acinaceous

grape

acinaciform
Shaped much like a scimitar.
Of a leaf with a curved blade that broadens towards the tip.

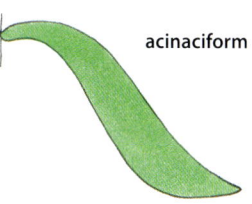

acinaciform

acinus, *pl.* **acini**
One of the small berries that make up some fruits, as the raspberry (*Rubus*). The stone or seed within the small berry. A grapeseed.

acinose Composed of acini. Resembling a bunch of grapes.

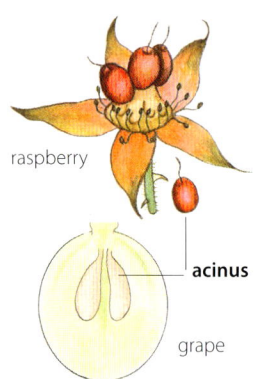
raspberry

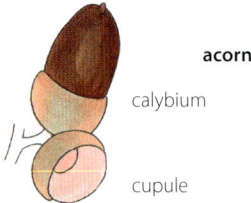
acinus

grape

acondylose, acondylous Having stems without joints or nodes.

acorn The fruit of the oak (*Quercus*) that is a nut (calybium) partly or completely enveloped at the base by a cupule.

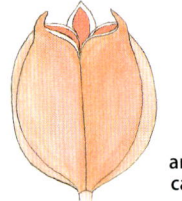
acorn
calybium
cupule

acotyledonous Having no cotyledons. Of flowering plants, embryos without cotyledons, as some orchids like cattleya (*Cattleya*) and dodder (*Cuscuta*).
cf. **cotyledonous**

acquired character A feature of an organism that occurs as a result of the environment and that is not transferred from one generation to the next.
cf. **inherited character**

acrocidal capsule
A capsule that splits open through slits at the tip, as bladdernut (*Staphylea*).

arocidal capsule

acrodromous Of leaves with two or more main veins running in arches towards the apex.

Acrodromous

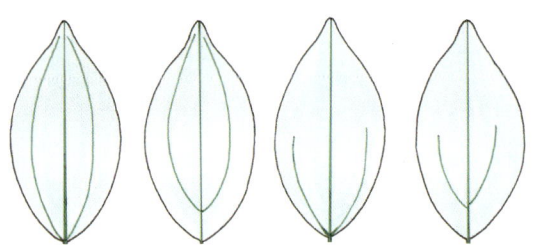
basal-perfect   suprabasal-perfect   basal-imperfect   suprabasal-imperfect

acropetal Developing in sequence from the base to the apex, as the racemose inflorescence of Persian fritillary (*Fritillaria persica*). Flowers at the bottom open first and buds near the top opening last.
= **ascending infloresence**
see also **centripetal, centrifugal**
cf. **basipetal**

acropetal

acrophyll
One of the mature fronds on the upper part of a climbing fern that is usually different from the basal fronds, as the climbing fern *Teratophyllum clemensiae*.
cf. **bathyphyll**

acrophyll

bathyphyll

acroscopic
Facing or directed towards the apex, as the pinnules of some ferns.
cf. **basiscopic**

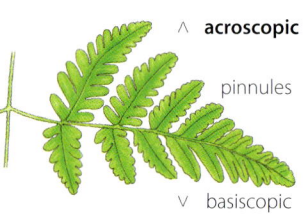
acroscopic
pinnules
basiscopic

acrospire In grasses (Poaceae), the coleoptile.

acrostichoid With sporangia covering the lower surface of a fern frond, usually densely so, as the genus *Elaphoglossum*.

acrostichoid

sporangia   fern frond

acrotonic
Having growth strongest in the upper part of the plant.
cf. **basitonic, mesotonic**

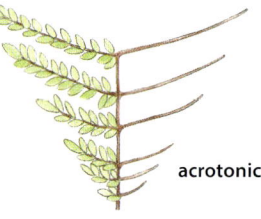
acrotonic

actino- A prefix meaning with radiating parts.

actinodromous
Of leaves with three
or more primary veins
diverging radially from
a single point.
cf. **palinactinodromous**

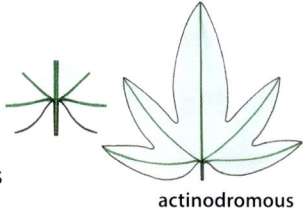

actinodromous

actinomorphic
Of flowers with radial
symmetry that divide
through the centre into
two or more like halves.
Regular.
see **polysymmetric**
cf. **zygomorphic**

actinomorphic

activator   A chemical substance that promotes a
growth process, as an auxin that caused the apex of
a stem to grow.
cf. **inhibitor**

aculeate   Having stiff
sharp prickles. Prickly, as
the margins of American
holly (*Ilex opaca*).
cf. **aculeolate**
aculeiform   Prickle-
shaped.

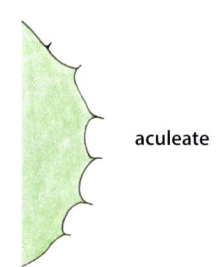

aculeate

aculeolate   Minutely prickly.
cf. **aculeate**

aculeus, aculei
A stiff, sharp-pointed
outgrowth, as a spine or
a prickle.

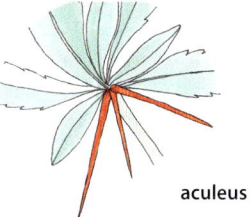

aculeus

acumen   A somewhat
elongated terminal point.
acuminate   Gradually
tapering to a sharp point
and forming concave
sides along the point.

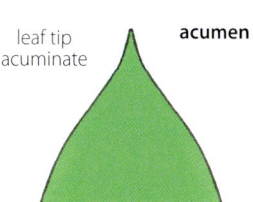

leaf tip
acuminate

acumen

acute   Tapering to a
pointed tip or base, with
the margins having an
angle of less than 90°.
cf. **obtuse**

leaf tip **acute**

acyclic   Arranged in
spirals and not in whorls,
as leaves on a stem.

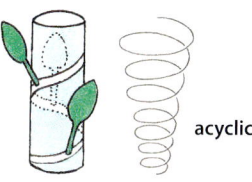

acyclic

ad-   A prefix meaning toward.
cf. **ab-**

adaptation   The process by which a plant's
genetic make-up is modified in order for it to better
fit its environment.

adaxial   The side of an
organ facing towards the
axis.
The upper surface of an
erect leaf with respect to
the vertical stem.
see also **resupinate**
cf. **abaxial**

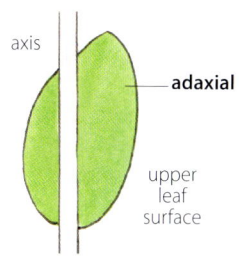

axis

adaxial

upper
leaf
surface

-adelphous
A suffix indicating the
number of fused stamens
in a flower. It is preceded
by an indication of the
number, as mona-, dia-,
tria- and poly-.

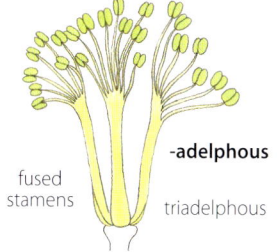

fused
stamens

-adelphous

triadelphous

adenosine triphosphate, ATP   A molecule,
found in all living things, that stores chemical
energy and releases it to fuel processes in a cell.
It is made up of adenine, the sugar ribose and three
phosphate groups.
see **dephosphorylation, mitochondrion**

adherent
Of unlike parts joined,
but only superficially,
and easily separated,
as the dorsal sepal
and lateral petals that
adhere to form a hood
in greenhood orchids
(*Pterostylis*).
cf. **coherent**

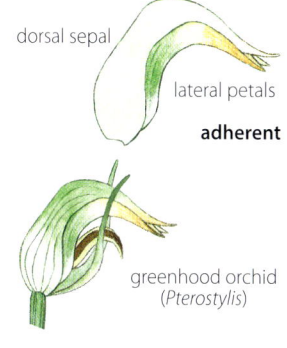

dorsal sepal

lateral petals

adherent

greenhood orchid
(*Pterostylis*)

adjacent   Of parts that are next to each other but
not overlapping or touching.
cf. **fused**

adjacent-ligular germination   Of palms, one of three types of germination, as Marquesas palm (*Pelagodoxa henryana*).

The cotyledonary petiole forms a button of tissue beside the seed and elongates very little. The radicle and plumular leaves emerge from the ligular sheath and develop next to the seed.

see also **remote-ligular germination, remote-tubular germination**

Adjacent-ligular germination

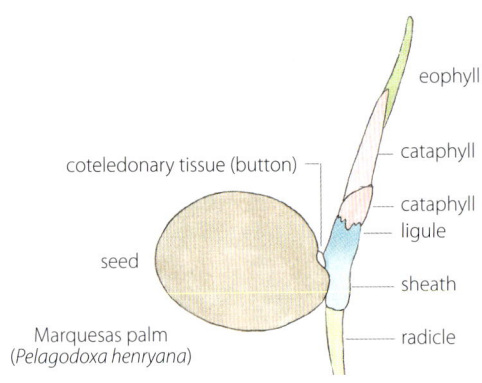

coteledonary tissue (button)

eophyll

cataphyll

cataphyll
ligule

seed

sheath

radicle

Marquesas palm
(*Pelagodoxa henryana*)

admedial, admedian
Of venation, running toward the midline of the leaf lamina.
cf. **exmedial**

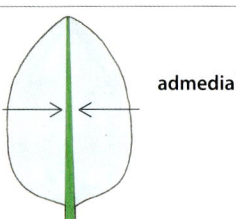

admedial

admissable   In nomenclature, a name allowed under the rules of the International Code of Nomenclature.

admixture   A collection containing plant material other than that intended by the collector.

adnate   Of unlike parts fused to one another, as the stamens fused to the corolla tube of common heath (*Epacris impressa*) or stipules fused to a petiole.
cf. **connate**

Adnate

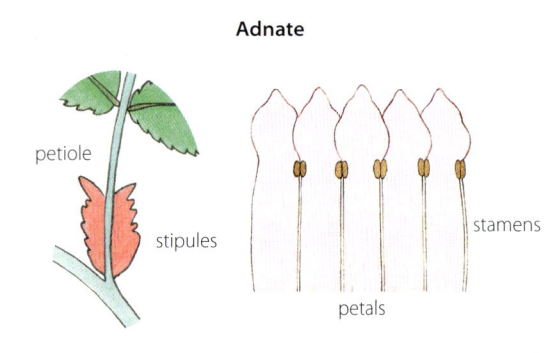

petiole

stipules

stamens

petals

adpressed   Pressed closely against but not fused, as leaves or hairs against a stem. Pressed at an angle of at most 15° from the vertical.
= **appressed**

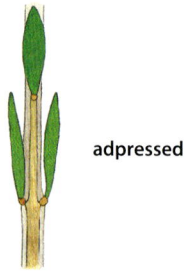

adpressed

adult   Fully grown.
The mature phases of a plant's life or plant part.
cf. **juvenile**

advanced   In phylogenetics, exhibiting a notable, but not necessarily superior, change from the original condition.
Apomorphy or derived are preferred terms.

adventitious   Of a plant part growing in an unusual location, as roots on a stem that originate from stem tissue.

adventitious embryony   Of flowering plants (angiosperms), the production of an embryo directly from cells of the nucellus or integuments in the ovule.
A form of agamospermy.
= **sporophytic apomixis**

adventitious roots   Of monocotyledons and some eudicots, roots that do not derive from the seed root, as a taproot. Adventitious roots grow in unusual places, as those on the stems of ivy (*Hedera helix*), on underground stems (rhizomes) and on the branches of strangler figs (*Ficus*)
see also **fibrous roots**

Adventitious roots

ivy stem

grass rhizome

strangler fig

adventive   Of a species new to a region that is not yet fully naturalised but is beginning to spread.

aerenchyma   A form of parenchyma in which the cells enclose large intercellular spaces that support gas exchange and buoyancy.
Found in the stems, roots and/or leaves of some aquatic plants.

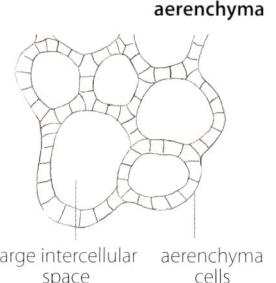

aerenchyma

large intercellular space   aerenchyma cells

aerial   Growing or borne above the ground or water, as the aerial roots of strangler figs and epiphytic orchids and the pneumatophores of mangroves.

Aerial

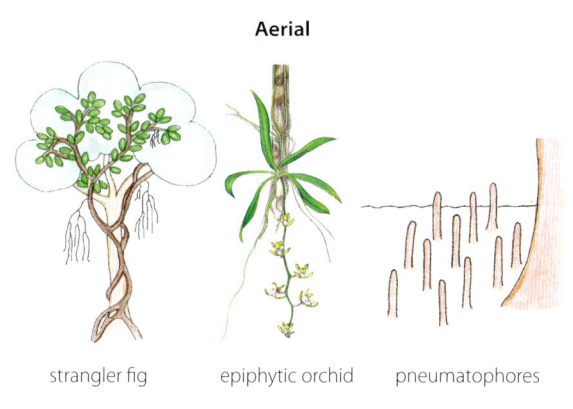

strangler fig   epiphytic orchid   pneumatophores

aero-   A prefix meaning of or related to the air.

aerobic respiration   The release of energy from organic compounds that takes place in the presence of oxygen.
cf. **anaerobic respiration**

aerophore   Of most ferns, a small outgrowth depression or line bearing stomata.

Aerophore

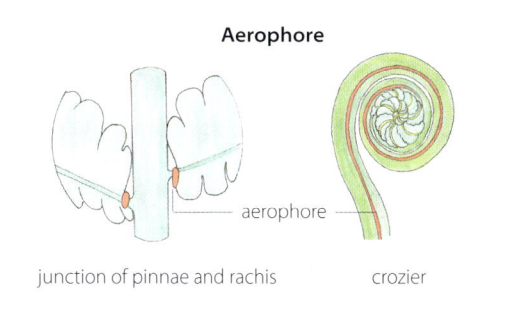

junction of pinnae and rachis   crozier

aerophore

aerophyte   A plant that grows on another plant for support but not for nutrients, as an orchid or staghorn on a tree.
= **air plant, epiphyte**
cf. **parasite**

aestival   Of or produced in summer.
cf. **autumnal, hibernal, vernal**

aestivation   Of flowers, the way in which petals and sepals are arranged relative to one another in the bud before it opens. Sepals and petals in the same flower can have different aestivation.
Dormancy in summer, as some grasses like tall fescue (*Festuca arundinacea*).
cf. **vernation**

aestivate   To pass the summer in a dormant state.

Some types of floral aestivation

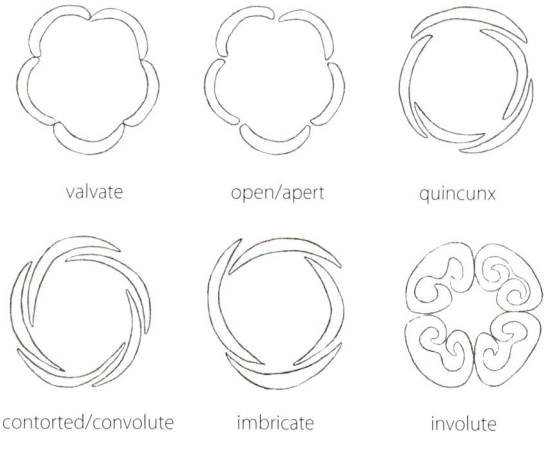

valvate   open/apert   quincunx

contorted/convolute   imbricate   involute

aff.   An abbreviation for affinis.

affinis, *abbr.* **aff.**   Having an affinity with but not identical to.
Of an undescribed species that is similar to or related to a known species, as the undescribed species *Geranium retrorsum* sp. aff. *retrorsum* and the described species *Geranium retrorsum*.

afoliate   Leafless.

agamospecies   A species that produces embryos and seeds in the absence of fertilisation.
A species in which sexual reproduction does not occur.

agamospermy   Any form of reproduction that involves cells in the ovule but takes place without fertilisation or meiosis.
The formation of embryos and seeds in the absence of fertilisation.
Agamospermy includes parthenogenesis, diplospory, apospory, adventitious embryony, apogamy and gametophytic apomixis.
see **apomixis**

agamospermous Exhibiting agamospermy.

ageotropic Of or relating to ageotropism. Of plant parts that would be expected to grow downwards but instead grow upwards against the influence of gravity, as the peg roots of mangroves.

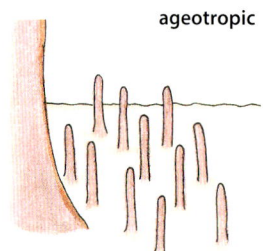

ageotropic

ageotropic peg roots

ageotropism Not reacting to gravity and instead turning upwards away from the earth. Negative geotropism.
see **gravitropism**
cf. **apogeotropism**

agglomerated Crowded in a dense cluster but not cohering, as pincushion flowers (*Brunonia*).
= **aggregated**

agglomerated

agglutinated Glued together, as pollen in the pollinia of some orchids (Orchidaceae).

aggregate fruit Fruit formed from the unfused carpels of a single flower, with the carpels becoming fruitlets.
A cluster of of berries is a baccacetum, of follicles a follicetum, of drupes a drupecetum, of achenes an achenecetum and of samaras a samaracetum.
= **etaerio**
see **compound fruit**
see also **apocarp**

Aggregate fruit

single flower

Raspberry (*Rubus*)

carpels

fruitlets

baccacetum
(*Phytolacca*)

samaracetum
(*Liriodendron*)

follicetum
(*Illicium*)

drupecetum
(*Rubus*)

achenecetum
(*Ranunculus*)

aggregated Crowded in a dense cluster but not cohering, as pincushion flowers (*Brunonia*).
= **agglomerated**

aggregated

air plant A plant that grows on another plant for support but not for nutrients, as an orchid or staghorn on a tree.
= **aerophyte, epiphyte**
cf. **parasite**

air spaces Intercellular spaces, between the cells of the spongy mesophyll in a leaf, that are saturated with water vapour. They allow the diffusion of carbon dioxide out of the leaf and oxygen into the leaf.
see also **stoma**

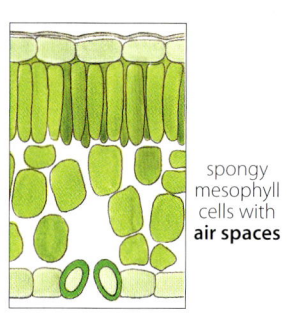

spongy mesophyll cells with **air spaces**

ala, *pl.* **alae** A wing-like structure, as the wing-like extension on the seeds of elms (*Ulmus*).
One of two clawed lateral petals of a pea flower.
alate Winged or having wing-like appendages.

Ala

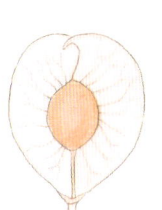

winged seed of elm (*Ulmus*)

lateral petal of pea flower

albumin Of plants, one of the proteins in the endosperm of a seed, as wheat (*Triticum*).
It is broken down during germination to provide nitrogen and sulfur for the developing seedling.
albuminous Of, like or containing albumin.

albuminous cell   A parenchyma cell associated with a sieve cell in the phloem of gymnosperms and lower vascular plants.

It moves sugar into and out of sieve cells and is considered to be the counterpart of a companion cell in angiosperms.

Albuminous cell

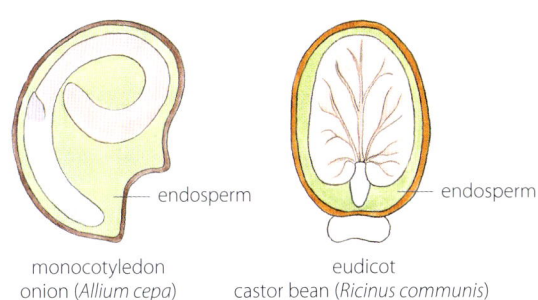

albuminous cells

sieve area

elongated sieve cell

albuminous seed   One having endosperm persisting in the mature seed as the main source of nourishment for the embryo, as most monocotyledons.

Some eudicots, as castor bean (*Ricinus communis*), have endosperm and are albuminous.

All gymnosperm seeds are albuminous.

= **endospermic seed**
cf. **exalbuminous seed**

Albuminous seed

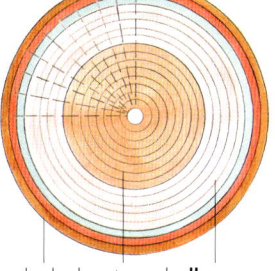

endosperm

endosperm

monocotyledon
onion (*Allium cepa*)

eudicot
castor bean (*Ricinus communis*)

alburnum
The younger layers of wood between the heartwood and the bark. It contains the functioning vascular tissue in which the sap flows.

= **sapwood**

bark   heartwood   **alburnum**

-ales   A suffix forming the names of orders of plants.

alete   Of a spore without laesurae.
cf. **inaperturate, monolete, trilete**

aleurone   A granular protein in the outermost layer of endosperm in grasses (Poaceae) and some eudicots.

aleurone layer
The outermost layer of endosperm in grasses (Poaceae) and some eudicots, as the legume fenugreek (*Trigonella foenum-graecum*).
It functions to release enzymes that digest starch in the endosperm for embryo and seedling growth.

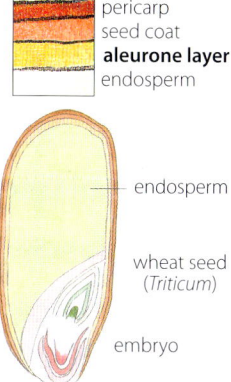

pericarp
seed coat
aleurone layer
endosperm

endosperm

wheat seed
(*Triticum*)

embryo

alga, *pl.* **algae**   Simple nonvascular plant-like organisms that lack roots, stems and leaves but have chlorophyll for photosynthesis.
They occur in fresh or salt water and range in size from microscopic and single-celled diatoms to macroscopic and multi-celled, as the seaweed kelp.
They are members of the taxonomic kingdom Protista (protists) and are a diverse collection of organisms containing many distinct groups that are not related as they do not share a common ancestor.

algal   Relating to or like algae.

Algae

brown algae
kelp
(*Macrocystis*)

red algae
laver
(*Porphyra*)

green algae
spaghetti algae
(*Chaetomorpha*)

Diatoms

algal bloom   The excessive multiplication of algae in either a saltwater or freshwater system.
It is harmful when it produces toxins or reduces oxygen in the water.
see **eutrophic**

alien   Of plants in a region where they are do not occur naturally.
Non-native species introduced from another place, often another country.
= **exotic**

aliferous   Having wings.
aliform   Wing-shaped.

alkali   An hydroxide of one of the alkali metals, as lithium, potassium, rubidium, caesium and sodium, that turns litmus paper from red to blue, indicating a pH of 8 to 14.
alkaline   Of a substance with a pH of 8 to 14.
see **litmus test**

alkalinity   Refers to the concentration of alkalis in a substance.

alkaloids   Organic compounds, containing nitrogen, that are often poisonous, as nicotine that is found in the nightshade family (Solanaceae).

allautogamy   A plant having both cross-pollinating and self-pollinating flowers.
allautogamous   Of or relating to allautogamy.

allele   Any one of the alternative forms of a particular gene that can be located at the same position (locus) on a chromosome.
see **heterozygous, homozygous**

alliaceous   Smelling or tasting like garlic.

allogamy   Pollination between flowers of the same species, especially between flowers on different plants.
see **cross-pollination, geitonogamy, outcrossing, xenogamy**
cf. **autogamy**

allopatric   Of distribution, occurring in separate non-overlapping geographical areas.
cf. **parapatric, sympatric**

allopolyploid, allopolyploidy   Having multiple chromosome sets from different species.
see **polyploidy**
cf. **autopolyploidy**

allotetraploid   A new species formed by having a diploid set of chromosomes derived from parents of two separate species.
= **amphidiploid**
see **ploidy**

alpine   Of high mountain areas above the natural treeline.

alternate   Of plant parts, as leaves and flowers, arranged in two rows and not opposite.
Of plants parts, as petals and sepals, occurring in succession one after another.
Of phyllotaxy, leaves or shoots occurring singly at a node and on opposite sides of a stem. The simplest form of a spiral arrangement.
cf. **distichous, opposite, superposed, whorled**

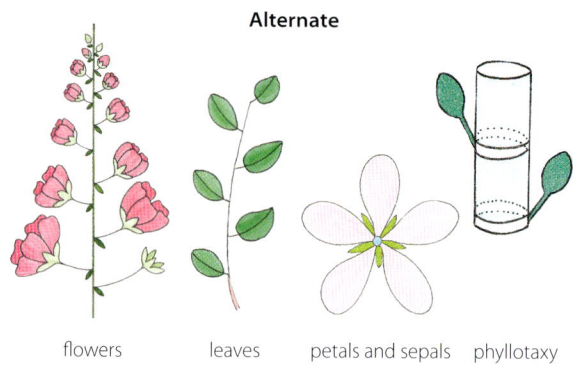

Alternate

flowers   leaves   petals and sepals   phyllotaxy

alternate vernation
Of young leaves in the unopened leaf bud arranged in alternating whorls of three.

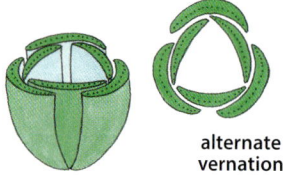

alternate vernation

alternate-bipinnate
Of a bipinnate leaf with the primary divisions arranged alternately on opposite sides of the rachis.
see also **pinnate**

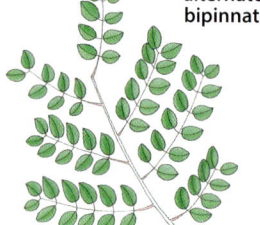

alternate-bipinnate

alternate-pinnate
Of a pinnate leaf with leaflets arranged singly on opposite sides of the rachis.

alternate-pinnate

alternation   One following another in turns.
see **alternation of generations**
Having parts between organs, as stamens borne between the petals of a flower.
cf. **superposition**

alternation of generations   In the life cycle of a plant, the alternation of a haploid gametophyte phase that reproduces sexually, with a diploid sporophyte phase that reproduces vegetatively.
see **angiosperms, ferns, gymnosperms, hornworts, liverworts, mosses**

alveola, *pl.* **alveolae, alveole, alveolus,** *pl.* **alveoli**
A small thin-walled angled cavity or pit on a surface. Often arranged in a honeycomb pattern.
alveolate   Pitted, as the receptacle of a daisy flower when the florets are removed.
cf. **foveate, foveolate**

alveola

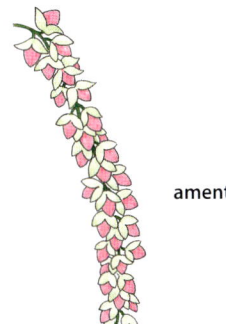
receptacle alveolate

AM   Arbuscular mycorrhiza.

amber   The fossilised resin of some species of extinct trees.
cf. **copal**

amensalism   An association between organisms of two different species in which one is inhibited or destroyed and the other is unaffected, as a shade tree that inhibits the growth of a plant beneath it.
see **symbiosis**
cf. **parasitism**

ament, amentum
A spike or raceme of unisexual, usually apetalous flowers on a pendulous flower stem, as the flowers of the walnut (*Juglans regia*).
A racemose inflorescence.
= **catkin**
amentaceous   Bearing, resembling or consisting of an ament or aments.
amentiferous   Bearing aments or catkins.

ament

amino acid   One of the small units that are the building blocks of proteins. They are variously attached to each other by peptides bonds, forming chains that make up proteins.

amorphic, amorphous   Having no defined shape.
Lacking halves that are mirror images on any plane.
see **asymmetric**
cf. **symmetric**

amphibious   Of plants living both in water and on land, as those in seasonally inundated habitats.
cf. **aquatic, terrestrial**

amphicarpic, aphicarpous   Of a plant with two kinds of inflorescence that produce two kinds of fruit, as the milkwort (*Polygala polygama*).
Aerial fruit ripens above ground and a self-fertilised subterranean fruit ripens below ground.
cf. **amphigeal, heterocarpous**

amphicarpic

amphicribal   Of a concentric vascular bundle with phloem encircling the xylem, as spike moss (*Selaginella*).
cf. **amphivasal**

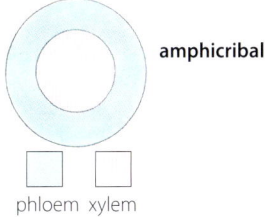
amphicribal
phloem xylem

amphidiploid   A new species formed by having a diploid set of chromosomes derived from parents of two separate species.
= **allotetraploid**
see **ploidy**

amphigeal   Of a plant with two kinds of flowers, the upper from the stem and the lower from the root or rootstock, as the milkwort (*Polygala polygama*).
cf. **amphicarpic**

amphigeal

amphimixis   Reproduction resulting from the union of a male and female gamete. Sexual reproduction.
cf. **apomixis**

amphipodium
A rhizome system that has both monopodial and sympodial branching.
amphipodial   Relating to an amphipodium.
cf. **leptomorph, pachymorph**

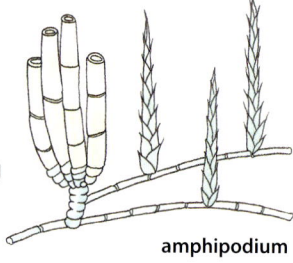
amphipodium

amphisarca   A pulpy fruit with a firm hard or woody rind, as Indian bael (*Aegle marmelos*). Derived from a many-carpelled septate superior ovary.
cf. **pepo**

amphisarca

amphistomatous   Of leaves with stomata on both the upper and lower surfaces.
cf. **epistomatous, hyperstomatous**

amphitony
Development of lateral growth on the upper and undersides of the main shoot.
cf. **epitony, hypotony**

amphitony

amphitropous
Of ovule orientation, with the ovule curved and at a right angle to the funicle. The micropyle and chalaza are near each other.
see **ovule orientation**

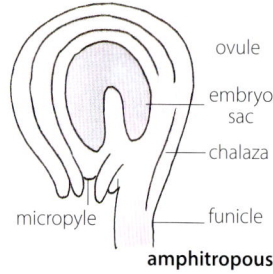
ovule
embryo sac
chalaza
micropyle
funicle
amphitropous

amphivasal
Of a concentric vascular bundle with xylem encircling the phloem, as begonia (*Begonia*).
cf. **amphicribal**

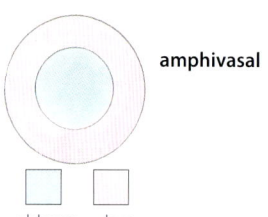
amphivasal
phloem xylem

amplectant
Clasping a support, as the twining petioles of mountain clematis (*Clematis aristata*).

amplectant

amplexicaul
Clasping the stem but not completely encircling it, as the base of some leaves.

amplexicaul

ampliate
Enlarged or swollen, as the corolla tube of eyebright (*Euphrasia*).

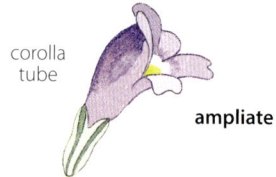
corolla tube
ampliate

ampulla, *pl.* **ampullae**
An organ shaped like a squat, rounded flask or bladder.
A small membranous float attached to the roots or leaves of some aquatic plants, as the waterwheel plant (*Aldrovanda vesiculosa*), containing a watery fluid and a small bubble of air.
ampullaceous
Inflated and swelling out towards the base like a short flask or bladder.
ampulliform   Shaped like an ampulla.

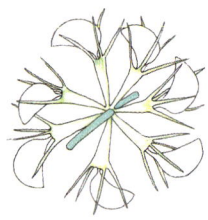
waterwheel plant
(*Aldrovanda vesiculosa*)

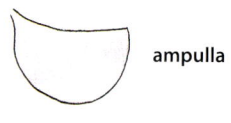
ampulla

amylaceous   Of, relating to or resembling starch.
see **amylum**

amylase   A group of enzymes that catalyse the hydrolysis of starch into sugars in plants, animals and microbes. Alpha amylase and beta amylase are present in plants.

amylolysis   The conversion of stored starch to sugars by the action of enzymes or acids.
see **amylum**

amyloplast   A leucoplast that converts glucose into starch and stores it as starch grains.

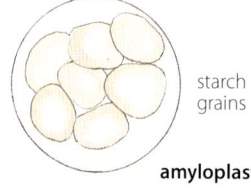
starch grains
amyloplast

amylum   Starch, a complex carbohydrate produced by most green plants as a form of energy storage. Found in seeds, fruits, tubers, roots etc.

anabolism   The use of energy to build larger complex molecules from simple molecules, as simple sugar molecules (monosaccharides) that are joined together to make dissaccharides and polysaccharides.
cf. **catabolism**

anadromous, anadromic   Of fern venation, having the the first lateral veins in a segment arising apically from the midrib.
cf. **catadromous, isodromous**

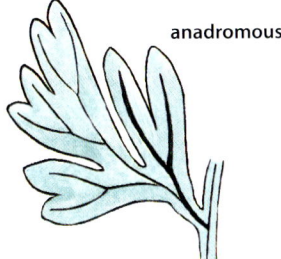
anadromous

anaerobic respiration   The release of energy from organic compounds that takes place in the presence of little or no oxygen.
cf. **aerobic respiration**

anagenesis   One of two main ways in which speciation occurs in response to the environment. Anagenesis is linear, with a slow accumulation of change. At a certain point, species A becomes species B, and species A becomes extinct.
cf. **cladogenesis**
anagenetic   Of or relating to anagenesis.

Anagenesis

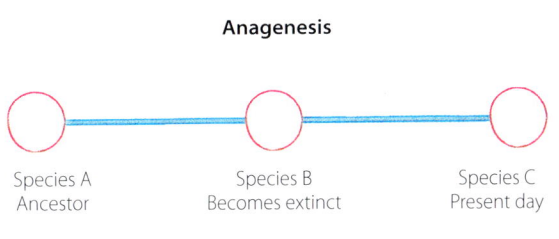

Species A
Ancestor

Species B
Becomes extinct

Species C
Present day

analogous   In phylogenetics, of similar characters that have evolved from different ancestral sources, as the wings of birds and bats.
see **homoplasy**
cf. **homologous**

anandrous   Having no stamens, as female flowers.

anantherous   Of stamen filaments that lack an anther.

anastomosis, *pl.* anastomoses
Of veins in a leaf connecting together to form a network of loops.
= **closed venation**

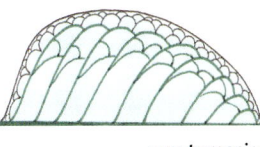
anastomosis

anatomy   The study of the internal structure of plant tissues and organs at a microscopic level.
= **phytotomy**
cf. **morphology**
anatomical   Of or relating to anatomy.

anatropous   Of ovule orientation, with the ovule turned at 180° so that it lies along its funicle and in line with the chalaza at the top and the micropyle, facing downwards, at the base. The ovule is fused to the funicle.
see **ovule orientation**

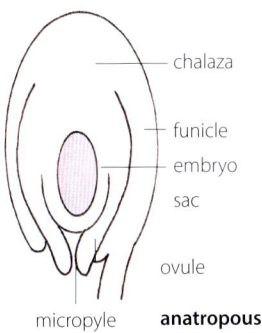
chalaza
funicle
embryo
sac
ovule
micropyle   **anatropous**

anauxotelic   Of an inflorescence axis having growth stop with an aborted vegetative bud, as the female flower of scrub sheoak (*Allocasuarina paludosa*).
cf. **auxotelic, monotelic**

ancestor   An actual or hypothetical entity from which another entity is descended.
see **phylogenetic taxonomy**
ancestral   Primitive. Of a characteristic belonging to or inherited from an ancestor.
cf. **derived character**

ancipital, ancipitous   Flattened and two-edged instead of round, as the stems of certain sedges.

androclinum   Another term for clinandrium.

androdioecious   Of a species with staminate (male) and bisexual flowers on different plants.
see also **andromonoecious**
cf. **gynodioecous**

androecious   Of a plant having only male flowers.
see also **staminate**
cf. **gynoecious**

androecium
The male reproductive organ of a flower.
The stamens collectively.
cf. **gynoecium**

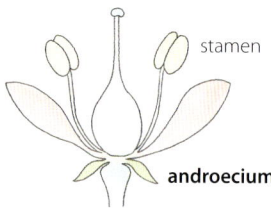
stamen
androecium

androgynomonoecious Of a species with staminate (male) flowers, pistillate (female) flowers and bisexual flowers on the same plant.
see also **andromonoecious, gynomonoecious**
cf. **trioecious**

androgynophore
An elongated stalk (stipe), inserted on the receptacle, that bears the stamens and pistil of a flower above the corolla and calyx.
= **gonophore, gynandrophore**

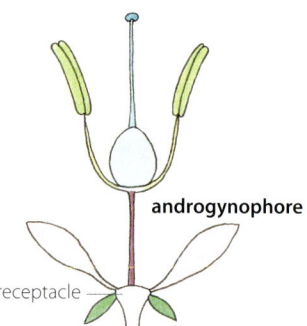
androgynophore
receptacle

androgynous Of an inflorescence with both male and female flowers.
see **perfect**
see also **bisexual**

andromonoecious Of a species with staminate (male) and bisexual flowers on the same plant.
see also **androdioecious**
cf. **gynomonoecious**

androphore A tube, formed by the united filaments of stamens. It bears the anthers above the petals and sepals and is inserted on the receptacle.
Typical of the mallow family (Malvaceae).

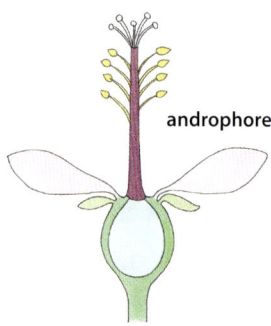
androphore

andropolygamous Of a species with staminate and bisexual flowers either on one plant or on a different plants within the same species.
see also **andromonoecious**
cf. **gynodioecous**

anemochore A plant whose seeds, spores or fruits are dispersed by wind.

anemochory Dispersal of seeds, spores and fruit by wind.
anemochorous Of or relating to anemochory.

anemogamy Dispersal of pollen and pollination by wind.
= **anemophily**
cf. **anemochory**
anemogamous Of or relating to anemogamy.

anemophile A plant that is pollinated by wind.

anemophily Dispersal of pollen and pollination by wind.
= **anemogamy**
cf. **anemochory**
anemophilous Pollinated by wind-blown pollen.

aneuploidy Having an abnormal number of chromosomes, as a cell that does not contain an exact multiple of the haploid number of chromosomes.
Having too many or too few chromosomes after cell division.

anfractuose
With twists and turns.
Convoluted, as the anthers of the gourd *Lagenaria*.

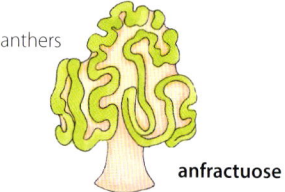
anthers
anfractuose

angiocarp
A fruit that is partially or entirely enclosed in a shell, involucre or husk, as an acorn in its cupule.
angiocarpous Of or relating to an angiocarp.

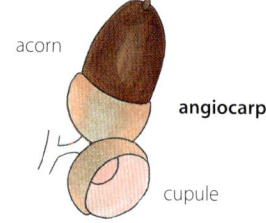
acorn
angiocarp
cupule

angiosperms *see* pages 16–17

angular, angulate
An outline having sharp angles.
Sharply pointed instead of curved, as a ridged stem.
A leaf having the sinuses and lobes with margins meeting at a point.

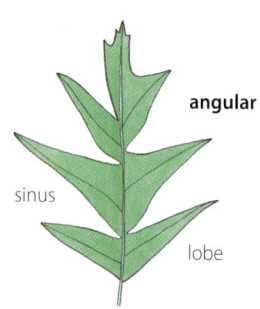
angular
sinus
lobe

angular collenchyma Collenchyma cells with thickenings located in the corners of the cells and no intercellular spaces.

angusti- A prefix meaning narrow.

angustiseptate
Of a fruit flattened with a narrow internal partition (septum), as the silicula of hairy shepherd's purse (*Microlepidium pilosulum*).

angustiseptate

septum

aniso- A prefix meaning unlike or not equal.

anisocotyly Of eudicots, having one cotyledon visibly larger than the other.
see also **macrocotyledon, microcotyledon**
cf. **isocotyly**

anisomerous Having whorls with a different number of parts, as a flower with a whorl of sepals, petals, carpels and/or stamens differing in number from the rest.
= **heteromerous**
cf. **isomerous**

anisophylly Having paired leaves that differ in shape and size.
cf. **isophyllous**
anisophyllous
Relating to anisophylly.

anisophylly

anisospory Having spores of two kinds produced in the same sporangium, smaller spores that produce male gametophytes and larger spores that produce female gametophytes. Occurs in bryophytes but is unusual.
cf. **heterospory, homospory**
anisosporous Producing both male and female spores in the same sporangium.

anisostemonous
With stamens different in number to the petals.
cf. **isostemonous**

anisostemonous

annual A plant living, reproducing and dying in one growing season.
cf. **biennial, perennial**

annular, annulate Ring-shaped. Having an annulus.

annular collenchyma Collenchyma having intercellular spaces and the cell wall with uniform thickening that is ring-like in cross-section.

annulus, *pl.* **annuli**
Of pollen, a ring-like thickening of the pollen wall surounding a pore or ulcus.
cf. **margo**
Of ferns, the thick-walled hygroscopic cells in the wall of the sporangium.

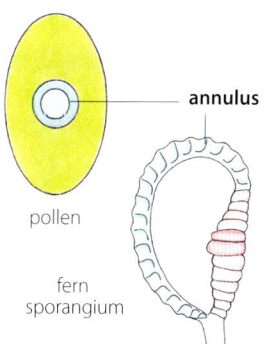
annulus
pollen
fern sporangium

anomalous secondary growth
The production of cambium in a monocotyledon. A condition that is not commonly seen and is present in only a limited number of families or genera, as yucca (*Yucca*).

anomaly A marked deviation from the normal standard.
anomalous Inconsistent with what is usual, normal or expected.
see **anomalous secondary growth**
cf. **atypical**

ante- A prefix meaning before.

antepetalous, antipetalous
Directly in front of the petals, as stamens that are opposite rather than alternate with the petals.

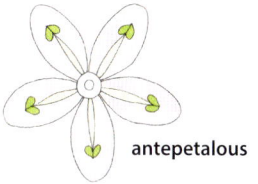
antepetalous

anteposition
The placement of one part above another on the same radius, as stamens borne above the petals of a flower.
anteposed Situated vertically on or above another part.
= **superposition**
cf. **alternation**

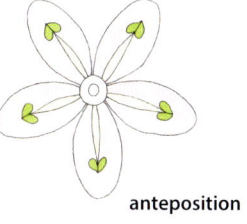
anteposition

angiosperms   Flowering, seed-bearing plants. They include herbaceous plants, shrubs, grasses and trees.
  Currently angiosperms are divided into several groups, the larger groups being Eudicots (75%), Monocotyledons (22%), Amborellales, Nymphaeales, Austrobaileyales, Chloranthales, Ceratophyllales and Magnoliids.
  The reproductive structure is the flower and comprises an ovary with ovules and anthers with pollen.
  Seeds develop enclosed in an ovary that becomes a fruit.
  Characteristics include double fertilisation and triploid nutritive material (endosperm) in the seed.
  Eudicot seeds have usually two seed leaves (cotyledons) and monocotyledons have one.
  Angiosperms have a life cycle alternating between a haploid sexual gametophyte generation and a diploid asexual sporophyte generation.
  The sporophyte generation is the larger familiar green plant.
  The gametophyte generation is microscopic and lives on the sporophyte.
  see also **seed**
  cf. **gymnosperm**

Alternation of Generations

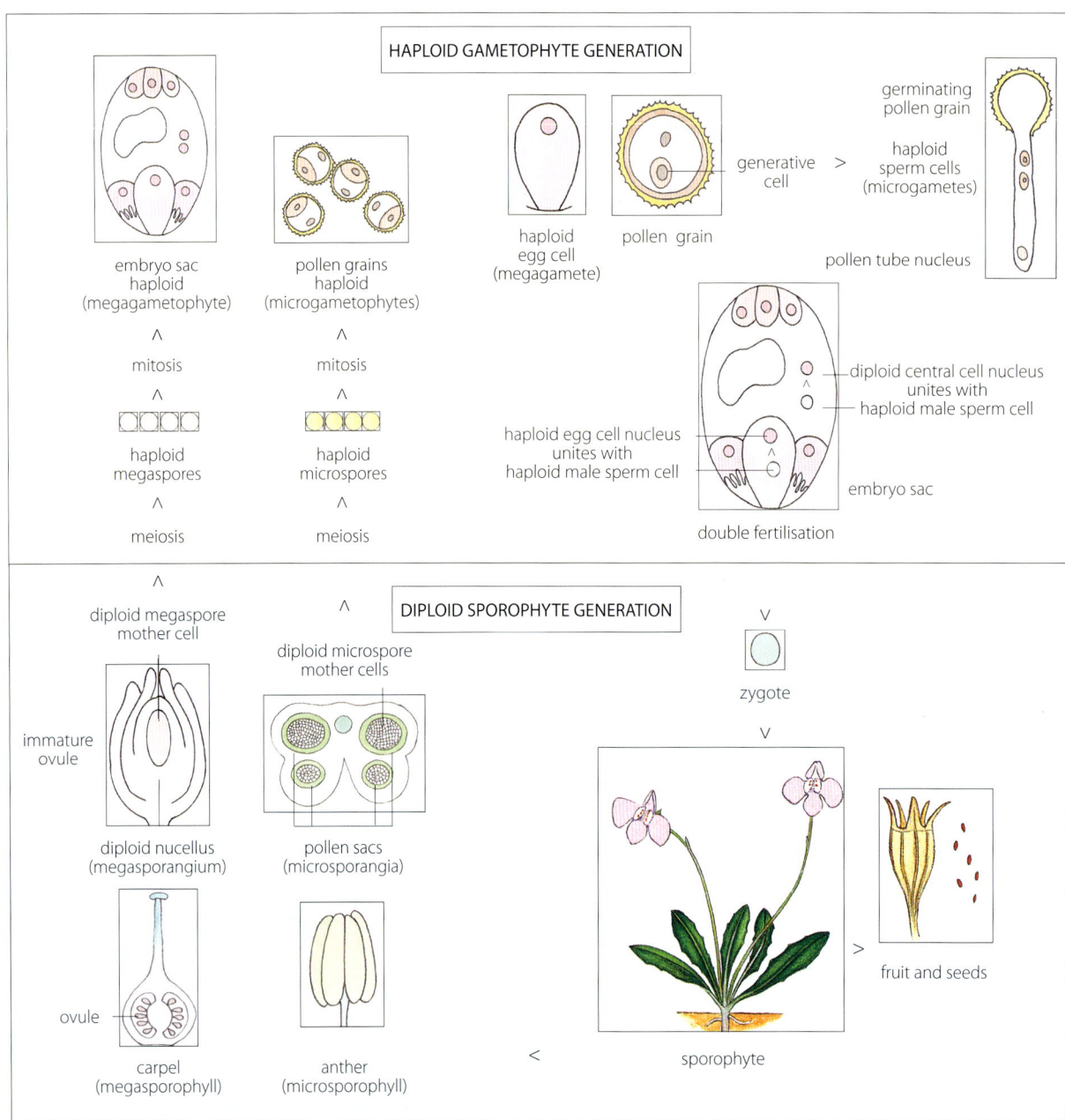

Angiosperms

MONOCOTYLEDONS

EUDICOTS

SEED

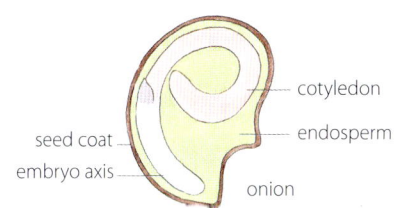

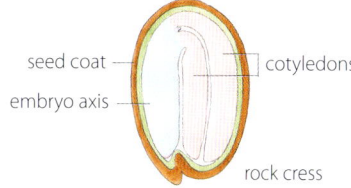

One seed leaf (cotyledon).
Endosperm that persists.

Two seed leaves (cotyledons).
Endosperm usually absorbed by developing seed.

FLOWER PARTS

water plantain   iris   lily

rapeseed   geranium   pea

Usually in multiples of three.
Sepals often same colour and shape as petals.

In multiples of four or five.
Sepals and petals usually distinct.

POLLEN

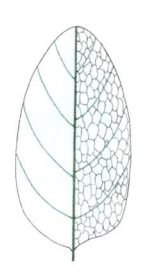

A single pore or furrow.

Typically with three pores or three furrows.

LEAF VENATION

Main veins parallel.

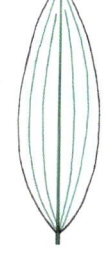

Veins reticulated.

STEM IN CROSS-SECTION

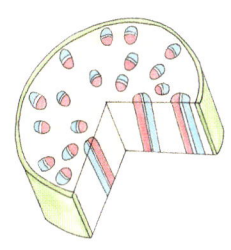

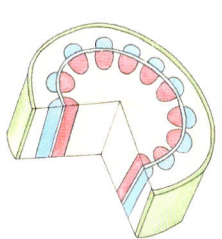

Vascular bundles scattered.

Herbaceous eudicot with vascular bundles in a ring.

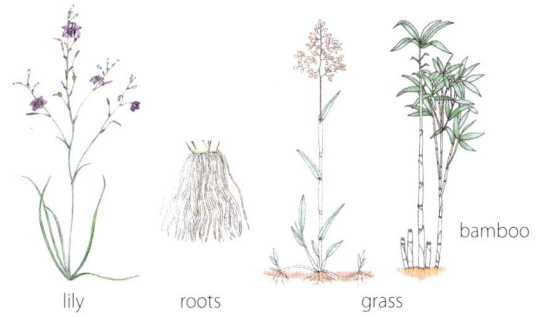

lily   roots   grass   bamboo

Usually herbaceous plants that lack side shoots.
Root system usually fibrous, never a taproot.

Herbaceous or woody plants.
Root system often a taproot.

anterior   Away from the axis.
The side of a flower facing away from the stem.
cf. **posterior**

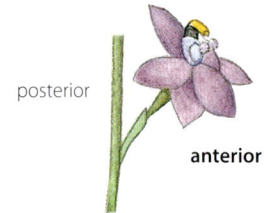

posterior

anterior

antesepalous, antisepalous
Directly in front of the sepals, as stamens that are opposite rather than alternate with the sepals.

antesepalous

anthecium, anthoecium   Of grasses (Poaceae), strictly the lemma and palea of a single floret, but often taken to include the reproductive parts and lodicules, and, occasionally, the attached rachilla. Particularly applied to members of the tribe Paniceae in which only the terminal floret is fertile.

anthela, *pl.* **anthelae**
An umbel-like inflorescence in some rushes (Juncaceae) and sedges (Cyperaceae) in which the primary branches arise more or less from the same point.
cf. **digitate inflorescence**
anthelate   Umbel-like.

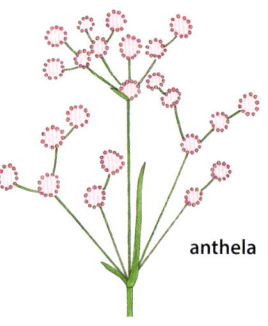

anthela

anthelodium
A corymb-like inflorescence in some sedges (Cyperaceae), with branches ending in spikelets rather than individual flowers.

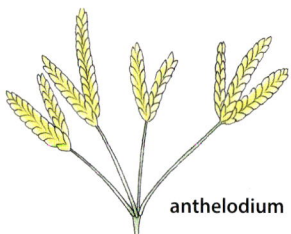

anthelodium

anther   *see* page 19

anther cap
Of some orchids, the cap-like anther that terminates the column and covers the pollinia, as spider orchids (*Caladenia*).

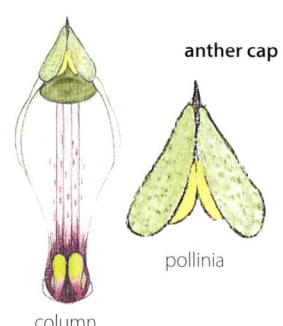

anther cap

pollinia

column

anther sac
The chamber (locule) in the anther of a flowering plant (angiosperm) in which pollen is produced.
= **microsporangium**
see **pollen sac**

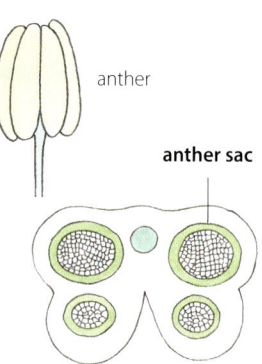

anther

anther sac

antheridium, *pl.* **antheridia**
The male, sperm-producing reproductive organ of bryophytes, ferns and fern allies.
cf. **archegonium**
antheridial   Relating to the antheridium.

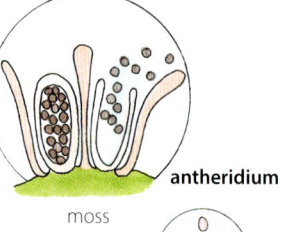

antheridium

moss

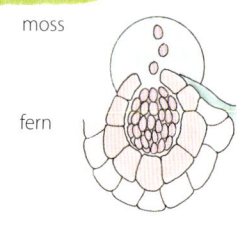

fern

antheriferous   Bearing anthers.

antherozoid, *pl.* **antherozoa**   A motile male gamete, that moves by means of whip-like hairs (flagellae).
In bryophytes and ferns it is produced in the antheridia.
Also found in some gymnosperms, where it is formed in the pollen tube prior to fertilisation.
= **spermatozoid**
cf. **oosphere**

anthesis   The period of flowering from opening of the bud to the withering of the stigma and stamens.

anthocarp
A dry indehiscent one-seeded fruit enclosed by one or more flower parts. Characteristic of the four o'clock family (Nyctaginaceae), as bougainvillea (*Bougainvillea*).
see **accessory fruit**

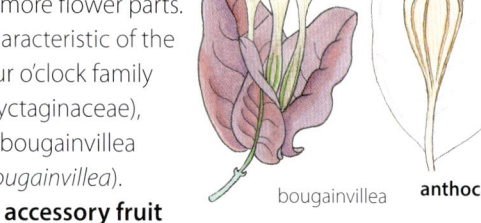

bougainvillea

anthocarp

anther   Of flowering plants, the part of the stamen in which pollen is produced.
It is usually on a stalk-like filament.
Of orchids, the structure housing the pollinia and situated apically or ventrally on the column and either
cap-like in form or two sacs opening by longitudinal slits.

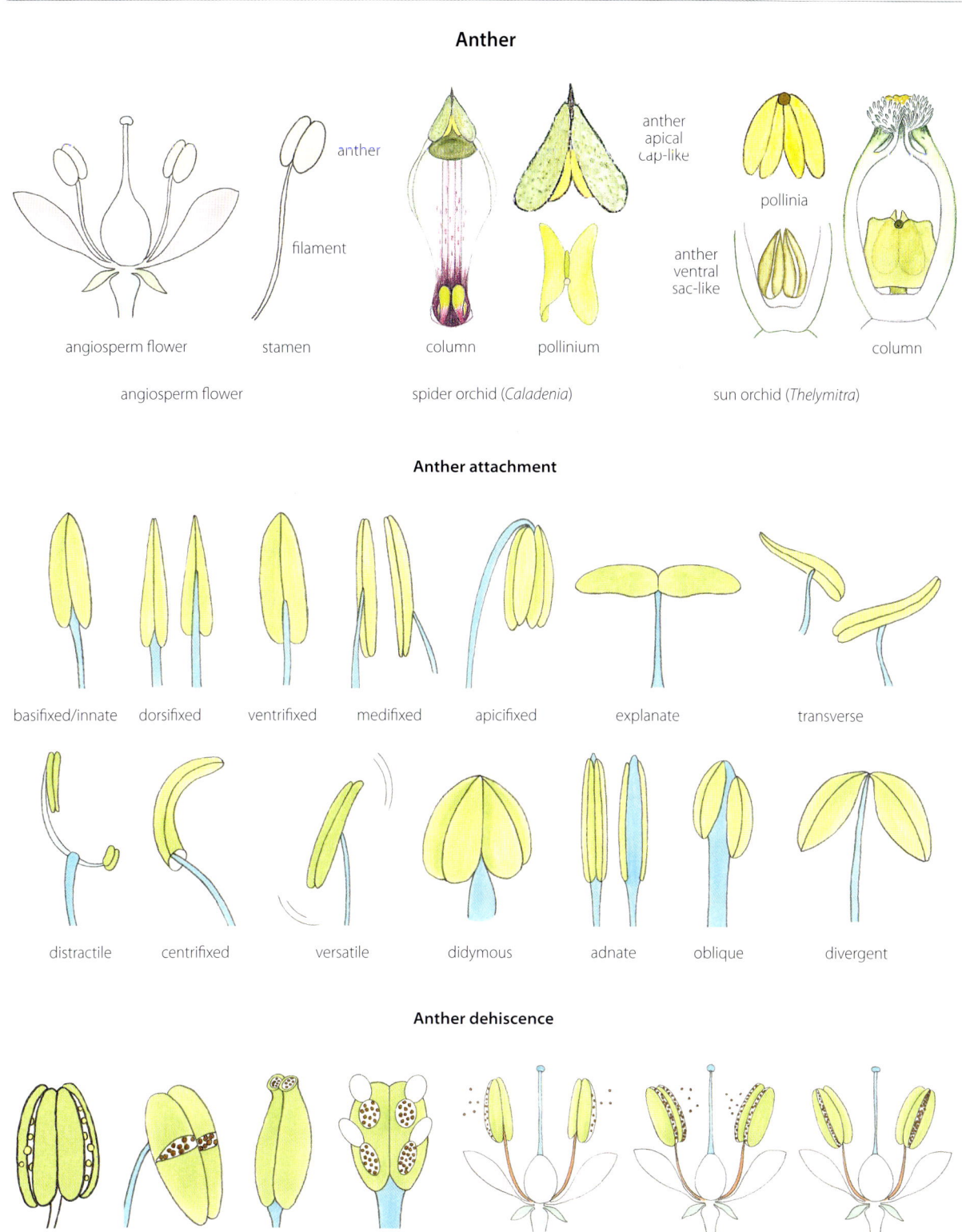

Anther

angiosperm flower   stamen

angiosperm flower

column   pollinium

spider orchid (*Caladenia*)

anther apical cap-like

pollinia

anther ventral sac-like

column

sun orchid (*Thelymitra*)

Anther attachment

basifixed/innate   dorsifixed   ventrifixed   medifixed   apicifixed   explanate   transverse

distractile   centrifixed   versatile   didymous   adnate   oblique   divergent

Anther dehiscence

longitudinal   transverse   poricidal   valvate   extrose/posticous   introrse/anticous   latrorse

anthocyanins   A group of plant pigments that give brilliant colours ranging from pink through to scarlet, purple and blue.
Found in fruits like purple grapes and blackberries. It acts as visible signals on flower petals and fruit to attract insects, birds and animals for pollination and seed dispersal.
see **flavonoids**

anthophily   Love of flowers.
anthophilous   Flower-loving. Applied to an animal that can be a pollinator.

anthophore
An elongated stalk (stipe) inserted on the receptacle, bearing the corolla, stamens and pistil of a flower above the calyx.

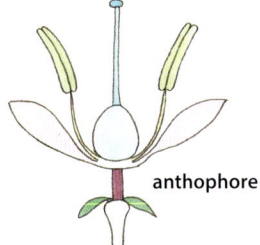
anthophore

anthotaxis, anthotaxy   The pattern of growth in an inflorescence. Growth may be either monopodial (racemose) or sympodial (cymose).

anthotelic   Of growth or branching with the tip of the stem ending in a flower or an aborted floral bud, as a determinate or definite inflorescence.
see **cymose inflorescence**
cf. **blastotelic**

anti-   A prefix meaning against, away or opposite.

anticlinal   Of the cell division plane, or any lines generally, at right angles to the surface of the plant body.
cf. **periclinal**

anticous   Facing towards the axis.
= **introrse**
cf. **extrorse, latrorse**

anticous dehiscence
Of anthers, facing inwards and opening longitudinally to release pollen towards the centre of the flower, as the daisy family (Asteraceae).
= **introrse dehiscence**
see also **anther dehiscence**

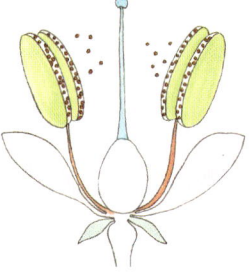
anticous dehiscence

antipodal cells
In angiosperms, three cells in the embryo sac positioned at the end opposite the egg cell. These cells degenerate prior to fertilisation and their function is uncertain.

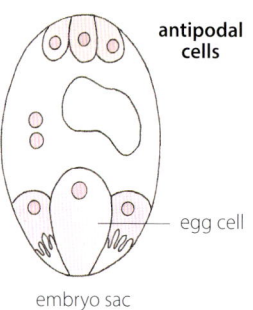
antipodal cells

egg cell

embryo sac

antrorse   Curved or bent forward or upward towards the apex, as hairs or spines.
cf. **retrorse**

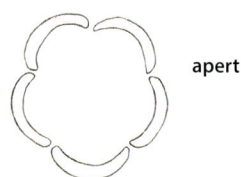
antrorse

apert   Without touching neighbouring parts, as petals in some buds.
see **aestivation**

apert

aperture   A variously shaped, thinner region on the wall of pollen grains in seed plants (angiosperms and gymnosperms) and of spores in lower plants (cryptogams).
see **germinal aperture, laesura, leptoma**
An aperture present in one layer of the wall (ectoaperture) is described as simple.
An aperture formed in more than one layer of the wall (endoaperture) is termed compound or composite.
aperturate   Having one or more apertures. Pollen wall apertures are recognised by their number (mono-, di-, tri-, tetra-, penta-, hexa-, poly-), their position (panto-, zono-, polar) and their shape (colpate, colporate, porate, sulcate).
cf. **inaperturate**

Aperture

Pollen grain

pore   colpus   colporus

sulcus   sulculus   ulcus   ulculus

Spore

monolete   trilete

apetalous
Of a flower having no petals, as the flowers of clematis (*Clematis*).

apetalous

sepals

apex   The tip of an organ or structure.

aphlebium *pl.* **aphlebia**
An atypical pinna at the base of the frond of some fossil ferns and the forest tree fern (*Hemitelia capensis*).

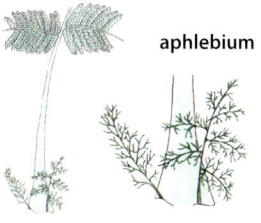

aphlebium

aphyllopodic
Referring to the annual flowering stems of sedges in the genus *Carex* that have bladeless basal sheaths.

aphyllopodic

bladeless basal sheath

aphyllous   Lacking leaves.

apical   At, of or on the apex or tip.
cf. **basal**

apical cell   Of eudicot embryogenesis, the first division of the zygote. The smaller apical cell rests on the larger basal cell. It gives rise to the hypocotyl, shoot apical meristem andcotyledons.
see **embryogenesis**

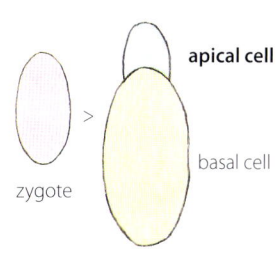

apical cell

zygote   basal cell

apical dominance   Plant growth that suppresses lateral growth from axillary buds and concentrates growth in buds at the apex or tip of the plant.

apical hook   The hook-like curve of the epicotyl found in seedlings that germinate buried in the soil.
It protects the shoot from damage as the seedling is pushing upwards towards light.
= **plumular hook**
see **hypogeal germination**

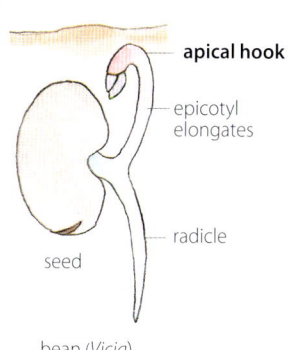

apical hook

epicotyl elongates

radicle

seed

bean (*Vicia*)

apical meristem   A region at the tips of roots and shoots with continuously dividing cells that elongate and increase the length of the plant. These cells can differentiate into three kinds of primary meristem. The protoderm that gives rise to the epidermis. Ground meristem gives rise to parenchyma, collenchyma and sclerenchyma. Procambium gives rise to primary xylem and primary phloem and produces two secondary meristems, the cork cambium and the vascular cambium.
cf. **intercalary meristem, lateral meristem**

Apical meristem

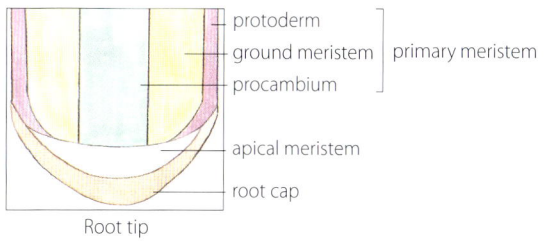

protoderm
ground meristem | primary meristem
procambium

apical meristem

root cap

Root tip

apical placentation
Having one or few ovules develop on a placenta at the top of a simple ovary or on placentas at the top of a compound syncarpous ovary.
see **placentation**

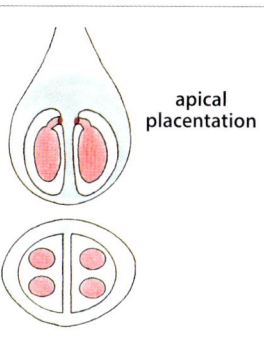

apical placentation

apicifixed   Of a stamen filament attached to the upper end of the connective of an anther.
cf. **basifixed**
see **anther attachment**

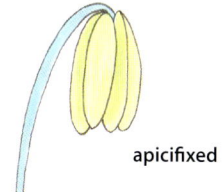

apicifixed

apicule, apiculus,
   pl. **apiculi**
   A short, pointed tip, as at
   the apex of a leaf or floral
   segment.
   apiculate   Tipped with
   an apicule.

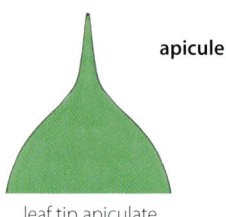

apicule

leaf tip apiculate

apo-   A prefix meaning lacking or separate.

apocarp   A fruit formed
   from the separate carpels
   of a single flower, as
   columbine (*Aquilegia
   vulgaris*).
   see **aggregate fruit,
   apocarpous**
   cf. **multiple fruit,
   syncarp**

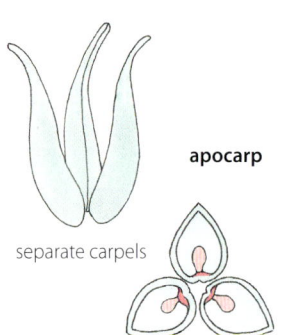

apocarp

separate carpels

apocarpous
   Of a compound
   gynoecium with two or
   more separate carpels.
   Of a fruit formed from
   two or more separate
   carpels, as columbine
   (*Aquilegia vulgaris*).
   = **choricarpous**
   see **apocarp**
   cf. **syncarpous**
   apocarpy   The
   condition of being
   apocarpous.

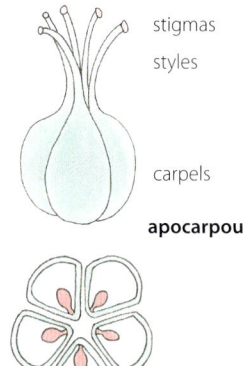

stigmas

styles

carpels

apocarpous

separate carpels

apogamy   Absence of sexual reproduction.
   A form of vegetative reproduction.
   In flowering plants (angiosperms), the formation of
   an embryo from cells in the embryo sac other than
   the egg cell.
   Of ferns and mosses, a haploid gametophyte
   (prothallus) producing buds asexually that develop
   into a haploid sporophyte, as lip ferns (*Cheilanthes*).
   cf. **agamospermy, apospory**
   apogamous   Exhibiting apogamy.

apogeotropism   Turning away from the earth, as
   stems and leaves that grow upwards from the soil.
   Negative geotropism.
   see **tropism**
   cf. **ageotropism**
   apogeotropic   Of or relating to apogeotropism.

apolar   Having no poles.
   Of pollen grains and spores without distinct polarity.

apomict   A plant that reproduces by apomixis.
   see **facultative ~**
   apomictic   Reproducing without meiosis or
   fertilisation of gametes.

apomixis   The production of viable seed from the
   ovule without fertilisation or meiosis.
   A form of cloning.
   see **gametophytic apomixis, sporophytic
   apomixis, vegetative apomixis**
   cf. **amphimixis**

apomorph, apomorphy   In cladistics, a derived
   character.
   A character that has evolved and become modified
   from one present in an ancestral form.
   cf. **autapomorph, autapomorphy,
   plesiomorph, plesiomorphy**
   apomorphic   In cladistics, derived.

apopetalous
   With petals free from
   each other.
   = **choripetalous,
   dialypetalous,
   polypetalous**
   cf. **gamopetalous,
   sympetalous**

apopetalous

apophysis,
   pl. **apophyses**
   A natural swelling or
   enlargement, as the
   enlarged neck between
   the base of the capsule
   and the top of the seta
   in mosses.
   = **hypophysis**
   Of some conifers, the
   part of the cone scale
   that is exposed when the
   cone is closed.

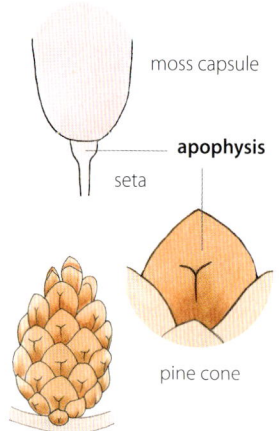

moss capsule

apophysis

seta

pine cone

apoplast   All of the cell walls and intercellular
   spaces in a plant.
   It allows for the rapid diffusion of water and solutes.
   The symplast together with the apoplast make up
   the whole plant.
   cf. **symplast**

apoplastic pathway   Diffusion of water, through the cellulose cell walls and intercellular spaces, from one cell to another.

One of the pathways of movement of water and solutes radially from the root epidermis through the endodermis to the vascular cylinder where it will be transported vertically in the xylem.

see also **apoplast, symplastic pathway**

aposepalous

With sepals free from each other.

= **chorisepalous, dialysepalous, polysepalous**

cf. **gamosepalous, synsepalous**

aposepalous

apospory

In flowering plants (angiosperms), the development of an embryo from a diploid somatic cell positioned next to the megaspore mother cell.

Of ferns and mosses, the development of a gametophyte directly from sporophyte tissue without meiosis and spore production so that both sporophyte and gametophyte have the same number of chromosomes, as severed parts of the fronds of bracken (*Pteridium aquilinum*) that reproduce vegetatively this way.

cf. **apogamy**

aposporous   Exhibiting apospory.

apostemonous

With stamens separate from one another.

cf. **diadelphous, monadelphous, syngenesious**

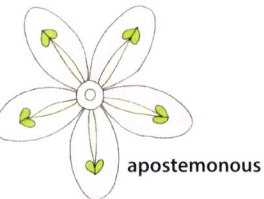

apostemonous

apotepalous

With tepals free from each other, as day lilies (*Haemerocallis*).

= **choritepalous, polytepalous**

cf. **gamotepalous, synsepalous**

apotepalous

appendage   A part attached to a main structure, as that below the anthers of chocolate lily (*Arthropodium strictum*).

appendicular   Relating to an appendage.

appendiculate   Bearing an appendage.

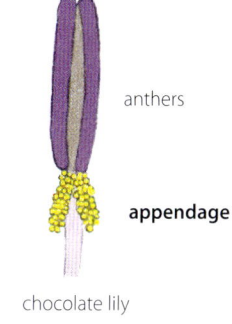

anthers

appendage

chocolate lily

applanate   Flattened, as the pods of some acacia (*Acacia*).

applanate vernation   Of pairs of leaves in bud flattened opposite each other, as some snowdrops (*Galanthus*).

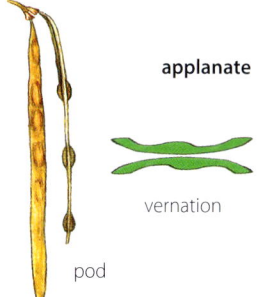

applanate

vernation

pod

appressed   Pressed closely against but not fused, as leaves or hairs against a stem.

Pressed at an angle of at most 15° from the vertical.

= **adpressed**

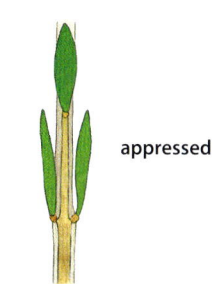

appressed

approximate   Close together but not united.

Closely resembling, almost exact or correct.

aquatic   Of or relating to water, as an aquatic habitat like a lake.

cf. **amphibious, terrestrial**

arachnoid   Covered with fine loosely entangled whitish hairs, resembling a spider's web, as some leaves.

= **cobwebbed**

arachnoid

arboreal   Pertaining to or growing on trees, as some mosses.

arboreous   Wooded, especially heavily wooded.

arborescent   Tree-like in growth structure or appearance, as tree ferns.

tree fern   **arborescent**

arbuscular mycorrhiza

A mutually beneficial symbiosis formed between Glomeromycota fungi and the roots of many vascular plants.

A network of hyphae where growth on the outside of the root is limited and hyphae penetrate the cells of the root cortex and form arbuscules. One of the endomycorrhizas.

see **mycorrhiza**

Arbuscular mycorrhiza

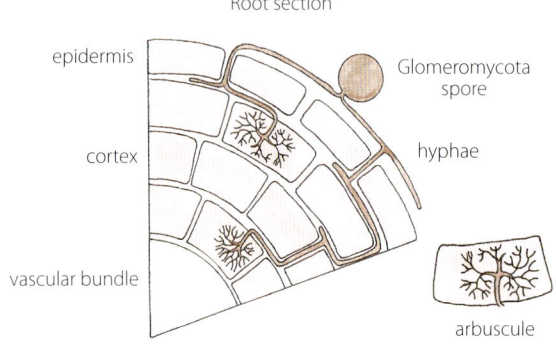

Root section

epidermis

Glomeromycota spore

cortex

hyphae

vascular bundle

arbuscule

arbuscule   A branching tree-like structure.

Of fungi, a tuft of branching hyphae, formed by some types of mycorrhiza, in the root cells of a host plant.

arbuscular   Branched like a tree.

arbuscule

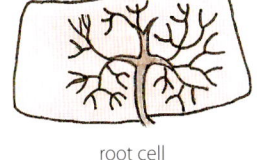

root cell

archegone, archegonium, *pl.* **archegonia**

The flask-shaped female, egg-producing reproductive organ of bryophytes, ferns and fern allies and some gymnosperms, like cycads and conifers.

see **venter**

cf. **antheridium**

archegonial   Relating to the archegonium.

Archegonium

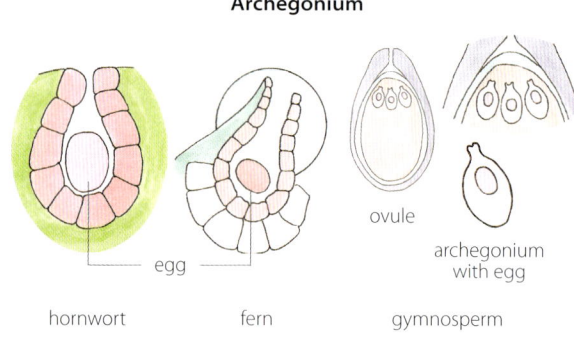

egg

ovule

archegonium with egg

hornwort

fern

gymnosperm

archespore, archesporium, *pl.* **archesporia**

One of the cells in the immature pollen sac of a developing anther that give rise to a microspore mother cell.

archesporial   Of or relating to an archespore.

archespore

immature pollen sac

immature anther

arching   Curving gently downwards, usually more freely than arcuate and less markedly than pendent, as the branches of arching shrubs.

arching

arcuate   Curved like a bow, as the veins of some leaves.

arcuate

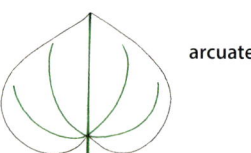

arenicolous   Growing in sand.

arenophile   A plant that thrives in or requires a sandy environment to grow.

areola, *pl.* **areolae, areole,** *pl.* **areoles**

A space between the lines of a net-like pattern.

A depression or a raised area on a cactus stem bearing flowers and/or spines.

One of the small spaces defined by a network of veins on a leaf.

One of the raised areas surrounded by grooves forming a network on a pollen grain.

areolate   Of the spaces between the lines of a net-like pattern.

Of a network with raised areas surrounded by grooves, as on some pollen grains.

cf. **faveolate, foveate, foveolate, reticulate**

areola

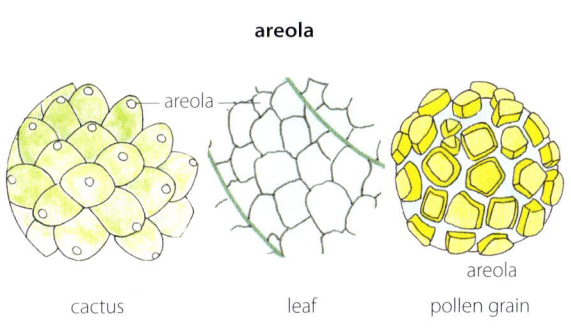

cactus   leaf   pollen grain

aril A fleshy appendage of various origins that partially or completely envelops the seed of some gymnosperms and angiosperms.
Specifically, the fleshy structure that develops from the funicle, as the pea family (Fabaceae), and attracts dispersal agents like ants.
arillate Bearing or related to an aril.
cf. **arillode**

seed   aril

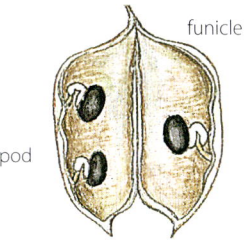

funicle

pod

golden tip (*Goodia lotifolia*)

arillode A fleshy structure, partially or completely enveloping a seed.
It develops from parts other than the funicle, as a caruncle that develops from the micropylar region of the seed coat or a strophiole that develops from the raphe.
= **false aril**
cf. **aril**

arista An awn or bristle.
aristate Tapered to a very narrow, elongated bristle-like point, as a leaf tip.
Bearing an awn or awns, as the beard of barley.

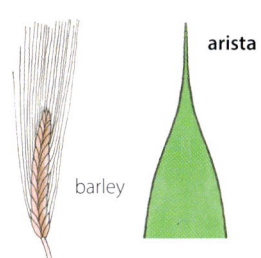

arista

barley

aristulate Having a small awn or bristle at the tip.

armature
A protective covering of thorns, spines, barbs or prickles

armature

armed Bearing protective spines, thorns and prickles that deter herbivory.
cf. **inermous, unarmed**

armed

aromatic Producing volatile oils that are distinctively fragrant or spicy.

article Part of an organ that breaks away easily at a joint or articulation. A segment of a jointed stem or fruit.

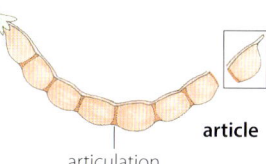

article

articulation

articulation A joint. A point of separation.
articulate, articulated Jointed. Separating freely at the joints (nodes).

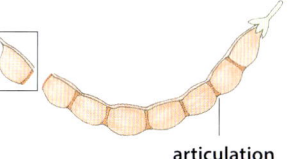

articulation

articulate-pinnate
Of a pinnate leaf with the rachis between the leaflets jointed.

articulate-pinnate

artificial selection The active breeding of traits that are considered desirable, as selective breeding of disease resistant crop plants.
Controlled crosses between parents result in desirable traits being passed on to offspring.
cf. **natural selection**

arundinaceous Resembling a reed or cane.

ascending Proceeding from a lower to a higher part. Gradually going upward.
Of leaves on a stem, spreading horizontally or obliquely then becoming erect.
cf. **descending**

ascending axis The stem of a plant along which the above-ground parts are arranged.

ascending inflorescence

Having the lower flowers on the axis and branches opening first.

= **acropetal**

cf. **descending inflorescence**

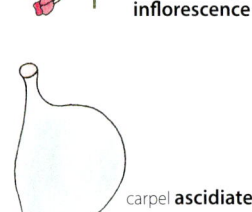

ascending inflorescence

ascidiate   Of carpels, shaped like a symmetrical or asymmetrical vase or an urn, as the pepper family (Piperaceae).

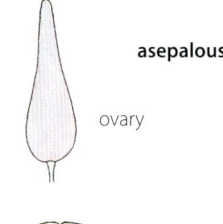

carpel **ascidiate**

Ascomycota   A phylum that accounts for 75% of all described fungi, commonly known as sac fungi, truffles or moulds. Some species form mycorrhizal relationships with the heath family (Ericaceae) and other plants.

see **fungus, mycorrhiza**

Ascomycota

asepalous   Of a flower having no sepals, as the female flower of willows (*Salix*) that lacks both sepals and petals.

stigma

asepalous

ovary

aseptate   Lacking partitions or walls dividing a cavity, as an aseptate ovary or fruit.

cf. **septate**

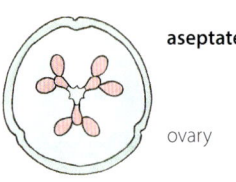

aseptate

ovary

asexual   Having no sex organs, as the asexual sporophyte of a fern.

Independent of sexual processes, as a plant that reproduces vegetatively.

asexual reproduction   The production of new individuals, without the fusion of gametes, by apomixis or vegetative reproduction.

Asexual reproduction can occur naturally, as new plants generated from agamospermy, stolons and rhizomes, or artificially, as new plants generated from cuttings, grafting and layering.

see also **micropropagation**

cf. **sexual reproduction**

asperity

A protuberance that gives a surface a rough texture.

= **scabrosity**

asperate, asperous

Rough with hairs or hard points.

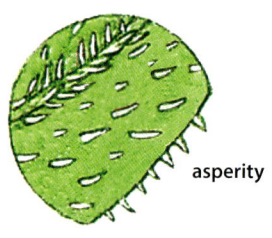

asperity

asperulate, asperulous   Minutely rough with short hard projections or points.

assurgent   Growing upward and curved, as the stems of balsam (*Impatiens assurgens*).

assurgent

astylocarpellous

Of a carpel without a style and supporting stalk (stipe).

cf. **stylocarpellous**

see **astylocarpepodic**

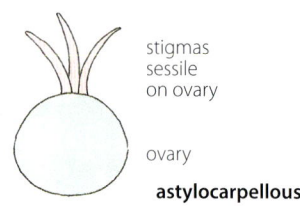

stigmas sessile on ovary

ovary

astylocarpellous

astylocarpepodic

Of a carpel without a style and with a supporting stalk (stipe).

cf. **stylocarpepodic**

see **astylocarpellous**

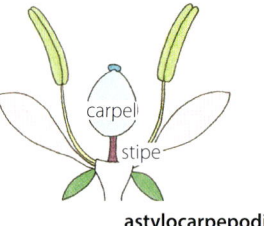

carpel

stipe

astylocarpepodic

astylous   Of a carpel lacking a style, with the stigma sessile on the ovary, as some members of the poppy family (Papaveraceae).

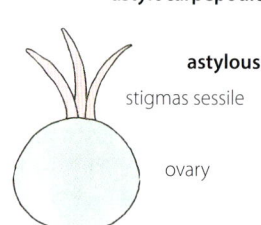

astylous

stigmas sessile

ovary

asymmetry   The quality of being made up of unlike parts facing each other.

cf. **symmetric**

asymmetric, asymmetrical   With any plane through the centre producing unlike halves, as a the leaf of a begonia (*Begonia*) and the flower of a canna lily (*Canna*).

see **amorphic**

cf. **symmetric**

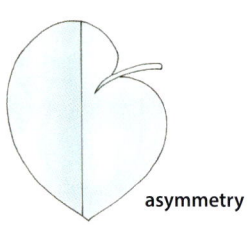

asymmetry

begonia (*Begonia*)

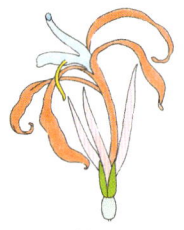

canna lily (*Canna*)

atactostele   Stele with vascular bundles usually scattered. Characteristic of the stems of monocotyledons.
cf. **eustele**

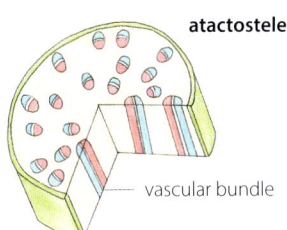

atactostele

vascular bundle

atavism   The reappearace of a character of a distant ancestor after several generations. A throwback.

ategmic   Of an ovule, lacking integuments surrounding the nucellus.
cf. **bitegmic, unitegmic**

atepalous   Of a flower having no tepals. Without a perianth, as the female flower of willows (*Salix*).
= **achlamydeous, naked flower**

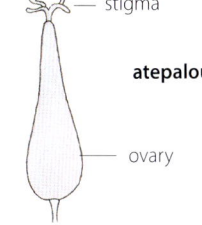

stigma

atepalous

ovary

ATP   Adenosine triphosphate.

atropous   Of ovule orientation, with the chalaza at the base, the micropyle at the top and facing upwards and the embryo sac between, with all aligned with the funicle on a straight axis. The most primitive ovule orientation.
= **orthotropous**
see **ovule orientation**

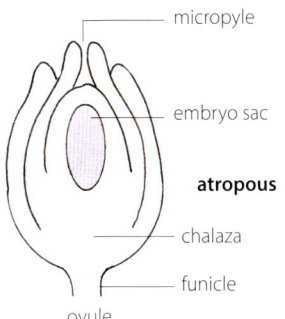

micropyle

embryo sac

atropous

chalaza

funicle

ovule

attenuate   Gradually tapered to a slender base or apex.

leaf base **attenuate**

attractant   Nectar, pollen or pseudopollen, but especially a substance that is a pheromone, that attracts an insect or other pollinator to a flower.

atypical   Not representative of the normal standard, unusual or irregular.
cf. **anomalous**

auct., auctoris, *pl.* **auctt., auctorum**
According to the author(s).

auct. non, auctoris non, *pl.* **auctt. non, auctorum non**
Placed after a plant name to show that it has been used incorrectly by one or more authors.

auctoris, *abbr.* **auct.** *pl.* **auctorum,** *abbr.* **auctt.**
According to the author(s).

auctoris non, *abbr.* **auct. non** *pl.* **auctorum non,** *abbr.* **auctt. non**
Placed after a plant name to show that it has been used incorrectly by one or more authors.

auricle
An ear or ear-shaped appendage or lobe, as that at the base of some leaves.
auriculate   Possessing an auricle or auricles, as the ear-like extension of the collar region between the leaf blade and the leaf sheath in grasses (Poaceae).

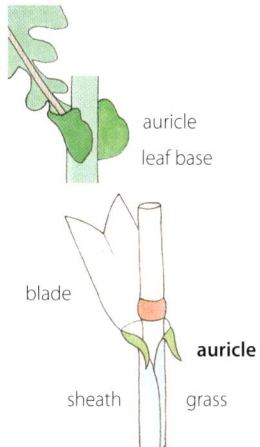

auricle

leaf base

blade

auricle

sheath   grass

autapomorph, autapomorphy   In cladistics, a character derived from an ancestor that is found only in one species.
An apomorphy that is restricted to just one species.
autapomorphic   Relating to an autapomorph.

author   In nomenclature, the first person to publish the name and description of a new taxon.
see **author citation**

author citation   Any reference to the author who first effectively published a taxon name.
In nomenclature, each author's name is given a unique abbreviation that is attached to the name of the taxon, as *Rubus* L.
The abbreviation L. refers to Carl Linnaeus who first described this genus.

authority   in nomenclature, the person whose name is abbreviated and cited after the taxon name, as *Rubus canadensis* L., with the L. refers to Carl Linnaeus who first described this species.

autochory
Of plants with a mechanism for self-dispersal of seeds, as the fruits of geraniums (*Geranium*) that dehisce explosively and release their seeds over a wide area.

autochory

fruit   seed dispersal

Geranium

autochthonous   Of the earliest known flora.

autogamy   Pollination of a flower with its own pollen.
= **self-pollination, selfing**
cf. **allogamy**
autogamous   Of a flower that is self-pollinating.

autonym   In nomenclature, an automatically created name for certain subdivisions of a genus or species that are based on the type specimen for that genus or species.
The genus name or specific epithet is retained, such as *Rubus* subgenus *Rubus* and *Geranium robertianum* subspecies *robertianum*.
Autonyms do not exist above the rank of genus.

autopolyploid, autopolyploidy   Having multiple chromosome sets derived from a single species.
see **polyploidy**
cf. **allopolyploid**

autotroph   An organism capable of synthesising its own food from inorganic substances using light or chemical energy. Plants are the most familiar autotrophs.
see **chemoautotroph, photoautotroph**
cf. **heterotroph**
autotrophic   Of or relating to an autotroph.
see **trophic**

autumnal   Of or appearing in autumn.
cf. **aestival, vernal**

auxins   Plant hormones produced in the growing tips of shoots and roots.
Their influences include water absorption, cell division, inhibition of side shoots above ground thus resulting in growth at the apices and promotion of lateral and adventitious root growth below ground.
see **phytohormone**

auxotelic   Of an inflorescence axis in which growth continues beyond the flowering region into a leafy shoot so that it appears to be inserted between the leaves, as most species of tea-tree (*Melaleuca*).
see **intercalary**
cf. **anauxotelic**

auxotelic

avowed substitute   New name. A replacement name.
= **nomen novum**

awl-shaped   Narrowly triangular and tapering gradually to a fine point.
= **subulate**

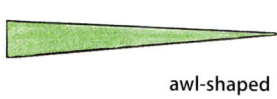

awl-shaped

awn   A slender stiff bristle-like appendage or prolongation.
Sometimes a continuation of the primary vein, as on the glume, lemma or palea of grasses (Poaceae).
awned   Aristate, bristled. Bearing one or more awns as a head of barley.
see **beard**

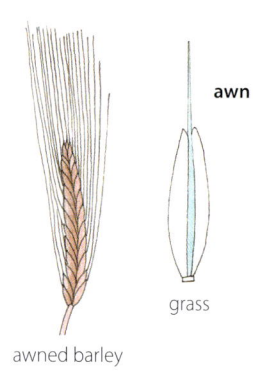

awn

grass

awned barley

axial   Of or belonging to an axis. In the direction of or along an axis. Situated on an axis.

vertical axis

axial

horizontal axis

axil   The upper angle between a leaf and a stem or a branch and a trunk.
axillary   Borne in an axil, as a bud, flower or inflorescence in the angle between a leaf and the stem.
cf. **intercalary, terminal**

axil

axillary inflorescence

axile   Attached to the axis of a structure, as ovules attached to placentas on the central axis of the ovary.
cf. **parietal**

axile placentation
With carpels fused creating a multilocular ovary and ovules arranged along the central axis in the bell peppers (*Capsicum annuum*).
see **placentation**

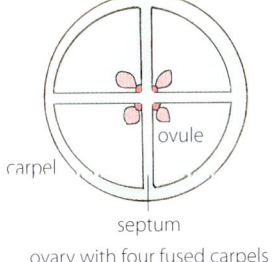

ovary with four fused carpels

axile placentation

axillary bud   A bud in the axil of a leaf that is capable of developing into a stem, branch or flower. On the crown of grasses, it gives rise to tillers, stolons or rhizomes.

axis, *pl.* **axes**   An imaginary or real straight line used as a reference to determine position, symmetry and rotation.
The stem along which the above-ground parts of a plant are arranged is commonly called the ascending axis and the root the descending axis.
The central part around which organs or plant parts are arranged, as ovules in axile placentation.
see **polar axis**
cf. **plane**

Axis

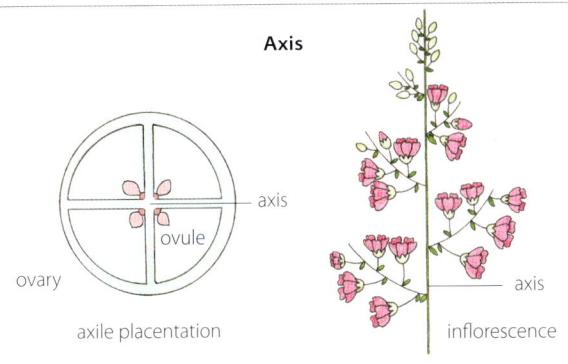

axile placentation

inflorescence

bacca   A fleshy indehiscent fruit with the seeds immersed in pulp, as grapes and tomatoes.
= **berry**
baccate   Berry-like. Bearing berries. Pulpy like a berry.

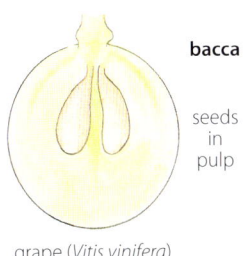

bacca

seeds in pulp

grape (*Vitis vinifera*)

baccacetum
An aggregate fruit composed of a cluster of berries, as pokeweed (*Phytolacca americana*).

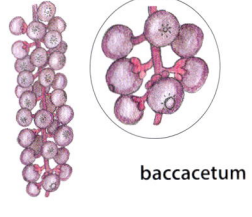

baccacetum

back bulb   Of some orchids, an old leafless, pseudobulb that remains after the terminal growth is finished.
If it is alive and has a bud at the base, it can be used for propagation.

back bulb

backcross   The cross between a hybrid and one of its parents.

bacterium, *pl.* **bacteria**   Microscopic single-celled organisms that reproduce by splitting into two. There is a cell wall but no nucleus and there are no organelles. DNA floats freely in the cytoplasm.
see **prokaryote**
bacterial   Of or relating to bacteria.

Bacterium

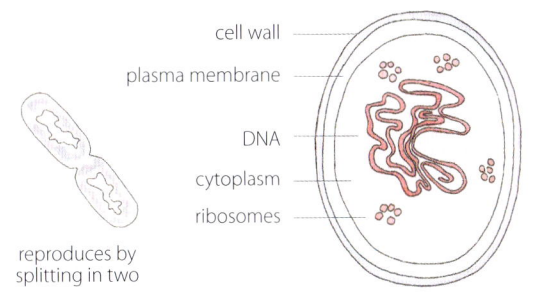

cell wall
plasma membrane
DNA
cytoplasm
ribosomes

reproduces by splitting in two

baculum, *pl.* **baculi**
A rod-shaped sculpturing element on a pollen grain or spore.
baculate   With rod-shaped projections.
baculiform   Rod-like.

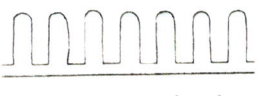

baculum

Bakerian mimicry   A female flower, lacking pollen or nectar rewards, mimics the male flower of the same species to attract a pollinator.
cf. **Dodsonian mimicry, Pouyannian mimicry, Vavilovian mimicry**

balausta   A fruit with a leathery pericarp that encloses a number of irregular cells containing seeds with a succulent testa.
Derived from an inferior ovary.
The fruit is tipped with the lobes of the persistent calyx, as pomegranate (*Punica granatum*).

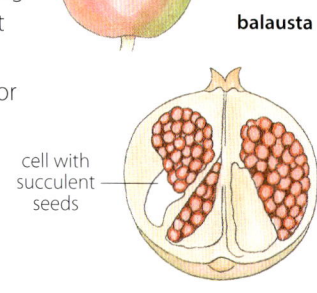

persistent calyx

balausta

cell with succulent seeds

balsam   A sweet-smelling oil or resin derived from various plants.
A plant producing this substance, as pine (*Pinus*).
balsamiferous   Producing balsam.

bamboo   A usually tall, tree-like grass in the subfamily Bambusoideae of the grass family (Poaceae), characterised by woody, hollow, jointed, cylindrical stems.

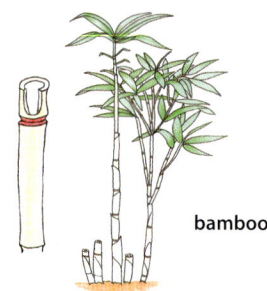

bamboo

banner   The large upper petal of a pea flower (Faboideae).
= **standard, vexillum**

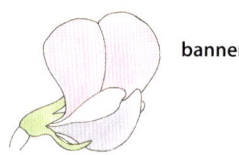

banner

barb   A short sharp hooked bristle or stiff hair-like projection, as on the spines of the fruit of sheep's burr (*Acaena*).
A beard.

barb

barbate   Bearded. Having long thin hairs, as the corolla lobes of beard-heath (*Leucopogon*).

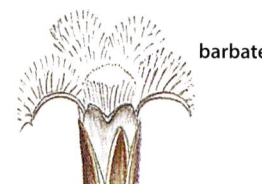

barbate

barbed   Jagged with hooks or points, as the pappus bristles of many daisies (Asteraceae).
Bearded.
cf. **barbate**

barbed

barbella, *pl.* **barbellae**   A short stiff hair.
barbellate   With short stiff hairs.

barbule   A minute barb or beard.
barbulate   Finely or minutely barbed or bearded.

bark   The tough outer covering of stems, branches and roots of woody plants that is composed of alltissue outside the ring of vascular cambium.

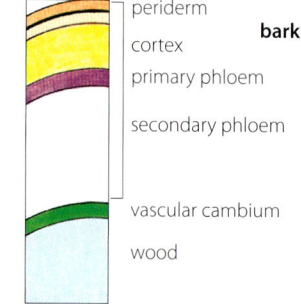

periderm
bark
cortex
primary phloem
secondary phloem
vascular cambium
wood

basal   At or near the base or arising from the base, as leaves at the base of a stem or veins at the base of a leaf.
= **proximal**
see also **suprabasal**
cf. **apical, terminal**

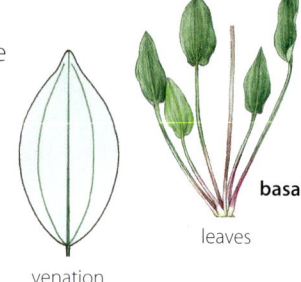

basal leaves

venation

basal cell   Of eudicot embryogenesis, the first division of the zygote in eudicots, the larger of the two cells below the smaller apical cell.
It gives rise to the hypophysis and suspensor.

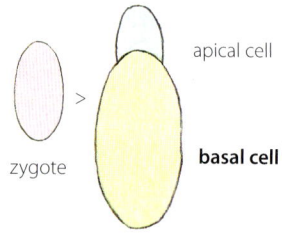

zygote

apical cell

basal cell

basal placentation
Having one or few ovules develop on a placenta at the base of a simple ovary or on placentas at the base of a compound syncarpous ovary.
see **placentation**

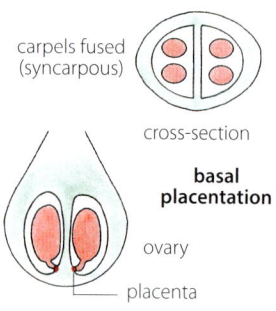

carpels fused (syncarpous)

cross-section

basal placentation

ovary

placenta

basal plate   The modified compressed stem at the base of a bulb or corm.
It has meristem tissue that gives rise to leaves, the flower bud, adventitious roots and offsets.

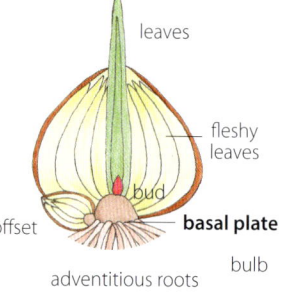

leaves

fleshy leaves

bud

offset

basal plate

bulb

adventitious roots

basal-imperfect

Having veins from the base of a leaf and lateral veins that extend for less than two-thirds of the leaf surface.

see **acrodromous**

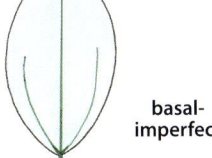

basal-imperfect

basal-perfect

Having veins from the base of a leaf and lateral veins that extend for at least two-thirds of the leaf surface.

see **acrodromous**

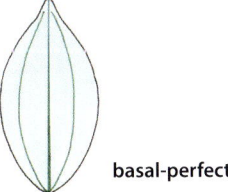

basal-perfect

basi-

A prefix meaning positioned at the base.

basicidal capsule

A capsule that splits at the base, as the birthwort genus (*Aristolochia*).

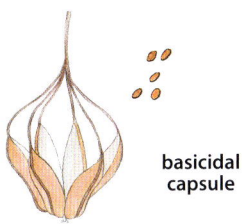

basicidal capsule

Basidiomycota

The phylum of fungi known as club fungi that produce spores in the cap on a microscopic club-shaped fruiting body called a basidium.

Includes mushrooms, puffballs and rusts.

They form mycorrhizal relationships with orchids and forest trees.

see **atropous, fungus, mycorrhiza**

Basidiomycota

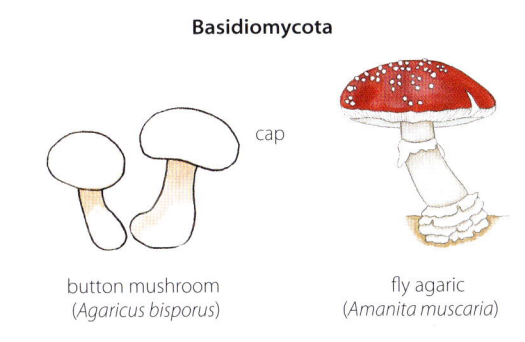

cap

button mushroom
(*Agaricus bisporus*)

fly agaric
(*Amanita muscaria*)

basifixed

Attached at or by the base, as a stamen filament attached to the base of the anther.

= **innate**

see **anther attachment**

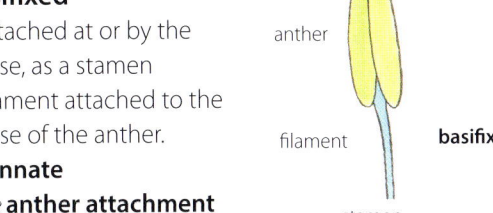

anther

filament   basifixed

stamen

basilar

Relating to or situated at the base.

= **basal**

basinerved

With veins arising from the base.

cf. **pinnate venation**

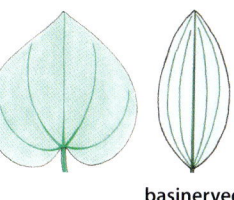

basinerved

basionym

In nomenclature, an earlier valid scientific name of a species that has since been renamed and from which the new name is partially derived, as the name *Geranium dissectum* var. *australe* (Nees) Benth. is derived from the basionym *Geranium australe* Nees.

basipetal

Developing, in sequence, from the apex to the base, as a cymose inflorescence. The flowers at the top open first and those at the base open last.

= **descending inflorescence**

see also **centrifugal, centripetal**

cf. **acropetal**

basipetal

basiphile

A plant that has a preference for, or grows exclusively on, alkaline soils that have a pH of more than 7.

basiphilous Thriving in an alkaline environment, as plants growing on some sand dunes.

see also **calcicole**

cf. **acidophilous**

basiscopic

Facing or directed towards the base, as the pinnules of some ferns.

cf. **acroscopic**

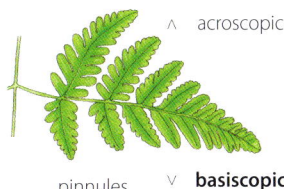

∧ acroscopic

pinnules   ∨ **basiscopic**

basitonic

Having growth strongest in the basal part of the plant.

cf. **acrotonic, mesotonic**

basitonic

bast

The commercial or trade name for phloem fibres used in manufacture of cords, ropes and weaving, as flax bast that is used to make linen. Also applied to any fibres from the outer parts of a plant.

bathyphyll   One of the first basal fronds of a climbing fern that is usually different from mature fronds produced on the upper part of the plant, as the climbing fern *Teratophyllum clemensiae*.
cf. **acrophyll**

bathyphyll

acrophyll

beak   A usually firm and slender tapering extension, as that between the cypsela and the pappus of some daisies (Asteraceae) and the persistent style on the achene of buttercups (*Ranunculus*).
see also **rostrum**

beaked   Having a beak.

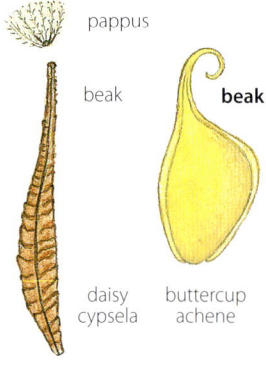

pappus

beak

beak

daisy cypsela

buttercup achene

beard   The long hair-like awns of barley and wheat.

bearded   In grasses (Poaceae), with long hair-like awns, as barley.
= **barbate**

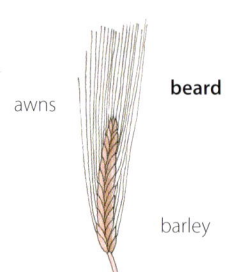

awns

beard

barley

Beltian body
A protein- and lipid-rich detachable tip on the leaflets of some *Acacia* species.
It acts as a food body for ants that in turn protect the plant.
see **myrmecophyte**

Beltian bodies

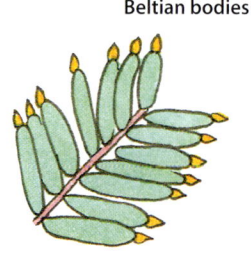

berry   A general term for a fleshy indehiscent fruit with the seed or seeds immersed in pulp, as grapes (*Vitis vinifera*) and tomatoes (*Solanum lycopersicum*).

berry

grape

betalains   A group of yellow and violet pigments that replace anthocyanins in most plant families of the Caryophyllales.
They act as visible signals on flowers and fruit to attract insects, birds and animals for pollination and seed dispersal.

bi-   A prefix meaning two.

bibacca   A fused double berry, as honeysuckle (*Lonicera*).
see **composite fruit**

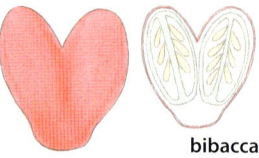

bibacca

bicarinate   Having two keels or ridges along the centre of the lower surface, as the palea of grasses.
cf. **carinate**

bicarpellary, bicarpellate
Of a flower having a gynoecium with two free or fused carpels, as the carrot family (Apiaceae).
cf. **polycarpellary, unicarpellate**

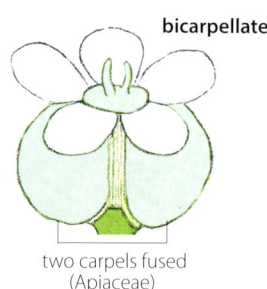

bicarpellate

two carpels fused (Apiaceae)

bicollateral vascular bundle   With phloem in two groups, one on the outside of the xylem and one on the inside of the xylem.
Characteristic of some angiosperm families, as the cucumber family (Curcurbitaceae).
A type of conjoint vascular bundle.

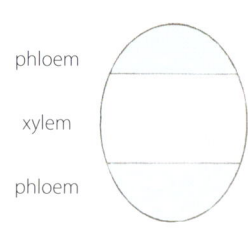

phloem

xylem

phloem

bicollateral vascular bundle

bicolour, bicoloured, bicolourous   Having two colours.

biconcave
Concave on both sides.

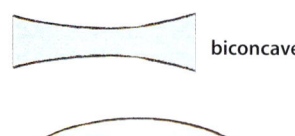

biconcave

biconvex
Convex on both sides.

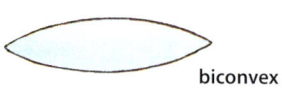

biconvex

bicrenate   Of a margin with crenate teeth that are themselves crenate, as the margins of some leaves.
= **doubly crenate**
cf. **crenate, crenulate**

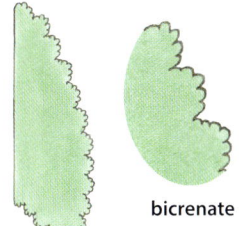

bicrenate

bidentate   Having two teeth.
Doubly toothed, as the margins of some leaves with tooth-like projections that are themselves toothed.
= **doubly dentate**

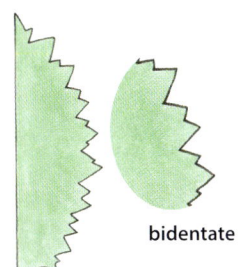

bidentate

biennial   A plant that completes its life-cycle in two years, producing only vegetative growth in the first year then flowering, fruiting and dying in the second year.
cf. **annual, perennial**

bifacial   Having structurally different upper (adaxial) and lower (abaxial) surfaces, as the leaves of most dicotyledons.
Typical of leaves that orient themselves at an angle to the main axis.
= **dorsiventral**
cf. **equifacial, unifacial**

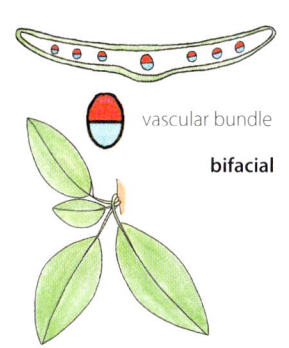

vascular bundle

bifacial

bifid   Split into two parts from the tip.
Split to about half its length to form two pointed lobes, as leaves of some palms of the species *Chamaedorea*.

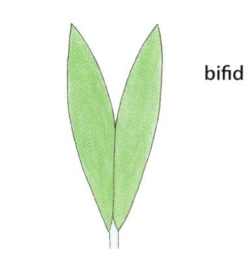

bifid

bifoliate   With two leaves, as the green bird orchid (*Chiloglottis cornuta*).
see **foliate**
cf. **trifoliate, unifoliate**

bifoliate

bifoliolate   Of a compound leaf with two leaflets. Leaflets may be sessile or petiolulate.
cf. **bifoliate**

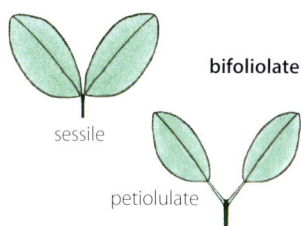

bifoliolate

sessile

petiolulate

bifurcate   Divided into two more or less equal branches or prongs. Forked.
see **furcate**

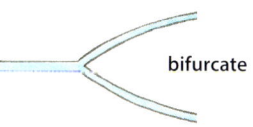

bifurcate

bigeminate   Of a compound leaf having a forked petiole and a pair of leaflets at the end of each branch.
see **geminate**

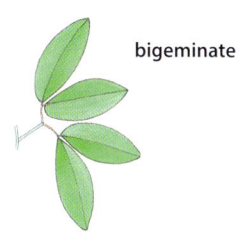

bigeminate

bigeneric hybrid   Of a hybrid resulting from a cross between species of different genera.
= **intergeneric hybrid**

bijugate   Of a pinnate leaf having two pairs of leaflets.
see **jugate**

bijugate

bilabiate   Having a tubular corolla with a two-lipped limb that may or may not be lobed, as the corolla of the mint family (Lamiaceae).

upper lip lobed

bilabiate

lower lip lobed

tube

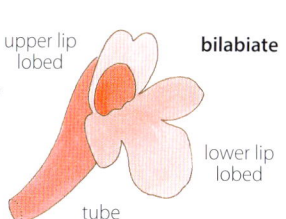

bilateral   Having identical parts on the left and right sides of an axis, especially when of equal size.
cf. **unilateral**

bilateral symmetry
The quality of being divisible through the centre into exactly similar halves on one plane only, as flowers in the pea family (Faboideae).
see **monosymmetric, zygomorphic**
cf. **actinomorphic**

pea flower

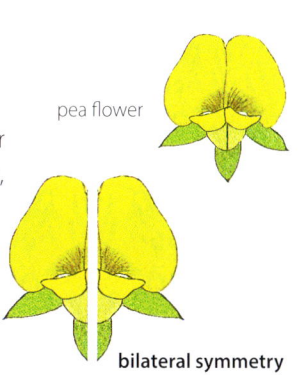

bilateral symmetry

bilobate, bilobed
Having two lobes, as the bilobed leaf of orchid trees (*Bauhinia*).

bilobate

bilocular Of an ovary, anther or fruit, having two locules or cavities for ovules, pollen or seeds.
cf. **plurilocular, unilocular**

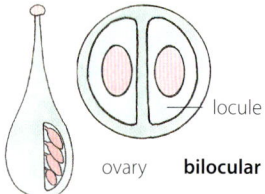
locule
ovary   bilocular

binary key A key that offers two contrasting characters or couplets at each step. By selecting one option each time an unknown plant can be identified.
= **dichotomous key**

binate Growing in pairs, as a petiole having two leaflets.
= **geminate**

binate

bine
The flexible twining stem of some plants, as hops (*Humulus lupulus*) and bindweed (*Convolvulus*). Any plant with such a stem.
cf. **vine**

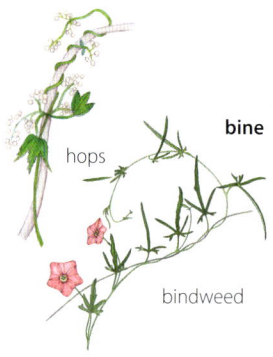
bine
hops
bindweed

binomial Having two names.

binomial nomenclature The system, devised by Linnaeus in 1753, in which plants have a two word name, the first is the genus name and the second the species name (specific epithet), as *Geranium robertianum* commonly known as herb Robert.
The basic unit of naming in botany.
The genus and species names are usually derived from Latin or Greek.

biodiversity The variety of plant and animal life in a particular place on earth or on earth as a whole.

biome *see* page 35

biota Flora, fauna and other forms of life, as fungi and microbes, that inhabit a given area.

biotic Living or once living, including plants, fungi and animals.
In an ecosystem, relating to or resulting from living organisms.
see also **community**
cf. **abiotic**

biotope The smallest region of a habitat characterised by uniform environmental conditions and populated by characteristic flora and fauna.

biotype A population in which all individuals have the same genotype.

biparous Of a cymose inflorescence forming two stems at each branching point, as a biparous cyme.
see also **dichasium, dichotomous**
cf. **multiparous. uniparous**

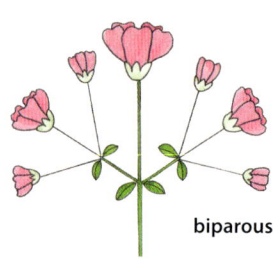
biparous

biparous cyme Another name for dichasium.

bipartite Divided almost to the base into two lobes, as a bipartite leaf.

bipartite

bipinnate Of a pinnate leaf with the primary divisions (pinnae) themselves divided into leaflets (pinnules).
A twice pinnately divided compound leaf.
see **alternate-bipinnate, opposite-bipinnate**
see also **pinnate**

bipinnate

bipinnatifid Twice pinnatifid.

bipinnatipartite Twice pinnatipartite.
Of a pinnately lobed leaf with lobes and lobules pinnatipartite.

bipinnatisect Twice pinnatisect.
Of a pinnately lobed leaf with lobes and lobules pinnatisect.

bipinnatisect

biome An ecological concept, closely related to an ecosystem but on a larger and often global scale, including aspects of climate, geography and the organisms living in it. There are terrestrial and aquatic biomes.
A biome includes examples across the globe. The tropical rainforest biome extends from South America to Southeast Asia, whereas the tropical rainforest of the Amazon Basin is an ecosystem.
cf. **community, ecosystem, habitat, population**

Biomes of the world

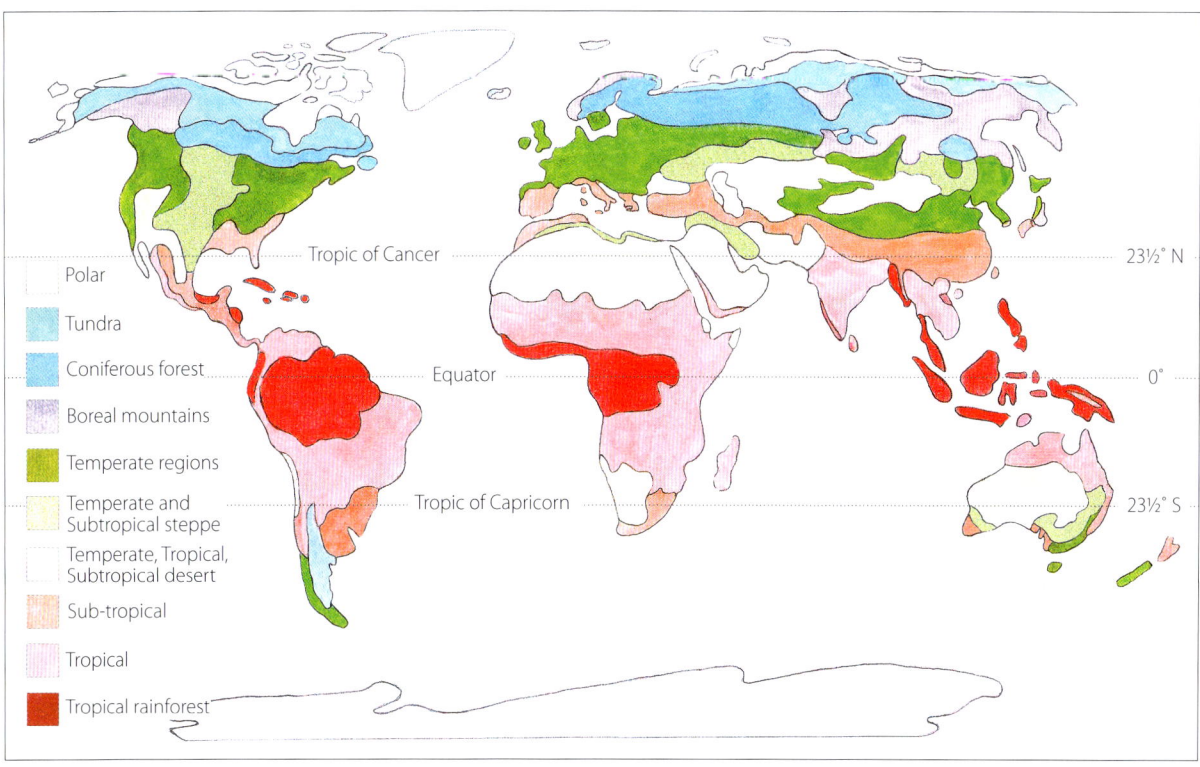

Polar

Tundra

Coniferous forest

Boreal mountains

Temperate regions

Temperate and Subtropical steppe

Temperate, Tropical, Subtropical desert

Sub-tropical

Tropical

Tropical rainforest

Tropic of Cancer — 23½° N

Equator — 0°

Tropic of Capricorn — 23½° S

---

biseriate Arranged in two rows or whorls.
see also **seriate**

biserrate Of a margin with saw-like teeth that are themselves toothed, as the margins of some leaves.
= **doubly serrate**
cf. **serrate, serrulate**

biserrate

bisexual Of a flower with both stamens and a pistil or pistils fertile.
= **hermaphrodite, perfect**
Of an inflorescence, having all flowers with both stamens and pistils fertile and/or flowers with either stamens or pistils fertile.
cf. **neuter, unisexual**

bispecific hybrid A hybrid between two different species belonging to the same genus. *Geranium* x *cantabrigiense* is a hybrid between *G. macrorrhizum* and *G. dalmaticum*.
= **interspecific hybrid**

bisporangiate Of an anther having a single lobe (monothecal anther) with two pollen sacs. The pollen sacs coalesce before dehiscence.
cf. **tetrasporangiate, unisporangiate**

Bisporangiate

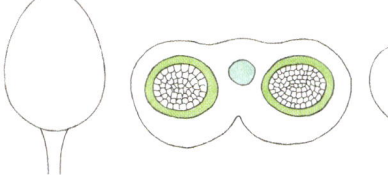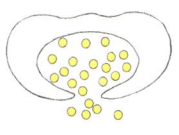

one-lobed anther   two pollen sacs   pollens sacs coalesce

bisymmetric, bisymmetrical

Divisible through the centre into two exactly similar halves that are at right angles to each other, as the flowers of bleeding heart (*Lamprocapnos*).
cf. **bilaterally symmetrical, monosymmetric, polysymmetric**

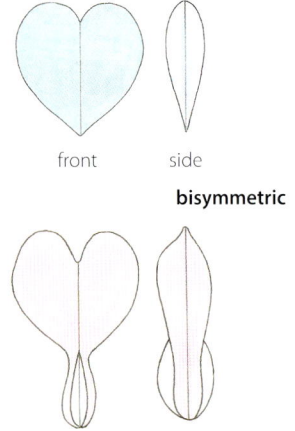

front   side

bisymmetric

bleeding heart (*Lamprocapnos*)

bitegmic

Of an ovule, with two integuments surrounding the nucellus, as monocotyledons and most eudicots.

Of a seed coat having two integuments, the outer integument being the testa and the inner integument being the tegmen, as monocotyledons and most eudicots.
cf. **ategmic, unitegmic**

Bitegmic

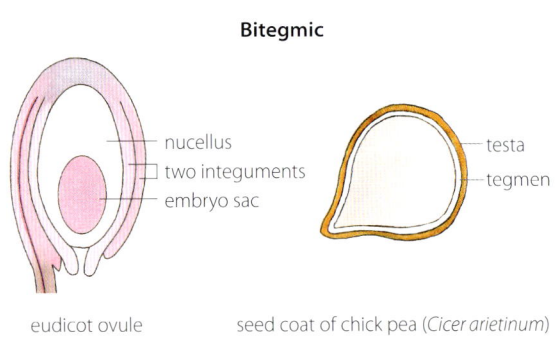

nucellus
two integuments
embryo sac

testa
tegmen

eudicot ovule   seed coat of chick pea (*Cicer arietinum*)

biternate

Consisting of three parts, with each part again divided into three, as a biternate leaf.

biternate

bivalent

In meiosis, said of a chromosome that is paired with its homologous chromosome during synapsis.

bladder

An inflated membranous sac-like structure, as the traps of bladderworts (*Utricularia*).

bladdery   Thin-walled and inflated like a bladder. Resembling a bladder, as the bladdery scales on the fruit of blue devil (*Eryngium ovinum*) or the bladder-like hairs on saltbush (*Atriplex*).

Bladder

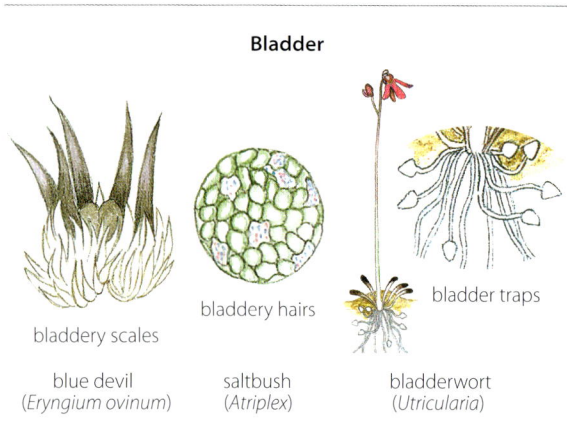

bladdery scales

bladdery hairs

bladder traps

blue devil
(*Eryngium ovinum*)

saltbush
(*Atriplex*)

bladderwort
(*Utricularia*)

blade

The flat expanded part of a leaf or petal.
= **lamina**

Of grasses (Poaceae), the long narrow part of the leaf above the sheath. The blade and shearth are connected by the collar.

Blade

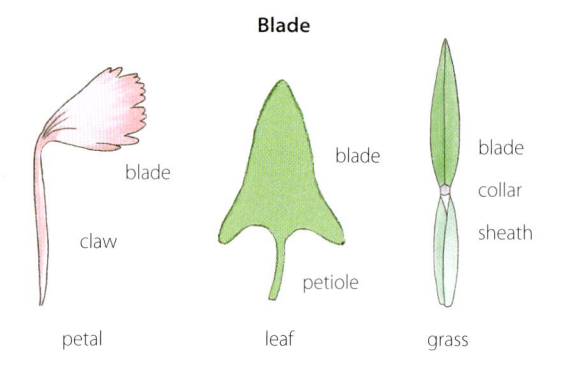

blade

claw

blade

petiole

blade

collar

sheath

petal   leaf   grass

blade meristem

Of grasses (Poaceae), a band of meristematic tissue between the leaf sheath and the leaf blade that generates growth of the leaf blade.
= **collar**

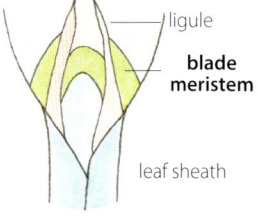

ligule

blade meristem

leaf sheath

blastotelic

Of growth or branching with the the tip of the stem ending in a vegetative bud so that it has the potential to grow indefinitely, as an indeterminate or indefinite inflorescence.
see **racemose inflorescence**
cf. **anthotelic**

blind

Lacking flower buds, or buds failing to develop into a flower.

bloom   A flower or the flowering time of a plant.
A white-grey coating on the surface of some fruits, leaves and stems.
see also **algal bloom, glaucous, pruinose**

blossom   A flower or mass of flowers. To bloom.

blotch   A patch of colour of an irregular shape.
blotched   Having distinct, irregularly shaped patches of colour.

bog   A freshwater wetland with poorly drained peaty soil and vegetation that typically includes peat mosses (*Sphagnum*), sedges and heaths.
cf. **marsh, swamp**

bole   The trunk of a tree below the lowest branch.
cf. **crown, trunk**

bole

boll   A dry, rounded capsule that splits into segments, especially of cotton (*Gossypium*) and flax (*Linum usitatissimum*).

boll

cotton

bony   Hard and tough, as the stone in the fruit of olives (*Olea europaea*).
= **osseous**

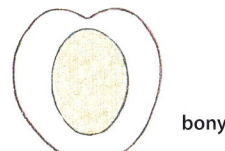

bony

boot stage   Of grasses (Poaceae), the transitional phase between the vegetative phase and the reproductive phase.
The swelling of the developing inflorescence inside the enclosing sheath of the flag leaf.
see **booting**

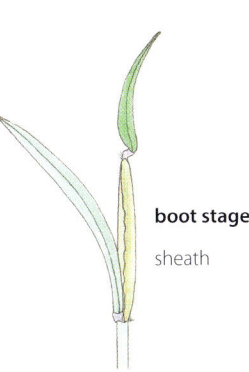

boot stage

sheath

booting
Of grasses (Poaceae), the inflorescence beginning to push through the uppermost sheath on the culm.
see **boot stage, heading**

booting

boreal   Relating to the northern biotic area that forms a nearly continuous belt across North America and Eurasia.
It is characterised by coniferous forests, long harsh winters and short summers.
see **biome, taiga**

boss   A rounded protruberance, as the prominent mound on the labellum of a helmet orchid (*Corybas*).

boss

bostryx   A flattened spirally coiled cymose inflorescence.
A single new stem develops repeatedly on the same side of the axis.
= **helicoid cyme**
see also **monochasium**
cf. **cincinnus**

bostryx

2
4   3
1

botryoidal, botryose   Having the form of a cluster of grapes.

botuliform   Shaped like a sausage.

brachiate   Applied to branches that are widely spreading and paired on alternate sides of the stem, as mint (*Mentha*).
cf. **decussate**

brachiate

brachy-   A prefix meaning short.

brachyblast
A short lateral branchlet, with very short internodes, often bearing leaves in clusters, as some pines (*Pinus*).
see **sterigma**

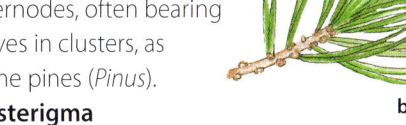

brachyblast

brachystylous

Of heterostylous flowers, those having short styles, as the cowslip (*Primula vulgaris*).
cf. **dolichostylous, isostylous**

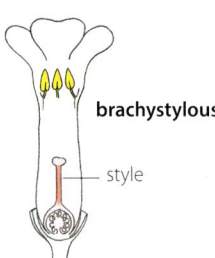

brachystylous

style

bracken

The bracken family (Dennstaedtiaceae) comprises twelve genera of mostly terrestrial ferns with a creeping or erect rhizomes and fronds that are usually large and much-divided.
see **fern**

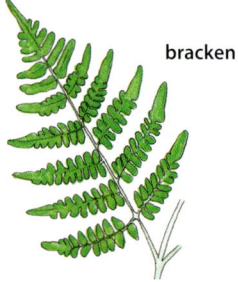

bracken

bracken (*Pteridium*)

bract

A modified or reduced leaf, typically differing in shape, size or colour from other leaves.

It may be associated with a flower or a compound inflorescence, a cone or a grass spikelet (the glumes, lemmas and paleas are bracts).

Bracts may protect an inflorescence in bud.

They form an involucre surrounding the inflorescence of daisies (Asteraceae).

Some are large and showy as the bracts of poinsettia (*Euphorbia pulcherrima*).

Fertile bracts subtend flowers, as those in a raceme.

Infertile bracts are those not associated with flowers.

Cones, as those of pines, have a bract, subtending the cone scale, that bears the ovules.

bracteate   Having bracts.

bractiform   Having the appearance of a bract.

Bract

bract — raceme

bract — compound inflorescence

bracts — protective

involucre

bracts — poinsettia

ovule — bract — cone scale — pine cone

lemma — palea — glume — grass spikelet

bracteole, bractlet

A small bract, borne singly or in pairs, on the peduncle of a solitary flower, or in a compound infloresence, or on the calyx of a flower.

A small bract subtending a cyme unit in a compound cyme.

bracteolate
Having bracteoles.

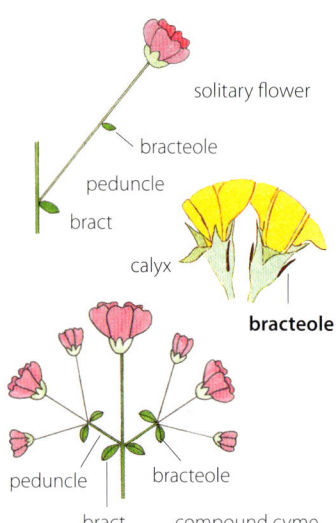

solitary flower

bracteole

peduncle

bract

calyx

bracteole

peduncle — bracteole

bract — compound cyme

bracteose

With many bracts.
With showy bracts, as the bracts of poinsettia (*Euphorbia pulcherrima*).

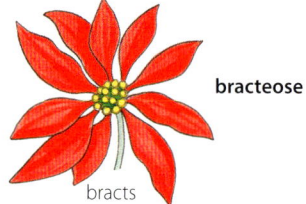

bracteose

bracts

bramble

Any plant in the bramble genus *Rubus*. Typically with prickly stems called canes and edible fruit, as blackberries, raspberries and boysenberries.
Any rough, usually wild, tangled or prickly shrub.

bramble

branch

A division or subdivision of a stem or axis, as a tree, shrub or inflorescence.

branching   The organisation of the branches, including branching patterns.
see **acrotonic, basitonic, mesotonic**

branchlet   A small branch.

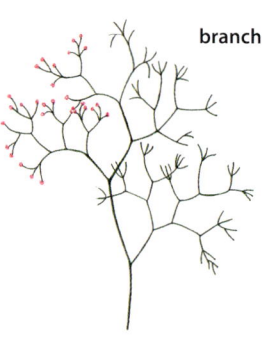

branch

branch   Of phylogenetics, the line connecting two nodes on a phylogenetic tree.
see **cladogram**

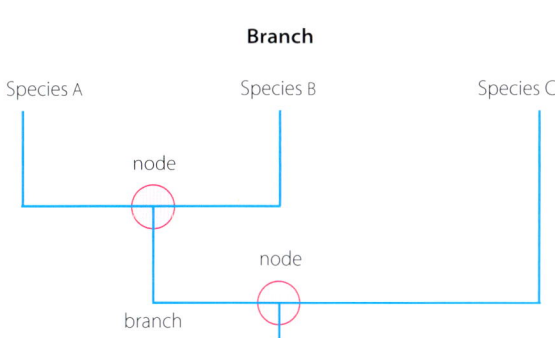

Branch

branch complement
Of bamboos, a branch or branch clusters arising from a mid-culm node. The number and arrangement of branches in a cluster can help identify a species.

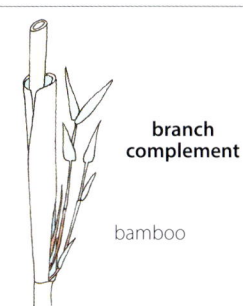

branch complement

bamboo

breakaway   A steep-sided rocky slopes, as those of some scarps, particularly those of mesas.

brevi-   A prefix meaning short.

breviaxe   Of a pollen grain that is oblong in shape due to having a polar axis that is shorter than its equatorial diameter.

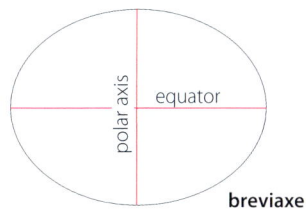

polar axis

equator

breviaxe

brevisulcate   A pollen grain with a very short sulcus.

bristle   A stiff hair.
In grasses, the upper part of a twisted awn.
bristled   Bearing bristles, as the margins of some leaves.

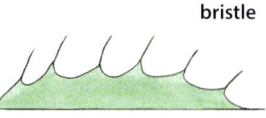

bristle

brochidodromous
Of leaves with secondary veins not reaching the margin but joining to form a series of prominent arches.

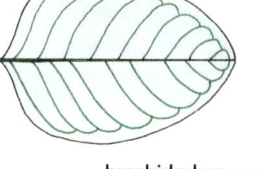

brochidodromous

bryophytes   Mosses, hornworts and liverworts. Nonvascular, simple, low-growing plants that lack true leaves, flowers and roots and produce spores rather than seeds.
They may be dioecious, with separate male and female plants, or monoecious, with male and female parts on the same plant.
A larger gametophyte generation alternates with a smaller dependent sporophyte generation.

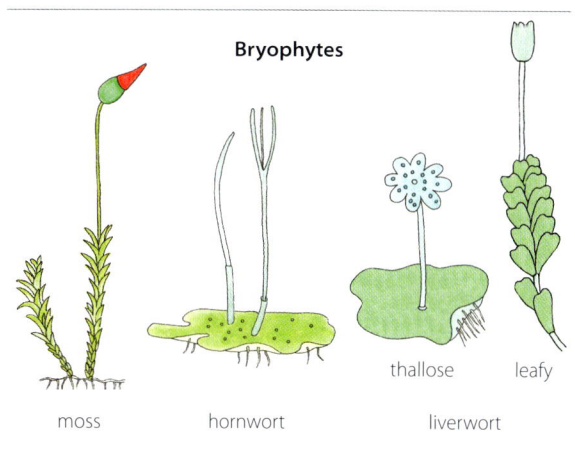

Bryophytes

moss   hornwort   liverwort

thallose   leafy

bud   An undeveloped shoot, usually in the axil of a leaf or on a stem. A vegetative bud develops into leaves or branches and a reproductive bud produces flowers.

bud

bud scale   A modified leaf that acts as a protective covering and tightly encloses the developing flower or leaf bud of some plants, as the leaf bud of sugar maple (*Acer saccharum*).
= **perule**

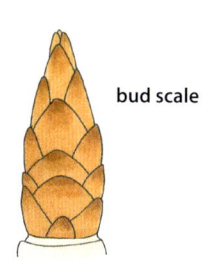

bud scale

budding
A form of graftage in which the scion inserted into the stock is a single bud.
cf. **grafting**

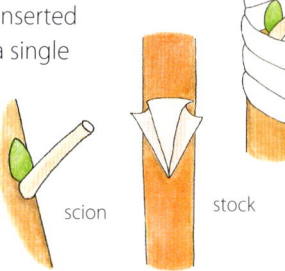

graft

scion   stock

budding

bulb A short thick underground stem (basal plate), with fleshy modified leaves (scales) surrounding next seasons bud. The scales develop from the base of the stem as do the roots.

The scale leaves have food reserves that support the bulb during dormancy and the resumption of active growth.

A bulb may be tunicate or imbricate.

Tunicate bulbs usually have fleshy cylinder-like leaves, arranged in concentric circles, that are covered by a sheath (tunic) of dry membranous scale leaves, as onions (*Allium*).

Imbricate or naked bulbs have fleshy overlapping leaves that lack a tunic, as the lily genus (*Lilium*).
cf. **corm**

bulbiferous, bulbose, bulbous Producing bulbs. Resembling a bulb.

bulbiform Having the shape of a bulb.

Bulb

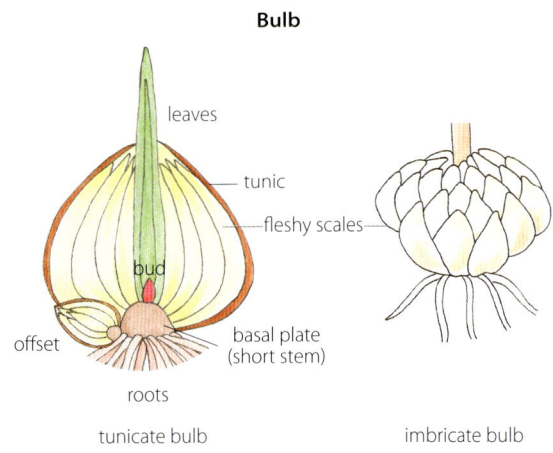

leaves
tunic
fleshy scales
bud
offset
basal plate (short stem)
roots

tunicate bulb   imbricate bulb

bulbel, bulbil, bulblet A small bulb, often arising from the axil of a leaf, as tiger lilies (*Lilium tigrinum*), or on an above-ground stem in place of a flower, as tree onion (*Allium cepa* var. *proliferum*), or at the base of a mature bulb.

All have the ability to develop into a new plant.

bulbiliferous Having or resembling a bulbil.

Bulbel

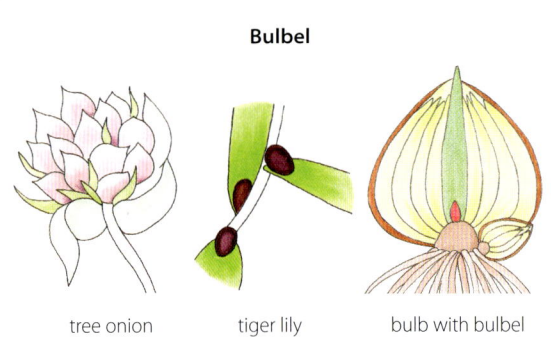

tree onion   tiger lily   bulb with bulbel

bulbotuber A corm.

bullate With blister-like swellings on the surface, as the leaves of some begonias (*Begonia*).

bullate

bulliform cell A bubble-shaped cell that is mostly water and may play a role in the hygroscopic opening and closing of leaves to prevent water loss in dry weather. Usually in the epidermis of the leaves of monocotyledons like grasses.

bunch grass Any of various grasses that grow in clumps or tufts rather than in a continuous mat.
cf. **mat grass**

bunch grass

bundle cap
Of herbaceous dicot stems, a cluster of sclerenchyma fibres towards the outside of the phloem in a vascular bundle that helps strengthen the stem.

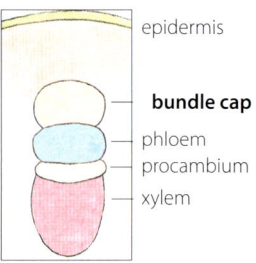

epidermis
bundle cap
phloem
procambium
xylem

bundle scar
The healing layer that forms on vascular tissue after a leaf falls.

The arrangement and number of bundle scars on a leaf scar is used for plant identification.
see also **abscission**

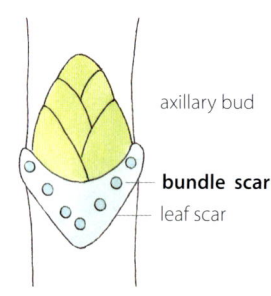

axillary bud
bundle scar
leaf scar

bundle sheath
A sheath of supporting sclerenchyma fibres that surrounds a vascular bundle in stems of monocotyledons.

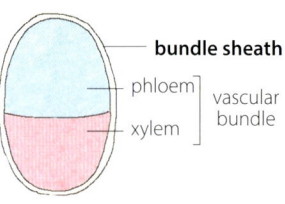

bundle sheath
phloem
xylem
vascular bundle

bur, burr A multiple fruit of achenes enclosed in a persistent prickly involucre, as cocklebur (*Xanthium*).
see **accessory fruit**

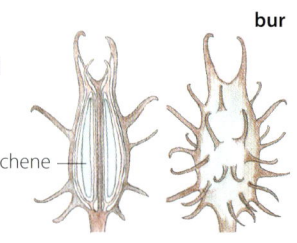

bur
achene

burl   A usually rounded knotty growth filled with dormant buds on the trunk, roots or branches of some trees.

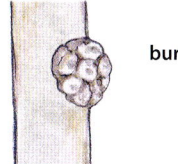
burl

bursicle   Of some orchids, a membranous covering over the viscidium that is easily broken when touched, thereby exposing the sticky viscidium, as in the genus *Orchis*.

bush   A low woody perennial plant with several stems and no distinct trunk.
= **shrub**

bush

bush   In Australia, New Zealand and South Africa, uninhabited mostly dry sclerophyll forests, woodland, or open grassland communities.

buttress   A thickened flared support at the base of a tree. It provides stability and may extend many metres from the trunk. Found in some mangroves and figs.

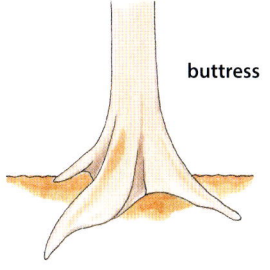
buttress

buzz pollination   Vibration from buzzing, as by a bee clasping an anther to shake pollen free from the flower of a flax lily (*Dianella*).

buzz pollination

caducous   Falling off prematurely or easily, as the petals of some flowers.

caducous

caespitose, cespitose   Forming small dense tufts, as some grasstrees (*Xanthorrhoea*).

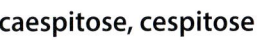
caespitose

grasstree

Of grasses (Poaceae), with tillers joined together at the base by very short stems or apparently stemless.
= **tufted**
cf. **mat grass**

grass

calcarate   Spurred, as the calyx of nasturtium (*Tropaeolum*).

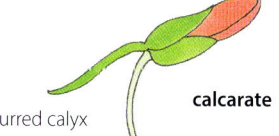
spurred calyx   calcarate

calcareous   Mostly or partly composed of calcium carbonate, as some soils.

calceolate, calceiform   Slipper-shaped. Resembling a round-toed shoe, as the labellum of the lady's slipper orchid genus (*Paphiopedilum*).

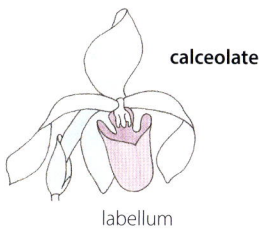
calceolate
labellum

calcicole, calciphile   A plant that cannot tolerate acidic soils and grows only on or mainly on alkaline soils containing lime.
cf. **calcifuge**
calcicolous   Growing in soils rich in lime.
cf. **acidophilous, calcareous**

calcifuge   A plant that cannot tolerate lime and grows only on or mainly on acidic soils.
cf. **calcicole, calciphile**

calciphyte   A plant tolerating a soil pH of 8 or higher.
calciphytic   Tolerating a soil pH of 8 or higher.

callose   A complex plant polysaccharide commonly associated with sieve areas of sieve elements and present in pollen tubes. Also produced in response to wounding and infection.

callosity   A raised hardened or thickened mass, as that on the labellum of an onion orchid (*Microtis*).
cf. **callus**
callose   Hard and thick in texture. Bearing callosities.

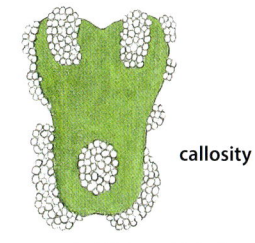
callosity

onion orchid (*Microtis*) labellum

callus, *pl.* **calli**   A projection or outgrowth of thickened tissue, as on the lip of the orchid genus *Caladenia*.

The toughened tissue that develops over a wound. In some grasses (Poaceae), a hard, sometimes bristly and/or sharp-pointed projection at the base of a floret, formed from the rachilla joint and/or the base of the lemma.

callused   Having a callus.

Calycle

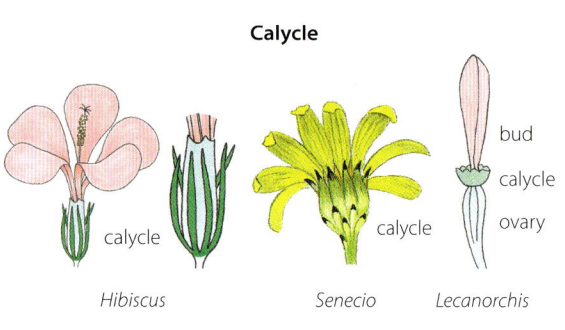

Hibiscus   Senecio   Lecanorchis

Callus

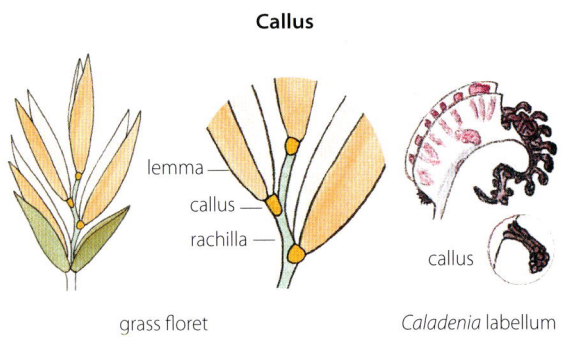

lemma

callus

rachilla

callus

grass floret   *Caladenia* labellum

calybium   A hard one-seeded nut derived from a one-loculed inferior ovary, as the nut part of an oak acorn (*Quercus*) that is partly surrounded by a cupule.

calybium

cupule

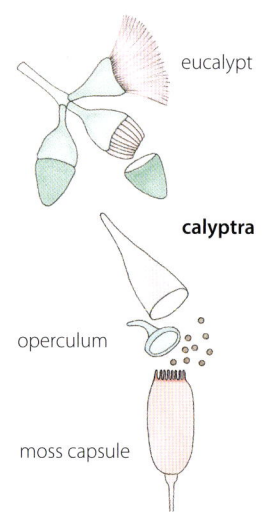

calycine   Relating to, attached to or resembling a calyx, as the calycine bracts on carnations (*Dianthus*). Like a cup.

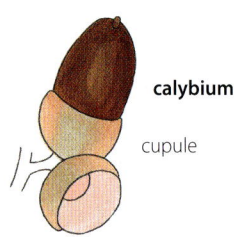

calyx   **calycine**

bracts

calycle, calicle, calyculus, *pl.* **calyculi**

A whorl of free or fused bracts that look like a second calyx, as hibiscus (*Hibiscus*).

A row of bracts immediately subtending the involucral bracts in some daisies, as fireweed (*Senecio*).

A small cup-shaped structure, as that below the calyx of the orchid *Lecanorchis*.

= **epicalyx**

calycular, calyculate   Cup-like.

Having a whorl of free or fused bracts that looks like a second calyx.

calyptra, *pl.* **calyptrae**

A hood or lid. Of some flowering plants, a deciduous cap, formed by fusion of the perianth, that covers the stamens and carpels in bud, as eucalypts.

A deciduous cap-like structure partly covering the capsule of mosses.

cf. **operculum**

calyptrate   Relating to, bearing or resembling a calyptra.

eucalypt

calyptra

operculum

moss capsule

calyx, *pl.* **calyces**

The sepals of a flower that surround the flower parts in bud. Sepals can be free or united.

The calyx is usually green, but in monocotyledons, like lilies, it is mostly the colour and texture of the petals.

calyciform, calycoid

Having the form or appearance of a calyx.

cf. **corolla**

calyx

geranium

sepals

lily

petal

calyx

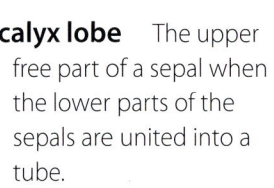

calyx lobe   The upper free part of a sepal when the lower parts of the sepals are united into a tube.

see **calyx tube**

calyx lobe

calyx

calyx tube

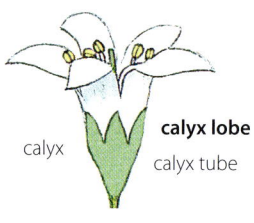

calyx tube   The part of a calyx below the lobes with the sepals united to some extent.
see **calyx lobe, gamosepalous**

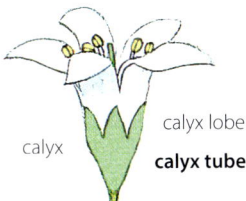

calyx

calyx lobe

calyx tube

CAM, crassulacean acid metabolism
A carbon fixation pathway that occurs when the stomata open at night to admit carbon dioxide while minimising water loss.
Found in many xeric plants like cacti.
Named after the stonecrop family (Crassulaceae) in which this pathway was first studied.
cf. **photosynthesis**

cambium, *pl.* **cambia, cambiums**   Meristematic tissue that is responsible for secondary growth.
There are two kinds: cork cambium and vascular cambium.
see **lateral meristem**

campanulate
Bell-shaped.
Of a corolla having a broad tube terminating in a flared limb, as the flowers of bluebells (*Wahlenbergia*).

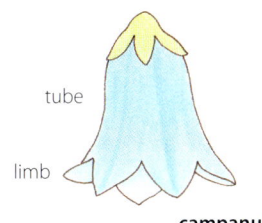

tube

limb

campanulate

camptodromous   Of leaves with secondary veins not terminating at the margins.

Camptodromous

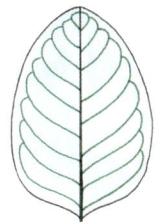

brochidodromous

cladodromous

eucamptodromous

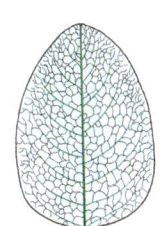

reticulodromous

campylodromous
Of leaves with several primary veins that originate at the base and run in strongly curved arches that converge towards the apex.

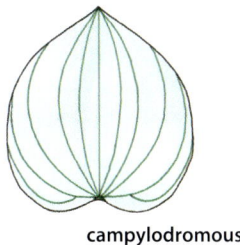

campylodromous

campylotropous
Of ovule orientation, with the ovule curved and at more or less 90° to the funicle, and the micropyle bent downwards.
see **ovule orientation**

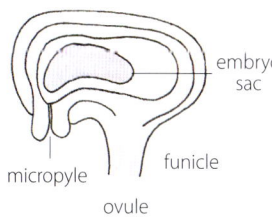

embryo sac

funicle

micropyle

ovule

campylotropous

canaliculate   Having one or more longitudinal grooves or channels, as some leaves.
= **channelled**

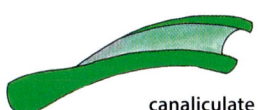

canaliculate

cancellate   Lattice-like, as the scales of the spleenwort fern (*Asplenium*).
= **clathrate**

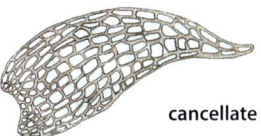

cancellate

candelabriform
Having the shape of a tall branched candlestick, as candelabra tree (*Euphorbia candelabrum*).

candelabriform

cane   A strong, slender, often jointed, stem of various plants such as sugarcane, bamboo, rattans, raspberries and grapevines.

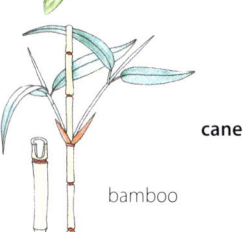

cane

bamboo

canescent   Pale silvery-grey.
Covered with a greyish to whitish layer of very short, closely interwoven fine hairs.
= **hoary**

canopy   Of a single
tree, its crown.
In a forest, the
uppermost layer
formed by the crowns
of the trees.

canopy

cantharophile   A plant pollinated by beetles.

cantharophily   Dispersal of pollen and pollination
by beetles.
cantharophilous   Pollinated by beetles.

capillary   Very slender
and hair-like.
Similar to filiform but
more delicate.

capillary

capitate   With a knob-
like head.
Of an inflorescence, with
flowers sessile and in a
dense cluster, as some
wattles (*Acacia*).
Of a stigma, shaped like
the head of a pin, as
that of lesser loosestrife
(*Lythrum hyssopifolia*).

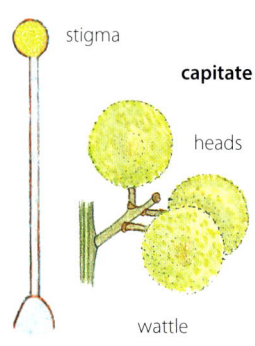

stigma

capitate

heads

wattle

capitellate   Minutely knob-shaped. Terminating in
a very small knob, as the stigmas of some flowers.
Grouped to form a capitulum.

capitular   Of or
relating to a capitulum
as the papery, petal-
like capitular bracts
surrounding the
capitulum of golden
everlasting (*Xerochrysum
bracteatum*).

capitular

bracts

daisy capitulum

capitulum, *pl.* **capitula**   A dense cluster of sessile
or almost sessile small flowers (florets) arranged on
a flattened or rounded receptacle.
A capitulum may be simple or compound.
It may be indeterminate (racemose), as the typical
inflorescence of the daisy family (Asteraceae), made
up of ray florets and/or tubular florets.
It may be determinate (cymose), as the bushmint
genus (*Hyptis*).
see **head**

Capitulum (Asteraceae)

 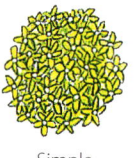 

Simple

all ray florets   all tubular florets   ray and tubular florets

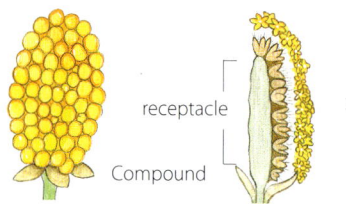 

receptacle

Compound

>

many capitula   capitula on a receptacle   capitulum of florets

caprification   Transfer of pollen from the native
caprifig flowers, by a fig wasp, to some cultivated
figs.

capsule
A mostly dry fruit that splits open to release seeds in
angiosperms and spores in bryophytes.
It is composed of cavities (locules), separated by
walls (septa), that contain seeds or spores.
In angiosperms, it is derived from a syncarpous
inferior or superior ovary with two or more carpels.
see **capsule dehiscence**
capsular   Of or pertaining to a capsule.

Capsule (angiosperms)

 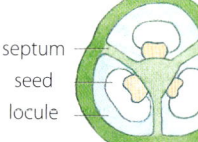

septum
seed
locule

coast bitterbush (*Adriana quadripartita*)

capsule dehiscence   A capsule is described
according to how it splits open at maturity.

Capsule dehiscence

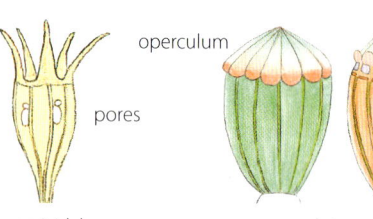 

operculum

pores

pores

poricidal

operculate-poricidal

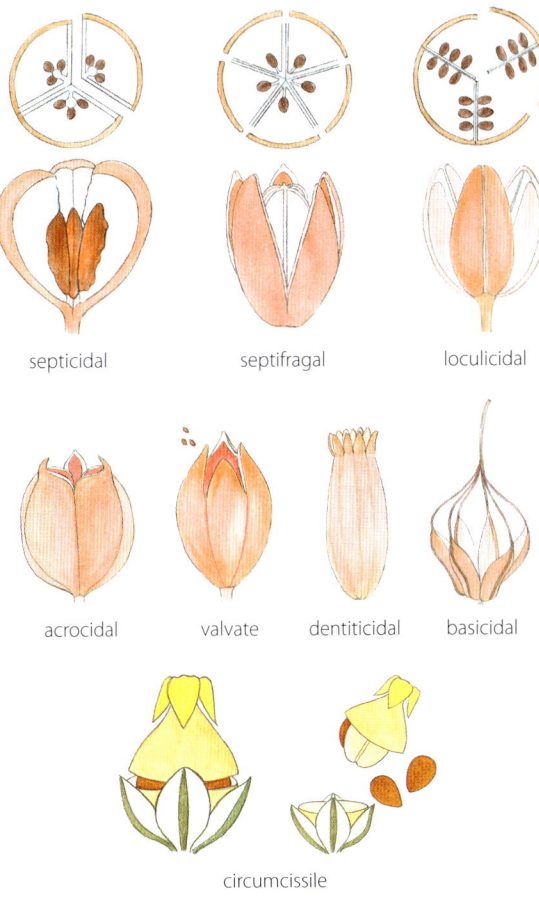

septicidal   septifragal   loculicidal

acrocidal   valvate   dentiticidal   basicidal

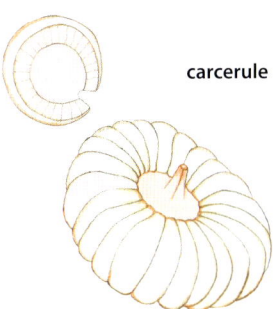

circumcissile

carbohydrate   Any member of a group of chemical compounds containing carbon, hydrogen and oxygen, including sugars, starches and cellulose. It is produced by green plants from carbon dioxide in the atmosphere and water during photosynthesis.

carbon, (C)   A non-metallic chemical element, having the symbol C, that occurs in many inorganic and all organic compounds.
Plants take carbon from the atmosphere in the form of carbon dioxide during photosynthesis.
It is converted into simple sugars to build starches, carbohydrates, cellulose, lignin and protein.

carcerule, carcerulus
A dry schizocarpic fruit that splits at maturity into four or more nutlets, as hollyhock (*Alcea*). Derived from a multicarpellary syncarpous superior ovary.

carcerule

carina   A ridge along the centre of the lower surface like that on the bottom of a boat.
The two lower petals of a pea flower, united along their lower margin to form a keel that encloses the stamens and pistil.
carinate   Shaped like the keel of a boat or having a keel, as some leaves.
= **keeled**

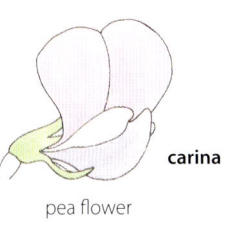

carina

pea flower

carina

leaf

carneous, carnose   With the texture of flesh. Fleshy, flesh-coloured, pale red.

carnivorous
Of plants, adapted to trap and digest small animals, especially insects, as insectivorous plants with modified leaves like Venus fly trap (*Dionaea muscipula*).
cf. **herbivorous**

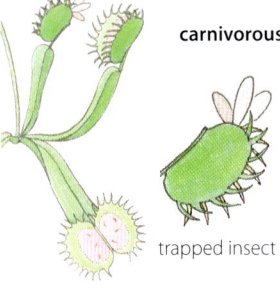

carnivorous

trapped insect

carotene
A red, yellow or orange pigment in the chromoplasts of plants cells that absorbs ultra-violet, violet and blue light in photosynthesis.
see **carotenoids**

carotenoids
A group of yellow, orange or red pigments synthesised by many plants that provide fruit and flowers with bright colours.
They are grouped into carotenes and xanthophylls. There are about 500 in all and examples are beta-carotene that gives carrots and other vegetables their orange colour and lycopene that gives tomatoes their red colour.
They act as visible signals on flower and fruit to attract insects, birds and animals for pollination and seed dispersal.
They interact with chlorophyll inside chloroplasts and help in light absorption for photosynthesis.

carpel The female reproductive unit of a flower, typically consisting of a stigma, style and ovary with ovules.
A simple pistil has only one carpel and the terms carpel, pistil and gynoecium are then synonymous.
A compound pistil has more than one carpel.
see **megasporophyll, ovuliferous scale**
see **gynoecium, pistil**
see also **apocarpous, syncarpous**
carpellary Of or like a carpel.
carpellate Bearing carpels.

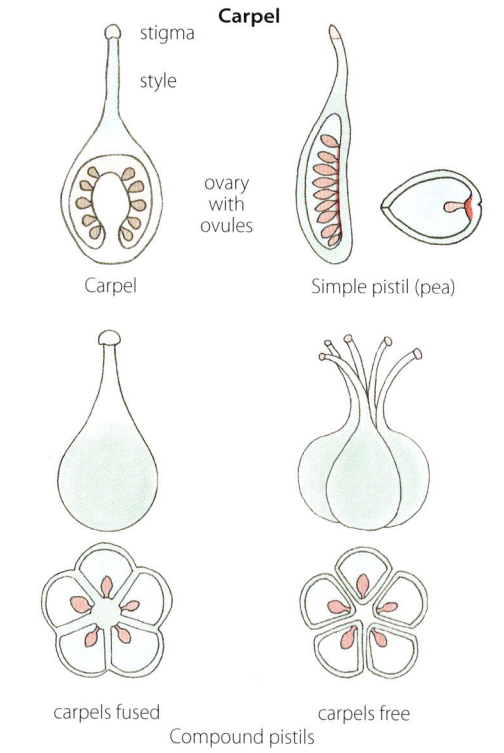

Carpel

stigma
style
ovary with ovules

Carpel   Simple pistil (pea)

carpels fused   carpels free
Compound pistils

carpellode
A sterile carpel, as in the male flower of coconut (*Cocos nucifera*).

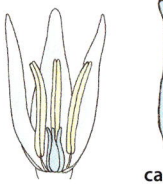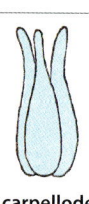

carpellode

carpellody The development of misshapen fruit caused by the fusion of the stamens to the ovary, as papaya (*Carica papaya*).

carpophore
A slender continuation of the flower stalk that bears the carpels in the parsley family (Apiaceae) and geranium family (Geraniaceae).

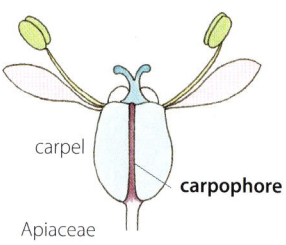

carpel

carpophore

Apiaceae

carpopodium In the daisy family (Asteraceae), the distinct short thickening at the base of some cypselas.
The point of attachment of the cypsela to the receptacle.

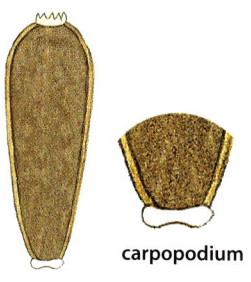

carpopodium

cartilagineous, cartilaginous Hard, tough and gristly.

caruncle A fatty food body on some seeds that attracts dispersal agents, as the appendage on the seed of coast bitter bush (*Adriana quadripartita*).
carunculate Bearing or related to a caruncle.

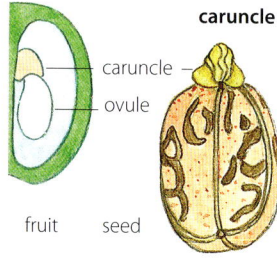

caruncle

caruncle
ovule

fruit   seed

caryophyllaceous
Of, relating to or belonging to the carnation family (Caryophyllaceae). Typically with a five-petalled corolla and each petal with a long erect claw and a spreading lamina.

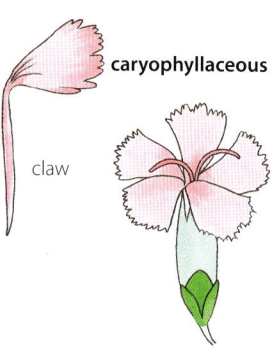

caryophyllaceous

claw

caryopsis A dry indehiscent fruit with one seed fused to the fruit wall (pericarp).
Derived from a one-carpelled superior ovary.
Characteristic of grasses (Poaceae), as wheat (*Triticum*).
cf. **achene, cypsela, diclesium**
= **grain**

Caryopsis

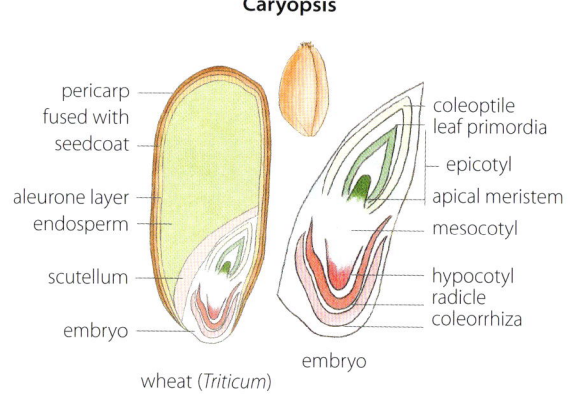

pericarp fused with seedcoat

aleurone layer
endosperm

scutellum

embryo

coleoptile
leaf primordia
epicotyl
apical meristem
mesocotyl
hypocotyl
radicle
coleorrhiza

embryo

wheat (*Triticum*)

Casparian strip Of the endodermis in roots, a waxy, waterproof band of material embedded in the sides of cells where they abut.

It prevents soil water and nutrients from diffusing through the endodermal cell walls and forces it to pass through the plasma membrane and cytoplasm to reach the vascular cylinder.

see also **apoplastic pathway, symplastic pathway**

Casparian strip

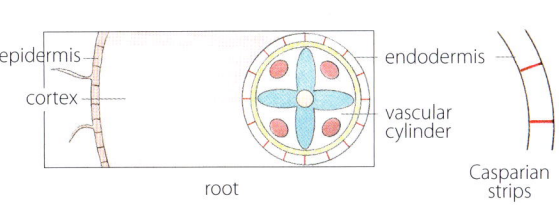

epidermis
cortex
endodermis
vascular cylinder
root
Casparian strips

castaneous A deep reddish-brown colour like that of a chestnut (*Castanea*).

casual Of an alien plant that has not become naturalised.

catabolism The release of energy during the breakdown of larger complex molecules into smaller molecules, as the conversion of glucose during respiration to produce carbon dioxide, water and energy that is stored as adenosine triphosphate.

cf. **anabolism**

catadromous, catadromic Of fern venation, having the first lateral veins in a segment extending toward the posterior margin.

cf. **anadromous, isodromous**

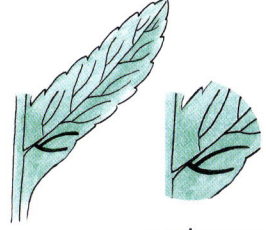

catadromous

catalyst A substance capable of initiating or speeding up a chemical reaction.

see also **enzyme**

cataphyll Any of several leaves that are not photosynthetic and function as protection, as the scale leaves that protect the buds of silver birch (*Betula pendula*), or for storage, as the fleshy scales of a bulb.

They may be present on aerial parts, as those at the base of rushes (*Juncus*) or underground parts, as those on the rhizomes of the turmeric family (Zingiberaceae).

When cataphylls occur on a seedling they are the first leaves after the cotyledon.

see also **plumular leaves**

Cataphyll

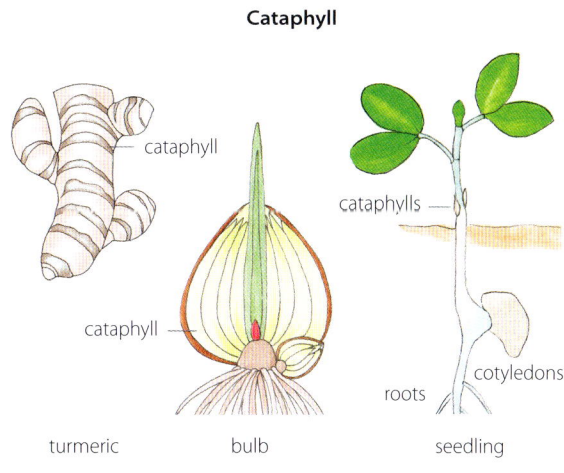

cataphyll
cataphylls
cataphyll
cotyledons
roots
turmeric
bulb
seedling

catkin A spike or raceme of unisexual, usually apetalous, flowers on a pendulous stem, as the flowers of the walnut (*Juglans regia*).

= **ament**

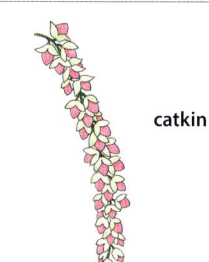

catkin

cauda, *pl.* **caudae** A slender tail-like appendage.
caudate Having a slender tail-like appendage.
caudiform Shaped like a tail.

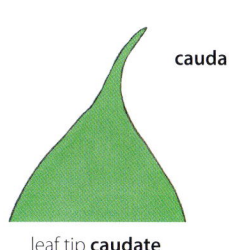

cauda

leaf tip **caudate**

caudex, *pl.* **caudexes, caudices** A stout swollen or succulent trunk-like stem, as that of pachycaul plants.

The main stem of a cycad, palm or tree fern. An underground stem that is the rootstock for some herbaceous perennials, as trillium (*Trillium*).
caudiciform Having the form of a caudex.

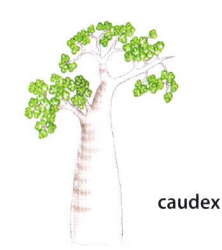

caudex

pachycaul (*Dendrosicyos*)

caudex
tree fern (*Cyathea*)
trillium (*Trillium*)

caudicle   Of orchids
(Orchidaceae)
and milkweeds
(Asclepiadaceae), the
slender elastic extension
of the pollen mass that
connects it to the sticky
viscidium.
It attaches to a pollinator
and helps separate the
pollen mass from the
anther.
cf. **stipe**

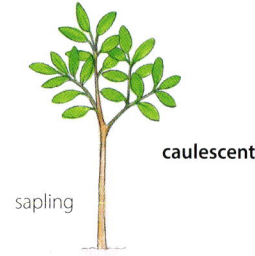

caulescent   Having a
well-developed above-
ground stem, as a
sapling. Includes shrubs,
trees and herbs with
aerial shoots and leaves.
cf. **acaulescent**

cauliflory   The
production of flowers
on the older branches or
trunks of woody plants,
as the watermelon tree
(*Syzygium moorei*).
cauliflorous   Exhibiting
cauliflory.

cauline   Growing on a
stem, having a stem.
Of leaves growing on a
stem.
cf. **basal, terminal**

caulome   The stem structure of a plant considered
as a whole.
cf. **phyllome**

cecidium, *pl.* **cecidia**   A gall.

cell   The microscopic structural and functional
unit of all living organisms, consisting of a nucleus,
cytoplasm, organelles, cell membrane and, in plants,
a cell wall.
see **eukaryote, prokaryote**
cellular   Of, relating to or consisting of cells.

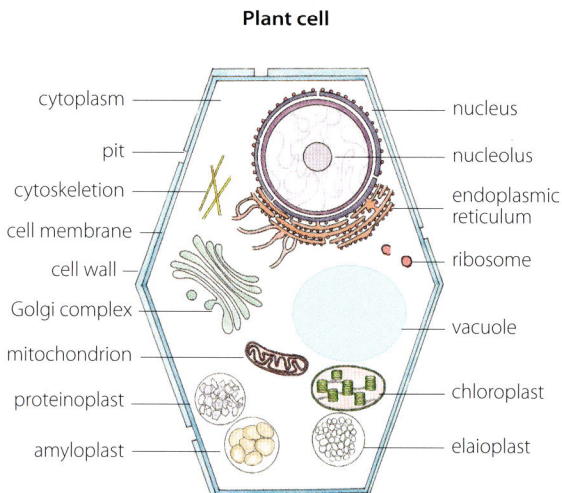

Plant cell

cytoplasm — nucleus
pit — nucleolus
cytoskeletion — endoplasmic reticulum
cell membrane — ribosome
cell wall —
Golgi complex —
mitochondrion — vacuole
proteinoplast — chloroplast
amyloplast — elaioplast

cell membrane
A thin semipermeable layer of tissue enclosing the
cytoplasm of a cell and, in plants, surrounded by the
cell wall.
It allows movement of some substances into and
out of the cytoplasm.
= **cytoplasmic membrane,
plasma membrane**

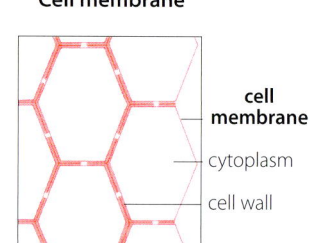

Cell membrane

cell membrane
cytoplasm
cell wall

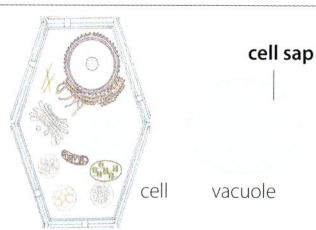

cell sap   The watery
fluid in a plant cell
vacuole that is made
up of water, salts and
sugar.

cell sap

cell   vacuole

cell wall
The outer surface of a plant cell that surrounds the
cell membrane.
It has up to three layers: a mainly cellulose layer
next to the cell membrane (the primary cell wall),
a pectin layer (the middle lamella) that is adhesive
and helps adjacent cell walls bind to one another
and, in some cells, a rigid lignin-rich strengthening
layer (the secondary cell wall).

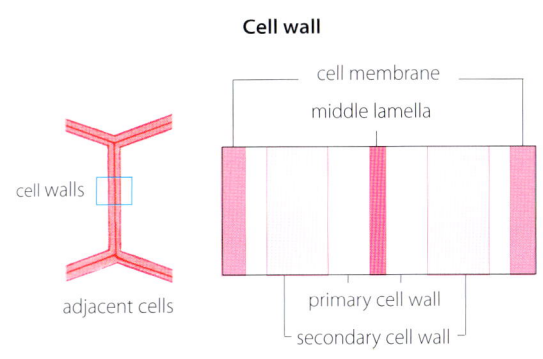

Cell wall

- cell membrane
- middle lamella
- cell walls
- adjacent cells
- primary cell wall
- secondary cell wall

cellular respiration   The pathway by which cells release energy from nutrients like glucose.
The energy released is trapped in the form of adenosine triphosphate for use in cell processes.
see also **aerobic respiration, anaerobic respiration, glycolysis, Krebs cycle**

cellulose   A complex carbohydrate, composed of glucose units, that is the main component of the cell walls of plants.

central cell   The largest cell in the embryo sac with two nuclei (the polar nuclei).
It is typically highly vacuolated and rich in food reserves.
see **double fertilisation**

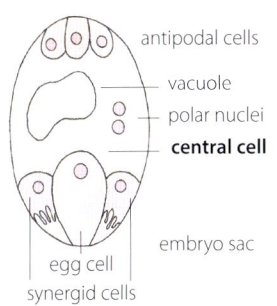

- antipodal cells
- vacuole
- polar nuclei
- **central cell**
- embryo sac
- egg cell
- synergid cells

central placentation
Carpels are fused but the internal walls (septa) are lacking, creating a unilocular ovary. The ovules are arranged along a central column (axis) that does not reach the top of the ovary, as primrose (*Primula*).
= **free central placentation**
see **placentation**

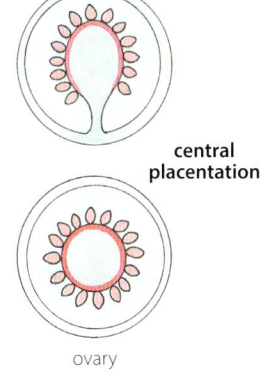

central placentation

ovary

centric   Having one usually cylindrical surface, as the leaf of an onion (*Allium cepa*) with no distinct upper and lower surfaces.
= **unifacial**
cf. **bifacial, equifacial**

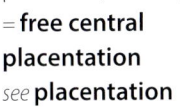
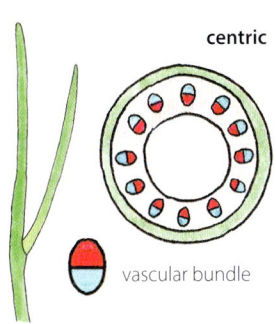

centric

vascular bundle

centrifixed   Of anther attachment in which the slender filament tip is inserted into a hollow or pit at the base of the anther.
see **anther attachment**

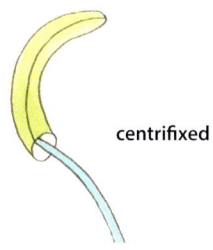

centrifixed

centrifugal
Flowering develops in sequence, from the centre towards the outer edge, as the cymose inflorescence of guelder rose (*Viburnum opulus*). Flowers at the centre open first and those on the outer side open last.
see also **acropetal, basipetal**
cf. **centripetal**

centrifugal

centripetal
Flowering develops, in sequence, from the outer edge towards the centre, as the racemose inflorescence of St. Peter's wort (*Hypericum tetrapterum*).
Flowers at the outer edge open first and those in the centre open last.
see also **acropetal, basipetal**
cf. **centrifugal**

centripetal

centromere
The point at which two chromatids touch.
Each chromosome is copied during cell division and the copy is joined to the original at a centromere.
A centromere is the attachment point for spindle fibres that separate the chromosome copies and pull them towards opposite ends of the cell before it divides.

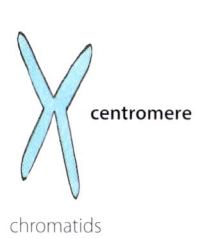

centromere

chromatids

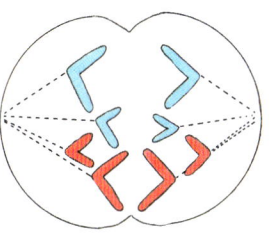

spindle fibres

cell division

cephalium, *pl.* **cephalia**
A woolly, densely bristled outgrowth at the top or side of some cactus species.
Flowers and fruit form on the cephalium.

cephalium

ceraceous, ceraceus Waxy in texture or appearance. A pale whitish-cream colour.

cereal Grass cultivated for grain for human or animal consumption, as maize, barley, wheat or oats.

cernuous Facing downwards, as some buds, flowers and fruit.

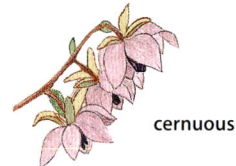
cernuous

cf. Compare with.

chaff Thin dry membranous scales or bracts, as the winnowed glumes, lemmas and paleas of grains and other grasses.
One of the bracts or scales on the receptacle of daisies (Asteraceae).
chaffy Covered with or consisting of chaff, resembling chaff.
see also **paleaceous**

chaff

wheat chaff

chalaza The region of an ovule opposite the micropyle where the integument(s) merge and are united with the nucellus.
The funicule, chalaza, and raphe, when present, form a continuous tissue.

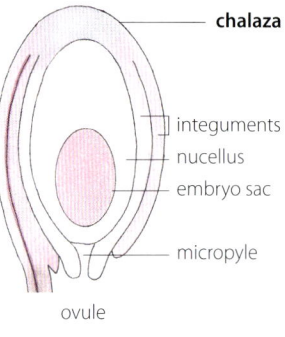
chalaza
integuments
nucellus
embryo sac
micropyle
ovule

chalazogamy
Entrance of the pollen tube through the chalazal tissue of the ovule.
cf. **mesogamy, porogamy**

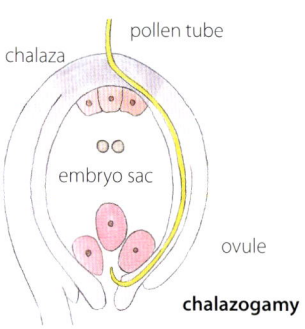
pollen tube
chalaza
embryo sac
ovule
chalazogamy

chamaephyte A plant to 25 cm high with perennating buds close to the soil surface and protected by leaf litter or snow in unfavourable conditions.
Includes subshrubs, cushion plants and plants with buds on stems that fall over, or ground covers on horizontal stems.
see also **cryptophyte, hemicryptophyte, phanerophyte, therophyte**

channelled Having one or more longitudinal grooves or channels, as some leaves.
= **canaliculate**

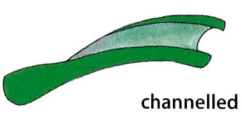
channelled

chaparral A dense semi-arid vegetation in California, with mainly tough woody evergreen shrubs to about two metres high. It has mild wet winters and summer droughts.
The chaparral is a Mediterranean-type ecosystem together with the garrigue and maquis in the Mediterranean Basin, the matorral in Chile, kwongan in southwestern Australia and fynbos in South Africa.

character, characteristic A structure, function or other attribute of a plant used to distinguish one taxon from another.
It may be inherited genetically or acquired in response to the environment.
see **acquired character, inherited character**

character state Two or more forms of a character, as leaf shape which may be lanceolate, ovate etc.

chartaceous
Thin and dry like paper, as the colourful leaf-like bracts of bougainvillea (*Bougainvillea*).
= **papery**

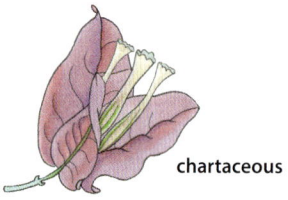
chartaceous

chasmogamy Flowers that open and expose the anthers and stigmas for pollination.
cf. **cleistogamy**
chasmogamous Of a flower that opens for pollination. Of a plant producing such flowers.

chasmophyte A plant that grows in rocky fissures and clefts.
cf. **chomophyte, lithophyte**

chemoautotroph, chemotroph An organism that oxidises chemical compounds to synthesise its own energy.
Typically a bacterium or a protozoan living in a hostile environment.
see **autotroph**

chemoautotrophic, chemotrophic Of or relating to a chemoautotroph or a chemotroph.
see **trophic**

chimaera, chimera An organism having a mixture of genetically different tissues that originate from two or more different zygotes.
Formed by processes like fusion of early embryos and grafting.
see **graft chimaera, mosaic**

chiropterochory Dispersal of seeds by bats.
cf. **chiropterogamy**
chiropterochorous Relating to chiropterochory.

chiropterogamy Dispersal of pollen and pollination by bats.
= **chiropterophily**
cf. **chiropterchory**
chiropterogamous Relating to chiropterogamy.

chiropterophile A plant species that is pollinated by bats.

chiropterophily Dispersal of pollen and pollination by bats.
= **chiropterogamy**
cf. **chiropterchory**
chiropterophilous Pollinated by bats.

chitin A tough substance found in the outer skeleton (exoskeleton) of arthropods, like crabs, insects and spiders, and in the cell walls of fungi.

chlamydeous Having a perianth consisting of a calyx and/or a corolla.
cf. **achlamydeous, dichlamydeous, heterochlamydeous, homochlamydeous, monochlamydeous**

chlorenchyma
Photosynthetic parenchyma cells that contain chloroplasts. Typically, it constitutes the mesophyll of a leaf, but it is also present in other green plant organs, as stems and unripe fruits.

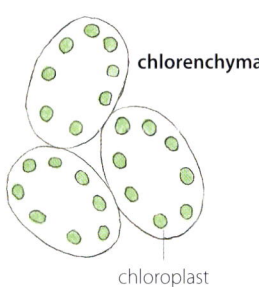

chlorenchyma

chloroplast

chlorophylls A group of greenish pigments two of which, chlorophyll *a* and chlorophyll *b*, are found in plants.
Chlorophyll *b* is found in green algae and chlorophyll *a* is found in the chloroplasts of other plants.
Chlorophylls absorb light energy for photosynthesis.
see **granum, thylakoid**
chlorophyllous Containing chlorophyll.

chloroplast The organelle in the cell that is responsible for photosynthesis in plants. It contains the greenish pigment chlorophyll in stacks of disc-shaped thylakoids that are the site of photosynthesis.
see **granum, stroma**

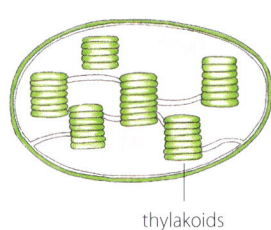

chloroplast

thylakoids

chloroplast DNA, cpDNA The small amount of DNA unique to chloroplasts.
Unlike nuclear DNA, cpDNA is arranged in rings.

chlorosis An abnormal condition resulting in green plants becoming yellowish due to a reduction of chlorophyll levels.
Its causes include mineral deficiency, disease or lack of light.
see also **etiolated**
chlorotic Of or relating to chlorosis.

chomophyte A plant that grows in soil on rocky ledges.
cf. **chasmophyte, lithophyte**

choricarpous

Of a compound gynoecium or fruit with two or more separate carpels.
= **apocarpous**
cf. **syncarpous**

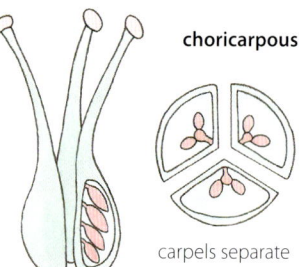

choricarpous

carpels separate

choripetalous

With petals free from each other.
= **apopetalous, dialypetalous, polypetalous**
cf. **gamopetalous, sympetalous**

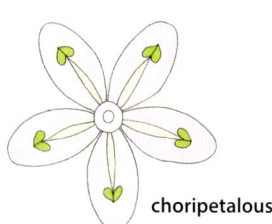

choripetalous

chorisepalous

With sepals free from each other.
= **aposepalous, dialysepalous, polysepalous**
cf. **gamosepalous, synsepalous**

chorisepalous

chorisis The division or splitting of a leaf or floral organ, as some stamens, into two or more parts. It may be a normal or abnormal development.

choritepalous

With tepals free from each other, as day lilies (*Haemerocallis*).
= **apotepalous, polytepalous**
cf. **gamotepalous, synsepalous**

choritepalous

chromatid Either of the two strands of DNA formed when a chromosome makes an exact copy of itself (replication) before cell division.
see **meiosis, mitosis**

Chromatid

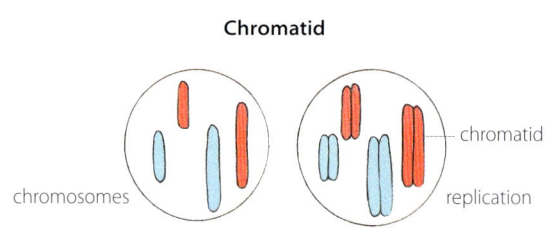

chromosomes   chromatid

replication

chromatin

Material composed of DNA and proteins in the nucleoplasm of the nucleus. It organises into visible chromosomes during cell division.

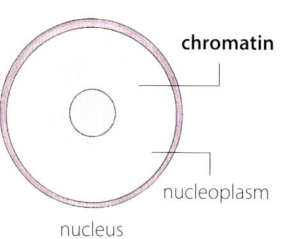

chromatin

nucleoplasm

nucleus

chromoplast

A plastid containing pigments other than chlorophyll, especially red, orange or yellow carotenoids. Found in fruits, flowers, roots and senescent leaves.

chromoplast

chromosome

One of the pairs of tightly coiled thread-like structures of DNA located in the nucleus of plant and animal cells. It carries genetic information.

In unicellular organisms that lack a nucleus (prokaryotes), like bacteria, there is no nucleus and the DNA floats freely in the cell.

The genetic information of each chromosome is transmitted from the parent cell to the daughter cells during cell division.

Each species has its own characteristic number of chromosomes.
see **chromatin**

chromosome set

The complement of chromosomes in the nucleus of a cell. It is constant for each species.

The basic set of chromosomes is called the monoploid set (x). A diploid species has two basic sets (2x), a triploid species has three basic sets (3x), and so on.

Chytridiomycota

The most primitive phylum of fungi known as chytrids. They are mainly aquatic and the only group having gametes with a flagellum that allows them to swim.

cicatrice, cicatrix, *pl.* cicatrices

A scar.
The mark left by the separation of one part from another, as by a leaf from a stem.

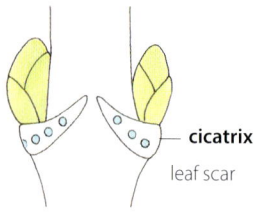

cicatrix

leaf scar

ciliola   A small, fine eyelash-like hair.
ciliolate   Bordered with small eyelash-like hairs, as the margins of some leaves.

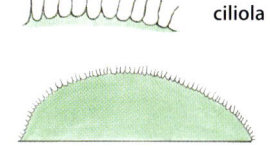

ciliola

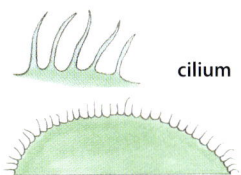

cilium, *pl.* **cilia**   A fine eyelash-like hair.
ciliate   Bordered with eyelash-like hairs, as the margins of some leaves.

cilium

cincinnus
A spirally coiled cymose inflorescence with a single new stem developing from one axil only. Branching continues to alternate from one axil to the other so that the axis is zigzagged.
= **scorpioid cyme**
see **monochasium**
cf. **bostryx**

cincinnus

cinereous   Resembling ashes. Grey with a coppery tint.

circinate   Coiled in a flat spiral with the tip innermost, as an unopened fern frond.
circinate ptyxis   Of a single leaf in bud that is flattened and spirally coiled, with the tip innermost.

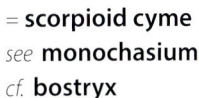

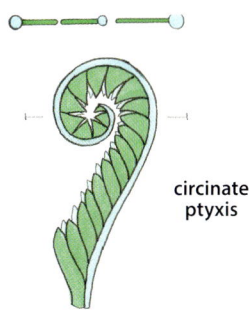

circinate ptyxis

circinotropous
Of ovule orientation, with the ovule turned at more than 360° so that the funicle becomes coiled around the ovule.
see **ovule orientation**

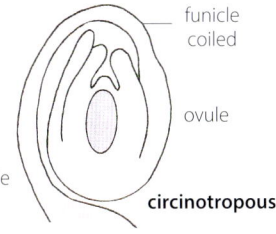

funicle coiled

ovule

funicle

circinotropous

circumcissile capsule
A capsule that dehisces around the circumference so that the upper part separates like a lid from the lower part, as plantain (*Plantago*).
= **pyxis**

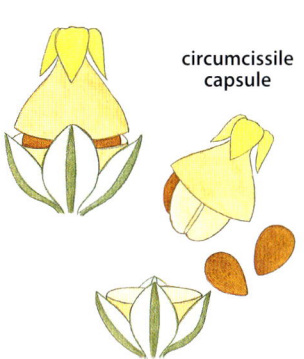

circumcissile capsule

circumscription   A description of what does and does not belong to a given taxon and of what sets it apart from others that are related to it.
see **description, diagnosis**

cirriferously-pinnate
Of a pinnate leaf with the terminal leaflet only, or the upper lateral leaflets also, reduced to tendrils.

cirriferously-pinnate

cirrus, cirrhus, *pl.* **cirri, cirrhi**   A tendril.
Of climbing palms (rattans), a barbed whip-like extension of a leaf midrib.
cirrose, cirrhose, cirrate   Ending in a long coiled tip.
cf. **flagellum**
cirrhiferous, cirriferous
Bearing a tendril or tendrils.
cirriform   Resembling a tendril.

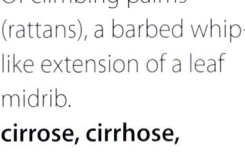

tendils

cirrus

cirrus

rattan

cirrate

cisterna, *pl.* **cisternae**
A flattened membrane-bound sac.
Occurs in the Golgi complex and the endoplasmic reticulum.
see **dictyosome**

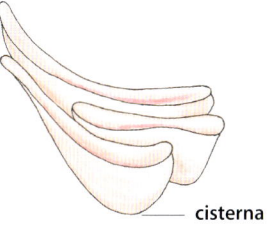

cisterna

clade A group of organisms that includes a hypothetical common ancestor taxon and its descendants.
see **cladogram**

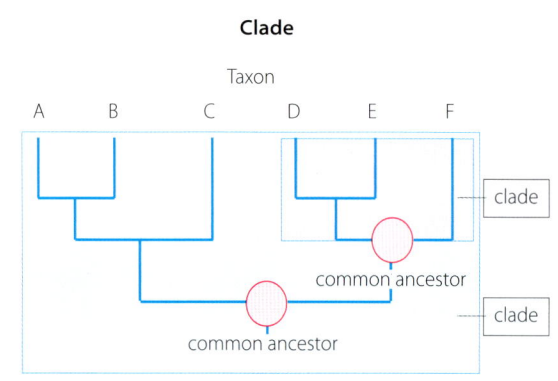

Clade

Taxon

A   B   C   D   E   F

clade

common ancestor

clade

common ancestor

cladistics A method of hypothesising relationships among organisms by reconstructing evolutionary trees. The tree is the result of analysis of data about characters or traits of the organisms, as those related to anatomy, physiology, behaviour or genetic sequencing.
see **clade, cladogram**

cladode One internode on a cladophyll. It is usually flattened and leaf-like but may also be needle-like on asparagus. Sometimes used synonymously with cladophyll and phylloclade.

cladode

Schlumbergera

cladophyll

Asparagus

cladodromous Of leaves with secondary veins branching repeatedly and becoming indistinct before reaching the margin.

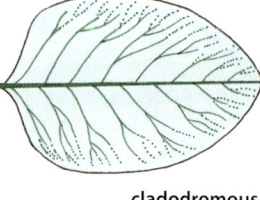

cladodromous

cladogenesis One of two main ways in which speciation occurs in response to the environment. Cladogenesis is branching, with an ancestral species separating into two or more new species.
see **cladogram**
cf. **anagenesis**

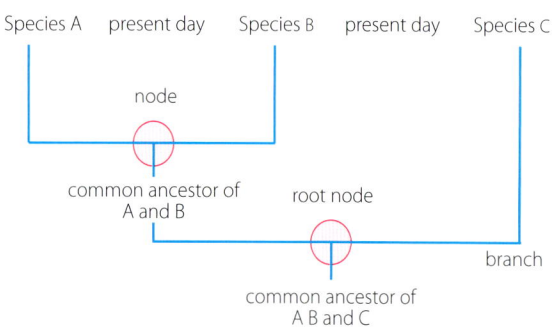

Cladogenesis

Species A   present day   Species B   present day   Species C

node

common ancestor of
A and B

root node

branch

common ancestor of
A B and C

cladogram A diagram showing evolutionary relationships within one or more clades. It shows a hypothetical ancestor, the lines of descent and the taxa that have evolved from it as the end-points.

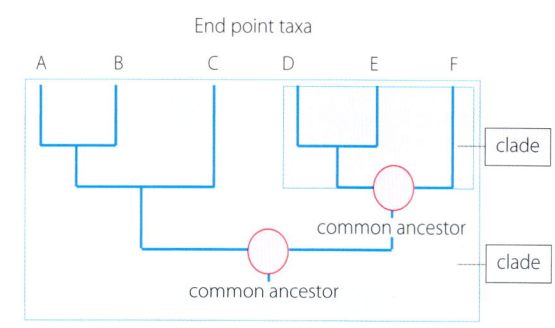

Cladogram

End point taxa

A   B   C   D   E   F

clade

common ancestor

clade

common ancestor

cladophyll A type of cladode that resembles leaf-like branches, as those in the cactus family. A term with variable definitions.
see **cladode**
= **phylloclade**

cladophyll

prickly pear
(*Opuntia*)

cladoprophyll A modified prophyll in some sedges, on the axis of lower paracladia, as the genus *Carex*. It may be tubular, scale-like or utricle-like and enclosing a flower.

cladoptosis Of the cypress family (Cupressaceae), with dead foliage falling simultaneously with branches and stems.
cladoptosic Relating to cladoptosis.

clambering Of a plant with thin weak stems that climbs or sprawls across objects or other plants without the use of tendrils or aerial roots, as black-eyed Susan (*Thunbergia alata*).
cf. **vine**

clambering

clasping Partially or completely surrounding an organ, as the leaf sheath of a grass clasps the culm.
cf. **amplexicaul**

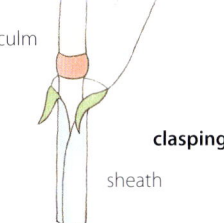
culm
clasping
sheath

class In taxonomic classification, a rank below division or phylum and above order.
Names of classes end in *-opsida*.
see **taxonomic hierarchy**

classification The systematic grouping of organisms, such as plants, animals, fungi and single-celled life forms, into categories according to their common attributes.
see **phenetics, phylogeny, taxonomy**

clathrate Lattice-like, as the scales of the spleenwort fern (*Asplenium*).
= **cancellate**

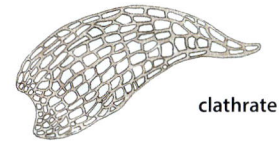
clathrate

clava, *pl.* **clavae** A club-shaped element.
clavate Having clavae, club-shaped.

clava

clavellate Club-shaped, but smaller than clavate.

clavuncle The expanded part of the style just below the stigma in the dogbane family (Apocynaceae).

claw The stalk-like base of a petal, sepal or bract.
= **unguis**
clawed Having a claw.
= **unguiculate**
see **caryophyllaceous**

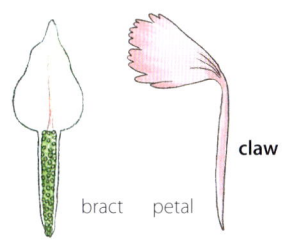
claw
bract petal

clay Very fine-grained soil that is sticky and greasy when wet and hard when dry. The particles are less than 0.002 mm in diameter.
Composed mainly of aluminium and silica.

cleft Split.
Of a pinnatifid or palmatifid leaf with sharp sinuses so that the lobes are pointed rather than rounded.
Of a leaf tip, split by a sharp sinus.
cf. **forked**

Cleft leaves

pinnately cleft   palmately cleft   leaf tip

cleistogamy
Self-pollination within an unopened flower.
see **autogamy**
cf. **chasmogamy**
cleistogamous Of flowers that self-pollinate in bud and remain closed, or self-pollinate before opening.
In orchids pollen falls from the anthers onto the stigmatic disc in the unopened flower.

Cleistogamy

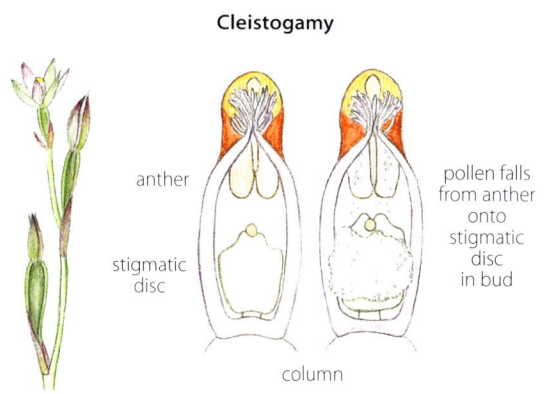

anther
stigmatic disc
column

pollen falls from anther onto stigmatic disc in bud

Slender sun orchid (*Thelymitra pauciflora*), self-pollination in bud.

cleistogene
In some grasses, a modified spikelet with florets that remain closed and are self-fertilised. Usually located within the leaf sheaths at the lower nodes.
see also **rhizanthogene**

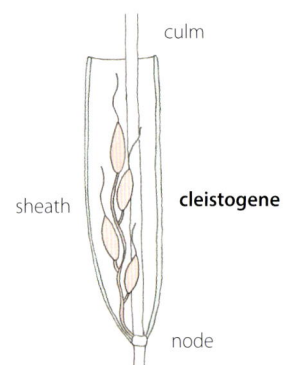
culm
sheath
cleistogene
node

climate   The average condition of the atmosphere, including temperature, precipitation, humidity, wind and seasonality, at a particular place over a long period of time.
cf. **weather**

climax community   The last stage in an ecological succession that sees a community of living things existing in equilibrium with its environment.
see also **sere**

climber   A plant, that grows vertically using tendrils or aerial roots, as ivy (*Hedera helix*), to attach itself to a support, usually another plant.
cf. **twiner**

climber
ivy

climbing fern
The climbing fern family (Lygodiaceae) comprises one genus (*Lygodium*). Plants have subterranean creeping rhizomes and climbing, twining fronds with a wiry rachis.
see **fern**

climbing fern
climbing fern (*Lygodium*)

clinandrium *pl.* **clinandria**   Of orchids, the depression or area on the column of a flower where the anther is situated.
= **androclinum**

clinanthium, clinum
The receptacle bearing the florets in the head (capitulum) of the daisy family (Asteraceae).

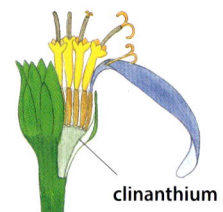
clinanthium

cline   A gradual change in a character or feature of a species across its geographical range.

clisere   A succession of climax communities in a given area over time. They are a result of changes in climate.

clone   Fragments of a genet that can survive separately, as a nodes with roots from a strawberry (*Fragaria*).
Individual fragments are genetically identical and are called ramets.
A set of organisms produced from a genet by vegetative reproduction.
see **genet**
clonal   Of or relating to a clone.

clonal colony   Of plants, a group of genetically identical individuals (clones) that grow in a given location. They all originate vegetatively from a single ancestor.
see **ramet**

closed vascular bundle   A vascular bundle that lacks a cambium layer between the xylem and phloem and has no secondary growth. Typical of monocotyledons.
cf. **open vascular bundle**

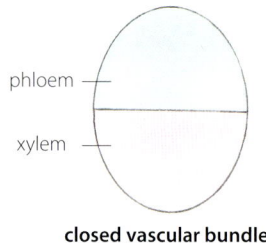
phloem
xylem
closed vascular bundle

closed venation
Of veins in a leaf connected together to form a network of loops, or that run parallel from base to apex, as grass blades and leaves of other monocotyledons.
see **anastomosis**
cf. **open venation**

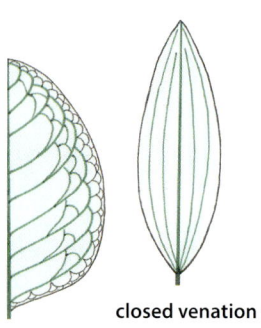
closed venation

club   Of orchids, the thickened apical part of a sepal or petal, often bearing osmophores, as that on some spider orchids (*Caladenia*).

club

clubmoss   The clubmoss family (Lycopodiaceae) comprises four genera of vascular plants that reproduce by spores rather than seeds.
They are terrestrial or epiphytic, with single-veined leaves and sporophylls sometimes arranged in cone-like strobili.
Spores are of only one kind (homosporous).
see **fern allies**

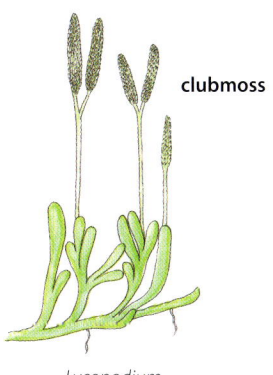

clubmoss

Lycopodium

clump   A group of similar things in a compact mass, as garlic (*Allium sativum*) or a clump of trees.
clumping   Not spreading.
Non-invasive, as clumping bamboos.
see **sympodium**

clump

garlic

clumping bamboo

cluster   A number of like things growing or collected together.
clustered   In a cluster, as leaves on a stem.

cluster

co-   A prefix meaning together, mutually or in common.

coalescent   Of plant parts fused or grown together to form a single unit, as the bracts of the cup of an acorn (*Quercus*).
cf. **contiguous**

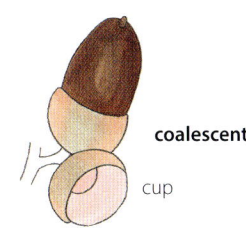

coalescent

cup

cobwebbed   Covered with fine loosely entangled whitish hairs, resembling a spider's web, as some leaves.
= **arachnoid**

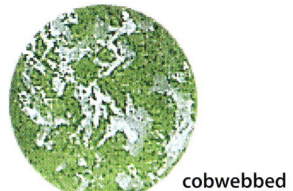

cobwebbed

coccus, *pl.* **cocci**   A single-seeded dry fruitlet of a schizocarp.
It may be indehiscent, as the carrot family (Apiaceae) or dehiscent, as geranium (*Geranium*).
see **carcerule, regma**
= **mericarp**

Coccus

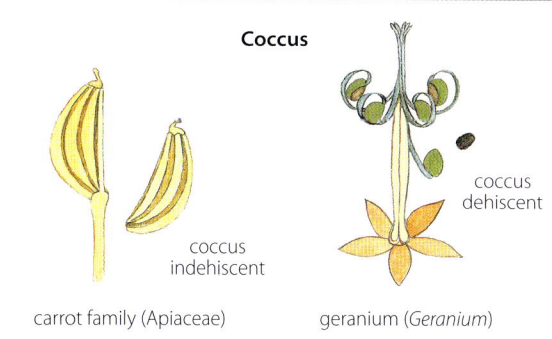

coccus
indehiscent

coccus
dehiscent

carrot family (Apiaceae)

geranium (*Geranium*)

cochlear, cochleate   Spoon-shaped.
cochlear vernation   Of young leaves in the unopened leaf bud, with one leaf larger than the others and shaped like a bowl, and exterior to the other leaves. A form of imbricate vernation.

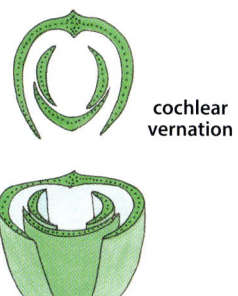

cochlear vernation

codominance   Of a heterozygous individual, having both alleles of a gene equally dominant and expressed in the phenotype.
cf. **epistasis, recessive**

codon   A sequence of three DNA or RNA nucleotides that corresponds with a specific amino acid.

coenanthium   An inflorescence with a nearly flat receptacle having margins that are slightly curved upwards.
The flowers are embedded in the receptacle itself, as the genus *Dorstenia*.

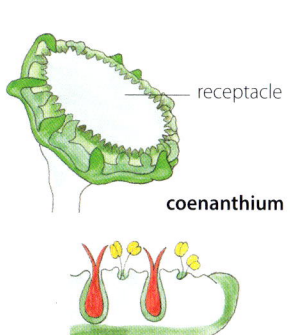

receptacle

coenanthium

receptacle with embedded
female and male flowers

coenocarp, coenocarpium   A fruit formed from an entire inflorescence and incorporating the bracts, perianth and axis, as pineapple (*Ananas comosus*) and white mulberry (*Morus alba*).
see **composite fruit**

Coenocarp

fruit   inflorescence   fruit

white mulberry (*Morus alba*)   pineapple (*Ananas comosus*)

coenocarpous   Bearing a coenocarp.
Of a gynoecium of two or more partly or entirely fused carpels.
see **paracarpous, semicarpous, syncarpous**

coenosorus, *pl.* **coenosori**
Of ferns, a group or line of sporangia resulting from many sori coalescing into one, as brake ferns (*Pteris*).

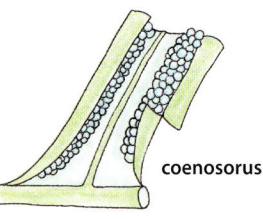

coenosorus

coetaneous   Maturing at the same time, as flowers developing at the same time as the leaves in some species of willow (*Salix*).

coevolution   The process in which two or more species evolve in response to the traits of the other(s).
The idea that interactions between species can drive adaptations, as the relationship between a hummingbird's bill shape and the shape of flowers it pollinates.

coflorescence
A lateral inflorescence in a synflorescence.
cf. **paraclade**

coflorescence

synflorescence

coherent   Of like parts joined, but only superficially, and easily separated, as the petals of some appleberries (*Billardiera*) cohere above the middle.
cf. **adherent**

coherent

cohesion   The fusion of similar plant parts, as petals or sepals, that are usually separate.

coil   A series of regularly spaced flat or cylindrical loops.
coiled   Curled or wound in concentric rings or spirals, as a tendril of the crosier of a fern.

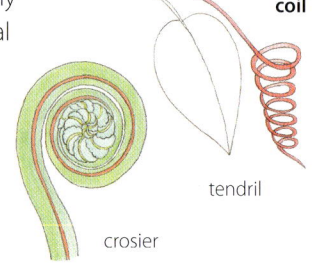

coil

tendril

crosier

coleoptile   Of the single seed in a caryopsis, the fruit of grasses (Poaceae), the sheath-like structure around the epicotyl in the embryo.
It provides protection during germination.
= **acrospire**
cf. **coleorrhiza**

Coleoptile

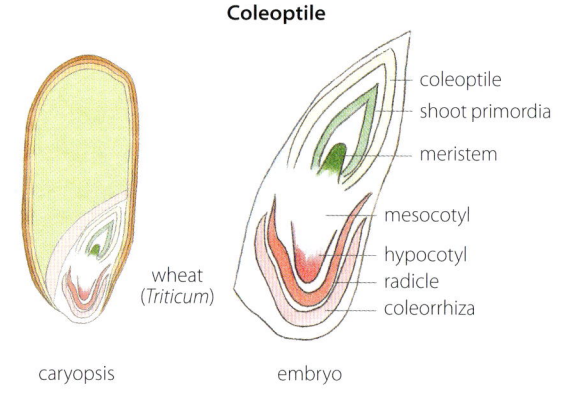

coleoptile
shoot primordia
meristem
mesocotyl
hypocotyl
radicle
coleorrhiza

wheat (*Triticum*)

caryopsis   embryo

coleorrhiza, *pl.* **coleorrhizae**
Of the single seed in a caryopsis, the sheath-like structure around the radicle in the embryo.
It provides protection during germination.
cf. **coleoptile**

Coleorrhiza

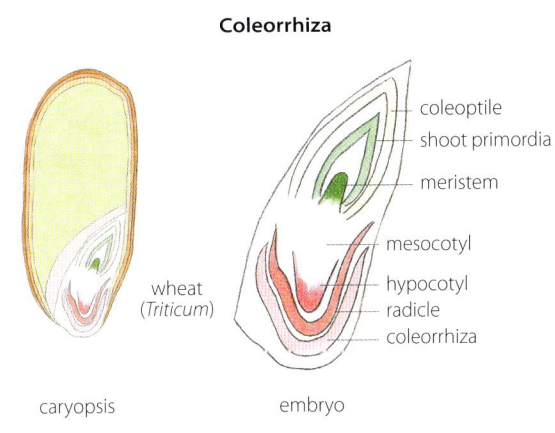

wheat (*Triticum*)

coleoptile
shoot primordia
meristem
mesocotyl
hypocotyl
radicle
coleorrhiza

caryopsis

embryo

collar   Of grasses (Poaceae), a band of meristematic tissue between the leaf sheath and the leaf blade that generates growth of the leaf blade.
= **blade meristem**
Of some deciduous terrestrial orchids, as spider orchids (*Caladenia*), the swollen region of the stem just below the soil surface that is colonised each growing season by mycorrhizal fungi.

Collar

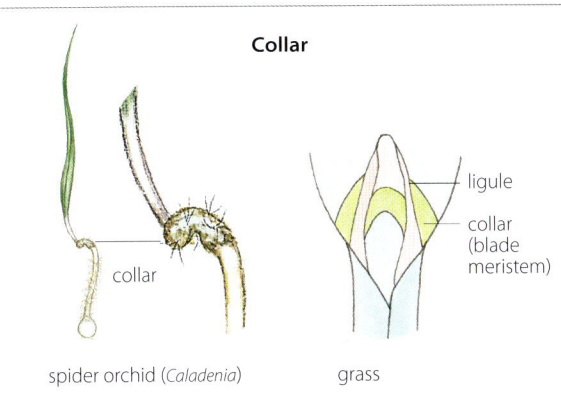

ligule
collar (blade meristem)

collar

spider orchid (*Caladenia*)

grass

collateral   Placed side by side.

collateral vascular bundles

With xylem arranged towards the interor and phloem arranged towards the exterior of the vascular bundle. They may be open or closed depending on the presence or absence of cambium.
A type of conjoint vascular bundle.
see **conjoint vascular bundles**

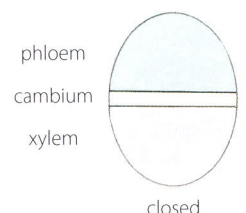

phloem
cambium
xylem

closed

collateral vascular bundles

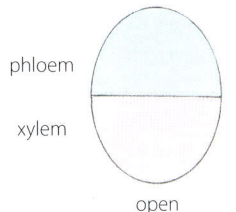

phloem
xylem

open

collection   A plant specimen collected for study and preservation and usually lodged in a herbarium.

collenchyma   Supporting tissue found in soft non-woody plants as they grow in length and is found mostly in leaves and stems.
It is composed of elongated cells with unevenly thickened walls and thick corners.
This tissue is flexible because it lacks lignin.
Different kinds of collenchyma are identified by the nature of the thickening of the cell wall.
see **angular ~, annular ~, lacunar ~, lamellar ~**

colleter   A specialised trichome, commonly in groups or tufts, that secretes mostly mucilage, in the madder family (Rubiaceae). Also at the base of the petiole in most of the dogbane family (Apocynaceae).

colliculate
Covered with low rounded projections, as the seeds of blackseed glasswort (*Tecticornia pergranulata*).

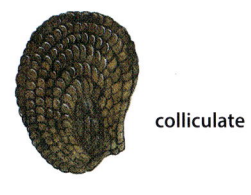

colliculate

colluviate   Covered in tiny bumps.

colonisation   The process by which a species spreads to a new area or establishes on denuded ground.

coloniser   A plant that establishes itself in a new area, as beach morning glory (*Ipomoea pes-caprae*) that colonises sand dunes.
see also **pioneer species**

coloniser

colony   A group of plants, usually of the same kind, that live and grow together, as many mosses that grow in compact colonies and some orchids, as the helmet orchid (*Corybas incurvus*).
see also **genet**

colony

helmet orchids

colpus, *pl.* **colpi**   An elongated, longitudinal aperture on a pollen grain evenly distributed over its surface or located at the equator.

colpate   Having a colpus.

cf. **colporus, pore, sulcus**

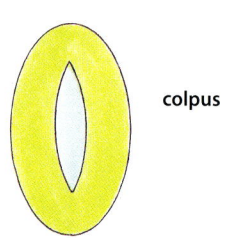

colpus

colporus, *pl.* **colpori**

An aperture on a pollen grain that is shaped like a colpus and has a circular pore-like region in the centre.

colporate   Having a colorpus.

cf. **colpus, pore, sulcus**

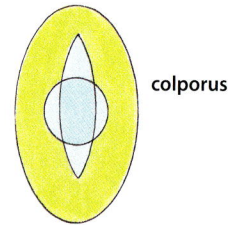

colporus

columella,
pl. **columellae**

A small column.
A central axis.
The axis to which the carpels of some fruit are attached, as geranium (*Geranium*).
A central column of tissue in a moss capsule.
One of the rod-like elements in the wall of a pollen grain that support the tectum layer.

see **pollen wall**

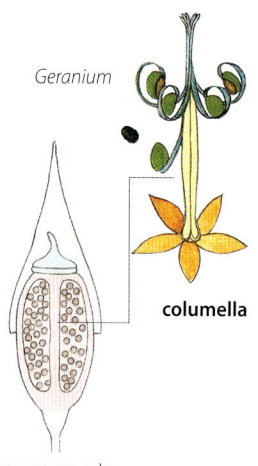

Geranium

columella

moss capsule

column   Of a flower, the male stamen and the female stigma and style fused together, as the central structure in an orchid flower (Orchidaceae), milkweed flower (Asclepiadaceae) and trigger plant flower (*Stylidium*).

= **gynandrium, gynostemium**

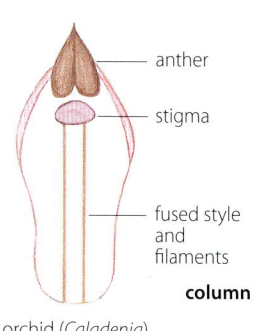

anther

stigma

fused style and filaments

column

orchid (*Caladenia*)

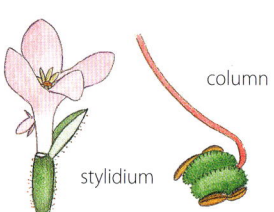

stylidium

column

column foot   Of many orchids, an extension at the base of the column to which the labellum, the bases of the lateral sepals and sometimes the bases of the petals are attached. Present in the greenhood genus *Pterostylis*.

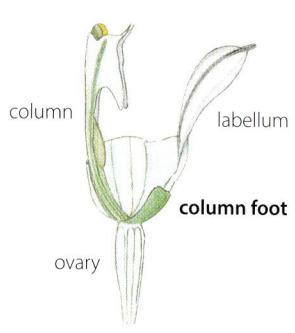

column

labellum

column foot

ovary

column wing
Of many orchids, an extension of the tissue of the column. Present in the spider orchid genus *Caladenia*.

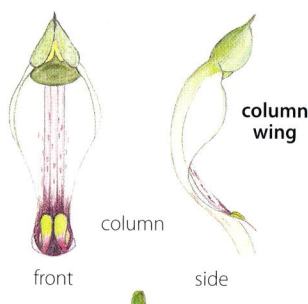

column wing

column

front

side

columnar   Growing in the shape of a vertical cylinder, as the Italian pencil pine (*Cupressus sempervirens*).

columnar

coma   A terminal tuft or cluster, as the silky hairs on the seeds of willow herbs (*Epilobium*).
The tuft of bracts on a pineapple (*Ananas comosus*).

comose   Bearing a coma or tuft.
Growing in tufts.

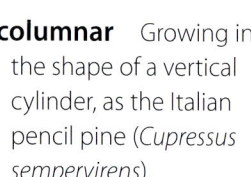

coma

willow herb

coma

pineapple

comb. nov., combinatio nova   A new name for a taxon that has the specific or infraspecific epithet used with a new genus or species name respectively.

= **new combination**

combinatio nova, *abbr.* **comb. nov.**
A new name for a taxon that has the specific or infraspecific epithet used with a new genus or species name respectively.

= **new combination**

commensalism   A relationship between two organisms where one benefits but the other is neither helped nor harmed, as an epiphytic orchid that is helped by being supported and brought closer to sunlight by the branch of a tree.
see **symbiosis**

commissure   The surface or face along which two structures are joined, as the carpels of the carrot family (Apiaceae).
see also **plicate**

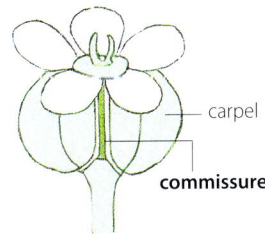

carpel

commissure

Apiaceae flower

commissural   Relating to or on the commissure.

common name   A local or popular name for a plant as opposed to the scientific name.
Caltha palustrus has many common names including marsh marigold and kingcup.

community   All living organisms (biotic components) in a given area.
Biotic components include the producers that are mainly green plants, the consumers that are mainly animals and the decomposers that are mainly bacteria and fungi.
cf. **biome, ecosystem, habitat, population**

compact
Closely clustered or packed together, as clusters of flower heads (capitula) in some daisies (Asteraceae).

compact

companion cell   A parenchyma cell, associated with sieve tube elements in angiosperms, that moves sugar into and out of sieve tube elements. Considered to be a more advanced counterpart of albuminious cells in gymnosperms.

Companion cell

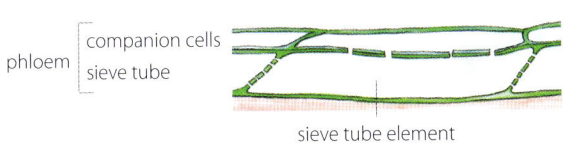

phloem
companion cells
sieve tube
sieve tube element

compatible   Capable of self-fertilisation or cross-fertilisation.
Capable of forming a successful graft.
see **cross-compatible**
cf. **incompatible**

complanate   More or less flattened in one plane, typical of a leaf lamina.

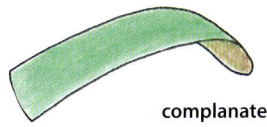

complanate

complete   Of a flower having all four whorls (sepals, petals, stamens and a pistil or pistils).
cf. **incomplete**

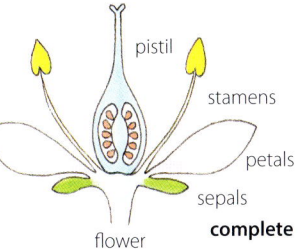

pistil
stamens
petals
sepals
complete
flower

complex tissue   Tissue composed of two or more different kinds of cell, as vascular tissue that is composed of xylem and phloem.
cf. **simple tissue**

complicate   Folded upon itself lengthwise, as some leaves.
= **conduplicate**

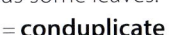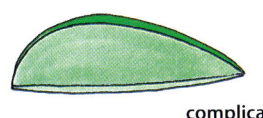

complicate

composite fruit   A fruit derived from an entire inflorescence with more than one flower, as mulberry and pineapple (both a sorosis), honeysuckle (bibacca) or fig (syconium).
It may incorporate parts of the flower other than the carpels.
= **multiple fruit**
see **coenocarp**
cf. **accessory fruit, aggregate fruit**

Composite fruit

Sorosis
pineapple
(*Ananas comosus*)

white mulberry
(*Morus alba*)

Bibacca
honeysuckle
(*Lonicera*)

Syconium
fig
(*Ficus*)

compound   Composed of two or more similar parts, as a cyme composed of cymules or a leaf composed of leaflets.
cf. **simple**

Compound

cymule

cyme
Compound inflorescence

leaflets

pinnate   palmate
Compound leaf

compound fruit
A fruit that develops from several carpels in a single flower, as buttercup (*Ranunculus*) or from several carpels in an infloresence, as pineapple (*Ananas comosus*).
see **aggregate fruit, composite fruit**
cf. **simple fruit**

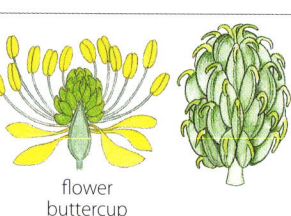

flower
buttercup

compound fruit

inflorescence
pineapple

compound gynoecium
A pistil with more than one carpel.
The carpels are either free (apocarpous) or variously fused (syncarpous).
cf. **simple gynoecium**

free

compound
gynoecium

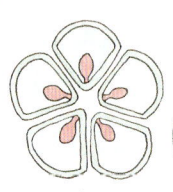
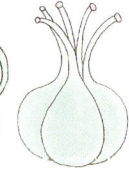

fused

compound leaf   *see* pages 63–64

compound ovary   An ovary resulting from two or more carpels that are fused together.
The ovary is multilocular if the carpel walls (septa) persist or unilocular if they break down.
cf. **simple ovary**

Compound ovary

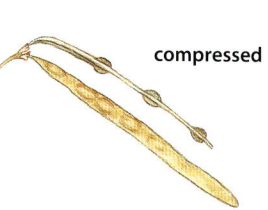

stigma
style

ovary

carpel   ovary aseptate   ovary septate

compressed   Flattened in one plane, either from side to side (laterally) or from top to bottom (dorsally), as some pods, stems etc.

compressed

compression wood   Structurally abnormal wood formed in response to stress.
Found on the underside of a lean, the side under compression, of stems in gymnosperms.
see also **reaction wood, tension wood**

con-   A prefix meaning uniform or the same.

concave   Of an outline or surface that curves inward.

concave

concentric vascular bundles
With either phloem encircling the xylem, as spike moss (*Selaginella*), or xylem encircling the phloem, as begonia (*Begonia*).
see **amphicribal, amphivasal, conjoint vascular bundles**

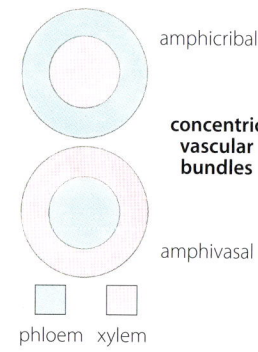

amphicribal

concentric
vascular
bundles

amphivasal

phloem xylem

conceptacle   A cavity having an outward opening pore in which sporangia develop, as in red and brown algae.
see **sporocarp**

concolorous   Having the same colour throughout.
Of a leaf, having the upper and lower surfaces the same colour.
cf. **discolorous, variegated**

compound leaf   A leaf divided into two or more leaflets on a common stalk.
  A pinnate leaf has leaflets attached along the rachis, which is an extension of the petiole.
  A palmate leaf has leaflets attached to the top of the petiole.
  Leaflets may be sessile or on a petiolule.

Compound leaf

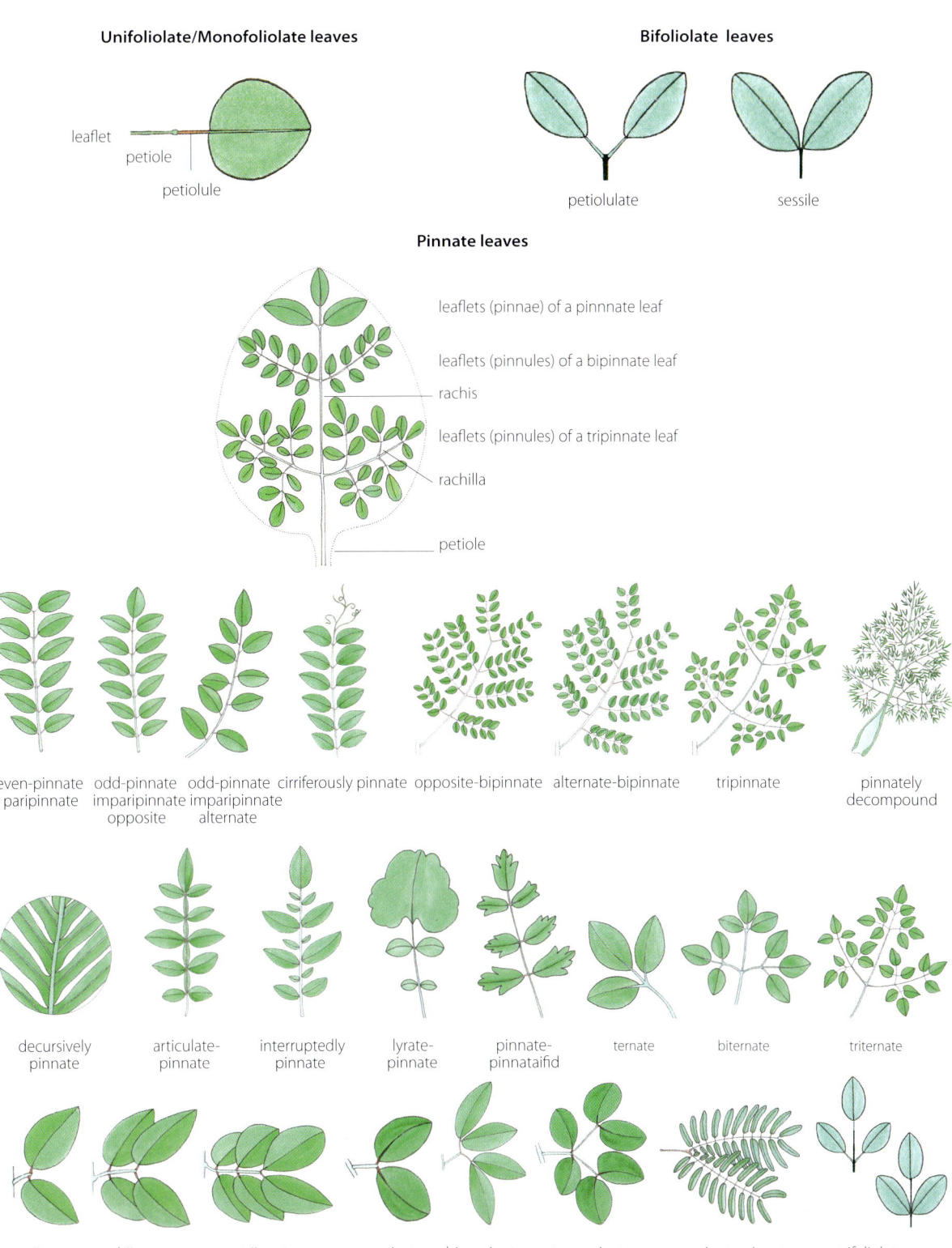

Unifoliolate/Monofoliolate leaves

leaflet
petiole
petiolule

Bifoliolate leaves

petiolulate

sessile

Pinnate leaves

leaflets (pinnae) of a pinnnate leaf

leaflets (pinnules) of a bipinnate leaf

rachis

leaflets (pinnules) of a tripinnate leaf

rachilla

petiole

even-pinnate paripinnate

odd-pinnate imparipinnate opposite

odd-pinnate imparipinnate alternate

cirriferously pinnate

opposite-bipinnate

alternate-bipinnate

tripinnate

pinnately decompound

decursively pinnate

articulate-pinnate

interruptedly pinnate

lyrate-pinnate

pinnate-pinnataifid

ternate

biternate

triternate

unijugate

bijugate

trijugate

geminate

bigeminate

tergeminate

geminate-pinnate

trifoliolate

Palmate leaves

leaflets

petiole

petiolulate

sessile

trifoliolate quinquefoliolate septenate fan-shaped digitate digitate-pinnate pedate

---

concrescent   Of the fusion of like or unlike parts.
see **adnate, connate**

condensation   The change of a substance from a gaseous state to a liquid state.
cf. **vaporisation**

condensed   Dense or compact, as opposed to lax or open, as the inflorescence of a pineapple (*Ananas comosus*).

condensed

conduplicate   Folded together lengthwise with the sides of the upper surface touching or almost touching, as the leaves of some grasses and palms.
conduplicate ptyxis
Of a single leaf in bud that is folded inwards lengthwise.

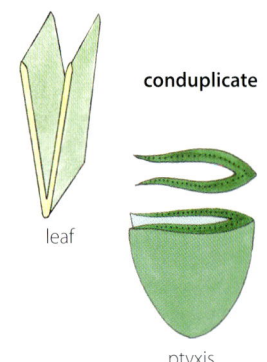

leaf

ptyxis

conduplicate

cone   A specific term for the unisexual reproductive structures of conifers that have a central axis with spirally arranged scales that bear ovules and pollen sacs.
A general term for a strobilus.
A three-dimensional shape or solid object with the base commonly a circle or ellipse and the sides tapering to a point.

conical, conoid, conoidal   Of or relating to a cone. Cone-shaped, as the canopy of the cypress, Alaska cedar (*Chamaecyparis nootkatensis*).

Cone

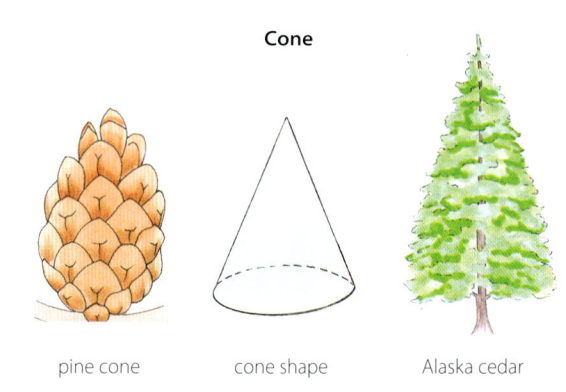

pine cone   cone shape   Alaska cedar

---

cone roots
A cone-shaped pneumatophore found in mangroves.
see also **knee roots, peg roots**

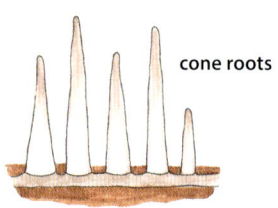

cone roots

cone scale
Of gymnosperms, one of the often woody scales that bears pollen sacs on the male cone and naked ovules or seeds on the female cone.
= **sporophyll**
see **megasporophyll, microsporophyll, ovuliferous scale**

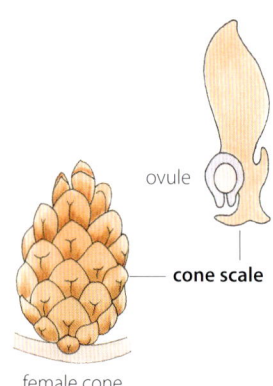

ovule

cone scale

female cone

pine (*Pinus*)

conflorescence A compound inflorescence composed of uniflorescences.
The uniflorescences differ substantially in structure from the overall structure of the conflorescence, as grevillea (*Grevillea*).

Conflorescence

conflorescence

uniflorescence

grevillea (*Grevillea*)

confluent
Running together.
Merging into one another, as the veins of some leaves.

confluent

conform Similar in shape, as the terminal pinna of some ferns is similar in shape to the lateral pinnae.

terminal pinna

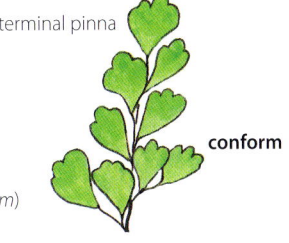

conform

maidenhair fern (*Adiantium*)

congeneric Belonging to the same genus.

conglutinated Joined together as if by glue.

congested
Extremely crowded together, as flowers in the spike of salt lawrencia (*Lawrencia spicata*).
cf. **lax**

congested

conglomerate
Densely clustered, crowded together as flowers in a glomerule of coast saltbush (*Atriplex cinerea*).

conglomerate

glomerule

conical taproot
A main, descending root that is broad at the top and tapers towards the base, as a carrot.
see **fusiform taproot, napiform taproot**

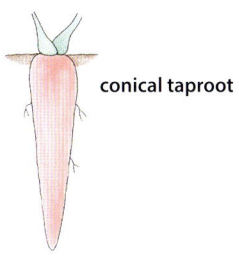

conical taproot

conidiospore
An asexually produced fungal spore that forms in a chain and breaks off when mature.

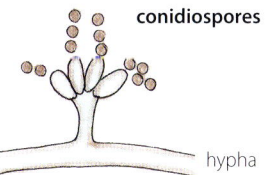

conidiospores

hypha

conifer Cone-bearing, usually evergreen trees or shrubs, such as pines and cypresses, with male and female cones usually on the same plant.
A member of Pinophyta, one of the four divisions of gymnosperms.
coniferous Producing or bearing cones.

conifer

conjoint vascular bundles With phloem and xylem arranged together in the same radius.
Found in stems and leaves.
There are three types: collateral vascular bundles, bicollateral vascular bundles and concentric vascular bundles.

Conjoint vascular bundles

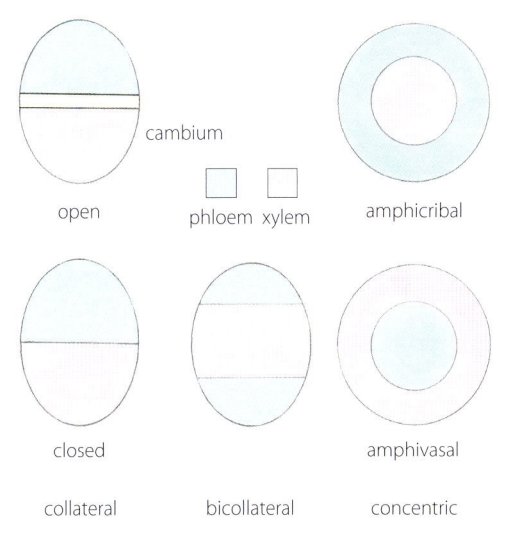

cambium

open

phloem xylem

amphicribal

closed

amphivasal

collateral

bicollateral

concentric

conjugate Joined together in pairs.
Of a pinnate leaf having only one pair of leaflets.
see **jugate**

conjugate

connate Of like parts fused to one another, as the petals of the bindweed family (Convolvulaceae) and the bases of some leaves.
cf. **adnate**

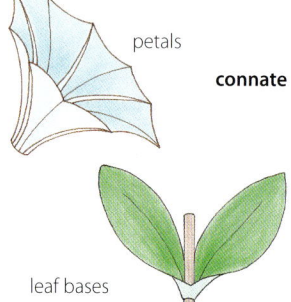

petals

connate

leaf bases

connate-perfoliate
With the broad bases of two opposite leaves joined together so that the stem appears to pass through a whole leaf.

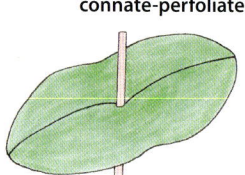

connate-perfoliate

connective Of an anther, tissue that joins the pollen sacs together and connects the anther to the filament.
It includes the vascular bundle.

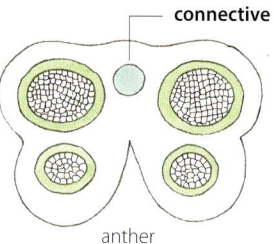

connective

anther

connivent Touching but not fused, as the petals of the keel of a pea flower.
Arched inward towards the tip, as the veins of some leaves.
= **convergent**

Connivent

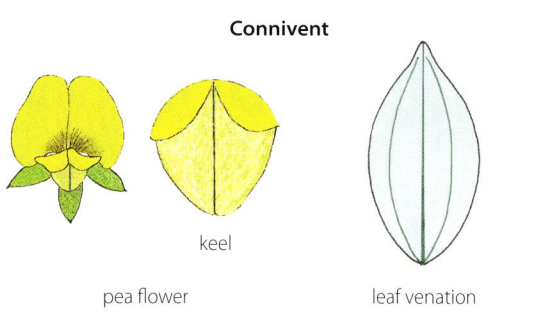

keel

pea flower

leaf venation

conservation status An indication of how likely a species is to survive now and into the future.
see **critically endangered, endangered, extinct, International Union for Conservation of Nature, relict, threatened, vulnerable**

conserved name A name to be kept even though it may formerly have been invalid.
= **nom. cons., nomen conservandum**

conspecific Belonging to the same species.

constricted Narrowed, especially by contracting at one place, as the jointed fruit of southern tick-trefoil (*Desmodium gunnii*).

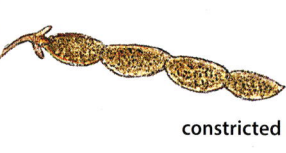

constricted

contiguous With parts touching and appearing fused but actually only in close contact, as the male zone of the spadix is contiguous with the female zone in the dragon arum (*Dracunculus*).
cf. **coalescent**
Of plant communities that are directly adjacent to one another.

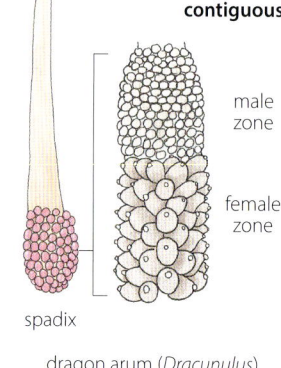

contiguous

male zone

female zone

spadix

dragon arum (*Dracunculus*)

continental drift
The fragmentation of Pangea into Laurasia and Gondwana, and the gradual movement of the continents across the earth's surface through geological time that is explained by plate tectonics.

Continental drift

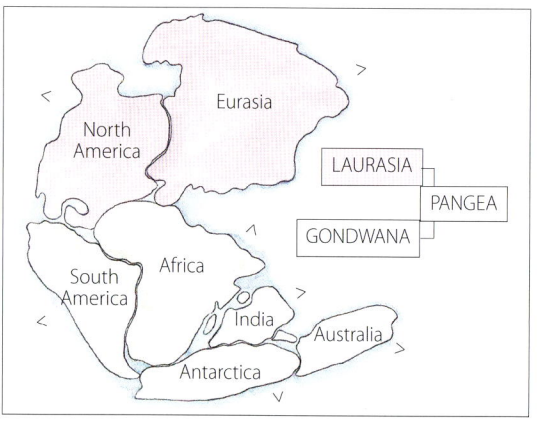

North America

Eurasia

LAURASIA

PANGEA

GONDWANA

South America

Africa

India

Australia

Antarctica

continuous
In grasses (Poaceae), of the jointed rachis from which the spikelets fall

grass rachis

leaving the stem entire.
Uninterrupted, as a stem
that lacks joints.
cf. **articulate, jointed**

stem   **continuous**

contorted   Twisted or bent. Convolute.
contorted aestivation   Of young petals, tepals or
sepals in the unopened bud, with one edge inside
and the other edge outside the adjacent one, as the
corolla of a hibiscus (*Hibiscus*).
= **convolute aestivation**
contorted vernation   Of young leaves in the
unopened leaf bud with margins overlapping on
one side. A form of imbricate vernation.
= **convolute vernation**

Contorted

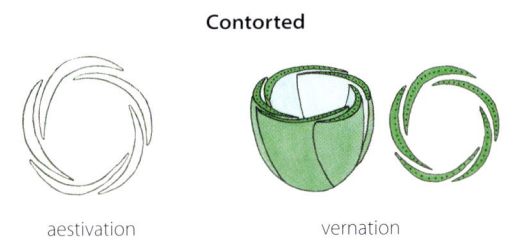

aestivation   vernation

contracted   Shortened
or narrowed and dense
as opposed to open, as
the spikelets of some
species of sedge (*Carex*).
cf. **lax**

contracted

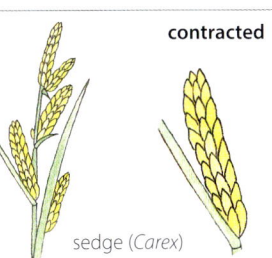

sedge (*Carex*)

contractile roots
Specialised fleshy
wrinkled roots that
contract and pull
the plant deeper
into the soil.
Found in many plants,
as bulbs, corms and
some taproots.

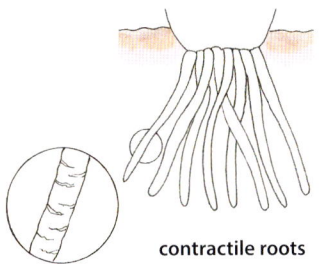

contractile roots

contraligule
A membranous flap on
the rim of the leaf sheath
opposite the blade, as
some nutrushes (*Scleria*).
see **ligule**

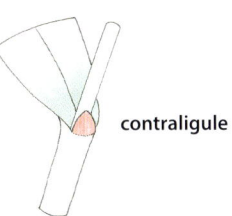

contraligule

convergent   Touching
but not fused, as the
petals of the keel of a
pea flower.
Arched inward towards
the tip, as the veins of
some leaves.
= **connivent**
cf. **divergent**

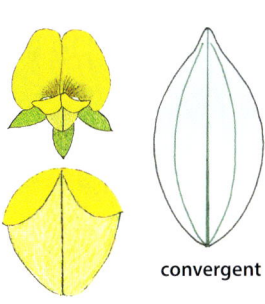

keel   **convergent**

convergent evolution   The independent
evolution of similar traits in two or more unrelated
or distantly related organisms.
= **parallel evolution, homoplasy**
cf. **divergent evolution**

converted clade name   Of phylogeny, a clade
name converted from a pre-existing name.
= **nomen cladi conversum**

convex   Of an outline
or surface that curves
outward, as the surface
of a circle or a sphere.

convex

convolute   Rolled longitudinally upon itself.
convolute aestivation   Of young petals, tepals or
sepals in the unopened bud, with one edge inside
and the other edge outside the adjacent one, as the
corolla of a hibiscus (*Hibiscus*).
= **contorted aestivation**
convolute ptyxis   Of a leaf rolled from one margin
to the other as banana (*Musa*) and aroids (Araceae).
convolute vernation   Of young leaves in the
unopened leaf bud, with margins overlapping on
one side. A form of imbricate vernation.
= **contorted vernation**

Convolute

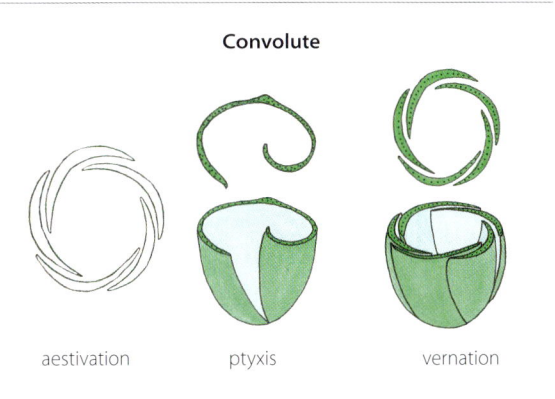

aestivation   ptyxis   vernation

copal   An aromatic, amber-like resin exuded from
various tropical trees.
cf. **amber**

coppice, copse   A thicket of small trees or shrubs. To form a coppice by periodic cutting back to ground level to stimulate new growth.

coralloid, coralliform   Like coral.

coralloid roots
Of cycads, a much-branched root mass that looks like coral and grows on the surface of the soil. The relationship between the plant and these roots is symbiotic. The roots host cyanobacteria that fix nitrogen used by the plant and the plant supplies products from photosynthesis that support the bacteria.

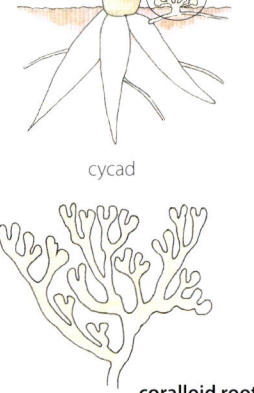
cycad

coralloid roots

cordate   Heart-shaped in outline with the notch at the base.
cf. **obcordate**
cordiform   Shaped like a heart.

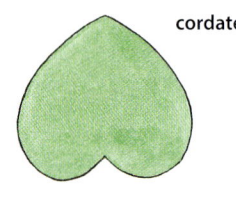
cordate

coriaceous   Leathery. Tough but somewhat flexible, as the phyllodes of some wattles (*Acacia*).

coriaceous leaves

cork   Non-living cells with waxy walls that are impervious to water and gases.
= **phellem**
see also **suberin**
corky   Of, like or consisting of cork. Having a light, compressible and resilient texture.

cork cambium   A layer of meristem that is responsible for secondary growth. It is part of the periderm in woody plants and some herbaceous plants. Cork cells (phellem) are produced on the side towards the surface of the plant and parenchymatous tissue (phelloderm) on the inner side.
= **phellogen**
see **cambium, lateral meristem**
see also **fusiform initials, ray initials**

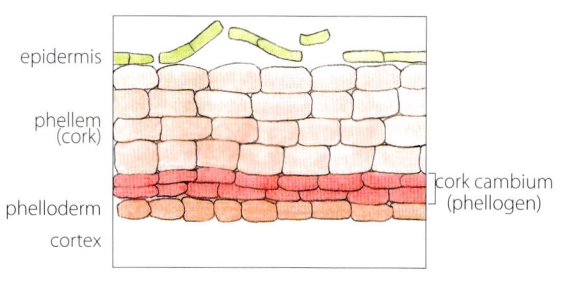
Cork cambium
epidermis
phellem (cork)
phelloderm
cortex
cork cambium (phellogen)

corm   A solid, rounded underground stem that stores food. It is covered with scale leaves and an outer tunic, as crocuses (*Crocus*) and gladioli (*Gladiolus*). It bears buds on the basal plate that can produce new plants.
see **cormel**
cormous   Bearing corms.

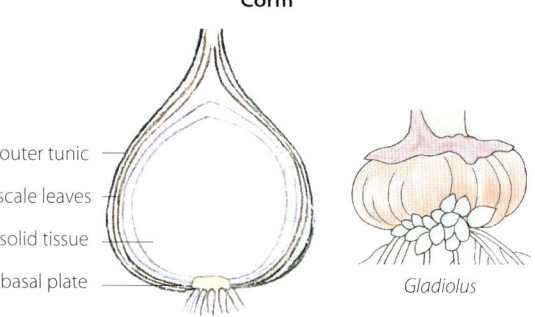
Corm
outer tunic
scale leaves
solid tissue
basal plate
Gladiolus

cormel, cormlet
A small corm, that can produce a new plant, growing at the base of a corm, as gladioli (*Gladiolus*).

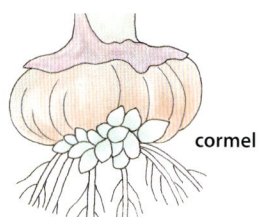
cormel

corneous   With a hard, smooth texture, as the leaf margin of some agaves. Having an incurved, tapering appendage like the horns of cattle.
= **horny**

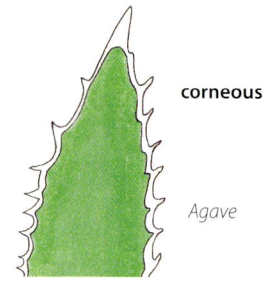
corneous
Agave

corniculate, cornute
With a small horn. Horn-shaped, as the capsule of goat's horn mangrove (*Aegiceras corniculatum*).

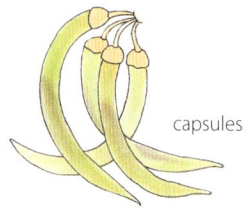
capsules

Bearing a horn-like spur or appendage, as the toadflax flower (*Linaria*).

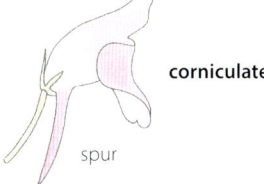

corniculate

spur

corolla   The petals of a flower that may be free or united.
it surrounds the reproductive organs (the ovary and the stamens), and is usually coloured to attract pollinators.
cf. **calyx**

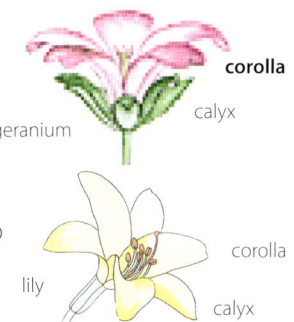

corolla

calyx

geranium

corolla

lily

calyx

corolla lobe
The upper free parts of a petal when the lower parts of the petals are united into a tube.
see **corolla tube**

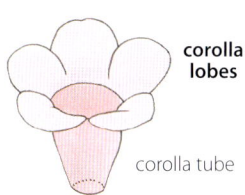

corolla lobes

corolla tube

corolla tube
The part of a corolla below the lobes, with the petals united.
see **corolla lobe, gamopetalous**

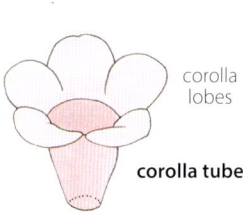

corolla lobes

corolla tube

corona   A crown.
A ring of tissue arising from the perianth of a flower, as the solid trumpet-shaped outgrowth at the centre of a daffodil (*Narcissus*), or the ring of horned lobes at the centre of the flower of milkweed (*Asclepias*).
The fringed membrane on the seed of Chilean needlegrass (*Nassella neesiana*).
coronate   Having a crown.
coroniform   Having the form of a crown.

Corona

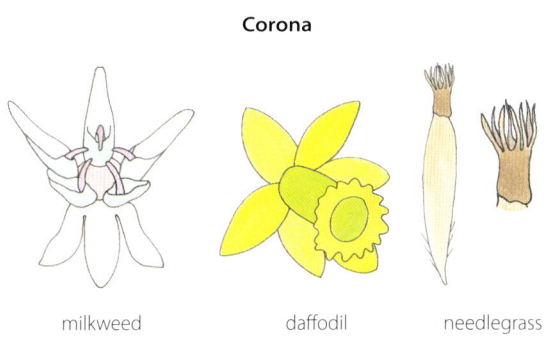

milkweed   daffodil   needlegrass

corpusculum
A sticky gland linked to the pollinia by translator arms in the milkweed family (Apocynaceae).

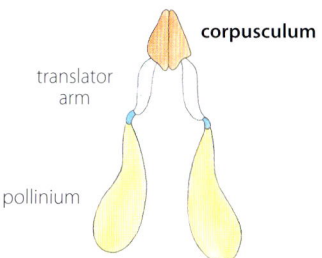

corpusculum

translator arm

pollinium

corr., correctus   Correct name.

correct name   In nomenclature, a name allowed under the rules of the International Code of Nomenclature.
= **correctus,** *abbr.* **corr.**

correctus, *abbr.* **corr.**   Correct name.

corrugated   Wrinkled into parallel ridges and troughs, as the stone of a peach (*Prunus persica*).

corrugated

cortex   Of the vascular cylinder, a region of unspecialised cells (ground tissue) lying between the epidermis and the vascular bundles in stems and roots.
cortical   Of or relating to the cortex.

Cortex

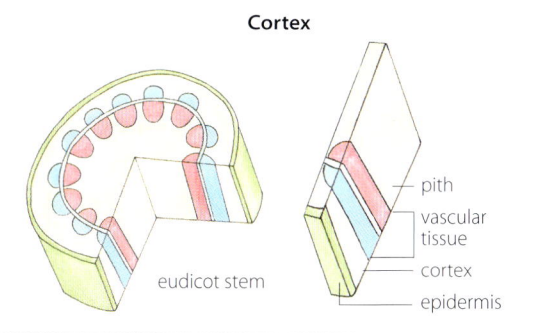

pith

vascular tissue

cortex

epidermis

eudicot stem

corticate   Having a cortex, bark or rind.

corticolous   Growing on bark, as a corticolous lichen.

corymb   A flat-topped or rounded racemose inflorescence with the stalks (pedicels) starting at different points on the peduncle.
Pedicels of the lower flowers are longer than those of the flowers above, bringing all flowers to about the same level.
A corymb can be simple or compound.
It is an indeterminate or indefinite inflorescence.
corymb page 70 (cont.)
cf. **umbel**

corymbiform   Having the form, but not necessarily the structure, of a corymb.

corymbose   Relating to or having the characteristics of a corymb.
see **acropetal, centripetal**
see also **cymose inflorescence**

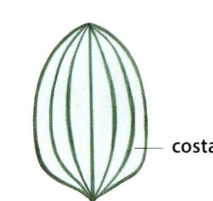

Corymb

simple corymb   compound corymb

peduncle   pedicel   peduncle

cosmopolitan   Having a worldwide distribution, as a species growing all over the world in habitats that are suited to it.

costa *pl.* **costae**   A rib or vein of a leaf, leaflet or frond. The midrib.

costate   Ribbed. Having one or more longitudinal veins or ribs.

costa

costapalmate
Of a palmate leaf with a short midrib (costa).

costa   **costapalmate**

costule   The midrib of a pinna or of a pinnule.

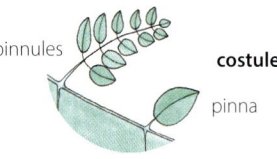

pinnules   **costule**

pinna

cottony   Soft, white and fibrous like the fluffy fibre of a cotton boll (*Gossypium*).

cottony

cotyledon   One of the leaves that originates directly from the tissue of the embryo of seed-bearing plants.
Eudicots, as legumes, mostly have two cotyledons, and monocotyledons, as grasses and lilies, have one cotyledon.
The scutellum of a caryopsis, the seed of the monocotyledonous grass family (Poaceae), is considered to be a modified cotyledon.
Gymnosperms have a variable number of cotyledons.

Commonly called a seed leaf.

cotyledonous   Having cotyledons.
cf. **acotyledonous**

Cotyledon

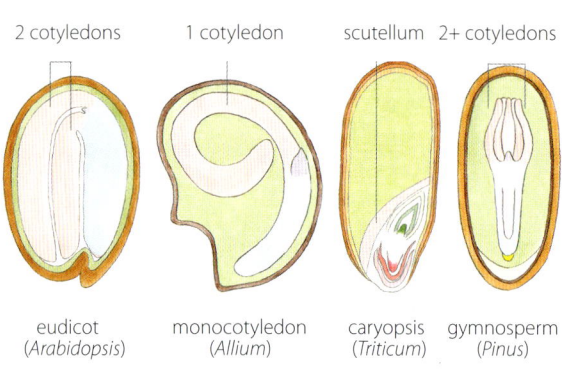

2 cotyledons   1 cotyledon   scutellum   2+ cotyledons

eudicot   monocotyledon   caryopsis   gymnosperm
(*Arabidopsis*)   (*Allium*)   (*Triticum*)   (*Pinus*)

cotyledon orientation   Various, including: straight, violet (*Viola hederacea*), coiled, hop (*Humulus lupulus*), folded and incumbent, mustard (*Erysimum cheiranthoides*) or folded and accumbent, bitter cress (*Barbarea sisymbrium*).

Cotyledon orientation

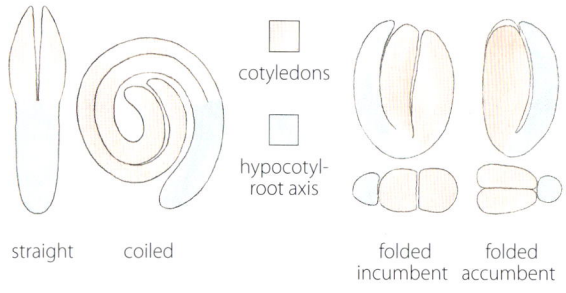

cotyledons

hypocotyl-root axis

straight   coiled   folded   folded
incumbent   accumbent

cotyledonary node
The point of attachment of the embryo axis to the cotyledon(s), as the eudicot broad bean (*Vicia faba*).

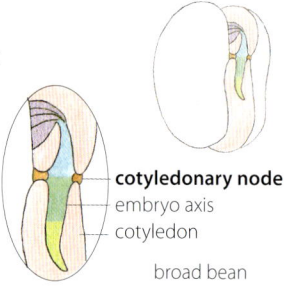

cotyledonary node
embryo axis
cotyledon

broad bean

cotyledonary petiole
Of palms with remote germination and some other monocotyledons, the first structure to emerge from the seed.

It grows down into the soil, forming a swelling from which the first seedling root (radicle) and the plumular leaves emerge.

= **hyperphyll**

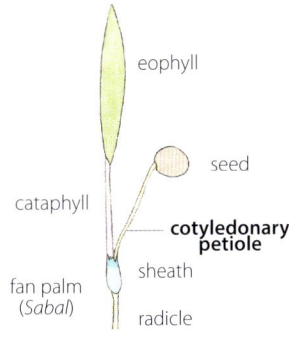

coumarin   A poisonous chemical compound found in many plants that has the sweet scent of new-mown hay.
Present in tonka beans (*Dipteryx odorata*) and sweet woodruff (*Galium odoratum*).

cpDNA, chloroplast DNA   The small amount of DNA unique to chloroplasts.
Unlike nuclear DNA, cpDNA is arranged in rings.

craspedodromous
Of leaves with veins starting from a point on the midrib and ending at the margin.

Craspedodromous

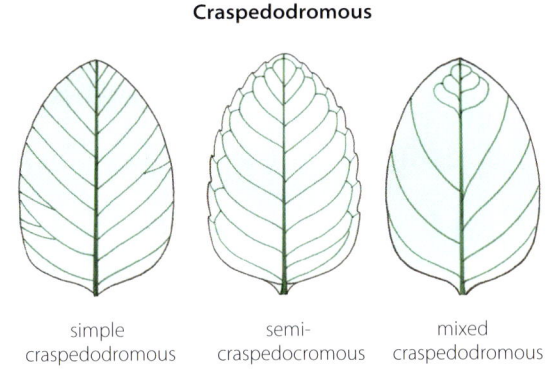

simple craspedodromous

semi-craspedocromous

mixed craspedodromous

crassi-   A prefix meaning thick.

crassinucellate   Of an ovule with several or numerous layers of cells in the nucellus.
Having a thick nucellus.
cf. **tenuinucellate**

crassulacean acid metabolism, CAM
A carbon fixation pathway that occurs when the stomata open at night to admit carbon dioxide while minimising water loss.
Found in many xeric plants like cacti.
Named after the stonecrop family (Crassulaceae) in which this pathway was first studied.
cf. **photosynthesis**

crateriform
Having the form of a shallow bowl, as the corolla of some species of *Jaltomata*.

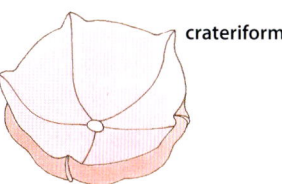

creeper   A plant that grows horizontally and sends out roots from the nodes on the stem, as creeping thyme (*Thymus serpyllum*).

creeping
Prostrate and producing roots at the nodes, as creeping Charlie (*Glechoma hederacea*).
= **repent, reptant**

cremocarp
A dry two-seeded schizocarpic fruit that separates from the central axis (carpophore) into two one-seeded mericarps.
Derived from a two-carpelled inferior ovary.
Characteristic of the carrot family (Apiaceae).

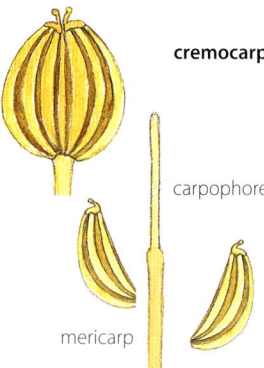

crenel   A round or convex flat tooth.
crenate   Scalloped.
of a toothed margin with regular, blunt or rounded teeth, as the margins of some leaves.
cf. **bicrenate, crenulate**

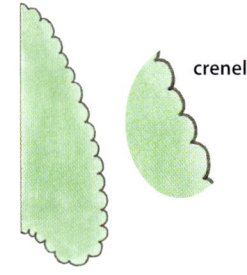

crenule
A small, rounded notch.
crenulate   Minutely scalloped.
Of a toothed margin with minute regular blunt or rounded teeth, as the margins of some leaves.
cf. **bicrenate, crenate**

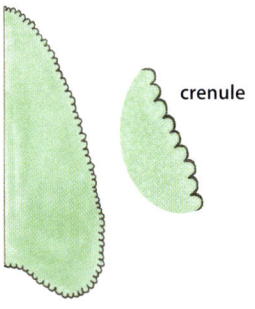

crepuscular   Relating to, occurring or active at dusk, as flowers that open or emit fragrance at this time.
cf. **diurnal, matutinal, nocturnal, vespertine**

crest An elevated line or
ridge on the surface or at
the summit of an organ.
An appendage
terminating an organ,
as the tuft on the keel-
shaped petal of some
milkworts (*Polygala*).
Of orchids (Orchidaceae),
complicated callus, as
that of the butterfly
orchid (*Oncidium*).
= **crista**

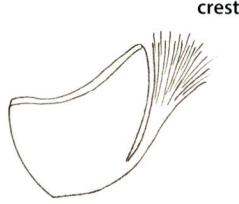

crest

petal of a milkwort flower

crested Having a crest.
= **cristate**

crispate, crisped
Curled or ruffled, as the
margins of some leaves.

crispate

crista Another term for crest.
  cristate Having a crest.
  = **crested**

critically endangered According to the IUCN,
a conservation status covering species that are
considered to be facing an extremely high risk of
extinction in the wild.

crosier, crozier
The coiled tip of a young
fern frond.
= **fiddlehead**

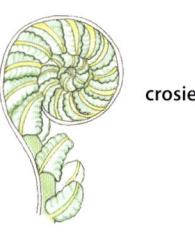

crosier

cross A hybrid.
The offspring of parents from two different genera
species or subspecies.

cross-compatible Of plants capable of
cross-fertilisation.
see **compatible**

cross-fertilisation Fertilisation by pollen from a
different plant, usually but not always, of the same
species.
see **cross-pollination**

cross-pollination Pollination between flowers on
different plants, usually but not always, of the same
species.
cf. **cross-pollination**

crossing over
In meiosis, the exchange
of genetic material,
during synapsis, between
the maternal and
paternal chromosomes
so that the haploid
chromosomes in the
sperm and the egg will
differ from the diploid
parent chromosomes.
= **recombination**

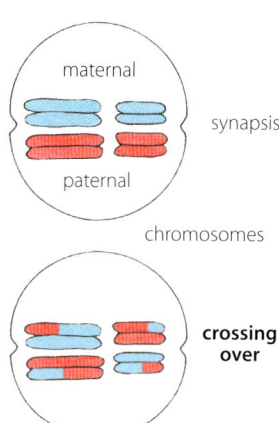

maternal

synapsis

paternal

chromosomes

crossing
over

crown
Of a tree, the branches and foliage above the bole.
Of a palms (Arecaceae), the cluster of fronds borne
at the tip of the stem.

Crown of a tree and of a palm

bole

tree

palm

Of grasses, the connecting tissue between the roots
and the shoots.
Leaves, flowering stems, tillers, rhizomes, stolons
and the adventitious or secondary roots arise from
the crown.
see **seminal root system**

Crown of grasses

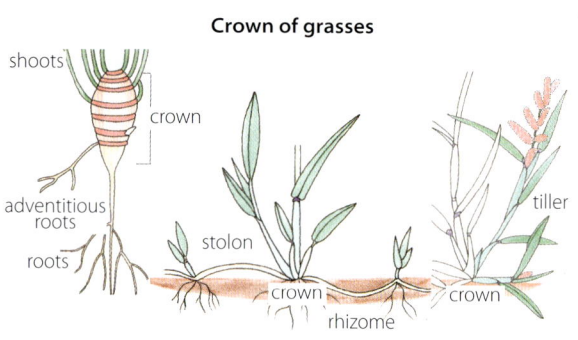

shoots

crown

adventitious
roots

tiller

roots

stolon

crown

crown

rhizome

crownshaft
A cylinder of clasping
leaf sheaths at the top
of the stem of some
pinnate-leaved palms
(Arecaceae).

crownshaft

crucifer   A cross.
cruciferous   Shaped like a cross, as the flowers of rapeseed (*Brassica napus*). Of or relating to plants of the mustard family (Brassicaceae, formerly Cruciferae).
cruciate, cruciform In the form of a cross.

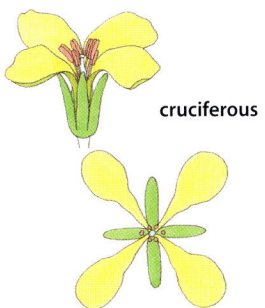
cruciferous

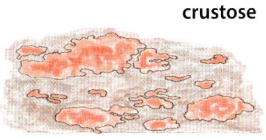
rapeseed (*Brassica napus*)

crustaceous   Hard, crust-like, thin and brittle.

crustose   Of lichens, forming a thin crust that adheres closely to the surface on which it grows.

crustose

lichen (*Caloplaca*)

crypto-   A prefix meaning hidden.

cryptocotyly   Of seed germination, having the cotyledons remain within the seed coat at germination.
cf. **phanerocotyly**

cryptogams   Organisms that reproduce by means of spores and have no true flowers or seeds. Includes ferns, bryophytes, algae, fungi and lichens.
cf. **phanerogam, spermatophyte**

cryptophyte   Plants with perennating buds underground and food reserves stored in other subterranean perennating organs like tubers. Cryptophytes are subdivided into geophytes, helophytes and hydrophytes.
see also **chamaephyte, hemicryptophyte, phanerophyte, therophyte**

cucullus   In the form of a hood or cowl, as the upper sepal of monkshood (*Aconitum*).
cf. **galea**
cucullate   Hooded, hood-shaped.
cf. **galeate**

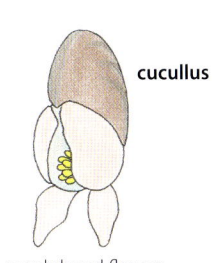
cucullus

monkshood flower

culm   Of sedges (Cyperaceae), commonly a 3-sided, solid or pithy stem that lacks nodes.

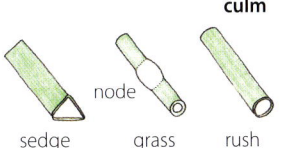
culm

node

sedge   grass   rush

An aerial stem of some monocotyledons that bears the inflorescence. Of grasses (Poaceae), a jointed, usually hollow stem, with solid nodes. Of rushes (Juncaceae), commonly a pithy terete stem with inconspicuous nodes.

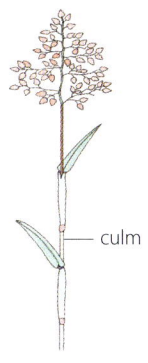
culm

culm leaf   Of bamboos, one of the protective leaves that wraps around new shoots and the young culm as it grows. It provides protection and gives the culm strength during the initial growing period.

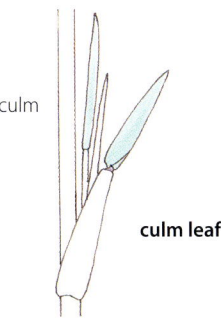
culm

culm leaf

cultigen   A plant known only in cultivation and whose origins are unknown or lost. Includes many cultivars and hybrids.

cultivar   An independent category for plants that are a result of selective breeding of natural species, as those in agriculture, forestry and horticulture.

cultivar epithet
The final element of a full cultivar name, enclosed in single quotation marks to distinguish it from the scientific name that precedes it. As genus – *Geranium* + species epithet – *robertianum* + cultivar epithet – 'Celtic White' gives the cultivar name *Geranium robertianum* 'Celtic White'.

cuneate
Narrowly triangular, broad above and tapering by straight lines towards the base.

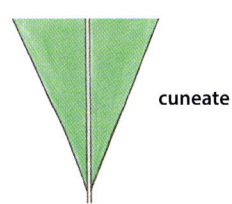
cuneate

cuneiform
Wedge-shaped.

cuneiform

cupule   A cup-shaped structure, as the cup-shaped involucre of an acorn and the indusium of some tree ferns in the genus *Cyathea*.
cupule page 74 (cont.)

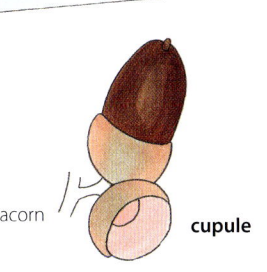
acorn

cupule

see also **calybium**

cupular   Relating to or shaped like a cupule.

cupulate   Bearing a cupule.

cupuliform   Shaped like a cupule.
= **cyatheoid, cyathiform**

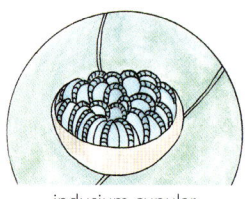

indusium cupular
Cyathea

curvinervate, curvinerved   Having the veins of leaves curved.

cushion plant   A very compact, low-growing, mat-forming plant with a long taproot and numerous small leaves and flowers over a layer of insulating dead and living material.
A plant of alpine, subalpine, arctic and subarctic regions.

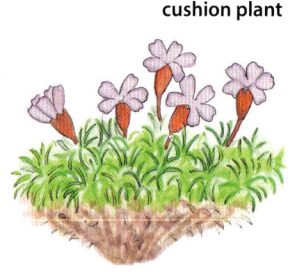

cushion plant

moss campion (*Silene acaulis*)

cusp   A short, sharp, rigid point.

cuspidate   Tipped with a cusp.

cusp

cuticle   The waxy layer of cutin on the epidermis of the aerial parts of plants that reduces water loss. It is typically thicker on plants like cacti and succulents that live in dry climates.

cutin   The wax-like, water-repellent material present in the walls of some plant cells.
see **cuticle**

cutting   Part of a stem, leaf or root that is cut off and grows roots and shoots to produce a new plant.
= **slip**
see also **grafting**

cutting

cyanobacteria   Bacteria that contain chlorophyll and generate oxygen through photosynthesis. They can live on land, as those that partner with fungi to form lichens, but most are aquatic.

cyatheoid   Of the tree fern genus *Cyathea*. Cup-like, as the a cup-shaped indusium of some *Cyathea*.

cyathoid, cyathiform   Shaped like a cup.
= **cupuliform**

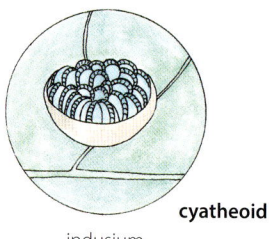

cyatheoid
indusium

cyathium, *pl.* **cyathia**   The characteristic inflorescence of the spurge family (Euphorbiaceae) that looks like a single flower. It consists of a cup-shaped involucre of fused bracts that encloses several male flowers, each reduced to one stamen, and one female flower reduced to a single pistil with a pedicellate ovary.
see also **pseudanthium**

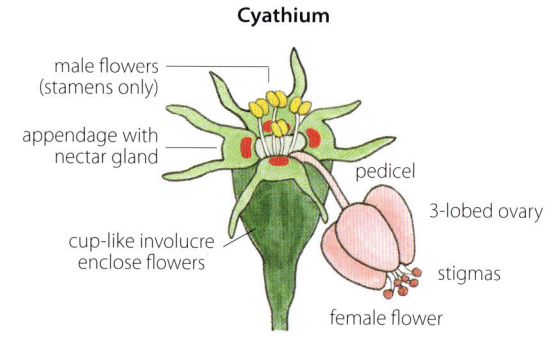

Cyathium

male flowers (stamens only)
appendage with nectar gland
pedicel
3-lobed ovary
cup-like involucre enclose flowers
stigmas
female flower

cyathophyll
Of the spurge genus (*Euphorbia*), bract-like structures below the tiny flowers that are leaf-like, reduced to tiny scales or may be showy and brightly coloured.

cyathophyll

poinsettia (*Euphorbia pulcherrima*)

cycad   Palm-like plants bearing large male or female cones on separate plants. Members of Cycadophyta, one of the four divisions of gymnosperms.

cycad

cyclic   Having identical parts arranged in whorls, as leaves on a stem.
cf. **seriate**

cyclic

cylinder　A solid or hollow body with parallel sides and a circular or oval cross-section.

cylinder

cylindric, cylindrical Like a cylinder.

cymbiform　Shaped like the bow of a boat, as the united lower petals (keel) of pea flowers (Fabaceae).
= **navicular**

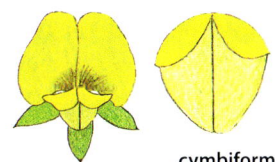

cymbiform

cyme　*see* page 76

cymose　Arranged in a cyme, bearing cymes.
cf. **racemose**

cymule　A three-flowered part of a compound cyme, with the oldest flower in the middle and the lateral ones younger.
A simple cyme, as a dichasium.
The ultimate division of a compound cyme.

Cymule

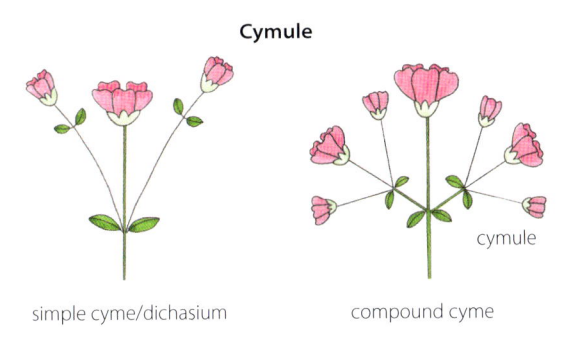

simple cyme/dichasium　　compound cyme

cymule

cynarrhodium
The fruit of the rose genus (*Rosa*).
The hollow hypanthium contains the achenes that are the true fruit.
Commonly called a hip.
see **accessory fruit**

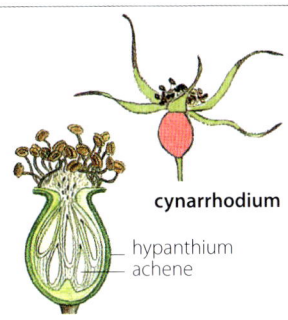

cynarrhodium

hypanthium
achene

cypsela, *pl.* **cypselae, cypselas**　A dry indehiscent fruit, with or without a pappus, having one seed attached to the fruit wall (pericarp) at one point only.

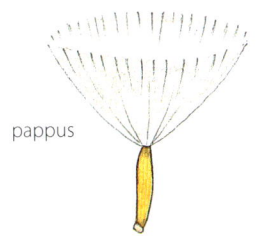

pappus

Derived from a two-carpelled inferior ovary in which only one ovule develops into a fruit and the other is aborted. Characteristic of the daisy family (Asteraceae).
cf. **achene, caryopsis, diclesium**

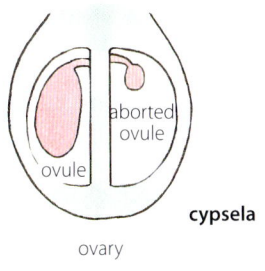

aborted ovule

ovule

cypsela

ovary

cystolith　A stone-like mass of calcium carbonate that forms in specialised epidermal cells of some plants.
The deposit is typically attached to the cell wall by a cellulose stalk or peg.
see **lithocyst**

cytokinesis　In meiosis and mitosis, the process of forming a cell wall that divides the cytoplasm of a parent cell into two or four daughter cells, each with its own nucleus.

Cytokinesis

cell wall

daughter cells

mitosis　　　meiosis

cytokinins　Plant hormones that stimulate seed germination, seed development, leaf expansion, induction of flowering and help delay senescence. They work with auxins to regulate cell division in roots and shoots.
see **phytohormone**

cytology　The study of the cell and its contents.
cytological　Of or relating to cytology.

cyme, cymose inflorescence   An inflorescence in which the main axis and lateral branches end in a flower. Each cyme unit has a pedicel that bears a single terminal flower. A lateral branch with a terminal flower then develops in the bracteole at the base of the pedicel. Subsequent branching follows the same pattern. Branching may be on one or both sides of the axis.

The simplest cyme is a solitary flower.

The first formed flower is at the tip of the peduncle so that flowering begins at the top in descending or basipetal succession.

The arrangement of flowers is centrifugal, with the oldest flower in the centre and the youngest flowers towards the outside.

The number of flowers is definite.

A determinate or definite inflorescence.

see **inflorescence**

cf. **raceme, racemose inflorescence**

Cymose inflorescence

Flowering descending/basipetal

Flowering centrifugal

solitary flower

cyme/dichasium

compound cyme/dichasium

pleiochasial cyme/pleiochasium

scorpioid cyme
cincinnus

helicoid cyme
bostryx

drepanium

rhiphidium

verticillaster

fascicle

glomerule

cytoplasm   Of a plant cell, the substance (cytosol) surrounded by the plasma membrane and all organelles suspended in it other than the nucleus.
see **cell**

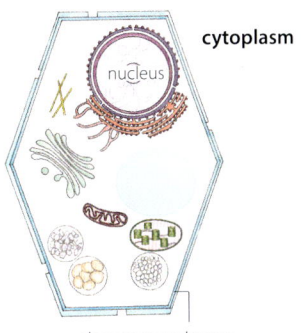

cytoplasmic   Of or relating to cytoplasm.

cytoplasmic membrane   A thin semipermeable layer of tissue enclosing the cytoplasm of a cell and, in plants, surrounded by the cell wall.
It allows movement of some substances into and out of the cytoplasm.
= **cell membrane, plasma membrane**

Cytoplasmic membrane

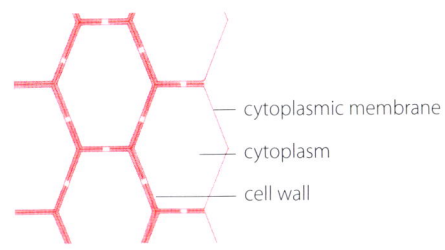

— cytoplasmic membrane
— cytoplasm
— cell wall

cytoplasmic streaming   The continuous movement of cytoplasm around the cell.

cytoskeleton   A system of microfilaments and microtubules in the cytoplasm that gives a cell its shape. It is involved in the motility of organelles and other components from one part of the cell to another.

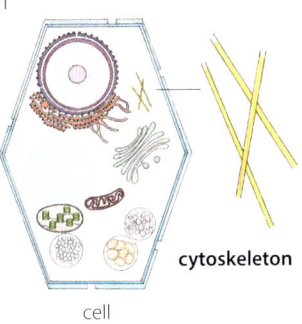

cytosol   A clear substance that makes up most of the volume of the cytoplasm.
= **ground substance, hyaloplasm**

damping off   Of plant seed and seedlings, a fungal or other pathogenic disease that causes the seed to rot or the seedling to collapse.

dauciform   Broad at the top and tapering towards the base. Carrot-shaped, as some roots.

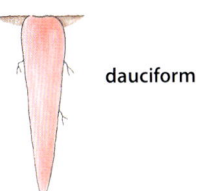

dealbate   Whitened.

deca-   A prefix meaning ten.

deciduous   Falling seasonally as a plant that sheds it leaves annually. Falling off or shed at a particular stage of development as the petals of a flower.
see also **caducous, fugaceous**
cf. **evergreen, persistent**

declinate   Bending downwards in a curve, as the stamens of cassia (*Cassia*).

decompound   Many times compound, as the leaf of fennel (*Foeniculum vulgare*).
see **pinnately decompound**

decorticate   To peel, to remove the outer layer of, as bark or a husk.

decumbent   Of stems lying or growing along the ground with the tips curving upwards, as matted St John's wort (*Hypericum japonicum*).
cf. **procumbent, prostrate**

decurrent   Having the base prolonged down the axis and adnate to it, as a leaf or leaflet with the base or the petiole extending down along the stem and attached to it.
= **decursive**

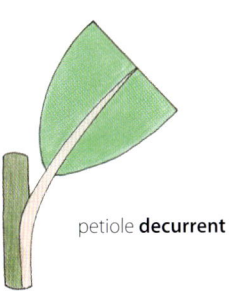

petiole **decurrent**

decursive   Having the base prolonged down the axis and adnate to it, as a leaf or leaflet with the base or the petiole extending down along the stem and attached to it.
= **decurrent**

petiole **decursive**

decursively-pinnate
Of a pinnate leaf with the base of the leaflets extending down along the rachis.

decursively-pinnate

decurved Curved gradually downward, as some leaves on a stem or the the surface of some leaves.
cf. **recurved**

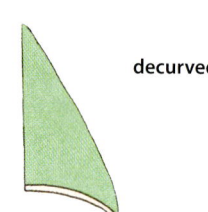
decurved

decussate Of opposite leaves with each pair arranged at right angles to the pair above and below.
cf. **brachiate**

decussate

decussate tetrad
A multiplanar tetrad with the four cohering members arranged in two pairs lying one across the other, more or less at right angles to each other.
see also **pollen tetrad**
see **multiplanar, viscin thread**

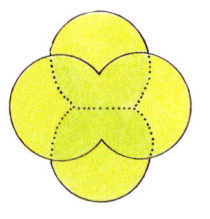
decussate pollen tetrad

definite Lacking a persistent terminal growing point and having growth occur in successive lateral branches, as elms (*Ulmus*).
The pattern of growth in which the apex of the main stem ceases to grow due to the abortion of the apical bud or the development of a flower or another structure, as a tendril.
Growth continues below the apex from a succession of axillary branches with a similar

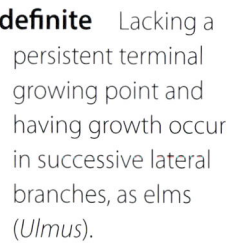
elm (*Ulmus*)

tendril

growth pattern.
Of sympodial or determinate growth, as a cymose inflorescence.
cf. **indefinite**

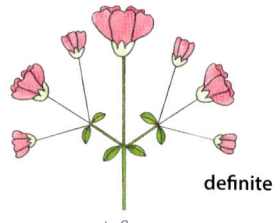
definite

cymose inflorescence

deflected, deflexed
Bent abruptly downwards, as margins of some fern fronds.

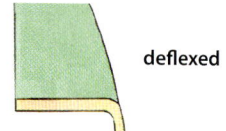
deflexed

defoliation Shedding of leaves. Loss of leaves by use of chemicals.

deforestation The removal of trees by humans, as forests that are cleared for growing crops or logged to make wood products.

dehiscence The splitting open of a plant part in order to release its contents, as a capsule releasing seeds or an anther releasing pollen.
dehisce To open spontaneously at maturity to release contents.
dehiscent Opening when ripe, as a capsule.
cf. **indehiscent**

Dehiscence

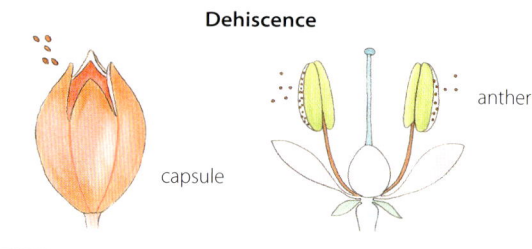

capsule

anther

delimitation Prescribing limits or boundaries, as those that define a genus or species.

deliquescence The condition of repeated divisions ending in fine divisions, as the venation of some leaves.
deliquescent Having a branching axis with the trunk soon lost or 'dissolved' into the successively divided branches, as most deciduous trees.
cf. **excurrent**

leaf venation

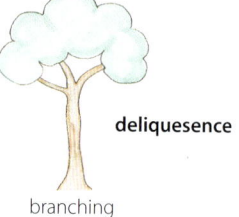
deliquesence

branching

deltate   Of a triangle with sides of about equal length and broad at the base.
= **deltoid**

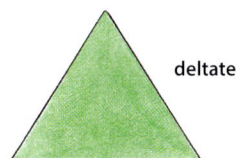
deltate

deltoid   Of a triangle with sides of about equal length and broad at the base.
= **deltate**
cf. **obdeltoid**

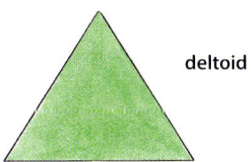
deltoid

demi-   A prefix meaning half.

dendriform   Tree-like in form, arborescent.

dendritic   Tree-like. Branched like a tree, as the veins in some leaves or a hair that branches at the apex like a tree.
= **dendroid**
see also **cladodromous, open venation**
cf. **anastomosis, closed venation**

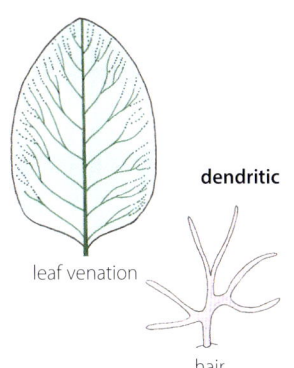
dendritic
leaf venation
hair

dendroid   Tree-like. Branched like a tree, as a hair that branches at the apex like a tree.
= **dendritic**

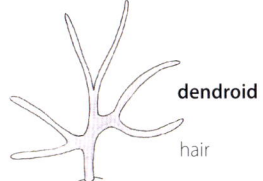
dendroid
hair

dentate   Having teeth. With shallow tooth-like projections, like an equilateral triangle, at right angles to the margin, as some leaves.
= **toothed**
cf. **crenate, edentate, serrate**

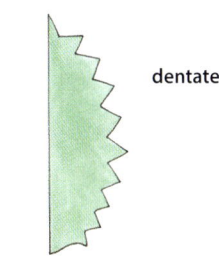
dentate

dentation
Referring to teeth on a margin, as a leaf or the petals of some flowers, as carnations (*Dianthus*).

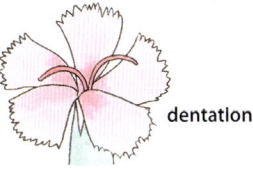
dentation

denticidal capsule   A capsule that splits open at the apex and forms a ring of teeth, as chickweed (*Cerastium*).

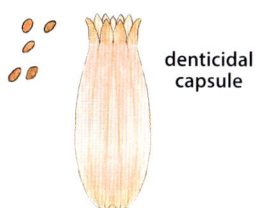
denticidal capsule

denticle   A very small tooth.
denticulate   Minutely toothed, minutely dentate. Having margins with minute tooth-shaped projections like an equilateral triangle, as the margins of some leaves.

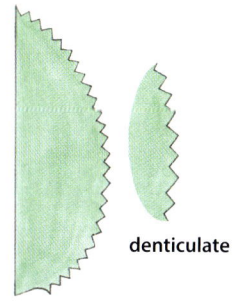
denticulate

dentiform   Tooth-shaped.

deoxyribose nucleic acid, DNA   A double-stranded, helically arranged acid in the nucleus of a cell that is the main constituent of chromosomes.
see also **ribose nucleic acid (RNA)**

depauperate   Of a plant that is stunted and arrested in growth and development due to poor growing conditions.
Of an ecosystem, lacking in species variety.

dependent   Hanging down.

dephosphorylation
Removal of a phosphate group from an organic compound through hydrolysis.
Adenosine triphosphate loses one or two phosphate groups when releasing energy for processes in the cell, thus becoming adenosine diphosphate or adenosine monophosphate.
The process can be reversed by phosphorylation.

depressed   Sunken or flattened as if pressed from above.

derived character
In phylogenetics, a character that has evolved from, but become modified and distinct from, the one present in an ancestor.
= **apomorphy**
cf. **ancestral**

dermal   Of the outer covering of a plant that includes the epidermis and the periderm.

dermal tissue   The epidermis (a primary tissue that is derived from the protoderm) and the periderm (a secondary tissue that is derived from cork cambium).

descending   Proceeding from a higher to a lower part. Gradually going downwards.

descending axis   The axis of a plant below ground, the root.
cf. **ascending axis**

descending inflorescence
Having the upper flowers on the axis and branches opening first.
= **basipetal**
cf. **ascending inflorescence**

descending inflorescence

descent   The transfer of genetic material from parents to offspring over time.

description   The assignment of features or attributes to a taxon.
A published account of the features of a taxon.
see **circumscription, diagnosis**

desert   An extremely dry landscape, with sparse vegetation that may be hot, as the Sahara Desert, or cold, as the Gobi Desert.
see **biome**

desertification   Land degradation in arid, semi-arid and dry subhumid areas resulting from human activities, like deforestation and grazing, and climatic variations.
see **ecoagriculture**

det. *abbr.,* **determinavit**   From the latin *determinavit*, meaning he/she determined.
Used on the label of a herbarium specimen and followed by the name of the person who identified the specimen.

determinate   Lacking a persistent terminal growing point and having growth occur in successive lateral branches, as elms (*Ulmus*).
The pattern of growth in which the apex of the main stem ceases to grow due to the abortion of the apical bud or the development of a flower or another structure, as a tendril.
Growth continues below the apex from a succession of axillary branches with a similar growth pattern.
Of sympodial or definite growth, as a cymose inflorescence.
cf. **indeterminate**

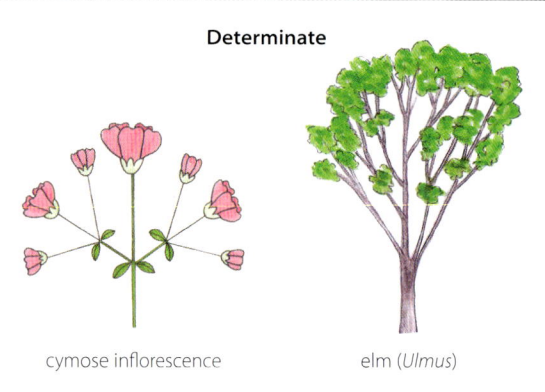
Determinate

cymose inflorescence   elm (*Ulmus*)

determination   Description and naming of an unknown species by comparing its characteristics with a known species, or recognising that it is new and warrants formal description and naming.
Specimens can be reidentified by adding a new determination and retaining a history of previous determinations that are termed 'unaccepted', though not necessarily wrong.
see **identification**

determinavit, *abbr.* **det.**   From the latin *determinavit*, meaning he/she determined.
Used on the label of a herbarium specimen and followed by the name of the person who identified the specimen.

Deuteromycota
An informal group of fungi known as imperfect fungi that are thought to reproduce only asexually by conidiospores.
They form ectendo-mycorrhizas with pines (*Pinus*), spruce (*Picea*) and larch (*Larix*).
see **fungus, mycorrhiza**

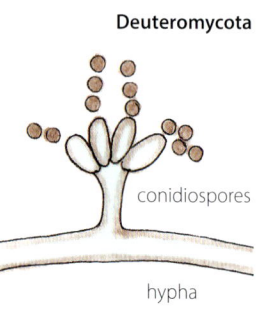
Deuteromycota

conidiospores

hypha

dextrorse Twining from the base in a spiral from left to right, as seen from the side. Twining in a clockwise direction, as seen from above.
cf. **sinistrorse**

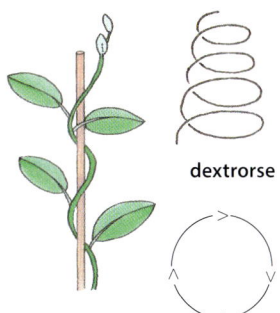

dextrorse

di- A prefix meaning two.

diad, dyad A group of two.
A pair of flowers in an inflorescence, as those of some mistletoes (*Amyema*).
Two united pollen grains.
cf. **monad, polyad, tetrad, triad**

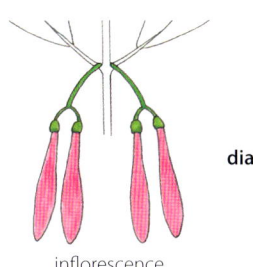

diad

inflorescence

pollen grains

diadelphous
Of stamens united by their filaments into two bundles, as some pea flowers that have nine stamens united and one free.
see **adelphous**

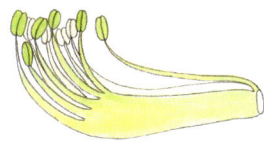

diadelphous

diagnosis A description of what distinguishes one taxon from another, based on selected characters.
see **circumscription, description**

dialypetalous
With a corolla of separate petals.
= **polypetalous**

dialypetalous

dialysepalous
With a calyx of separate sepals.
= **polysepalous**

dialysepalous

diandrous Having two stamens, as the flowers of derwentia (*Derwentia derwentiana*).

cf. **monandrous, pentandrous, polyandrous, tetrandrous, triandrous**

derwentia (*Derwentia derwentiana*)

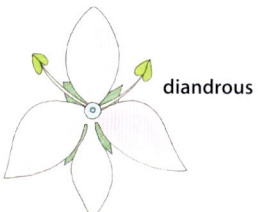

diandrous

diaphanous Delicate and transparent, or nearly so.

diaspore The dispersal unit of a plant.
It may be a seed, as flowering plants, gymnosperms and pteridosperms, a spore as ferns and lycopods, a plant fragment as bryophytes, a whole plant as tumbleweed (*Salsola*), a seedling as mangrove (*Rhizophora*) or a fruit as beet (*Beta vulgaris*).
= **disseminule**
see also **propagule**

diastase Any one of a group of enzymes that causes the transformation of starch into maltose.

diatom
A minute single-celled photosynthetic alga with a cell wall composed of silica. Found in vast quantities in fresh and marine waters.

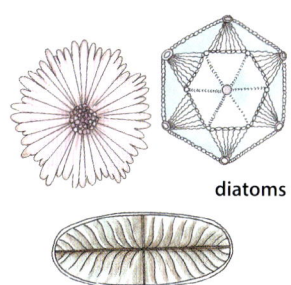

diatoms

diatropism Tendency of a plant to grow at right angles to the direction of the orienting growth stimulus, as lateral branches that grow at right angles to the stimulus of gravity.
see **plagiotropism, plagiogeotropism, tropism**
diatropic Of or relating to diatropism.

dichasium, *pl.* **dichasia, dichasial cyme**
A cymose inflorescence with the main axis bearing a terminal flower. A lateral axis may develop in each of the subtending bracts or bracteoles and bears a terminal flower.
see also **biparous, dichotomous**
cf. **monochasium, pleiochasium, polychasium**
dichasial Of a dichasium.

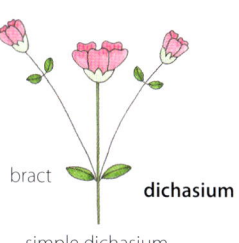

bract

dichasium

simple dichasium

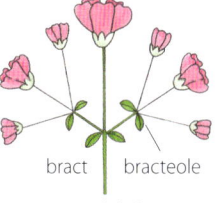

bract   bracteole

compound dichasium

dichlamydeous
Having a perianth of two whorls, that is, with both a calyx and a corolla.
cf. **chlamydeous**

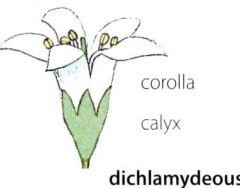
corolla
calyx
dichlamydeous

dichogamy   Of a flower, a way of preventing self-fertilisation by having stamens and stigmas maturing at different times.
see **protandry, protogyny**
cf. **herkogamy**
dichogamous   With stamens and stigmas maturing at different times.

dichotomous   Forking once or several times, each time into two equal branches.
cf. **trichotomous**

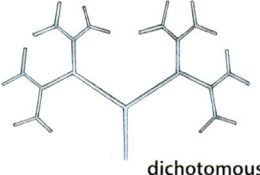
dichotomous

dichotomous cyme   Another name for dichasium.

dichotomous key   A key that offers two contrasting characters or couplets at each step so that by selecting one option each time an unknown plant can be identified.
= **binary key**

dichotomous venation
Forking and dividing always into two more or less equal branches, as veins of some leaves.
see also **bifurcate**

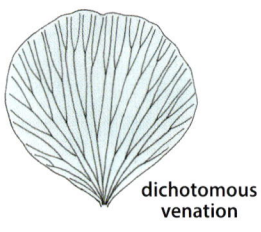
dichotomous venation

diclesium   A small dry indehiscent fruit.
An achene, surrounded by a free but persistent hardened perianth, as marvel of Peru (*Mirabilis jalapa*).
cf. **achene, caryopsis, cypsela**
see **accessory fruit**

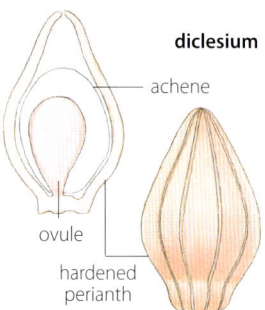
diclesium
achene
ovule
hardened perianth

diclinous   Having stamens and pistils in separate flowers.
see **androgynous, dioecious, monoecious, trioecious**

dicotyledon   Angiosperms were formerly divided into monocotyledons, with one seed leaf (cotyledon) in the embryo, and dicotyledons, mostly with two seed leaves in the embryo. Dicotyledons are no longer regarded as a natural grouping and angiosperms are now divided into several groups, the larger groups being eudicots (with pollen grains having three pores or furrows) and monocotyledons (with pollen grains having one pore or furrow).

dicotyledonous
Of a plant embryo having two cotyledons, as most eudicots.
Of a plant producing such embryos.
cf. **monocotyledonous, polycotyledonous**

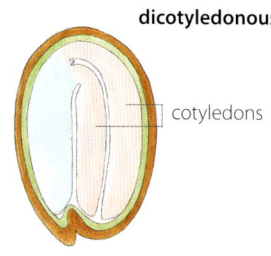
dicotyledonous
cotyledons
rock cress (*Arabidopsis*)

dictyosome
A stack of flattened membrane-bound cisternae that make up the Golgi complex.

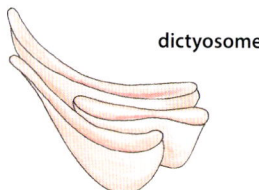
dictyosome

didymous   Of anthers where the connective is almost absent, as those of the genus *Clinosperma*.
cf. **distractile**
see **anther attachment**

didymous

didynamous   Having four stamens, two long and two short, as flowers of some members of the mint family (Lamiaceae).
cf. **tetradynamous, tridynamous**

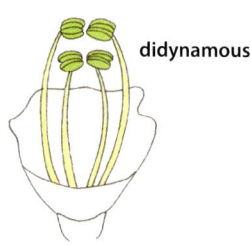
didynamous

differentiated   With visible distinctive characteristics, as a stamen divided into an anther and a filament or the petals and sepals of most flowers.

differentiation   The process by which cells become specialised to form a particular function, as amorphous meristem cells that undergo various changes to form the different tissues of the plant body.

diffuse   Of open
spreading growth that
is loosely branched,
as the inflorescence
of spreading hogweed
(*Boerhavia diffusa*).

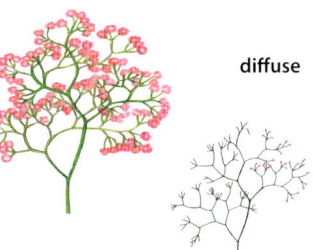

diffuse

diffusion   Passive movement of molecules from
a region of higher concentration to a region of
lower concentration in order to equalise the
concentration.
Movement is random and occurs in the absence of
a semipermeable membrane. It takes place in both
liquids and gases.
cf. **osmosis**

Diffusion

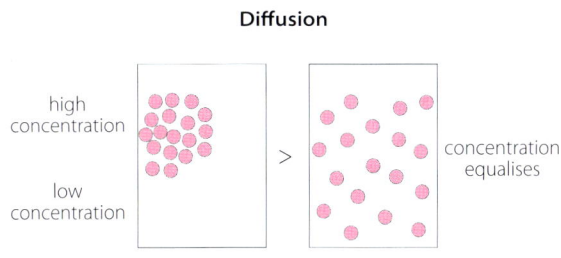

high
concentration

low
concentration

concentration
equalises

digiform   Shaped like a finger.
cf. **digitate**

digitate   With the
segments spreading
from a common point
like the fingers of a hand,
as some palmate leaves.

digitate

digitate inflorescence
An arrangement of spikes
or racemes, (rather than
flowers), radiating like
fingers from a common
point at the top of the
peduncle. Found in some
grasses (Poaceae) and
sedges (Cyperaceae).
cf. **anthela, panicle,
raceme, spike**

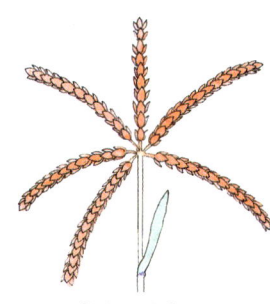

digitate inflorescence

digitate-pinnate   Of
a leaf with the segments
spreading from a
common point like the
fingers of a hand and
each segment divided
pinnately.

digitate-pinnate

digynous
Of a flower having two
carpels, whether free or
variously fused.

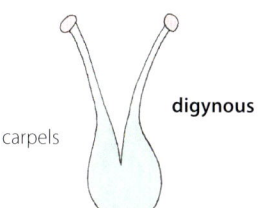

digynous

carpels

dilated   Widened.
Expanded, as the tube
of some flowers or the
funnel-shaped corolla of
bindweed (*Convolvulus*).
cf. **distended**

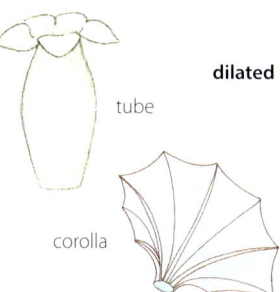

dilated

tube

corolla

dimerous   Having flower parts, such as petals,
sepals and stamens, in whorls of two or multiples of
two. 2-merous.
see **-merous**

dimidiate   Of a leaf or
leaflet, with the leaf on
one side of the midrib so
reduced that it appears
to be lacking.

midrib

dimidiate

dimorphic   Having two distinct forms, as the styles
of cowslip (*Primula officinalis*) and the leaves of small
river buttercup (*Ranunculus amphitricus*).
see **heterophylly, pin, thrum**
cf. **monomorphic, polymorphic, trimorphic**

Dimorphic

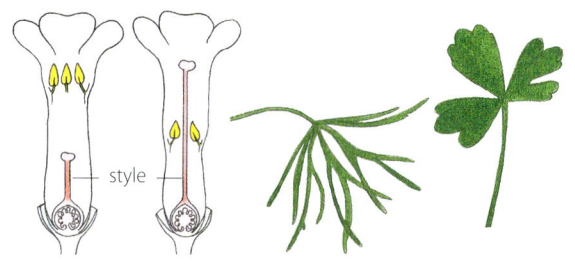

style

cowslip (*Primula officinalis*)   buttercup (*Ranunculus amphitricus*)

dioecious   Of a species with unisexual flowers on
different plants.
see **androgynous, diclinous, monoecious,
trioecious**

diplo-   A prefix meaning double, in pairs or
twofold.

diploid, diploidy   Having two complete sets (2x) of chromosomes in each somatic cell.
see **ploidy**

diplospory   In flowering plants (angiosperms), the formation of an embryo from the diploid megaspore mother cell before it undergoes meiosis. A form of agamospermy.
cf. **apospory**

diplostemonous
With two whorls of stamens, the inner whorl opposite the petals and the outer whorl alternate with the petals.
cf. **obdiplostemonous**

diplostemonous

dipterous   Of fruit or seed, having two wing-like expansions.

disarticulate
To separate at the joints (nodes), as the rachilla of some grasses.
cf. **articulate**

disarticulate

disc, disk   A thin, flat and circular object.
The central part of a daisy capitulum.
disciform, discoid
Resembling a disc. Circular and flattened, as the inflorescence of most button weeds (*Cotula*).

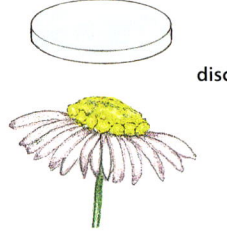

disc

disc floret
Of the daisy family (Asteraceae), one of the tubular florets that form the central portion of a head, as distinct from the surrounding ray florets.
= **tubular floret**
cf. **ligulate floret, ray floret**

head

disc floret

discolorous   With two different colours.
Of a leaf having the upper and lower surfaces unlike in colour.
cf. **concolorous, variegated**

discolorous

discrete   Individually separate and distinct.
cf. **coalescent**

disjunct distribution   Separated geographically, as populations of a species located in different areas due to habitat fragmentation.

dispersal   Movement from the parent plant to a new location, usually for reproduction, as the scattering of fruit, seeds and pollen grains by wind, water, insects and so on.
see **dispersal mechanism, dispersal unit**

dispersal mechanism   The means by which a diaspore, or some other structure like pollen or a propagule, is removed from the vicinity of the parent plant.
Includes dispersal by wind (anemochory), animals (zoochory), water (hydrochory) and ants (myrmecochory).

dispersal unit   The entity that is shed from the parent plant for reproduction, as seeds, spores or pollen grains.
see **disseminule**

dissected   Deeply divided.
Of a leaf having lobes with incisions that extend almost, but not quite, to the midrib or almost but not quite to the top of the petiole.

Dissected

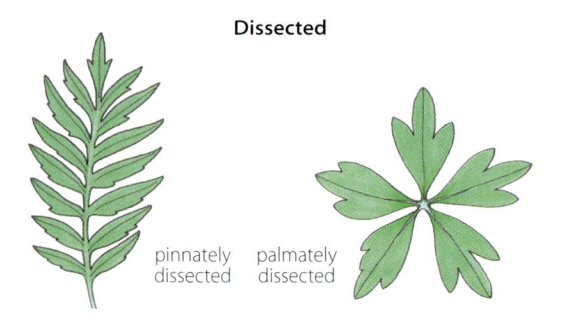

pinnately dissected   palmately dissected

dissemination   The spreading of disseminules.

disseminule   The dispersal unit of a plant. It may be a seed, as flowering plants, gymnosperms and pteridosperms, a spore as ferns and lycopods, a plant fragment as bryophytes, a whole plant as tumbleweed (*Salsola*), a seedling as mangrove (*Rhizophora*) or a fruit as beet (*Beta vulgaris*).
= **diaspore**
see also **propagule**

dissepiment   A partition or wall separating two cavities. In an ovary or fruit, usually formed by the fusion of adjacent carpel walls.
= **septum**

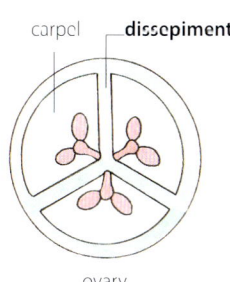

carpel   **dissepiment**

ovary

distal   Near the free end as opposed to the attached (proximal) end.

Distal

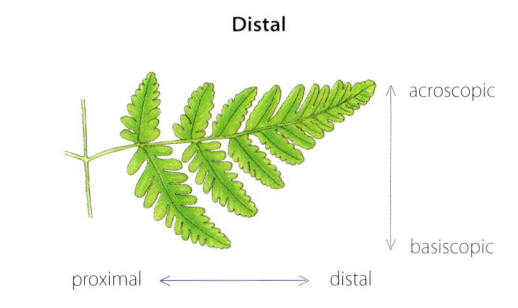

acroscopic

basiscopic

proximal ←→ distal

distal pole   Of a pollen grain in a tetrad, that part of the polar axis orientated towards the outside.
cf. **proximal pole**

Distal pole

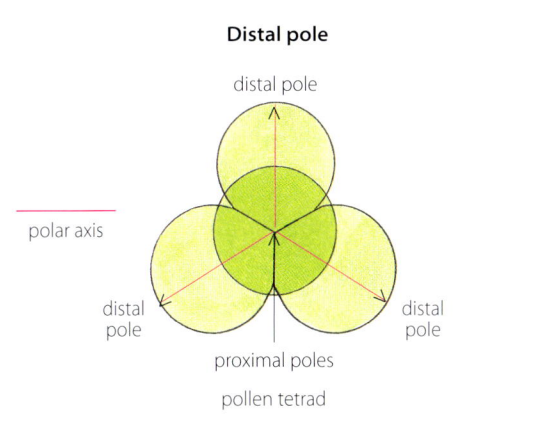

distal pole

polar axis

distal pole

distal pole

proximal poles

pollen tetrad

distant   Separate or apart in space, far removed. Widely spaced, as leaves on a stem.
= **remote**

distant

distended   Expanded or swollen from internal pressure, as a turgid cell that is swollen from excess water.
cf. **dilated**

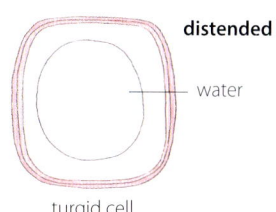

distended

water

turgid cell

distichous   Arranged in two vertical rows on opposite sides of a stem, as some leaves, with any third leaf above the one below it.
= **two-ranked**
see **orthostichy**

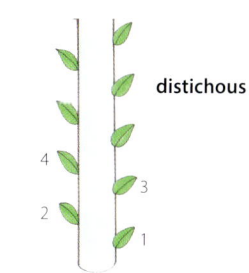

distichous

4

2

3

1

distinct   With like parts separate and not fused, as the petals and sepals of a flower.
= **free**
cf. **fused**

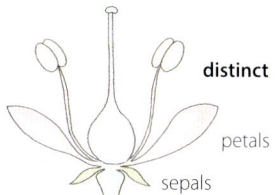

distinct

petals

sepals

distractile   Borne widely apart. Of anthers separated by a very long, narrow connective, as sage (*Salvia officinalis*).
cf. **didymous**
see **anther attachment**

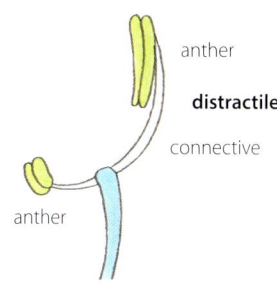

anther

distractile

connective

anther

distribution   The pattern of a species as it occurs over a particular geographical area.

distyly   Having styles of two different lengths in flowers on the same plant, as the cowslip (*Primula vulgaris*).
cf. **heterostyly, tristyly**
distylous   Exhibiting distyly.

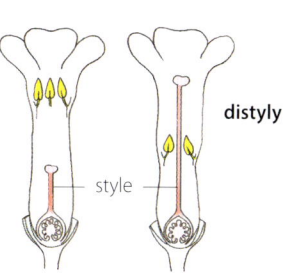

distyly

style

disulcate   Of a pollen grain having sulci arranged in pairs, either on opposite sides of the equator or at the distal pole.

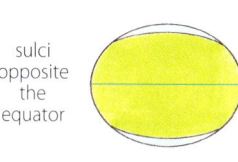

sulci opposite the equator

disulcate

sulci at the distal pole

disymmetric
Divisible through the centre into exactly similar halves on two planes of symmetry, as the flowers of rapeseed (*Brassica napus*).
cf. **bisymmetric, zygomorphic**

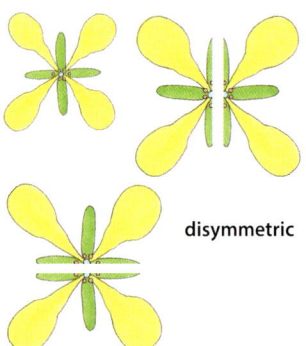

disymmetric

dithecal, dithecous
Of a stamen having two anther lobes and four pollen sacs (microsporangia), two in each lobe.
= **tetrasporangiate**
cf. **monothecal**

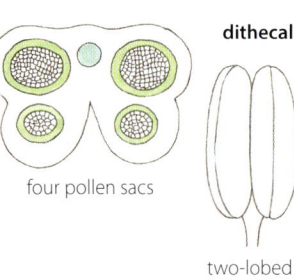

dithecal

four pollen sacs

two-lobed anther

diurnal   In the daytime. Of flowers, opening only during daylight.
cf. **nocturnal**

divaricate   Widely spreading, almost at right angles to the axis, as the pedicels of marsh speedwell (*Veronica scutellata*).

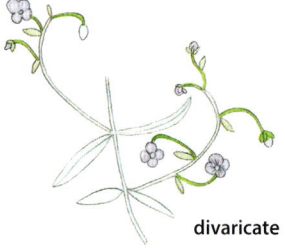

divaricate

divergent   Spreading broadly from the centre as the petals of many donkey orchids (*Diuris*).
cf. **convergent**
Of anthers, spreading from the centre.
cf. **explanate**

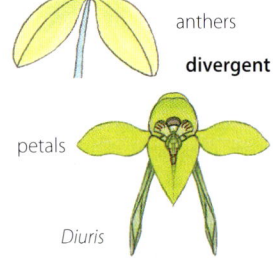

anthers

divergent

petals

Diuris

divergent evolution   The process whereby groups with a common ancestor accumulate differences over time in response to environmental pressures.
Divergent evolution leads to speciation.
cf. **convergent evolution**

diversity   Variety, as the number of different species in a given community or the variety of communities within an ecosystem.

divided   Separated into part.
Of a leaf that is divided into leaflets.

divided

division   A taxonomic classification between kingdom and class.
The name of a division in the plant kingdom ends in -*phyta*, as Magnoliophyta (flowering plants).
= **phylum**
see **taxonomic hierarchy**

DNA   Deoxyribose nucleic acid.
see also **RNA**

DNA sequencing   Determination of the order of the four building blocks (adenine, guanine, cytosine or thymine) that make up a DNA molecule.
= **gene sequencing**

Dodsonian mimicry
Imitation by the flower of one species that has no nectar, by the flower of another species that has nectar, in order to attract a pollinator, as the flowers of some donkey orchids (*Diuris*) that mimic bush peas (*Pultenaea*) that have nectar.
see **Bakerian mimicry, Pouyannian mimicry, Vavilovian mimicry**

Dodsonian mimicry

donkey orchid

bush pea

dolabriform   Having the shape of the head an axe or hatchet, as some leaves.

dolabriform

dolichostylous
Of heterostylous flowers, those having long styles, as the cowslip (*Primula vulgaris*).
cf. **brachystylous, isostylous**

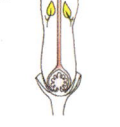

style

dolichostylous

domatium, *pl.* **domatia**   Structures produced by
a plant that provide shelter for tiny athropods. They
may feed on herbivorous pests harmful to the plant.
Examples of domatia are tiny pits, pockets or tufts of
hairs that house mites on the undersurface of some
leaves, or hollow stems and thorns that house ants.

Domatium

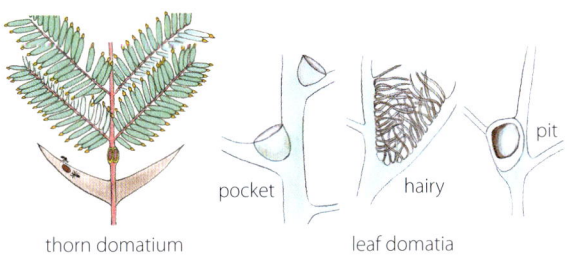

thorn domatium   leaf domatia

pocket   hairy   pit

domesticate   To breed, develop and stabilise a
desirable characteristic in a wild plant so that it can
be cultivated.

dominant   Of a heterozygous individual, an allele
of a gene that masks or conceals the expression of
the other allele.
cf. **epistasis, recessive**

dormancy
A state in which a plant is synchronised with the
external rhythms of the seasons and does not grow
at certain times of the year, as herbs that die back to
bulbs, or rhizomes and trees that are deciduous.
A period of time when a plant, or its seeds, does not
grow due to environmental conditions.
Dormancy allows a plant to wait for favourable
conditions before starting growth.
   dormant   Alive but temporarily inactive.

dormant buds   Usually applied to buds that do
not grow the following season.
Also sometimes applied to resting buds.
   = **latent buds**

dorsal   The back. The outer
side facing away from the
axis, as a leaf on a stem.
cf. **ventral**

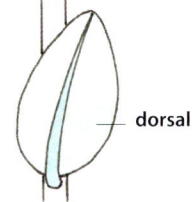

dorsal

dorsal sepal
The uppermost sepal
of a resupinate orchid
flower, as *Caladenia,* or
the lowermost sepal of
a non-resupinate orchid
flower, as *Prasophyllum.*

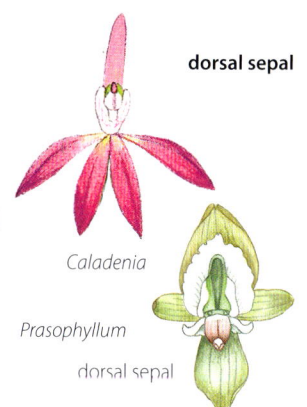

dorsal sepal

Caladenia

Prasophyllum

dorsal sepal

dorsifixed   Attached
on or by the back, as a
stamen filament attached
to the connective
somewhere along the
back of an anther.
cf. **medifixed,
ventrifixed**
see **anther attachment**

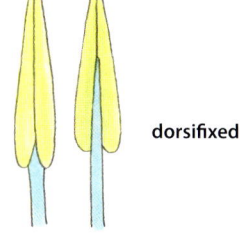

dorsifixed

dorsiventral   Having
structurally different
upper (adaxial) and
lower (abaxial) surfaces,
as the leaves of most
dicotyledons.
Typical of leaves that
orient themselves at an
angle to the main axis.
   = **bifacial**
cf. **isobilateral**

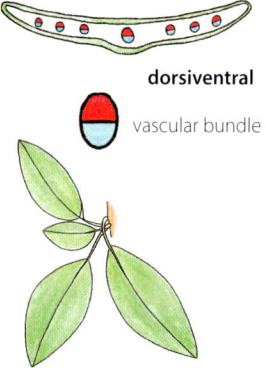

dorsiventral

vascular bundle

double   Of a flower with more than the usual
number of petals or with petal-like sepals, stamens
or bracts, as paeony (*Paeonia*).
   see **hose-in-hose**

double fertilisation
Of angiosperms, the two separate unions that occur
in an ovule.
After the two male sperm cells enter the embryo
sac, one sperm cell fertilises the egg cell nucleus,
forming the diploid (2*n*) zygote.
The other sperm cell combines with the two fused
polar nuclei to form a triploid (3*n*) nucleus in the
centre of the large central cell. This cell will give rise
to the endosperm of the seed.
   double fertilisation page 88 (cont.)
   cf. **simple fertilisation**

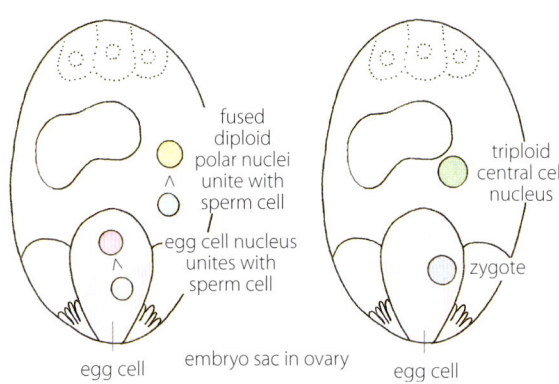

Double fertilisation

fused diploid polar nuclei unite with sperm cell

triploid central cell nucleus

egg cell nucleus unites with sperm cell

zygote

egg cell

embryo sac in ovary

egg cell

doubly crenate

Of a margin with crenate teeth that are themselves crenate, as the margins of some leaves.

= **bicrenate**

cf. **crenate, crenulate**

doubly crenate

doubly dentate

Having tooth-like projections that are themselves toothed. Doubly toothed, as the margins of some leaves.

= **bidentate**

cf. **biserrate**

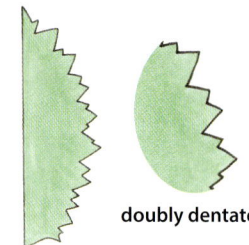

doubly dentate

doubly serrate

Of a margin with saw-like teeth that are themselves toothed, as the margins of some leaves.

= **biserrate**

cf. **serrate, serrulate**

doubly serrate

down   A covering of short soft hairs, as on the buds, leaves and fruit of some plants.

= **pubescence**

downy   Covered with down.

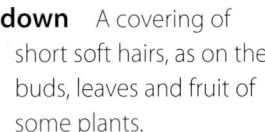

down

drepanium,
pl. **drepania**

A sickle-shaped cymose inflorescence that is flattened, with successive branches on one side only.

see also **monochasium**

cf. **rhiphidium**

drepanium

1 2 3 4 5

-dromous   A suffix meaning moving or running.

drooping   Hanging downwards limply, as leaves on a stem.

drooping

dropper   A root that grows downward from a bulb or corm that bears a replacement bulb or corm.

Of orchids, the short root that bears the replacement tuber.

= **sinker**

dropper

replacement tuber

leek orchid (*Prasophyllum*)

drupe   A fleshy fruit with the ovary wall ripening into three layers, the skin (exocarp), the flesh (mesocarp) and the stone (endocarp) that encloses the seed. Derived from a one-carpelled superior ovary, as the stone-fruit genus *Prunus*.

drupaceous   Of or like a drupe.

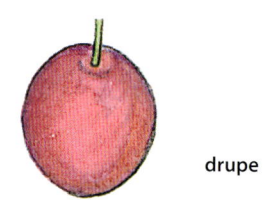

drupe

exocarp

mesocarp

endocarp

drupecetum

An aggregate fruit composed of a cluster of drupes, as the raspberry (*Rubus*).

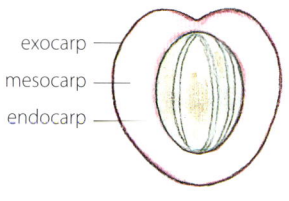

drupecetum

drupelet   A small drupe. One segment of an aggregate fruit, as raspberries and blackberries (*Rubus*).

drupelet

duct   A tube formed by a row of cells that have lost their adjacent cell walls.

duplicate specimen   Part of a single plant collection preserved as a separate specimen.

duramen   The central usually darker inactive central wood of a trunk or branch. It is made up of non-functioning sapwood cells infiltrated with other substances like lignin.
= **heartwood**

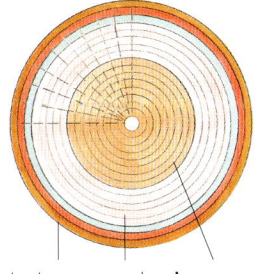

bark   sapwood   **duramen**

dystrophic   Of a body of brownish acidic water, with a high concentration of dissolved humus. It is low in oxygen and supports little life.
see **trophic**
cf. **eutrophic, mesotrophic, oligotrophic**

e-   A prefix meaning without.

ear   An auricle. An ear-shaped appendage at the bottom of a leaf. The grain-bearing spike or head of a cereal plant, as wheat or barley.

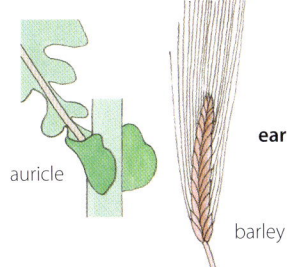

auricle

ear

barley

early wood   Wood in a growth ring, with large thin-walled cells, that is produced in spring. It is less dense than late wood that is produced later in the growing season when growth is slower. Early wood and late wood usually appear as two distinct bands.

ebracteate   Lacking bracts.

ebracteolate   Lacking bracteoles.

ecalcarate   Without a spur.

eccentric, excentric   Off-centre. Having the axis or other part not centrally placed, as the style of the buttercup genus (*Ranunculus*).

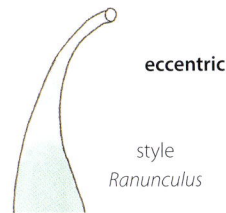

eccentric

style
Ranunculus

echinate   Of a surface with stout spine-like projections.

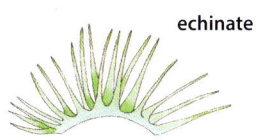

echinate

echinulate   Bearing small spines.

ecoagriculture   A means of land and natural resource management that conserves native biodiversity, provides agricultural products on a sustainable basis and supports viable livelihoods for local people.
It recognises that there are unprecedented demands on the world's finite resources.
see **desertification**

ecological succession   In ecology, the process of change in species structure as a community establishes over time.
Primary succession occurs on previously uncolonised areas like lava flows.
Secondary succession occurs on disrupted or disturbed areas.
= **succession**
see also **climax community**

Ecological succession

1. Colonisers
fungi, mosses, lichens

2. Pioneers
groundsels, grasses

3. Herbs

5. Trees

4 . Shrubs

ecology   The study of the distribution and abundance of living organisms in relation to the environment and in relation to each other.
In particular, the distribution and geographic range of a particular population.

ecostate   Of a leaf without a midrib, as some mosses.

ecostate

ecosystem   All living organisms (biotic components) and their physical surroundings (abiotic components) in a given area.
Biotic components include plants, animals, fungi, algae and microbes.
Abiotic components include temperature, water availability and soil.
An ecosystem may be small (a pond) or large (the tropical rainforest of the Amazon Basin), it may be aquatic, as a lake, river or coral reef, or terrestrial, as a grassland, forest or desert.
cf. **biome, community, habitat, population**

ecotone   A transitional area between two ecosystems, as that between a forest and a grassland.

ecotype   A group of organisms within a species that has adapted genetically to its particular environmental conditions. It can still reproduce with other members of its species from other areas that have not undergone these changes.

ectendomycorrhiza, *pl.* **ectendomycorrhizae, ectendomycorrhizas**
A mutually beneficial symbiosis formed between Deuteromycota fungi and the roots of some conifers.
Ectendomycorrhiza has characteristics of both ectomycorrhiza (that has a mantle of hyphae, around the outside surface of the root, ramify ing through the intercellular spaces in the root cortex) and endomycorrhiza (that has hyphae penetrating the cells and forming pelotons).
see also **mycorrhiza**

Ectendomycorrhiza

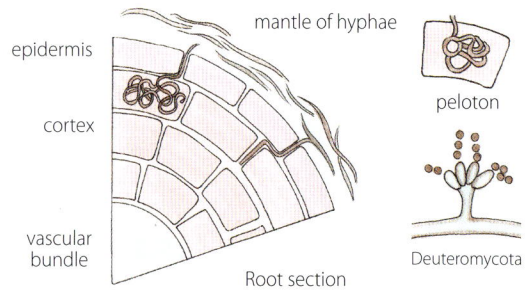

epidermis

cortex

vascular bundle

mantle of hyphae

peloton

Deuteromycota

Root section

ectexine, ektexine   The outer of two layers of the exine in the wall of a pollen grain, usually consisting of the tectum, infratectum and the foot layer.
see **pollen wall**

ecto-, ekto-   A prefix meaning outer.
cf. **endo-**

ectoaperture   An aperture present in one layer of the wall of a pollen grain. A simple aperture.

ectocarp   Another term for epicarp and exocarp.

ectomycorrhiza, *pl.* **ectomycorrhizae, ectomycorrhizas**   A mutually beneficial symbiosis formed between Basidiomycota fungi and the roots of the majority of forest trees.
A network of hyphae forms a sheath (a mantle) around the outside surface of a root and penetrates the cells of the root cortex by ramifying through the intercellular spaces, forming a Hartig net, rather than extending into the cells themselves.
The sheath of hyphae also extends into the surrounding soil.
= **ectotrophic mycorrhiza**
see **Hartig net, mycorrhiza**
cf. **ectendomycorrhiza, endomycorrhiza,**

Ectomycorrhiza

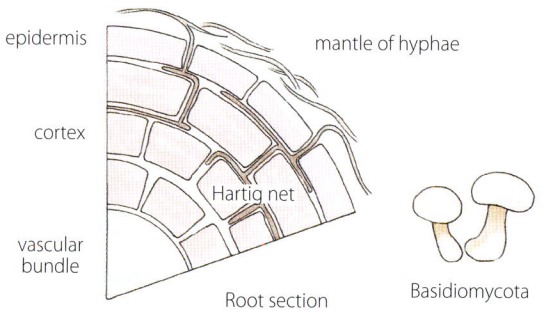

epidermis

cortex

vascular bundle

mantle of hyphae

Hartig net

Root section

Basidiomycota

ectophyte   A parasitic plant that lives on the surface of its host.
ectophytic   Relating to ectophytes.
cf. **endophyte**

ectotrophic   Obtaining nourishment from outside.
see **trophic**
cf. **endotrophic**

ectotrophic mycorrhiza
= **ectomycorrhiza**

edaphic   Relating to the soil.

edentate   Without teeth.
cf. **dentate**

effectively published   Published in accordance with the International Code of Nomenclature.
= **validly published**

efflorescence   The time or act of flowering.

egg   The female reproductive cell of plants and animals.
= **egg cell, megagamete, ovum**

egg apparatus
In angiosperms, the egg cell plus the two synergid cells at the micropylar end of the embryo sac.

Egg apparatus

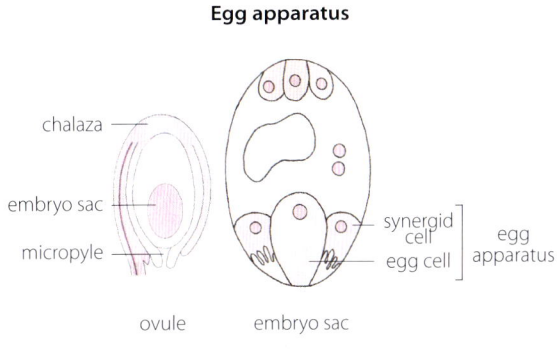

chalaza
embryo sac
micropyle
ovule
synergid cell
egg cell
egg apparatus
embryo sac

egg cell   The female sex cell that unites, at fertilisation, with a male sperm cell to form a zygote.
= **megagamete, ovum**
see **megagamophyte**
see also **egg, egg apparatus, embryo sac, oosphere**
cf. **sperm cell**

eglandular   Without glands.

elaiophore   Floral glands that secrete oils as a reward for pollinators, as found in some monocot families like Orchidaceae, Iridaceae and some eudicot families like Curcurbitaceae, Plantaginaceae and Primulaceae.

elaioplast   A leucoplast that stores lipids as rounded oil droplets.

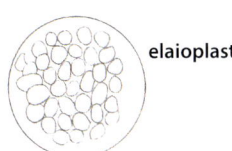

elaioplast

elaiosome   A fleshy lipid and protein-rich food body on some seeds that attracts ants.
It is eaten by the ant and the seed is discarded and dispersed underground in the ant's nest.

elaiosome
seed
golden wattle (*Acacia pycnantha*)

elaminate   Without a blade, as a sheath that lacks a blade in some sedges.
see **aphyllopodic**
cf. **laminate**

sheaths
elaminate

elater   One of the spiral cells among the spores in the capsule of a liverwort.
A slender appendage attached to the spores of horsetails.
Both change shape in response to moisture and are involved in the ejection of spores from the capsule.

eligulate   Without ligules.

ellipsoid   A three-dimensional shape, oval in outline and widest at the middle, with equally rounded ends, and elliptical in all sections through the long axis.
cf. **elliptic**

elliptic   A two-dimensional shape, oval in outline and widest at the middle, with equally rounded ends.
= **oval**
cf. **ellipsoid**

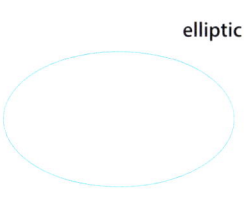

elliptic

elongate   Lengthened, as if stretched or extended.
To lengthen.
Of a plant part that is longer than wide, as grass leaves.

elongation   The process whereby a plant part is lengthened, as the the internodes in a grass stem or the elongation of a fruiting stem for seed dispersal, as *Corybas*.
see also **gibberellins, jointing**

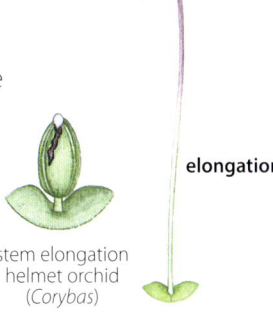

elongation
stem elongation
helmet orchid
(*Corybas*)

emarginate Having a distinct, broad, shallow notch at the apex, as with some leaves and petals.

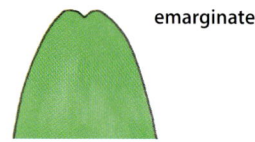

emarginate

embedded Enclosed or fixed firmly in a surrounding mass, as the flowers of glassworts (*Salicornia*).

flower

embedded

glasswort (*Salicornia*)

embryo The rudimentary plant within the seed before germination.
It consists of a plumule, a hypocotyl-root axis and one, two or more cotyledons.
embryonic In an early stage of development.
see **embryogenesis**

Embryo

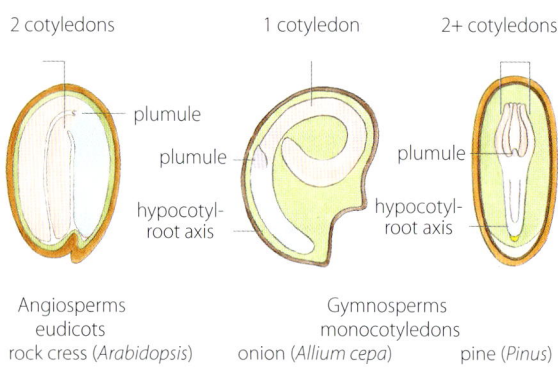

2 cotyledons

— plumule

— plumule

— hypocotyl-root axis

1 cotyledon

plumule —

2+ cotyledons

plumule —

hypocotyl-root axis

Angiosperms
eudicots
rock cress (*Arabidopsis*)

Gymnosperms
monocotyledons
onion (*Allium cepa*)   pine (*Pinus*)

embryo axis, embryonic axis
The plumule, epicotyl, hypocotyl and radicle together form the embryo axis.
It is attached to the cotyledon at the cotyledonary node.
It represents the axis of the future plant.
= **tigellum**

Embryo axis

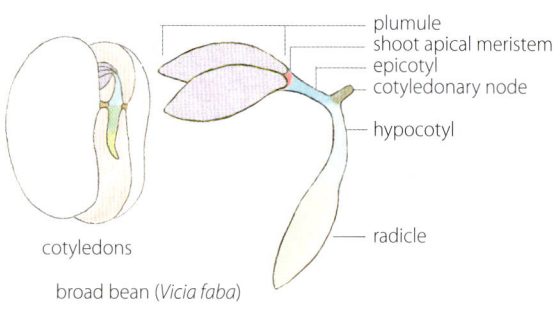

— plumule
— shoot apical meristem
— epicotyl
— cotyledonary node
— hypocotyl

— radicle

cotyledons

broad bean (*Vicia faba*)

embryo sac
Of angiosperms, the female gametophyte that develops within each ovule of the ovaries.
It commonly consists of seven cells, three antipodal cells, two synergid cells, one egg cell and one central cell with two polar nuclei.
= **megagametophyte**
see also **egg apparatus**

Embryo sac

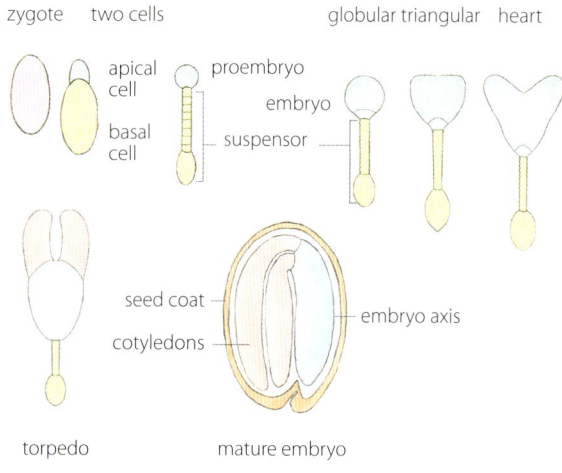

antipodal cells

vacuole
polar nuclei
central cell
egg cell
synergid cell
filiform apparatus

ovule

ovary
with ovules

embryo sac

embryogenesis, embryogeny
The process whereby the fertilised zygote undergoes a sequence of divisions to form an embryo.

Embryogenesis

zygote   two cells   globular triangular   heart

apical cell
basal cell

proembryo
embryo
suspensor

seed coat
cotyledons

embryo axis

torpedo

mature embryo

embryology The study of embryo formation and development.

embryony Having or producing an embryo.
see **monembryony, polyembryony**

Embryophyta A subkingdom of plants that develop embryos in specialised reproductive tissue and have a life cycle with alternating diploid and haploid generations.
Includes bryophytes, ferns, gymnosperms and angiosperms.
embryophyte Any member of the subkingdom Embryophyta.

embryotega Of some seeds, a cap-like covering over the micropyle that, at germination, lifts like an operculum and allows the radicle to emerge, as the spiderwort family (Commelinaceae).

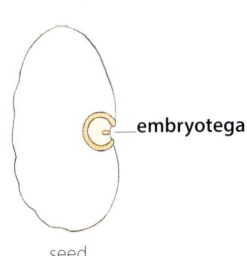

seed

emend., emendavit In nomenclature, a correction or amendment to the description of a taxon followed by the name of the person who made the change.

emendavit, *abbr.* **emend.** In nomenclature, a correction or amendment to the description of a taxon followed by the name of the person who made the change.

emergent Of an aquatic plant rooted in the soil below the water but with the stem and leaves rising out of the water, as arrowheads (*Sagittaria sagittifolia*).
= **emersed**
cf. **submerged**

emergent

emergent layer
The storey in a rainforest that extends above the canopy and is composed of very tall trees and may include woody climbers and epiphytes.

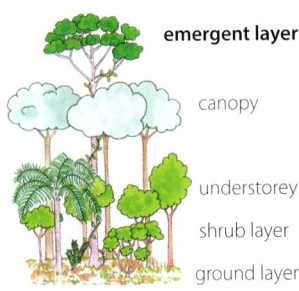

emergent layer

canopy

understorey

shrub layer

ground layer

emersed Of an aquatic plant rooted in the soil below the water but with the stem and leaves rising out of the water.
= **emergent**

enantiostyly
Deflection of the style to either the left side or to the right side of the floral axis, as the *Cassia* genus.
enantiostylous
Exhibiting enantiostyly.

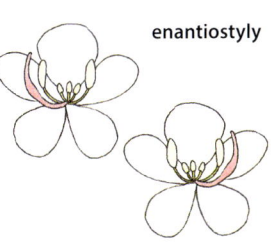

enantiostyly

enation An outgrowth from the surface of a plant part, as on the leaf of some members of the succulent genus *Eriospermum*.

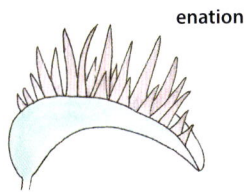

enation

endangered According to the IUCN, a conservation status covering species that are considered to be facing a high risk of extinction in the near future.

endarch Describes radial differentiation of xylem according to the relative position of protoxylem and metaxylem, in which protoxylem is positioned closest to the inside of the stem and metaxylem is positioned closest to the outside of the stem.
Found in the more specialised stems and leaves of seed plants.
cf. **exarch, mesarch**

endemic Native to a particular geographical region and not found elsewhere.

endexine The inner of the two layers of the exine in the wall of pollen grain.
see **pollen wall**

endo- A prefix meaning inner.
cf. **ecto-, ekto-**

endoaperture An aperture formed in more than one layer of the pollen wall. A compound or composite aperture.

endocarp The innermost layer of the fruit wall that may be tough or hard.
It surrounds the seed, as that of a stone fruit.
see **pericarp**

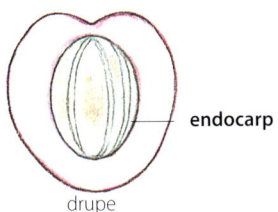

endocarp

drupe

endodermis   In roots and some herbaceous stems, a single layer of living cells around the vascular cylinder that, together with the pericycle, separates it from the cortex.
Each cell has a waxy Casparian strip that regulates water flow from the cortex into and out of the vascular cylinder.
see also **exodermis**

Endodermis

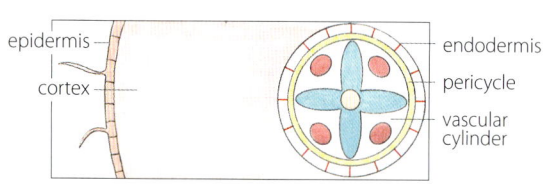

epidermis
cortex
endodermis
pericycle
vascular cylinder

endogenous growth   Growing from or originating from within.
Developing from inner tissue, as lateral roots that originate from the pericycle deep inside the root.
cf. **exogenous growth**

endomitosis   Chromosomes that duplicate but fail to separate, resulting in the doubling of the chromosomes in the nucleus.
see also **polyploidy**

endomycorrhiza, *pl.* **endomycorrhizae, endomycorrhizas**
The most common mutually beneficial symbiosis formed between a fungus and the roots of plant.
A network of hyphae with growth on the outside of the root limited and hyphae penetrating into the cells of the root cortex.
Examples include arbuscular mycorrhiza, ericoid mycorrhiza and orchidaceous mycorrhiza.
= **endotrophic mycorrhiza**
see **mycorrhiza**
cf. **ectendomycorrhiza, ectomycorrhiza**

Endomycorrhiza

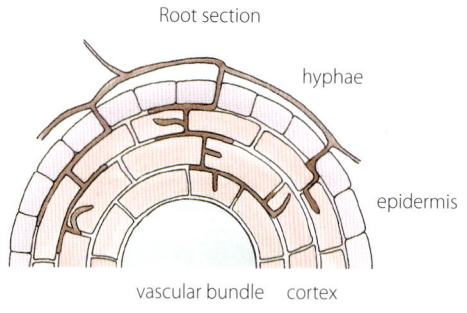

Root section
hyphae
epidermis
vascular bundle   cortex

endophyte   An organism, often a fungus or bacterium, that lives inside a plant, usually without causing harm.
endophytic   Relating to endophytes; infiltrating or invasive.
cf. **ectophyte**

endoplasmic reticulum
A cell organelle that is a network of flattened sacs and narrow tubes attached to the nuclear membrane.
It is composed of the smooth endopasmic reticulum and the rough endoplasmic reticulum.

Endoplasmic reticulum

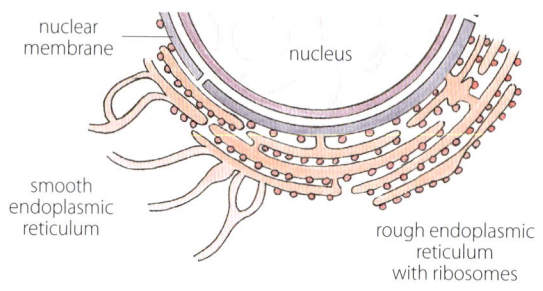

nuclear membrane
nucleus
smooth endoplasmic reticulum
rough endoplasmic reticulum with ribosomes

endosperm
Nutritive tissue in a seed that is not part of the embryo and is usually mostly starch with some lipids and proteins like albumin.
In angiosperms the tissue is derived from one male gamete and two female polar nuclei and is triploid.
In gymnosperms the tissue is derived from the female gametophyte and is haploid.
In gymnosperms and most monocots the endosperm persists and is the main source of food in the mature seed.
In most eudicots the endosperm is absorbed by the developing embryo and food is subsequently stored in the cotyledons that are produced by the embryo.
Orchids (Orchidaceae) lack a food supply in the seed and rely on a relationship with fungi to provide nourishment during germination.
Endosperm types include mealy, as wheat, oily, as poppies, bony, as some palms, and liquid, as coconuts.
see **albuminous seed, endospermic seed, exalbuminous seed, non-endospermic seed**
see also **cotyledon, perisperm**
endospermic   Of or relating to the endosperm. having endosperm, as endospermic seed.

Endosperm

endosperm absorbed   endosperm   endosperm   endosperm

eudicot (*Arabidopsis*)   monocotyledon (*Allium*)   caryopsis (*Triticum*)   gymnosperm (*Pinus*)

endospermic seed

One having endosperm persisting in the mature seed as the main source of nourishment for the embryo, as most monocotyledons.

Some eudicots, as castor bean (*Ricinus communis*), have endosperm and are albuminous.

All gymnosperm seeds are albuminous.

= **albuminous seed**
cf. **non-endospermic seed**

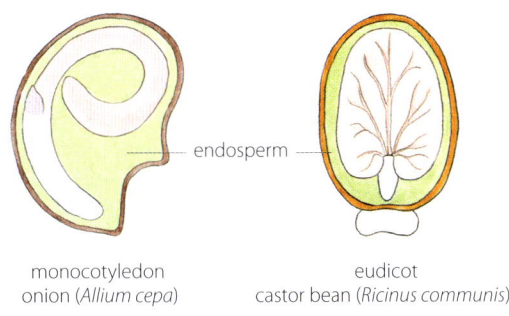

Endospermic seed

monocotyledon onion (*Allium cepa*)   eudicot castor bean (*Ricinus communis*)

— endosperm —

endospore

The innermost layer of a spore wall.
see **sporoderm**
cf. **intine**

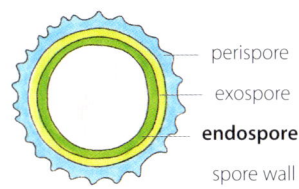

perispore
exospore
endospore
spore wall

endospory

Development of a gametophyte (protonema) within a spore, as in some liverworts and the spike moss genus *Selaginella*.

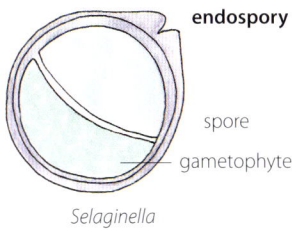

endospory

spore
gametophyte

Selaginella

endotegmen

Of the bitegmic seed coat of angiosperms, the inner epidermis of the inner integument (tegmen).
cf. **exotegmen, mesotegmen**
see also **endotesta, exotesta, mesotesta**

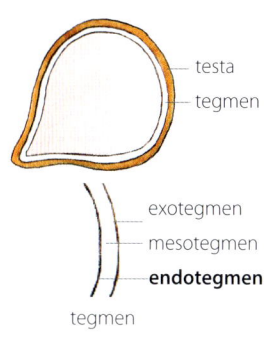

testa
tegmen

exotegmen
mesotegmen
endotegmen

tegmen

chick pea seed (*Cicer arietinum*)

endotegmic seed
Of angiosperms, having the mechanical layer in the endotegmen.
see **tegmic seed**

endotesta, *pl.* endotestae

Of the unitegmic seed coat of gymnosperms, the innermost parenchymatous layer, the other two layers being the outer sarcotesta and the middle sclerotesta, as pine (*Pinus*).

Of the bitegmic seed coat of angiosperms, the inner epidermis of the outer integument (testa).

cf. **exotesta, mesotesta**
see also **endotegmen, exotegmen, mesotegmen**
endotestal   Of, relating to or having a endotesta.

Endotesta

seed coat

endotesta
sclerotesta
sarcotesta

unitegmic seed coat pine seed (*Pinus*)

testa
tegmen

exotesta
mesotesta
endotesta
testa

bitegmic seed coat chick pea seed (*Cicer arietinum*)

endotestal seed   Of angiosperms, having the
mechanical layer in the endotesta.
see **testal seed**

endothecium   Of angiosperms, a fibrous
layer of the wall of an immature pollen sac (microsporangium), situated next to the epidermis, with weak areas that are linked to anther dehiscence.
see **microsporangial wall**

endotrophic   Obtaining nourishment from within.
  see **trophic**
  cf. **ectotrophic**

endotrophic mycorrhiza
  = **endomycorrhiza**

endozoochory   Dispersal of seeds and fruit by passing unharmed through the digestive system of an animal.
  endozoochorous   Of or relating to endozoochory.

ensiform   Shaped like the blade of a sword. Long and narrow with a sharp tip.
  = **gladiate**

ensiform

entire   Having a smooth margin that is not toothed, lobed or divided, as the margins of some leaves and petals.

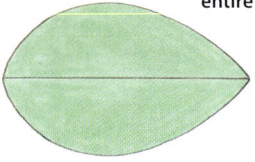

entire

entity   Something that exists as a discrete unit. It may be abstract, as a taxon in phylogenetic classification, or real, as a genome.

entomochory   Dispersal of pollen, spores, seeds or fruit by insects.
  entomochorous   Of or relating to entomochory.

entomogamy   Adapted to attract insect pollinators, pollination by insects.
  entomogamous   Of or relating to entomogamy.

entomophile   A plant that is pollinated by insects.

entomophily   Pollination by insects.
  entomophilous   Pollinated by insects.

enzyme   A substance in living organisms that regulates the rate at which chemical reactions proceed, without itself being altered.
  see also **catalyst**
  see **amylase, diastase**

eophyll   The first photosynthetic leaves of the seedling above the

the cotyledons. They are transitional leaves that are usually simpler in shape and smaller in size than adult leaves.
  see also **plumular leaves**

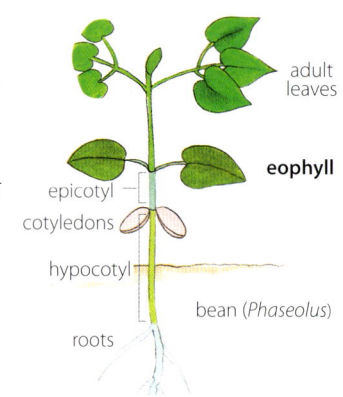

adult leaves

eophyll

epicotyl

cotyledons

hypocotyl

bean (*Phaseolus*)

roots

epaleate   Lacking a palea or paleae.

epedicellate   Lacking a pedicel.

ephemeral   Very short-lived. Lasting for a very short time, as the petals of guinea flowers ((*Hibbertia*).
  cf. **fugaceous, evanescent**

ephemeral

epi-   A prefix meaning upon, over or outer.

epiblast   Of a caryopsis, a scale-like appendage opposite the scutellum in the embryo of the seed, as wheat and rice. Some interpret it as a vestigial second cotyledon.

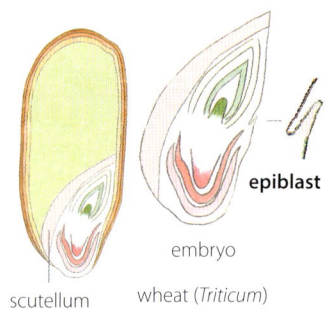

epiblast

embryo

scutellum

wheat (*Triticum*)

epiblem, epiblema   The epidermis of the root. It is of different origin to the epidermis of shoots but is continuous with it.
  = **rhizodermis**

epicalyx   A small cup. A whorl of free or fused bracts that looks like a second calyx, as hibiscus (*Hibiscus*).
  = **calicle, calycle, calyculus**

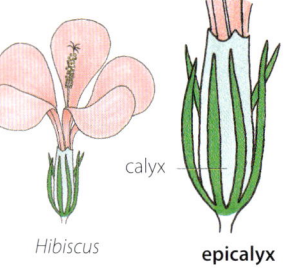

calyx

Hibiscus

epicalyx

epicarp   Another term for ectocarp and exocarp.
  see **exocarp**

epichile   Of orchids, the apical portion of the labellum, as pink fairies (*Caladenia latifolia*).

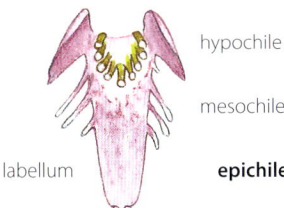

hypochile

mesochile

labellum

epichile

epicormic   Of growth from a dormant bud on the trunk or limb of some trees.
Usually triggered after injury, as fire or coppicing but also by stress or decline.

epicormic growth

epicortical   On top of the bark, as some mistletoes that produce epicortical runners that grow in a vine-like manner along the outside of the host branch.

epicortical runner

mistletoe

epicotyl   In grasses (Poaceae), that part of the embryo axis above the mesocotyl that is enclosed in the protective sheathing coleoptile.
Of a seed, in eudicots and most monocotyledons, the part of the embryo axis that is above the cotyledonary node and below the plumule, with the apical meristem at its tip.
Of a seedling, the part of the axis above the cotyledon(s) and below the plumular leaves.
cf. **hypocotyl, hypocotyl-root axis, mesocotyl**

Epicotyl

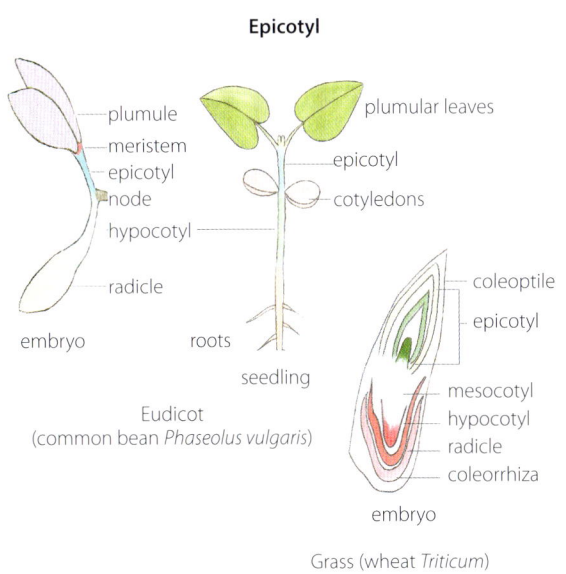

plumule
meristem
epicotyl
node
hypocotyl

radicle

embryo

plumular leaves
epicotyl
cotyledons

roots

seedling

Eudicot
(common bean *Phaseolus vulgaris*)

coleoptile
epicotyl

mesocotyl
hypocotyl
radicle
coleorrhiza

embryo

Grass (wheat *Triticum*)

epidermis   The outermost protective layer of cells on the surface of young shoots, roots and other organs of a plant.
It is a primary tissue derived from the protoderm in the apical meristem and is replaced by periderm in woody plants.
see **cuticle**
see also **epiblem**

epigeal, epigeous   On or out of the ground.
see **hypogeal**

epigeal germination, epigeous germination
Of seed germination, the radicle elongates and penetrates the soil and the elongating hypocotyl pushes the cotyledons out of the ground.
Common in eudicots.
cf. **hypogeal germination, viviparous germination**

Epigeal germination

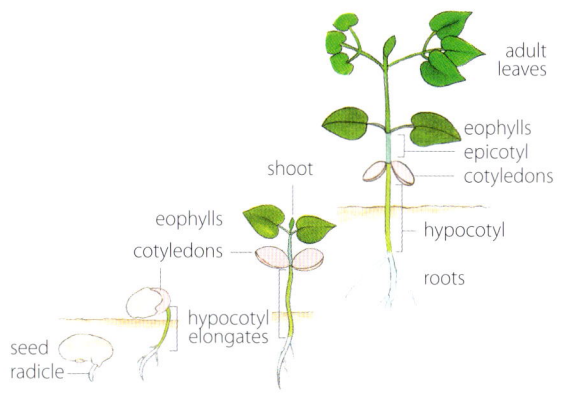

adult leaves

eophylls
epicotyl
cotyledons

shoot

hypocotyl

eophylls
cotyledons

roots

hypocotyl
elongates

seed
radicle

bean seed (*Phaseolus*)

epigenous   Growing on the surface of an organ, as fungus on a leaf.

epigynous berry   Sometimes used to distinguish a berry-like fruit derived from an inferior ovary and the hypanthium, as gooseberry (*Ribes uva-crispa*).

Epigynous berry

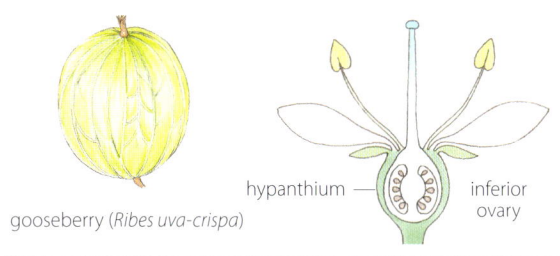

gooseberry (*Ribes uva-crispa*)

hypanthium

inferior ovary

epigyny Having the ovary embedded in a hypanthium.
The whorls of stamens, petals and sepals are borne above the ovary on the rim of the hypanthium.
epigynous Above the ovary.
cf. **hypogynous, perigynous**

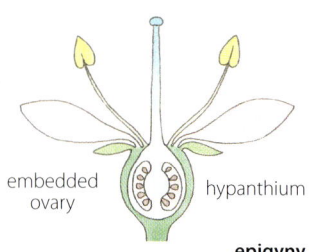

embedded ovary

hypanthium

epigyny

epilith Growing on the surface of rocks
epilithic Of or relating to an epilith.

epinasty Increased growth along the upper surface of a plant part causing it to bend downward.
cf. **hyponasty**
epinastic Of or relating to epinasty.

epipetalous Inserted on the corolla, as the stamens of many species.
cf. **episepalous, epitepalous**

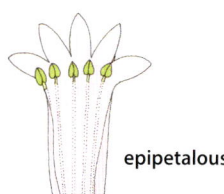

epipetalous

epiphloedal Growing on the bark of trees, as a lichen.

epiphylly Having epiphytes on leaves.
epiphyllous, epiphyllic Growing on a leaf, as the flowers of twisted stalk helwingia (*Helwingia*).

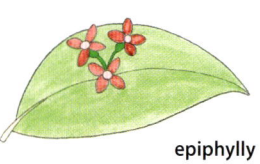

epiphylly

epiphyte A plant that grows on another plant for support but not for nutrients, as a staghorn or an epiphytic orchid on a tree.
= **aerophyte, air plant**
cf. **hemiepiphyte, parasite, saprophyte**
epiphytic Of or relating to an epiphyte.

epiphytic

orchid

episepalous Inserted on the sepals, as the nectaries of Burma lancewood (*Blackwellia tomentosa*).
cf. **epipetalous, epitepalous**

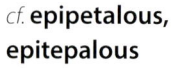

episepalous

nectary

sepal

epistase In some plants, cells modified from the nucellar epidermis.
They have thick cutinised walls and are situated near the micropylar region of the ovule, as agave (*Agave*).
cf. **hypostase**

epistasis Having an allele of one gene masking or concealing the output of an allele of another gene.
cf. **dominant**

epistomatous Of leaves with stomata only on the upper surface.
= **hyperstomatous**
cf. **amphistomatous, hyperstomatous**

epitepalous Inserted on the tepals, as the anthers of banksia (*Banksia*).
cf. **epipetalous, episepalous**

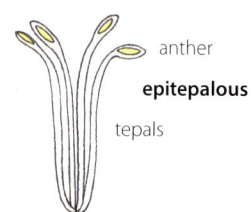

anther

epitepalous

tepals

epithet The word following the name of the genus and denoting a species, variety or other division, as *Geranium robertianum* and *Geranium sanguineum* var. *striata*.

epitony Development of lateral growth on the upper side of the main shoot.
cf. **amphitony, hypotony**

epitony

epitropous Of ovule orientation, with the micropyle proximal with reference to the funicle, as anatropous, campylotropous and amphitropous.
cf. **hypotropous, pleurotropous**

epizoochory Dispersal of seeds and fruit on the bodies of animals.
epizoochorous Of or relating to epizoochory.

equational division

The second division in meiosis.

The two haploid non-identical daughter cells divide again to produce four non-identical haploid cells.

The chromosome number after division is equal to that in the two daughter cells.

see **meiosis**

cf. **reduction division**

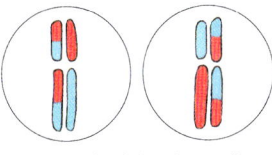

two haploid daughter cells

equational division

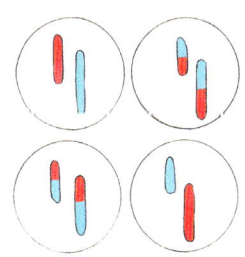

four haploid cells

equator
A real or imaginary circle dividing a sphere or other surface into two usually equal parts.

cf. **axis**

equatorial On or relating to the equator.

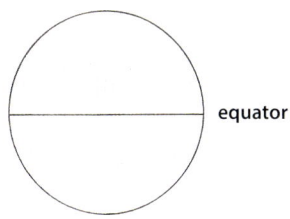

equator

equatorial plane
An imaginary flat surface dividing an object at the equator.

cf. **polar axis**

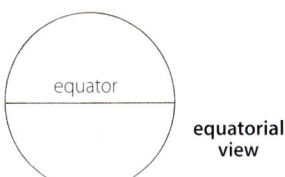

equatorial plane

equatorial view
An object as it appears when the equator is in the line of sight.

cf. **polar view**

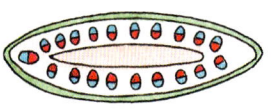

equator

equatorial view

equifacial
With two similar sides having no evident distinction internally or externally between the upper and lower surfaces, as the leaves of an iris (*Iris*).

Typical of leaves that orient themselves parallel to the main axis, as most monocotyledons.

= **isobilateral**

cf. **bifacial, unifacial**

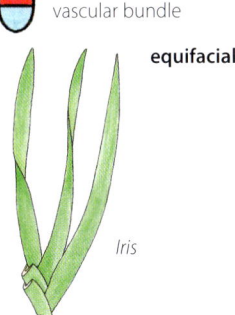

leaf cross-section

vascular bundle

equifacial

Iris

equilateral
Having all sides equal, as an equilateral triangle.

equilateral triangle

equinoctial
Of flowers that open and close regularly at specific times during the day, as some species of bindweed (*Convolvulus*).

equisetoid
Drooping in habit like the horsetail genus (*Equisetum*), as members of the sheoak family (Casuarinaceae).

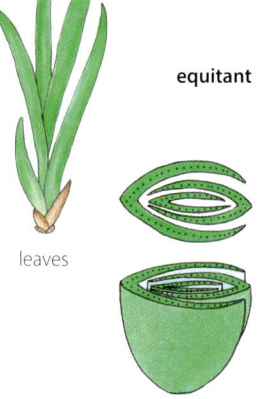

equisetoid

sheoak

equitant
Of leaves, folded lengthwise and cohering except at the base. The outermost leaf encloses the next to form two ranks, as the iris genus (*Iris*).

equitant vernation
Of young leaves in the unopened leaf bud with the margins folded to embrace both margins of another leaf.

equitant

leaves

vernation

erect
Of a plant habit or plant part, upright, as the vertical trunk of a tree.

erect

ergastic substances
Non-living substances in the protoplasm of a cell that are the product of various metabolic reactions.

Reserve materials of food synthesised by the cell, as carbohydrates stored in amyloplasts, proteins stored in proteinoplasts and fats stored in elaioplasts.

Secretory materials, as enzymes, nectar and chlorophyll, that are stored in cell protoplasm and are useful to the plant.

Excretory materials (waste products) stored in some cells of leaves fruit and bark, as latex and alkaloids.

cf. **protoplasm**

ericaceous   Belonging to or related to the heath family (Ericaceae).
Also used to describe plants that like acid soil.
cf. **ericoid**

ericoid
Of or resembling the genus *Erica*.
Of leaves, small and tough like those of heather.
cf. **ericaceous**

ericoid

ericoid mycorrhiza
A mutually beneficial symbiosis formed between Ascomycota fungi and the roots of plants mainly from the heath family (Ericaceae).
One of the endomycorrhizas with hyphae that penetrate the cells of the root cortex and form pelotons.
see **mycorrhiza**

Ericoid mycorrhiza

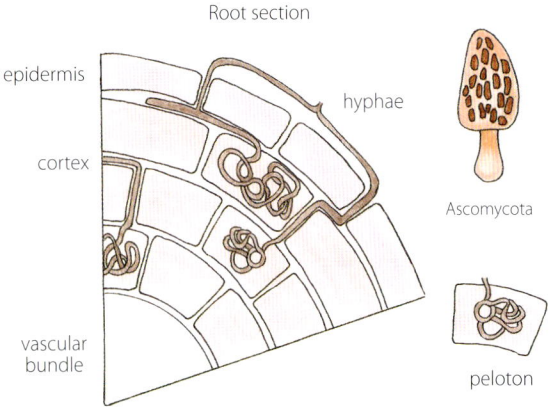

Root section

epidermis

hyphae

cortex

Ascomycota

vascular bundle

peloton

erose   Irregularly toothed. Of a margin that appears jagged or gnawed, as the margins of some leaves.
= **irregularly dentate**

erose

erostrate   Lacking a beak.
cf. **rostrate**

escape, escapee   A garden or crop plant that has become naturalised and is often weedy and invasive.

essential organs
The pistil and stamens of a flower that are necessary for the production of seeds that reproduce the plant.
cf. **accessory organs**

essential organs

flower

stamens

pistil

estipellate, exstipellate   Of leaflets that lack stipels at the base of the petiolules.
cf. **exstipulate**

estipulate, exstipulate   Of a leaf that lacks stipules at the base of the petiole.
cf. **exstipellate**

etaerio   Fruit formed from the unfused carpels of a single flower, with the carpels becoming fruitlets.
A cluster of of berries is a baccacetum, of follicles a follicetum, of drupes a drupecetum, of achenes an achenecetum and of samaras a samaracetum.
= **aggregate fruit**
see **compound fruit**
see also **apocarp**

Etaerio

Raspberry (*Rubus*)

single flower

carpels

fruitlets

baccacetum
(*Phytolacca*)

samaracetum
(*Liriodendron*)

follicetum
(*Illicium*)

drupecetum
(*Rubus*)

achenecetum
(*Ranunculus*)

ethylene   A plant hormone that influences other hormones in the transition from vegetative to reproductive stages and senescence.
It plays a key role in fruit ripening.
see **phytohormone**

etiolated   Of a plant that is pale, weak and elongated due to a lack of light.
see also **chlorosis**

eu-   A prefix meaning true.

eucamptodromous
Of leaves with secondary veins upturned and becoming indistinct before reaching the margin.

eucamptodromous

euhydrophile   A plant requiring submergence in fresh water.
euhydrophilous   Thriving submerged in fresh water.

euhydrophyte   A plant that is completely submerged (except for the infloresence), or a plant that is anchored to the substratum with floating leaves or floating and submerged leaves, or is submerged and free-floating.

eukaryote   An organism, other than cyanobacteria and bacteria, that has its DNA within a well-defined nucleus surrounded by a nuclear membrane as well as other organelles including mitochondria, Golgi apparatus and lysosomes.
cf. **prokaryote**
eukaryotic   Of a cell that has a clearly defined nucleus and organelles.

eudicot
The largest divisions of flowering plants (75% of all angiosperms).
Characterised by a seed with two cotyledons (seed leaves) in the embryo. Flower parts are often in multiples of four or five. Pollen has three furrows or pores. Leaves have reticulated veins. A stem in cross-section has vascular bundles in a ring. Typically the root is a taproot. There can be secondary, often woody, growth.
cf. **dicotyledon, monocotyledon**

Eudicots

seed coat — cotyledons
embryo axis — endosperm

rock cress

rapeseed

geranium

pea

Seed with two seed leaves (cotyledons). Endosperm usually absorbed by developing seed.

Flower parts usually in multiples of four or five. Sepals and petals usually distinct.

Pollen typically with three pores or furrows.

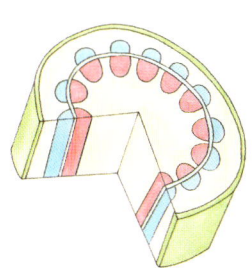

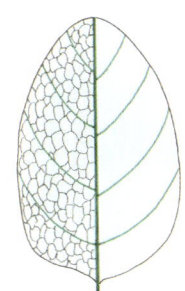

Herbaceous stem with vascular bundles in a ring.

Leaves with veins reticulated.

Herbaceous or woody plants.

Root system often a taproot.

---

euphyllophytes   A group of plants within the vascular plants (tracheophytes) that includes seed-bearing plants (spermatophytes) and spore-bearing ferns (monilophytes). They possess true leaves (megaphylls), usually with many veins.
cf. **lycophytes**

euryhaline   Of plants and animals able to tolerate a wide range of salinity.

eurythermal   Tolerant of a wide range of temperatures.

eusporangiate   Of spore-bearing vascular plants, that include some ferns, horsetails (*Equisetum*), whisk ferns (*Psilotum*), clubmosses (*Lycopodium*), spike mosses (*Selaginella*) and quillworts (*Isoetes*), that have sporangia arising from a group of epidermal cells.
True ferns (leptosporangiate ferns) have sporangia that arise from a single epidermal cell.
see **fern**

eustele   Stele with separated strands of vascular tissue. In cross-section they appear as a ring of discrete bundles with phloem to the outside and xylem to the inside.
Characteristic of eudicot stems.
cf. **atactostele**

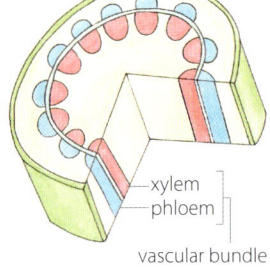

xylem
phloem] vascular bundle
eustele

eutrophic   Of a body of water rich in nutrients, like nitrogen and phosphate, that support abundant plant life. As the plant life decays, it can deplete the oxygen supply that sustains other organisms.
see **algal bloom, trophic**
cf. **dystrophic, mesotrophic, oligotrophic**

evanescent   Fleeting, remaining only a very short time.
cf. **fugaceous**

evaporation   The transition from a liquid to a vapour (gas) at temperatures below boiling point. It occurs on the surface of the liquid, compared with boiling, in which vapour forms below the surface as bubbles.
see **vaporisation**

even-pinnate
Of a pinnate leaf with leaflets arranged in pairs and terminating with a pair of leaflets.
= **abruptly-pinnate, paripinnate**

even-pinnate

evergreen   A plant that retains green leaves throughout the year.
cf. **deciduous**

evolution   Changes in the genetic makeup of a population, species or lineage over time by which they are thought to have developed and diversified from earlier forms.
Natural selection, mutation, genetic drift and gene flow (gene migration) are the basic mechanisms of evolution.
see also **macroevolution, microevolution**

ex-   A prefix meaning from, according to, without or former.

exalbuminous seed
One having endosperm absorbed by the growing embryo, with the embryo itself having cotyledons that store food for germination.
Found in most eudicots, as the bean family (Fabaceae).
= **non-endospermic seed**
cf. **albuminous seed**

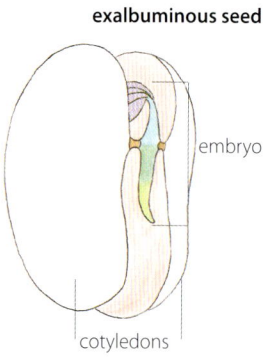

exalbuminous seed

embryo

cotyledons

broad bean (*Vicia faba*)

exarch   Describes radial differentiation of xylem in which protoxylem is positioned towards the outside of the stem and metaxylem towards the inside of the stem.
Found in some lower vascular land plants, as clubmosses (*Lycopodium*).
cf. **endarch, mesarch**

exarillate   Of a seed that has no aril.

excrescence   An outgrowth or protuberance that may be normal, as a nodule on the roots of legumes, or abnormal, as a gall.

excretion

The discharge of waste matter, such as carbon dioxide or salt, as some halophytic plants that excrete salt in droplets from their leaves.

cf. **secretion**

excretory Relating to excretion.

excretion

salt droplets

excurrent

Having the axis prolonged to form an undivided main stem or trunk, as firs and spruces. Running through to the apex and beyond, as the midrib of certain leaves.

cf. **deliquescent**

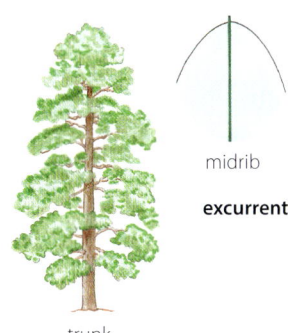

midrib

excurrent

trunk

exfoliating

Peeling off in layers or flakes, as the bark of paper birch (*Betula papyrifera*).

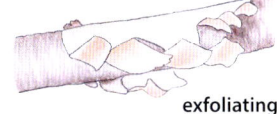

exfoliating

exindusiate

Without an indusium, as the sorus of the polypody fern (*Polypodium*).

cf. **indusiate**

exindusiate

sorus

exine

The outer layer of the pollen wall that is composed mainly of sporopollenin.

see **pollen wall**

exmedial, exmedian

Of venation, running away from the midline of the leaf lamina.

cf. **admedial**

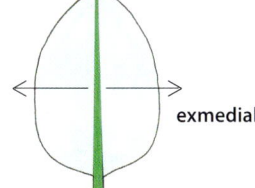

exmedial

exo-

A prefix meaning on the outside or external.

exocarp

The outer layer of the fruit wall, as the soft skin of a stone fruit, or the leathery rind of an orange.

= **ectocarp, epicarp**

see **pericarp**

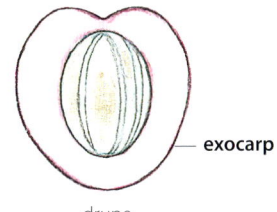

exocarp

drupe

exodermis

The outer layer of one or more cells with Casparian strips, that surrounds the cortex in some roots. It can supplement the function of the endodermis when plants are suffering a water deficit.

exogenous growth

Growing from or originating from without.

Developing from outer tissue, as root hairs that are outgrowths of epidermal cells.

cf. **endogenous growth**

exospore

The layer between the outer perispore and the inner endospore of a spore wall.

see **sporoderm**

cf. **exine**

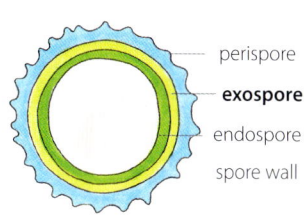

perispore

exospore

endospore

spore wall

exotegmen

Of the bitegmic seed coat of angiosperms, the outer epidermis of the inner integument (tegmen).

cf. **endotegmen, mesotegmen**

see also **endotesta, exotesta, mesotesta**

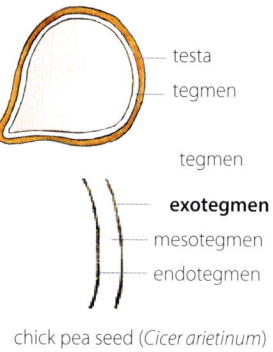

testa

tegmen

tegmen

exotegmen

mesotegmen

endotegmen

chick pea seed (*Cicer arietinum*)

exotegmic seed

Of angiosperms, having the mechanical layer in the exotegmen.

see **tegmic seed**

exotesta, *pl.* exotestae

Of the bitegmic seed coat of angiosperms, the outer epidermis of the outer integument (testa).

cf. **endotesta, mesotesta**

see also **endotegmen, exotegmen, mesotegmen**

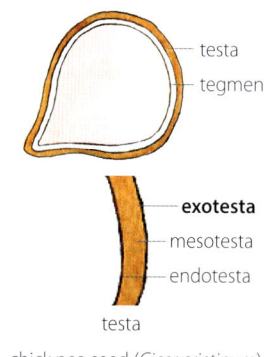

testa

tegmen

exotesta

mesotesta

endotesta

testa

chick pea seed (*Cicer arietinum*)

exotestal seed

Of angiosperms, having the mechanical layer in the exotesta.

see **testal seed**

exotic   Of plants in a region where they are do not occur naturally.
Non-native species introduced from another place, often another country.
= **alien**

explanate   Spread out and flattened. Of anthers spread at right angles to the filament, as some species of beardtongue (*Penstemon*).
cf. **transverse**
see **anther attachment**

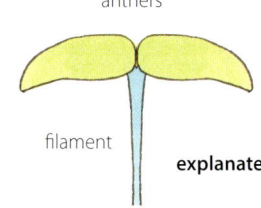

anthers

filament

explanate

explicative vernation
Of pairs of leaves in bud that are flattened opposite each other but with the edges folded back or sometimes rolled, as some snowdrops (*Galanthus*).

explicative vernation

exserted   Projected beyond.
Protruding, as stamens extending beyond the corolla tube.
cf. **included**

exserted stamens

extant   Still in existence.

extinct   Of conservation status, said of a species that is extinct in the wild or no longer existing at all.

extra-   A prefix meaning outside or beyond.

extrafloral   Located other than in the flower, as the nectary of the cherry (*Prunus avium*) that is on the leaf petiole rather than in the flower.

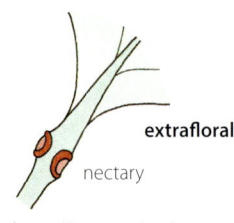

extrafloral

nectary

cherry (*Prunus avium*)

extrastaminal
Situated outside the whorl of stamens, as the glands on the male flower of spurge (*Phyllantus*).

extrastaminal glands

gland

extravaginal
Of branching, with the young shoot breaking through the base of the leaf sheath, as in some grasses (Poaceae).
cf. **infravaginal, intravaginal**

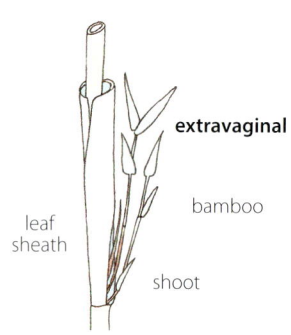

extravaginal

bamboo

leaf sheath

shoot

extrorse   Facing away from the axis.
= **posticous**
cf. **introrse, latrorse**

extrorse dehiscence
Of anthers, facing outwards and opening longitudinally to release pollen away from the centre of the flower, as the dayflower family (Commelinaceae).
= **posticous dehiscence**
see also **anther dehiscence**

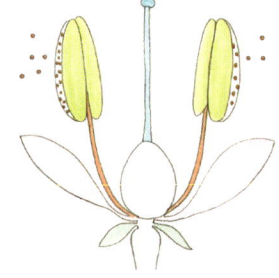

extrorse dehiscence

exudate   A substance produced and discharged by a plant.
It may be through pores or through a wound, as resins, mucilage, gums and latex.
Roots exude chemicals into the rhizosphere.
Nectar is sometimes considered to be an exudate.

eye   A small depression on a stem tuber.
Each represents a node bearing one or more buds, subtended by a scale leaf, from which a new plant can grow, as potatoes (*Solanum tuberosum*).
see **perennating bud, tuber**
cf. **slip**

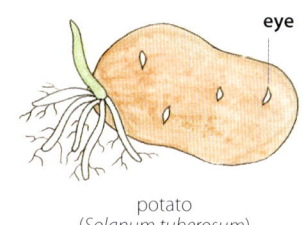

eye

potato (*Solanum tuberosum*)

f.   An abbreviation for form or forma.

F1 generation   The first filial generation.
the hybrid offspring of a cross between two true breeding individuals.
see **F2 generation, P generation**

F2 generation   The second filial generation.
the result of a cross, either by self-pollination or
cross-pollination, between two F1 hybrids.
see **F1 generation, P generation**

facultative   Optional, as an orchid flower that
can cross-pollinate or self-pollinate according to
environmental conditions.
cf. **obligate**

facultative apomict   A plant that can reproduce
either asexually (apomixis) or sexually.

falcate, falciform
Curved like the blade of
a sickle.

falcate

fall   The spreading or
drooping sepals of the
flower of an iris (*Iris*) as
distinct from the three
more or less upright
petals or standards.

standards

fall

false aril   An arillode.

false berry   A loosely
applied term referring
to a pulpy indehiscent
accessory fruit with one
to many seeds.
Derived from an
inferior ovary and
the hypanthium, as
gooseberry (*Ribes uva-
crispa*).
see **accessory fruit**

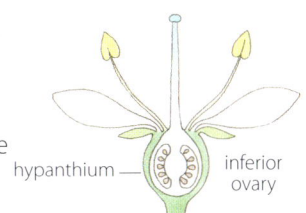

hypanthium

inferior
ovary

false
berry

gooseberry

false fruit
A fruit derived from a simple ovary or compound
ovary and some additional non-ovarian tissue like
the receptacle.
A strawberry has the true fruits (achenes derived
from the ovaries) embedded in the fleshy
receptacle.
Other accessory fruits include hips, pomes and
pineapples.
= **accessory fruit, pseudocarp**
cf. **true fruit**

False fruit

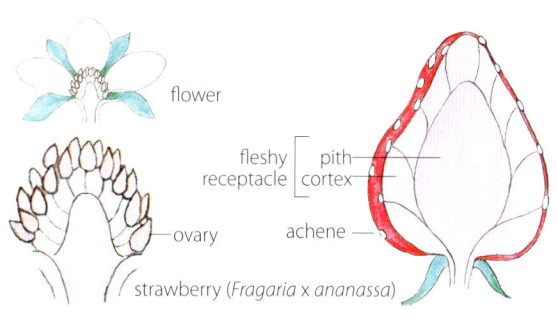

flower

fleshy
receptacle

pith
cortex

ovary

achene

strawberry (*Fragaria* x *ananassa*)

false indusium
A protective covering
over the sporangia
formed by the reflexed
margin of a fern frond, as
brake ferns (*Pteris*).

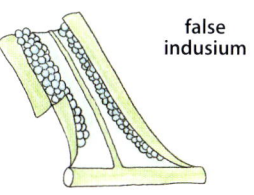

false
indusium

false septum
A projection from the
ovary wall into the
locules forming
a partition, as flax
(*Linum*) that has five
locules that are futher
divided by five false
septa.
see also **replum**

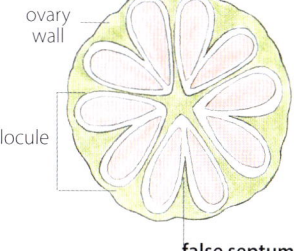

ovary
wall

locule

false septum

false vein   Of fern
fronds, a vein-like strand
with no vascular bundle.
= **pseudovein**

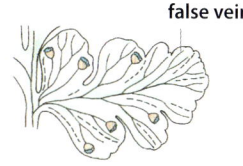

false vein

family   In taxonomic classification, a rank below
order and above genus.
Names of orders end in *-acae*, as Ochidaceae
(orchids).
see **taxonomic hierarchy**

fan palms   Palms with
palmate or costapalmate
leaves divided shallowly
or deeply into a variable
number of segments that
often split at the tips.
cf. **feather palms**

fan palms

farina   A powdery flour-like covering, as found on the stems and leaves of some plants.
farinaceous   Resembling flour.
Having a mealy texture or appearance.
Containing starch, as some seeds.
farinose   Covered with a mealy powder-like flour, as the lower leaf surface of some goosefoots (*Chenopodium*).

fasciation   The abnormally broad and flattened growth of a stem or stems so that it resembles several stems fused together side by side, as occurs in dandelion (*Taraxacum*).
fasciated   Showing abnormally flattened fusion of parts or organs.

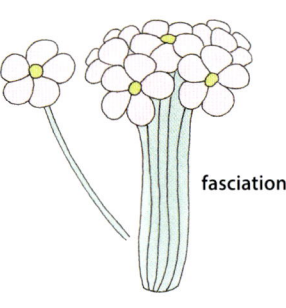

fasciation

fascicle   A tight bundle. The elements are almost always independent, but appear to arise from the same point.
fascicled, fasciculate
Arranged in bundles, as needles of the pine genus (*Pinus*).

fascicle

Pinus

fascicled cyme
A cymose inflorescence composed of a small bundle of pedicellate flowers arising from more or less the same point, as toro (*Myrsine salicina*).
cf. **glomerule, verticillaster**

fascicled cyme

fastigiate   Of a tree or shrub having a very narrow canopy, with the branches about the same length and more or less parallel to the main stem, as Lombardy poplar (*Populus nigra* var. *italica*).

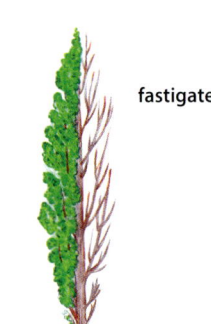

fastigate

faucal   Relating to the throat.

fauna   All of the animal life occurring in a particular area or geological period.
cf. **flora**

faveolate, favose   With cavities like honeycomb, honeycombed.
Pitted, alveolate.

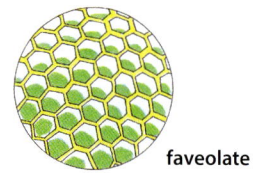

faveolate

feather palms
Palms with pinnate or, rarely, bipinnate leaves, consisting of a series of individual leaflets along an extension of the petiole called the rachis.
cf. **fan palms**

feather palms

felted   Intertwined, matted and compressed together, as some hairs.

female flower   A flower with functional carpels but no functional stamens.

fenestrate   Having openings or translucent areas like windows, as a fenestrate leaf.

fenestrate

fern   *see* page 107

fern allies   *see* page 108

ferrugineous, ferruginous   Rust-coloured, reddish-brown.
Of, relating to or containing iron, as some soils.

fertile   Able to reproduce sexually.
Of flowers with viable pollen and ovules.
Producing viable seeds and fruit.
Of shoots, branches, bracts etc. that bear flowers.
cf. **infertile, sterile**

fertile bract   A bract that bears a flower.
cf. **sterile bract**

fertile bracts

sterile bract

fertilisation   The union of gametes, a male sperm cell and a female egg cell, to form a zygote.
see **angiosperms, ferns, gymnosperms, hornworts, liverworts, mosses**
see also **double ~, simple ~**

fern   A vascular plant that reproduces by spores and has true leaves (fronds) usually arising from an underground rhizome, sometimes a stolon, or borne on an erect trunk (tree ferns).

Examples of true ferns from some of the main groups include: bracken (*Pteridium*), sword fern (*Nephrolepis*), maidenhair fern (*Adiantum*), tree fern (*Cyathea*), flowering fern (*Osmundia*), forked fern (*Gleichenia*), filmy fern (*Hymenophyllum*) and climbing fern (*Lygodium*).

see also **leptosporangiate**

cf. **fern allies**

Fern

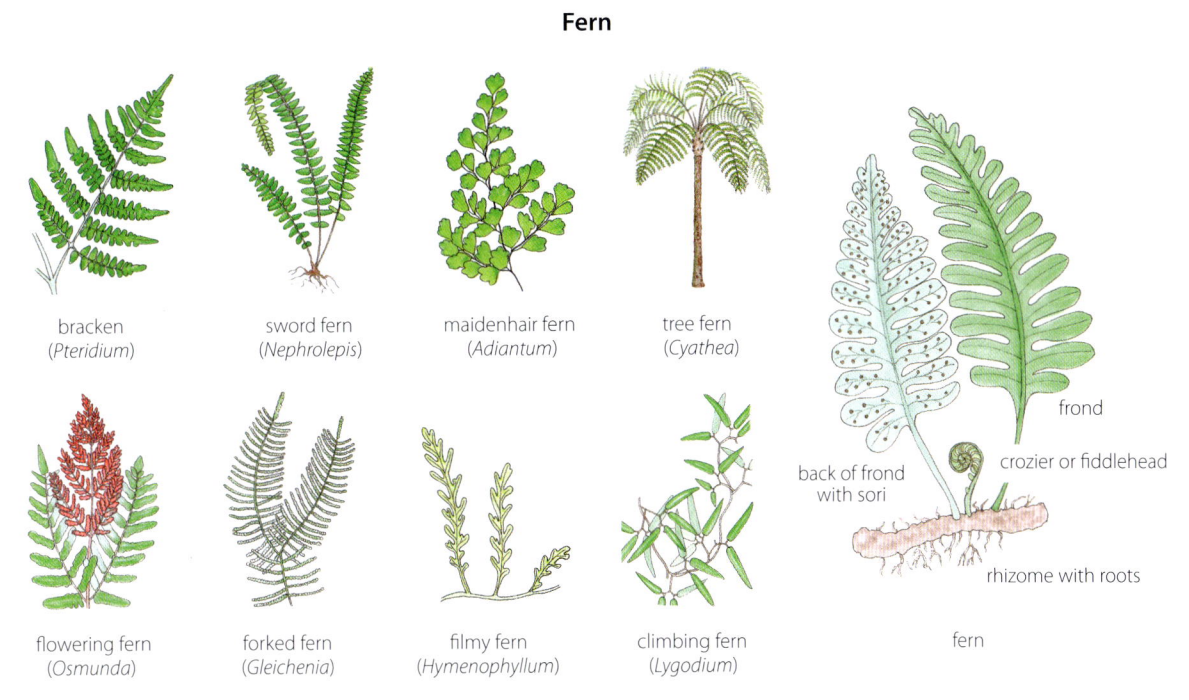

bracken
(*Pteridium*)

sword fern
(*Nephrolepis*)

maidenhair fern
(*Adiantum*)

tree fern
(*Cyathea*)

back of frond
with sori

frond

crozier or fiddlehead

rhizome with roots

flowering fern
(*Osmunda*)

forked fern
(*Gleichenia*)

filmy fern
(*Hymenophyllum*)

climbing fern
(*Lygodium*)

fern

ALTERNATION OF GENERATIONS

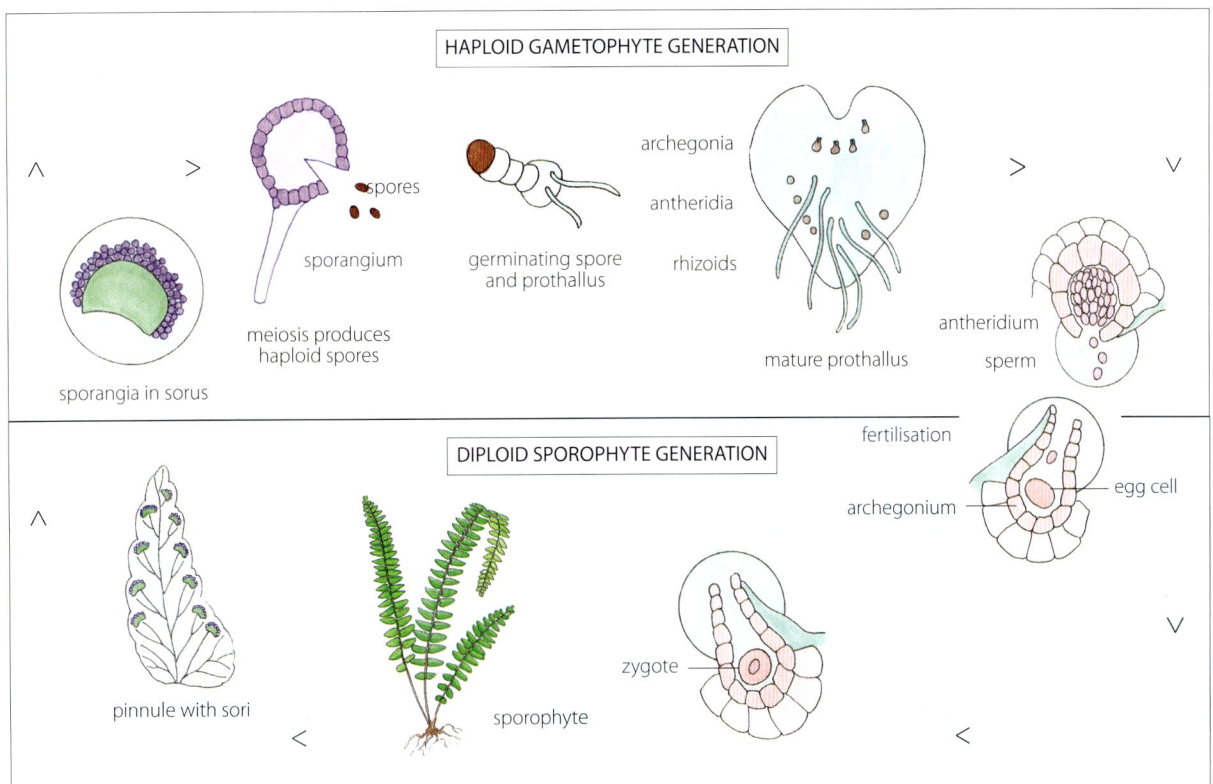

HAPLOID GAMETOPHYTE GENERATION

spores

sporangium

germinating spore
and prothallus

archegonia

antheridia

rhizoids

meiosis produces
haploid spores

mature prothallus

antheridium

sperm

sporangia in sorus

DIPLOID SPOROPHYTE GENERATION

fertilisation

archegonium

egg cell

pinnule with sori

sporophyte

zygote

fern allies Not a natural grouping but generally refers to vascular spore-bearing plants thought to be closely related to ferns. Includes horsetails (*Equisetum*), whisk ferns (*Psilotum*), quillworts (*Isoetes*), spike mosses (*Selaginella*) and clubmosses (*Lycopodium*).

Fern allies

horsetail (*Equisetum*)   whisk fern (*Psilotum*)   quillwort (*Isoetes*)   spike moss (*Selaginella*)   clubmoss (*Lycopodium*)

festucoid Resembling the perennial tufted grass fescu (*Festuca*).

fetid, foetid With an offensive stinking odour, as of decay.

fibre A long strand of thread-like material, as one of the fibres on a cotton seed. It may be hard or soft.
fibrose, fibrous Thread-like. Having, consisting of or resembling fibres.

fibre cell An elongated cell with tapered ends and thick lignified cell walls enclosing a lumen. Fibre cells are dead at maturity and function as support tissue. Together with sclereids, they form sclerenchyma that is associated with phloem and more commonly with xylem.

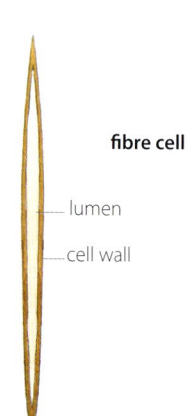

fibre cell

lumen

cell wall

fibril A small fibre, a delicate thread-like filament or hair, as the hairs on the rootlets of some plants.
fibrillar, fibrillate, fibrillose Bearing fine fibres or slender strands that are parallel to each other and not matted. Finely striated.

fibrillous Composed of small fibres.

fibrous roots
Of monocotyledons and some eudicots, a mass of adventitious roots of similar size that sprout from nodes on an underground stem, as at the base of some bulbs. Typical of most grasses.
cf. **taproot**

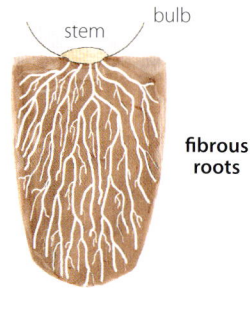

stem   bulb

fibrous roots

-fid A suffix meaning split or divided into parts.

fiddlehead The coiled tip of a young fern frond.
= **crosier**

fiddlehead

filament A slender, thread-like structure. Of a stamen, the stalk, not necessarily slender or thread-like, bearing the anther.
filamentous Bearing or resembling a filament.

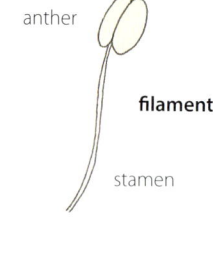

anther

filament

stamen

filantherous Of a typical stamen with a distinct anther and a filament.
cf. **laminar**

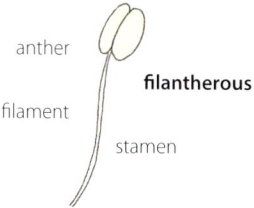

anther

filantherous

filament

stamen

filiferous   Bearing thread-like attachments, as the margins of some leaves.
cf. **filiform**

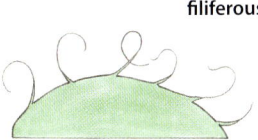
filiferous

filiform   Slender and thread-like.
cf. **capillary**

filiform

filiform apparatus   Numerous finger-like projections in the cytoplasm of a synergid cell.

Filiform apparatus

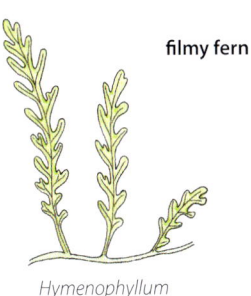

chalaza

embryo sac

mycropyle

ovule

filiform apparatus

synergid cell

embryo sac

filmy fern   Delicate small commonly ephytic ferns in the family Hymenophyllaceae, with the fronds typically one cell thick (*Hymenophyllum*).
see **fern**

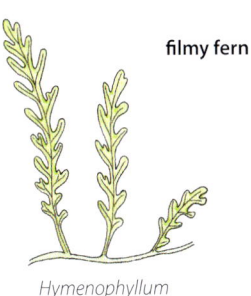
filmy fern

Hymenophyllum

fimbria, *pl.* **fimbriae**
A fine outgrowth, usually derived from the same material as the organ itself, as the hair-like fringe on the margins of some leaves.
fimbriate   Having fimbria, fringed.
cf. **ciliate, fimbrillate, lacerate, laciniate**

fimbria

fimbriate leaf margin

fimbrilla, *pl.* **fimbrillae**   A single division or tooth of a minute, fine fringe.
fimbrillate   Minutely fringed.
cf. **fimbriate**

fishtail   Of palms, leaflets shaped like a fish tail, with the tip appearing chewed, as the leaflet of the toddy palm (*Caryota urens*).
see also **praemorse**

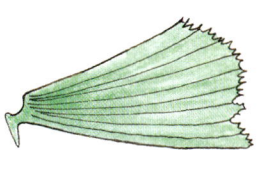
fishtail

fissum   Cleft or split, as the leaves of a stoneflower (*Argyroderma fissum*).

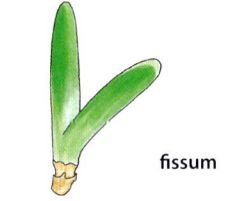
fissum

fistular, fistulose, fistulous
Hollow and cylindrical, as the leaf of an onion.

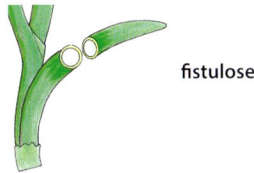
fistulose

fitness   Of natural selection, the degree of increased or decreased survival or reproduction of individuals in a population.

five-ranked   Of leaves arranged in five vertical rows, with any sixth leaf above the one below it.
= **pentastichous**
see also **orthostichy**

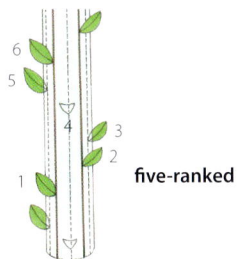
five-ranked

flabellate, flabelliform
Fan-shaped.

flabellate

flaccid   Drooping or lacking stiffness through lack of water; caused by water loss in cells and the shrinking of the cell contents away from the cell wall.
cf. **tumid, turgid**
flaccidity   The state of being flaccid, loss of turgour.

Flaccid

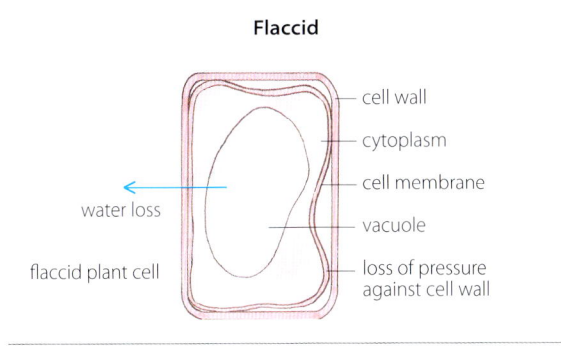

cell wall

cytoplasm

cell membrane

water loss

vacuole

flaccid plant cell

loss of pressure against cell wall

flag leaf   Of grasses (Poaceae), uppermost leaf of the culm.
see **boot stage**

flag leaf

flagelliflorous   Having an inflorescence that projects down beyond the crown of the tree on a long hanging pedicel, as the sausage tree (*Kigelia africana*) and some some species of fig (*Ficus*).

flagelliflorous

flagellum, *pl.* **flagellae**
A lash-like appendage, as a slender, flexible shoot or runner.
Of palms (Arecaceae), a barbed whip-like extension of the inflorescence found in climbing palms (rattans).
cf. **cirrus**
flagellate   With a whip-like extension, bearing flagella.
flagelliform   Elongate and slender like a whip.

flagellum

rattan

flavescent   Yellowish, turning yellow.

flavins   A group of orange-yellow pigments in plants that can only be seen when they are purified. Functions include control of phototropism.

flavonoids   A group of mostly yellow pigments found in plants that are usually only visible under ultraviolet light.
Functions include ultraviolet nectar guides on flowers that are visible to pollinators and ultraviolet-absorbing flavones in some leaves that protect against harmful radiation.
Includes anthocyanins that are brightly coloured red, purple and blue pigments.

fleshy   Composed of a firm juicy pulp, as the mesocarp of a drupe, such as a prune. Succulent.

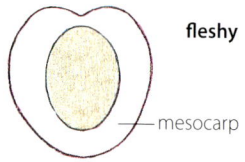

fleshy

mesocarp

flexible   Pliable, easily bent without breaking.

flexuose, flexuous
Bending or curving alternately in different directions, as the axis of a grass spike.

flexuose

floating   Borne on the surface of water, as the leaves of an aquatic plant. The plant may be rooted in the mud below, as yellow floating heart (*Nymphoides peltata*) or free-floating, as frogbit (*Hydrocharis morsus-ranae*).
see also **natant**

Floating

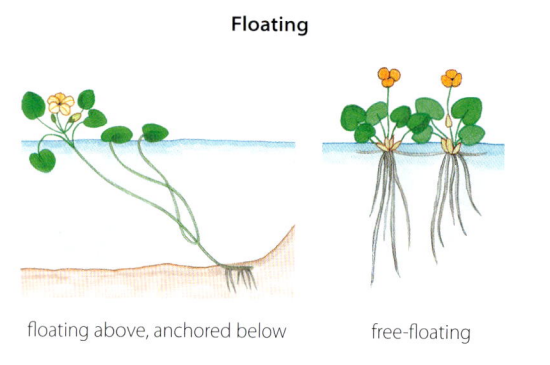

floating above, anchored below   free-floating

floccule   A small clump of material that resembles a tuft of wool.
flocculent   Having or resembling tufts of wool.

floccus, *pl.* **flocci**   A woolly tuft.
floccose   With tufts of woolly soft hairs.

flora   All of the plant species occurring in a particular area or geological period.
Classifications include indigineous flora, agricultural flora, horticultural flora and weed flora.
A taxonomic botanical publication documenting the flora of a particular area, usually with keys for plant identification.
cf. **fauna, vegetation**

floral   Belonging to or associated with a flower.

floral diagram

A stylised drawing of the cross-section of a flower showing the number and relative position of the various parts.

see also **floral formula**

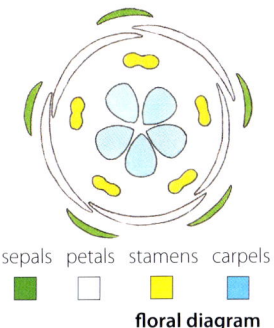

sepals petals stamens carpels

floral diagram

floral envelope

A collective term for the calyx and corolla of a flower, especially when both are similar, as the day lily (*Hemerocallis*). The sterile parts of a typical flower.

= **perianth, perigone**

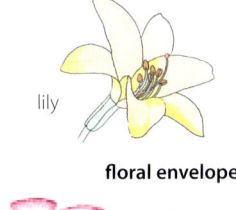

lily

floral envelope

corolla

calyx geranium

floral formula

A system for representing the structure of a flower using numbers, letters and symbols to convey information about a flower in a compact form.

Typically K = calyx, C = corolla, A = androecium, G = gynoecium.

Each letter is followed by a numeral to indicate the number of parts, as C5 indicating 5 separate petals and C(5) indicating 5 united petals.

see also **floral diagram**

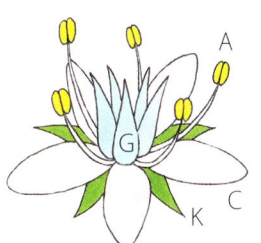

A5	androecium	5 stamens
G5	gynoecium	5 carpels
C5	corolla	5 petals
K5	calyx	5 sepals

floral formula

floral symmetry

The planes of symmetry that give a mirror image in a flower.

With only one plane of symmetry (mirror image) the flower is monosymmetric (zygomorphic, bilaterally symmetrical), as the flowers in the pea family (Faboideae).

With two planes of symmetry the flower is disymmetric, as the flowers of rapeseed (*Brassica napus*).

With two planes of symmetry at right angles to each other the flower is bisymmetric, as the flowers of bleeding heart (*Lamprocapnos*).

With three or more planes of symmetry the flower is polysymmetric (actinomorphic, radially symmetric,

regular), as the flowers of geraniums (*Geranium*) and some lilies.

When there are no planes of symmetry the flower is asymmetric (amorphic), as the flowers of the canna lily (*Canna*).

Floral symmetry

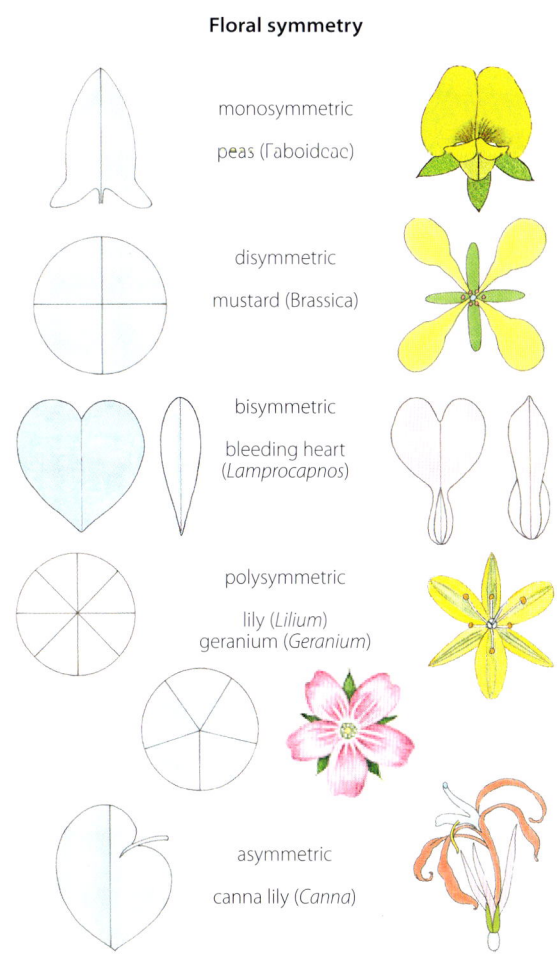

monosymmetric

peas (Faboideae)

disymmetric

mustard (Brassica)

bisymmetric

bleeding heart (*Lamprocapnos*)

polysymmetric

lily (*Lilium*) geranium (*Geranium*)

asymmetric

canna lily (*Canna*)

floral tube

A tubular or cup-shaped enlargement of the receptacle and/or the bases of the floral parts. It bears the petals, sepals and stamens.

It is either above the ovary, below the ovary and free or variously united with it.

= **hypanthium**

cf. **receptacle**

Floral tube

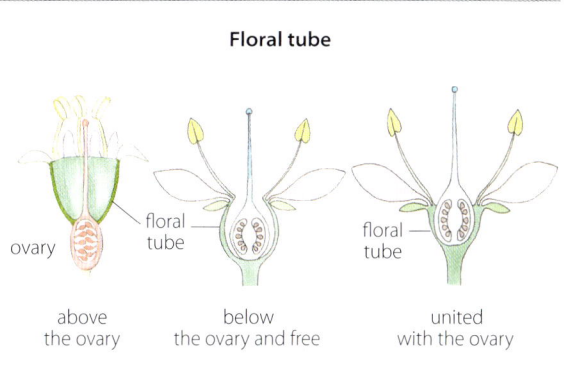

ovary

floral tube

floral tube

above the ovary

below the ovary and free

united with the ovary

floral whorl   Any one of the four flower parts, sepals, petals, stamens or carpels, borne at the same level on an axis.

stamens
carpels
petals
sepals

floral whorl

florescence   Flowering or blossoming. The flowering period. In grasses, a spikelet.

floret   A small flower, especially when part of a larger inflorescence, as the ray florets and tubular florets of daisies (Asteraceae).
Of grasses, each flower in a spikelet (locusta) together with the lemma and palea that enclose it.
see **floscule**

Floret

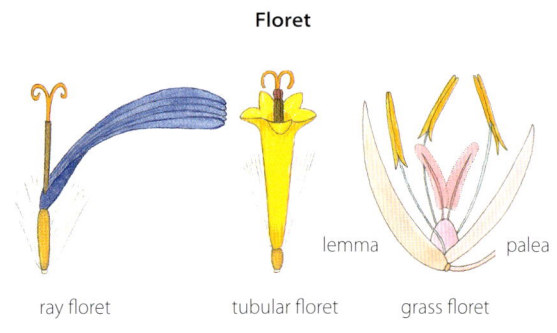

lemma   palea

ray floret   tubular floret   grass floret

floricane   A second-year cane of a bramble that bears flowers and fruit.
cf. **primocane**

floricane

floriferous   Bearing numerous flowers.

floristics   Study of the distribution of plant species and the relationships between them in a geographic area.

floscule   A little flower or floret.
  flosculose   Consisting of florets as the flower-like inflorescence of daisies (Asteraceae).

flower   The structure for sexual reproduction in angiosperms. Typically arranged in whorls, with a non-fertile calyx and corolla and fertile stamens and pistil(s) that are all inserted on a receptacle at the tip of the flower stem (pedicel).
see also **dioeceious, monoecious**

Flower

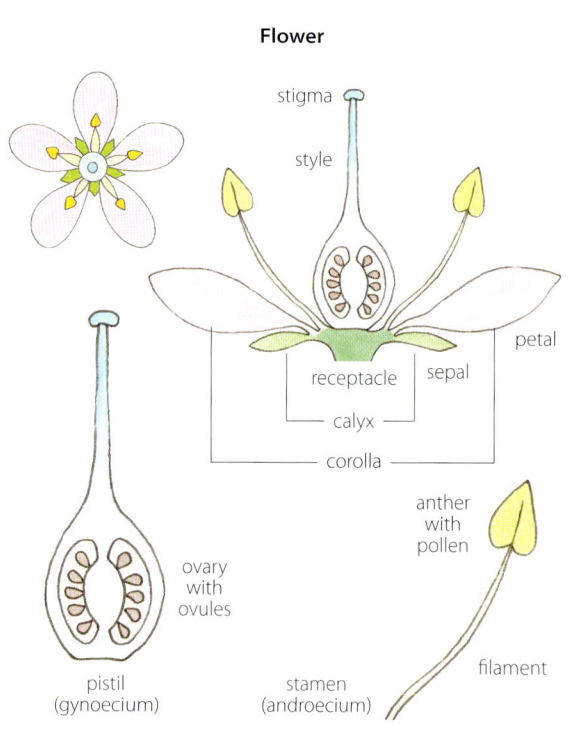

stigma
style
petal
receptacle   sepal
calyx
corolla
anther with pollen
ovary with ovules
pistil (gynoecium)
stamen (androecium)
filament

flowering fern
Common name of members of the genus *Osmunda*, so called because of the appearance of the fertile fronds that have spores located in rusty-coloured tassel-like clusters at the tips of the fertile fronds.
see **fern**

flowering fern

flowering fern (*Osmunda regalis*)

fluted   With longitudinal channels, grooves or furrows, as the stems of some cacti.

fluted

foliaceous   Of or resembling a leaf. Bearing numerous leaves.
= **foliose**

foliage   The photosynthetic leaves of a plant collectively.
see also **leaf**

foliage leaf   The main photosynthetic organ of a plant, usually consisting of a blade and a petiole.
see **leaf**

blade
petiole

Of bamboos, leaves on branches and at the tip of mature culms, with large blades attached to the sheath by a pseudopetiole.
cf. **culm leaf**

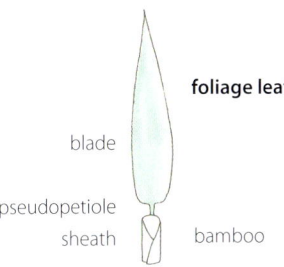

foliage leaf

blade

pseudopetiole

sheath

bamboo

foliar　Of or relating to leaves.
Leaf-like, as foliar stipules.

foliar stipules

foliate　Of or relating to leaves. Bearing leaves. Shaped like a leaf. Leaf-like.
see **bifoliate, trifoliate, unifoliate**
cf. **foliolate**

-foliate　A suffix denoting the number of leaves.

-foliolate　A suffix denoting the number of leaflets.

foliole　A leaflet in a compound leaf.
foliolate　Having leaflets.
see **bifoliolate, trifoliolate, unifoliolate**
cf. **foliate**

foliose　Resembling a leaf. Having many leaves.
=**foliaceous**
Of lichens, having a thallus with lobed, leaf-like extensions, as *Parmelia*.

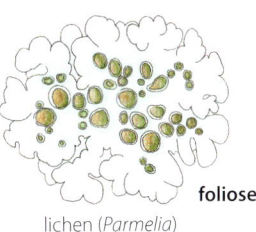

foliose

lichen (*Parmelia*)

follicetum
An aggregate fruit composed of a cluster of follicles, as star anise (*Illicium verum*).

follicetum

follicle　A dry dehiscent fruit with one or more seeds. It derives from a single carpel and splits open lengthwise along one suture only, as boronia (*Boronia*).
follicular　Of or pertaining to a follicle.

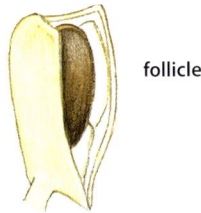

follicle

foot　Of an orchid, an extension of the base of the column.
see **column foot**
Of bryophytes, as mosses, the location at which the sporophyte is anchored to the gametophyte and through which nutrients are transferred from the gametophyte to the sporophyte.

Foot

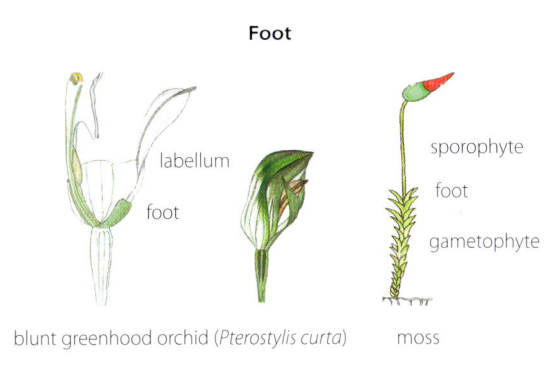

labellum

foot

sporophyte

foot

gametophyte

blunt greenhood orchid (*Pterostylis curta*)　　moss

foot layer　The inner layer of the ectexine in the wall of a pollen grain.
see **pollen wall**

foramen, *pl.* **foramina**　A minute opening, as the micropyle of the ovule through which the pollen tube usually enters; a very small hole.
foraminate　Having small holes or perforations.

forb　A broad-leaved herbaceous dicotyledonous plant, as distinct from grasses, sedges, shrubs and trees. Especially one growing in grasslands, prairies or meadows.

forest　A vegetation type with the tallest layer composed of trees.

fork　To separate from a common point into two branches or prongs. One of the branches or prongs into which anything is so divided. A Y-shaped or V-shaped branch.
forked　Branched to form a Y-shape or a V-shape, as the veins of some leaves and fronds.
=**furcate**
cf. **cleft**

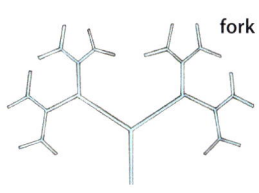

fork

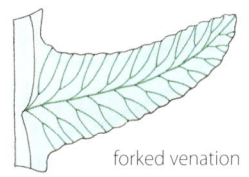

forked venation

forked fern

The forked fern family (Gleicheniaceae) has long-creeping branched rhizomes and most species have fronds that fork at the tips indefinitely. They typically form large masses that cover the ground and other vegetation.

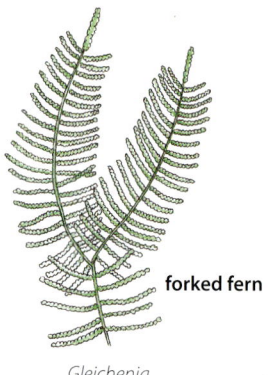

forked fern

Gleichenia

see **fern**

form, forma, *abbr.* **f.**   A subdivision of species below variety, usually differentiated by a minor characteristic like colour variation.
The lowest rank in taxonomic classification, as *Geranium maculatum* forma *albiflorum*.

see **taxonomic hierarchy**

fornix, *pl.* **fornices**   One of the little arched scales in the throat of a corolla, as hound's tongue (*Cynoglossum*) in the forget-me-not family (Boraginaceae).

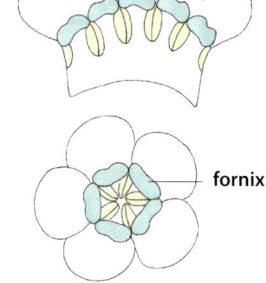

fornix

fornicate   Arched, bending over. Having fornices.

fossulate

With irregularly shaped narrow grooves.

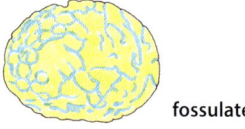

fossulate

fovea, *pl.* **foveae**   A small pit or depression.
foveate   Pitted.

foveola, *pl.* **foveolae**   A minute pit or depression.
foveolate   Minutely pitted.

fragmentation   Growth of a new plant from a piece of the parent plant, as occurs with some 'leafy' bryophytes like mosses.
A form of vegetative reproduction.

free   Separate, not joined to each other or to another organ. Usually referring to separateness of similar

free

petals

parts, as petals, or of dissimilar parts or organs, as stamens.

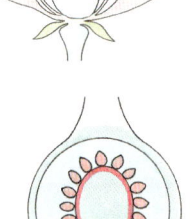

stamens

= **distinct**
cf. **fused**

free central placentation

With carpels fused but the internal walls (septa) lacking, creating a unilocular ovary, as primrose (*Primula*).
The ovules are arranged along a central axis that does not reach the top of the ovary.

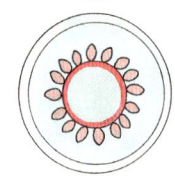

free central placentation

= **central placentation**
see **placentation**

free-floating

Unattached to the bottom substrate of a water body, as the aquatic plant frogbit (*Hydrocharis morsus-ranae*).

free-floating

see also **natant**

fringe   A border. A line of appendages along a border.
fringed   With a border of hairs, bristles, glands etc. along a margin.

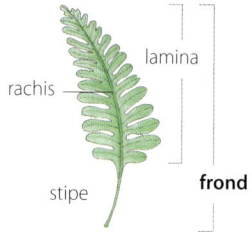

petal

fringe

fringe lily (*Thysanotus tuberosis*)

frond   The leaf of a fern, palm or cycad.
Fern fronds may be compound (pinnate or bipinnate) or simple and usually lobed (pinnatifid or bipinnatifid).

lamina

rachis

stipe

frond

fructan   A polymer of fructose located in the vacuoles that stores energy as soluble carbohyrate.

fructiferous   Bearing fruit.

fructification   The fruiting process of a plant.

fruit   The seed-bearing structure of flowering plants (angiosperms).

Formed after flowering from one or more ovaries and, sometimes, other floral parts.

A fruit may be dehiscent (a pod or capsule) or indehiscent (an achene or samara), dry (a capsule) or succulent (a drupe).

see **accessory ~, aggregate ~, multiple ~, schizocarp, simple ~**

Fruit

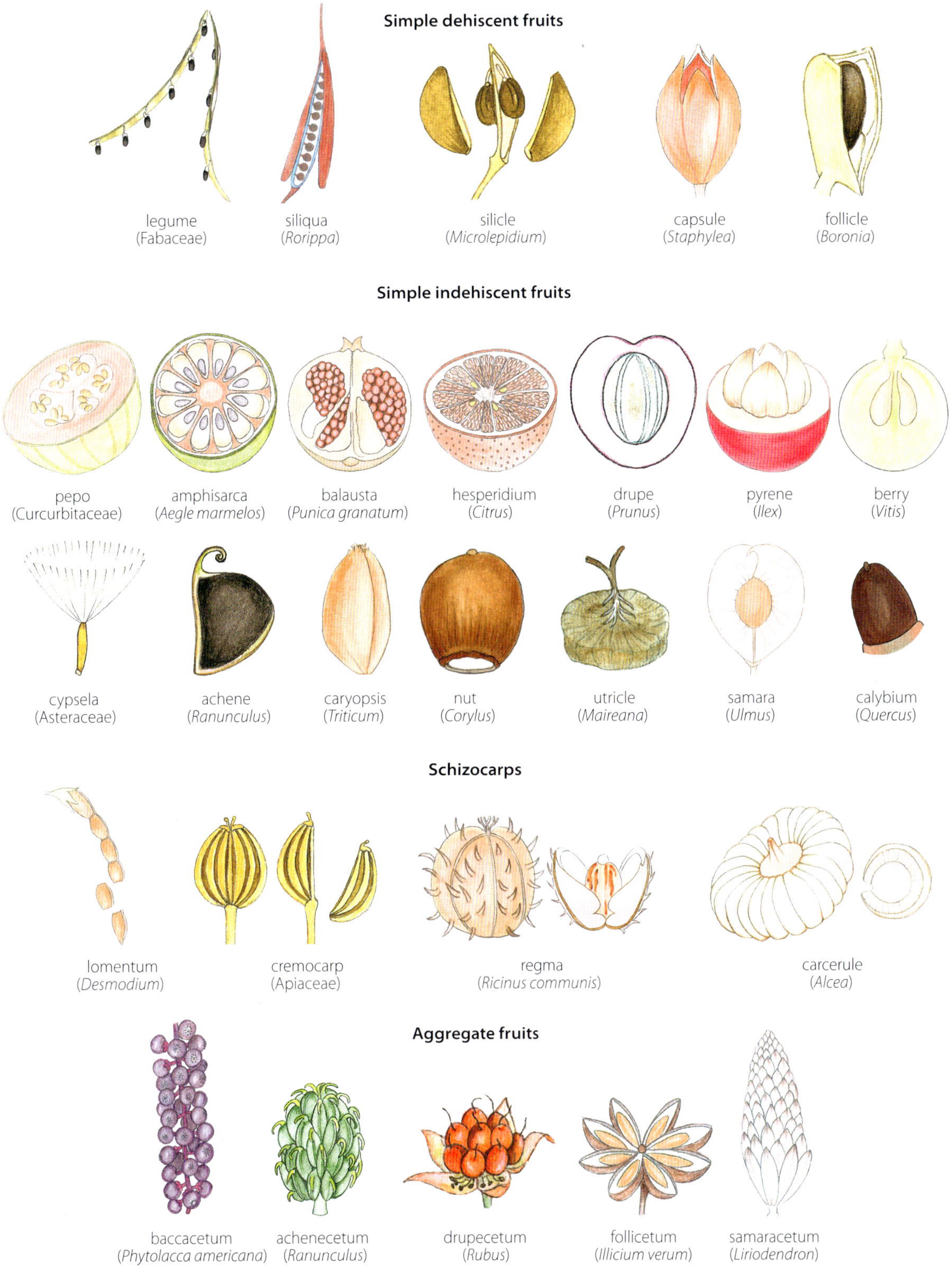

Simple dehiscent fruits

legume
(Fabaceae)

siliqua
(*Rorippa*)

silicle
(*Microlepidium*)

capsule
(*Staphylea*)

follicle
(*Boronia*)

Simple indehiscent fruits

pepo
(Curcurbitaceae)

amphisarca
(*Aegle marmelos*)

balausta
(*Punica granatum*)

hesperidium
(*Citrus*)

drupe
(*Prunus*)

pyrene
(*Ilex*)

berry
(*Vitis*)

cypsela
(Asteraceae)

achene
(*Ranunculus*)

caryopsis
(*Triticum*)

nut
(*Corylus*)

utricle
(*Maireana*)

samara
(*Ulmus*)

calybium
(*Quercus*)

Schizocarps

lomentum
(*Desmodium*)

cremocarp
(Apiaceae)

regma
(*Ricinus communis*)

carcerule
(*Alcea*)

Aggregate fruits

baccacetum
(*Phytolacca americana*)

achenecetum
(*Ranunculus*)

drupecetum
(*Rubus*)

follicetum
(*Illicium verum*)

samaracetum
(*Liriodendron*)

115

Composite or multiple fruits

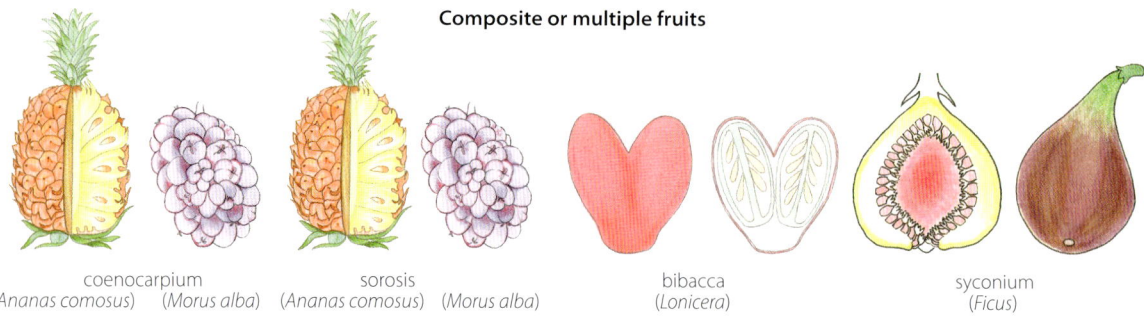

coenocarpium
(*Ananas comosus*) sorosis
(*Morus alba*) sorosis
(*Ananas comosus*) (*Morus alba*) bibacca
(*Lonicera*) syconium
(*Ficus*)

Accessory fruits

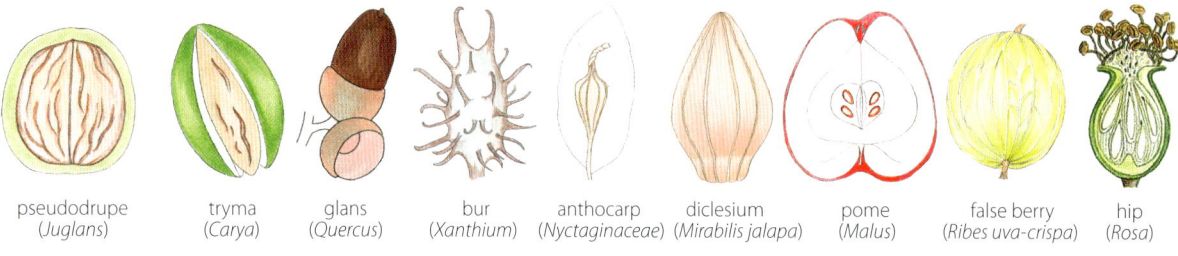

pseudodrupe
(*Juglans*) tryma
(*Carya*) glans
(*Quercus*) bur
(*Xanthium*) anthocarp
(*Nyctaginaceae*) diclesium
(*Mirabilis jalapa*) pome
(*Malus*) false berry
(*Ribes uva-crispa*) hip
(*Rosa*)

fruitlet   A small fruit. One of the individual parts of an aggregate fruit, as a raspberry (*Rubus*).

fruitlet

frutescent   Becoming shrubby, as rosemary (*Rosmarinus*) and lavender (*Lavandula*) that become woody at the base and remain herbaceous above.

frutescent

woody base

fruticose   With stems or branches but without a single main axis, shrub-like.
Of lichens, having upright or pendulous branches.

Fruticose

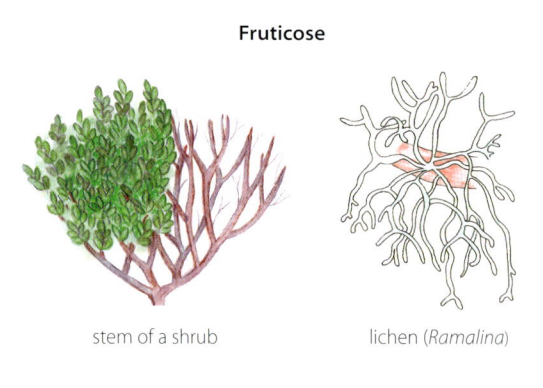

stem of a shrub   lichen (*Ramalina*)

fruticulose   Somewhat shrubby, like a small shrub.

fugaceous, fugacious
Falling or fading early as the petals of guinea flowers (*Hibbertia*).
cf. **ephemeral, evanescent, caducous**

fugaceous

fugitive species   A colonising species of disturbed sites that will usually be overgrown by a stronger competing species, as plants of deserts and ephemeral ponds that die when water evaporates.

fulvous   Reddish-brown.

fungus, *pl.* **fungi**   Unicellular organisms, or multicelluar organisms made up of hyphae, some of which form fruiting bodies that are mushrooms or toadstools.
Hyphae invade a food source, digest it externally, then absorb it.
Cell walls contain chitin, a substance found in the exoskeleton of arthropods and insects.
Reproduction is by spores.
Fungi are divided into five phyla: Ascomycota, Basidiomycota, Chytridiomycota, Glomeromycota and Zygomycota. There is an additional informal group, Deuteromycota.
see **mycelium, mycorrhiza, osmotrophy**
fungal   Relating to or like fungi.

Fungus

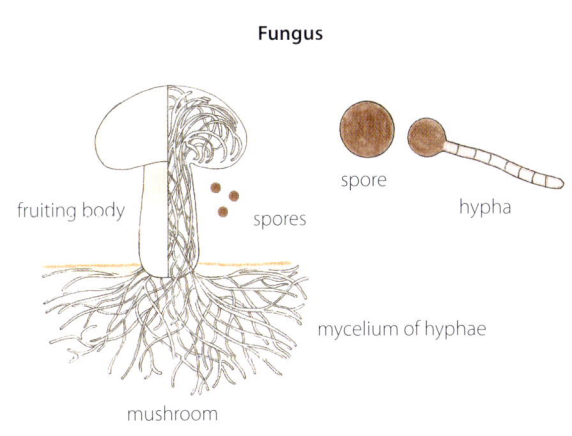

fruiting body

spore

spores

hypha

mycelium of hyphae

mushroom

---

funicle, funiculus
The stalk connecting an ovule to the placenta or attaching the seed to the wall of the pod, as some pea genera (Fabaceae).
funicular Relating to a funicle.

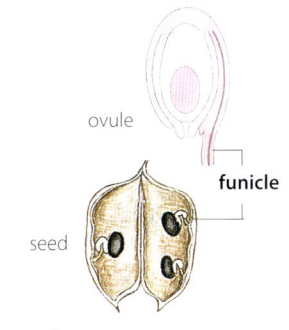

ovule

funicle

seed

funnelform Funnel-shaped, as the corolla of bindweed (*Convolvulus*).
see **infundibular**

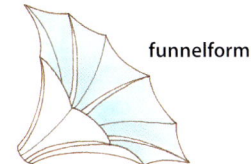

funnelform

furcate Dividing into two prong-like branches. Branched to form a Y-shape or a V-shape, as the veins of some leaves and fronds.
= **forked**
see **bifurcate, trifurcate**

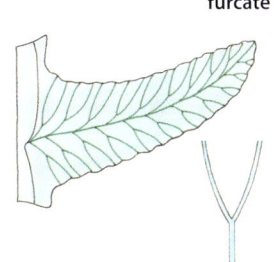

furcate

furfuraceous Covered with soft easily displaced bran-like scales. Scurfy.

furrow A groove or channel, as the indentation along the side of the culm in the bamboo genus *Phyllostachys*.
furrowed With a longitudinal groove.
= **sulcate**

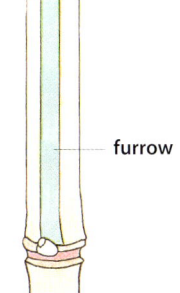

furrow

fuscous A dusky brownish-grey colour.

---

fused A general term meaning united, as the lateral sepals of greenhood orchids (*Pterostylis*) that are largely united.
see also **adherent, adnate, coalescent, coherent, connate**
cf. **contiguous, distinct, free**

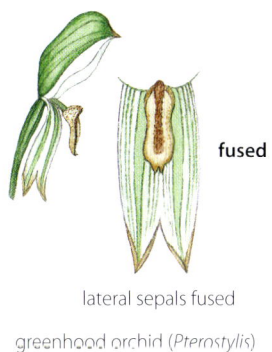

fused

lateral sepals fused

greenhood orchid (*Pterostylis*)

fusiform A three-dimensional shape that tapers at both ends and is widest in the middle. Spindle-shaped.

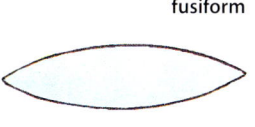

fusiform

fusiform initials One of two types of initial cells (the other being ray initials) in cambium that are the meristematic tissue responsible for secondary growth in plants.
Fusiform initials are elongated tapering cells that give rise to vertical growth and produce the conducting cells of both secondary xylem and secondary phloem in wood. They transport food and water vertically.

fusiform taproot
A main, descending root that is broad in the middle and tapers towards the apex and the base, as some radishes.
see **conical taproot, napiform taproot**

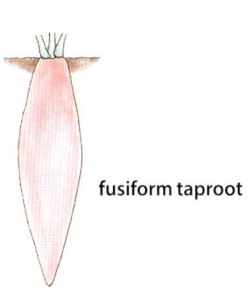

fusiform taproot

fynbos A shrubby vegetation in South Africa, with few trees or grasses, growing on poor soils and having cool wet winters and hot dry summers. It is a Mediterranean-type ecosystem together with the garrigue and maquis in the Mediterranean Basin, the chaparral in California, matorral in Chile and kwongan in southwestern Australia.

Gaia hypothesis A theory put forward by James Lovelock in the 1960s that proposed that the earth is a vast self-regulating organism that seeks an environment optimal for life.

galbulus, *pl.* **galbuli**
A modified cone that becomes fleshy and berry-like, typical of the juniper genus (*Juniperus*).

galbulus

galea   A helmet.
cf. **cucullus**
   galeate   Helmet-shaped, as the helmet-like structure formed by fusion of the petals and dorsal sepal of greenhood orchids (*Pterostylis*).
cf. **cucullate**

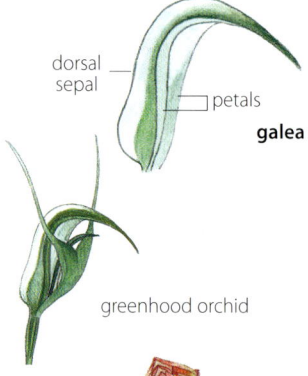
dorsal sepal
petals
galea
greenhood orchid

gall   An abnormal growth form induced in a plant in response to the presence of fungi, bacteria, irritation or insects, as the larvae of the gall midge that cause flower-like galls on beaded glasswort (*Sarcocornia quinqueflora*).
cf. **domatium**

gall

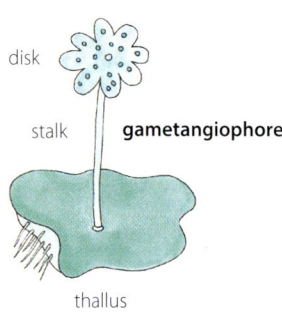
gall midge lava

gametangiophore
The gamete-bearing structure, including a disk and a stalk, that grows out of the thallus, in thallose liverworts. It is either male (an antheridiophore) or female (an archegoniophore).

disk
stalk
gametangiophore
thallus

gametangium, *pl.* **gametangia**   A reproductive cell or organ in which gametes are formed. The term is usually restricted to the sex organs of algae, fungi, bryophytes, ferns and fern allies.
see **antheridium, archegonium**

gamete   Of sexual reproduction, one of the haploid male or female sex cells that unite at fertilisation to form a zygote.
The sperm cells are male gametes and the egg cells are female gametes.

gametogenesis   The formation of the sex cells (male sperm and female eggs) in the reproductive organs of a plant.
In angiosperms, the male sperm cells form in a pollen grain and the female egg cell forms in the embryo sac.
In gymnosperms, the male sperm cells form in a pollen grain and the female egg cell forms in an archegonium.
In ferns, mosses, liverworts and hornworts, the male sperm cells form in an antheridium and the female egg cell forms in an archegonum.
see **megagametogenesis, microgametogenesis**

gametophore
In bryophytes, a structure bearing gametangia.
In mosses and hornworts it is a green perennial 'leafy' stem.
In liverworts it is either a green perennial 'leafy' stem or a thallus. It may be dioecious or monoecious.
cf. **gametophyte**

gametophore
hornwort
moss
liverwort

gametophyte   *see* page 119

gametophyte generation   The haploid phase of a plant's life cycle that produces the male and female gametes that unite to form a zygote.
see **alternation of generations, gametophyte**
cf. **sporophyte generation**

gametophytic apomixis   In flowering plants (angiosperms), the production of an embryo directly from the megaspore mother cell or a cell next to the megaspore mother cell.
A form of agamospermy.
see also **apospory, diplospory**
cf. **sporophytic apomixis**

gammate   Shaped like the greek letter gamma (Γ), as the calli on the labellum of the orchid *Caladenia capillata*.

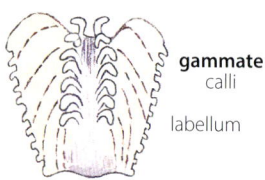
gammate
calli
labellum

gamo-   A prefix meaning united or fused.

gametophyte   All plants have a life cycle alternating between a haploid gametophyte generation and a diploid
sporophyte generation. The sporophyte produces haploid spores by meiosis that germinate and grow into a
haploid gametophyte.

A gametophyte produces male and female gametes (sperm and eggs) and may be bisexual or unisexual.
In nonvascular plants (bryophytes), the gametophyte is a thallus or a 'leafy' stem that is larger than the tiny
sporophyte. The thallus bears antheridia that produce sperm and archegonia that produce eggs.
Vascular plants that produce spores (ferns and fern allies), have a prothallus that is much smaller than the larger
sporophyte. The prothallus bears antheridia that produce sperm and archegonia that produce eggs.
Vascular plants that produce seeds, (gymnosperms and angiosperms) have microscopic dependent gametophytes
(male pollen and female ovules) that live in or on the larger photosynthetic sporophyte. Gymnosperms have male
and female gametophytes borne on cones and in angiosperms the male gametophyte (pollen grain) is borne in an
anther and the female gametophyte (ovule) is enclosed in an ovary.

see **alternation of generations**
cf. **gametophore, sporophyte**

Gametophyte

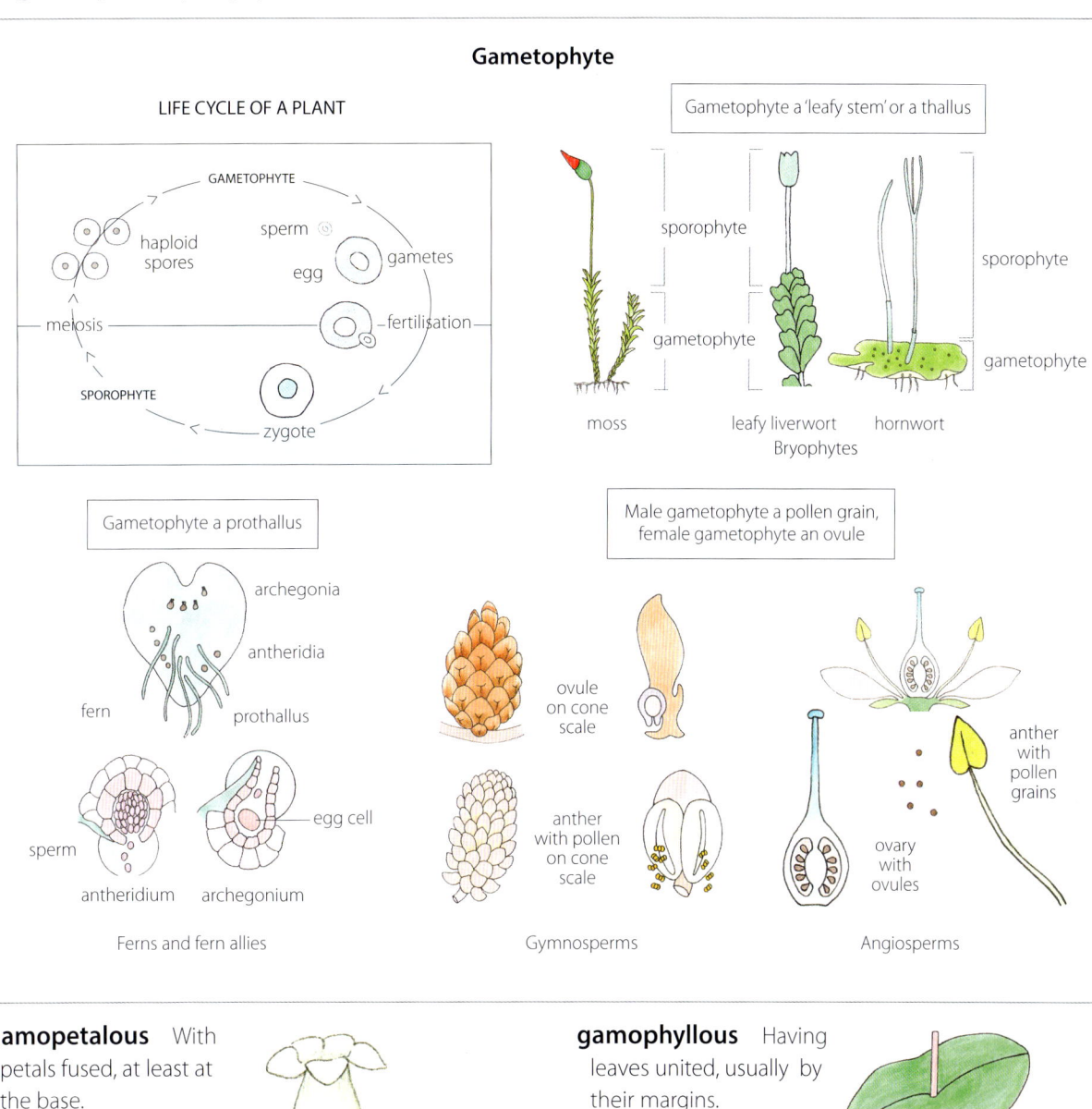

LIFE CYCLE OF A PLANT

GAMETOPHYTE

haploid spores
sperm
egg
gametes
meiosis
fertilisation
SPOROPHYTE
zygote

Gametophyte a 'leafy stem' or a thallus

sporophyte
gametophyte
sporophyte
gametophyte

moss   leafy liverwort   hornwort
Bryophytes

Gametophyte a prothallus

archegonia
antheridia
fern
prothallus
egg cell
sperm
antheridium   archegonium
Ferns and fern allies

Male gametophyte a pollen grain,
female gametophyte an ovule

ovule on cone scale
anther with pollen on cone scale
Gymnosperms

anther with pollen grains
ovary with ovules
Angiosperms

---

gamopetalous   With
petals fused, at least at
the base.
= **sympetalous**
cf. **apopetalous,
choripetalous,
polypetalous**

gamopetalous

gamophyllous   Having
leaves united, usually by
their margins.
having leaf-like parts
united, as a calyx.

gamophyllous

gamosepalous   With sepals fused, at least at the base.
= **synsepalous**
cf. **aposepalous**

gamosepalous

gamotepalous   With tepals fused, at least at the base, as flowers of lily of the valley (*Convallaria majalis*).
= **syntepalous**
cf. **apotepalous**

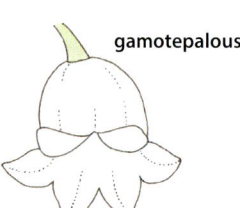

gamotepalous

garrigue   An open low shrubby vegetation less than a metre high, growing on limestone soils in low rainfall areas in the Mediterranean Basin.
It is a Mediterranean-type ecosystem together with the maquis, also in the Mediterranean Basin, the chaparral in California, matorral in Chile, kwongan in southwestern Australia and fynbos in South Africa.

geitonogamy   Pollination between flowers on the same plant.
see **allogamy, xenogamy**
cf. **autogamy**

gel   A semi-solid jelly-like substance in which a liquid is dispersed in a solid.

gelatinous   A wet, sticky state between solid and liquid, as the seed of an unripe almond (*Prunus dulcis*).

gelatinous seed

geminate   Growing in pairs.
Of a petiole having only two leaflets.
= **binate**

Geminate

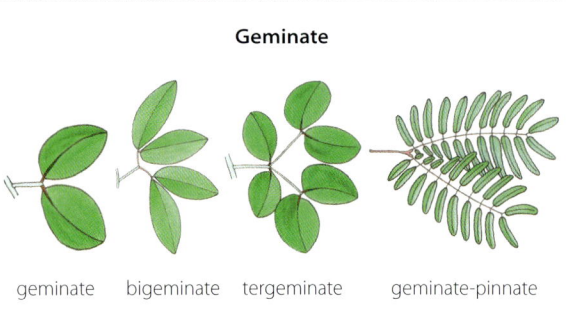

geminate   bigeminate   tergeminate   geminate-pinnate

geminate-pinnate   Of a compound leaf with a pair of leaflets and each leaflet being divided pinnately, as the red powder puff plant (*Calliandra haematocephala*).

geminate-pinnate

gemma, *pl.* **gemmae**   A bud or bud-like structure that separates from the parent plant and serves as a means of vegetative propagation, as tiny sundew (*Drosera pygmaea*).
cf. **turion**
An asexually produced cell or group of cells developed on the leaves of some bryophytes from which new plants may develop. In some thallose liverworts they are borne in specialised 'gemma cups'.
gemmate, gemmiferous
Having gemmae. Reproducing by gemmae.

Gemma

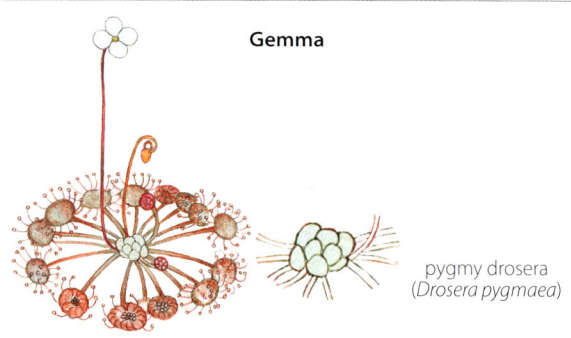

pygmy drosera (*Drosera pygmaea*)

gene   A unit of heredity consisting of a sequence of DNA in a specific location on a chromosome. It determines a particular characteristic of an organism.
see also **nucleotide**

gene expression   The process by which a DNA sequence is converted into instructions for protein synthesis.
cf. **genetic code**

gene flow   The exchange of genes between individuals within a population, or between different populations of the same species, as a result of dispersal of pollen, spores or seeds.
One of the basic mechanisms of evolution together with natural selection, mutation and genetic drift.
= **gene migration**
see **dispersal mechanism**

gene migration   The exchange of genes between individuals within a population, or between different populations of the same species, as a result of dispersal of pollen, spores or seeds.
One of the basic mechanisms of evolution together with natural selection, mutation and genetic drift.
= **gene flow**
see **dispersal mechanism**

gene pool   The sum of all genes and combinations of genes (alleles) in a population of a single species.

gene sequencing   Determination of the order of the four building blocks (adenine, guanine, cytosine or thymine) that make up a DNA molecule.
= **DNA sequencing**

generative cell   One of the two cells, lacking a cell wall, in a pollen grain.
It floats in the cytoplasm of the tube cell and divides, either before pollen is shed or in the pollen tube, to form two male gametes.
see also **tube cell**

Generative cell

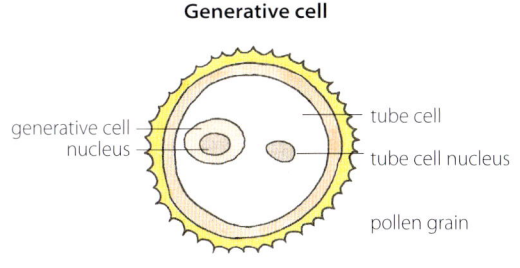

generative cell nucleus — tube cell
— tube cell nucleus
pollen grain

genet   A single genetic individual that is produced from the seed of a particular zygote.
A result of sexual reproduction.
cf. **ramet**

genetic code   The set of rules by which information encoded in genetic material (DNA or RNA sequences) is translated into amino acid sequences (proteins) by living cells.
see also **codon**

genetic drift   Change in allele frequencies in a population over generations due to chance.
One of the basic mechanisms of evolution together with natural selection, mutation and gene flow (gene migration).

genetic engineering   The manipulation of DNA to alter one or more of an organism's characteristics by adding, removing or replacing genes.
see also **vector**

genetic marker   A DNA sequence with a known physical location on a chromosome.

geniculate   Bent abruptly like a knee joint, as some stems.

pedicel
peduncle
geniculate

genome   Generally, the complete genetic information of an organism.
In particular, the DNA found in the nucleus of a cell. Chloroplasts and mitochondria have their own DNA and have their own separate genomes.

genotype   The genetic makeup of a cell or an organism inherited from its parents.
cf. **phenotype**

genus, *pl.* **genera**   In taxonomic classification, a rank below family and above species.
see **taxonomic hierarchy**
generic   Of or relating to a genus.

geocarpy   The ripening of fruit below ground from flowers borne above gound, as the peanut (*Arachis hypogaea*). The gynophore of the spent flower grows downward and pushes the fertilised ovule into the soil where it matures into the fruit.
geocarpic   Pertaining to geocarpy.

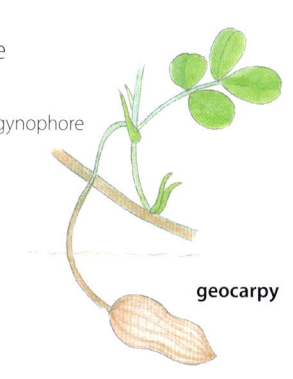

gynophore
geocarpy

geophilous   Of a plant having adaptations for surviving by disappearing underground when protection is needed.

geophyte   A plant with perennating buds underground in dry conditions, as those having rhizomes and tubers.
see also **cryptophyte**

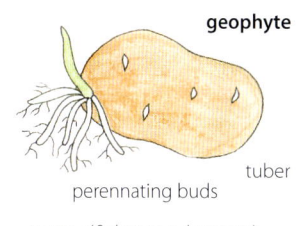

geophyte
tuber
perennating buds
potato (*Solanum tuberosum*)

geotropism   The response of a plant to gravity, either tending to grow downwards (positive geotropism), as roots, or tending to grow upwards (negative geotropism), as shoots.
= **gravitropism**
see **tropism**
geotropic   Of or relating to geotropism.

germinal aperture   The variously shaped, thinner region of the pollen wall through which the pollen tube emerges or the spore wall through which the protonema emerges.

germinate, germination   *see* page 123

gibberellins   A group of plant hormones responsible for growth and development, including seed germination, stem elongation due to increased cell size, transition to flowering and development of flowers, fruits and seed.
see **phytohormone**

gibbosity   A swelling or protuberance.
gibbous   Swollen or enlarged on one side, as the flower tube of common valerian (*Valerian officinalis*). With a pouch-like swelling.
see also **ventricose**

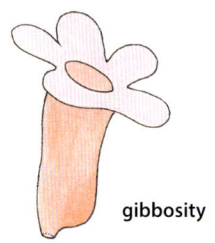
gibbosity

ginkgo   The deciduous maidenhair tree (*Ginkgo biloba*), so called because of its fan-shaped leaves like those of some maidenhair ferns.
It bears male or female cones on separate plants. The sole surviving species of Ginkgophyta, one of the four divisions of gymnosperms.

girth   Measurement around the circumference of something, as of a tree at a certain distance above ground level.

glabrate, glabrescent   Becoming glabrous. Almost glabrous.

glabrous   Lacking hairs, scales or other indumentum, smooth.

gladiate   Shaped like the blade of a sword, as the leaves of gladiolus and iris.
= **ensiform**

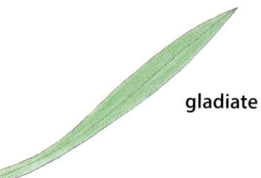
gladiate

gland   A cell or group of cells that secrete or excrete a substance, such as a hormone, enzyme, wax, nectar, salt or water. It may be on or near the plant surface and discharging the substance externally, as glandular hairs, nectaries and hydathodes, or internal and releasing the substance into a canal or cavity, as resin canals in pines and schizogenous cavities or ducts in Asteraceae.
glandular, glanduliferous   Having or relating to glands.

glandular hair   Tipped with a gland, as those on the leaf of a sundew (*Drosera*) that produce a secretion that traps and digests insects.

glandular hair

glandular-ciliate   With gland-tipped eyelash-like hairs, as the margins of some leaves.

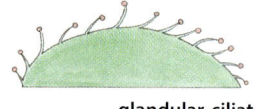
glandular-ciliate

glandular-punctate   With dot-like depressions, due to translucent or coloured glands, as the leaves of mints (*Mentha*).

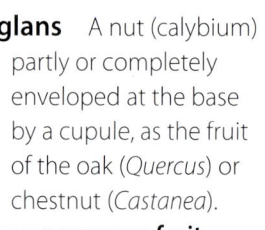
glandular-punctate

glans   A nut (calybium) partly or completely enveloped at the base by a cupule, as the fruit of the oak (*Quercus*) or chestnut (*Castanea*).
see **accessory fruit**

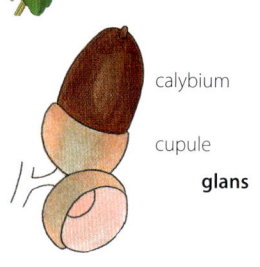
calybium
cupule
glans

glaucescent   Somewhat glaucous, becoming glaucous.

glaucous   Covered with a greyish, bluish or whitish bloom of fine white powder or wax.
see also **ceraceus, pruinose**

germination   Of seed plants, the resumption of growth of the embryo in the seed, resulting in the rupture of the seed coat, usually at the micropyle, and the emergence of the young root or radicle.

In epigeal germination, the radicle elongates and penetrates the soil and the elongating hypocotyl pushes the cotyledons out of the ground.

In hypogeal germination the epicotyl grows upward and the cotyledons remain underground.

In viviparous germination the seed germinates before becoming detached from the parent plant.

Of seed plants, the rupture of the pollen grain on the stigma of a flower and growth of the pollen tube.

Of cryptogams, the rupture of the spore and emergence of a multicellular protonema that will give rise to a small gamete-producing plant (gametophyte).

germinate   To sprout, shoot or produce buds.

Germination

Seeds

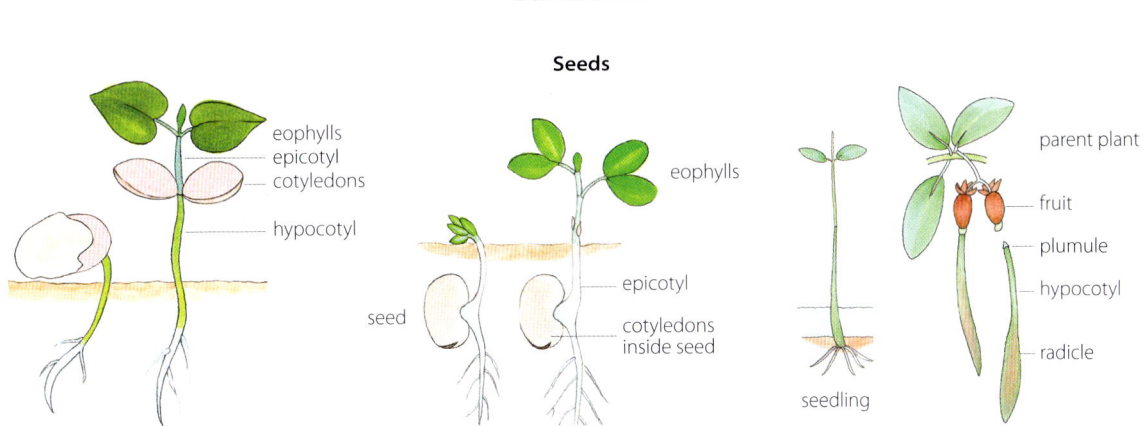

epigeal germination   hypogeal germination   viviparous germination

Pollen grains

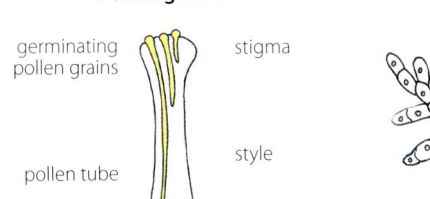

Spore (moss)

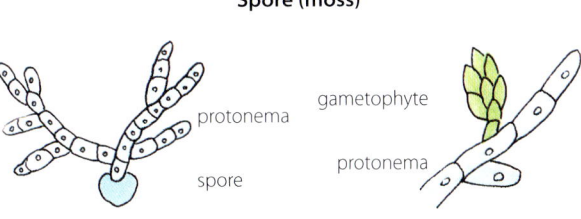

globose, globular
Spherical or nearly so, as the flower heads of some wattles (*Acacia.*)

globose

glomerate   Collected into a dense, spherical mass, as the flowers of some pennyworts (*Hydrocotyle*).

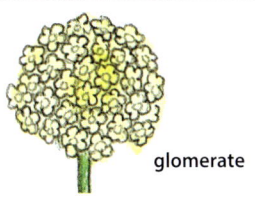

glomerate

glochid, glochidium,
pl. **glochidia**
A fine barbed detachable spine derived from the epidermis and occurring in the areoles of the prickly pear genus (*Opuntia*).
cf. **prickle, spine, thorn**
glochidiate   With fine barbed spines.

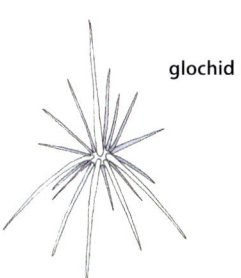

glochid

Glomeromycota
A phylum of fungi dependent on a symbiotic relationship with plant roots.
Its hyphae invade cells of the root cortex and form arbuscules.
Thought to reproduce asexually by spores.
see **fungus, mycorrhiza**

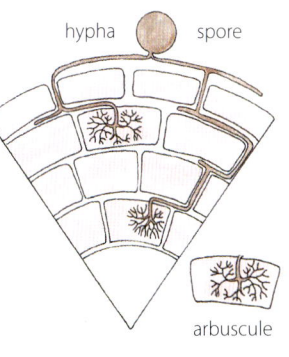

Glomeromycota

glomerule, *pl.* **glomeruli**
A dense cluster.
A dense head-like cymose inflorescence of almost sessile and usually small flowers. Commonly part of a compound inflorescence.
see **fascicled cyme, verticillaster**

glomerulate Having glomeruli. Arranged in small compact clusters, as the male flowers of some saltbushes (*Atriplex*).

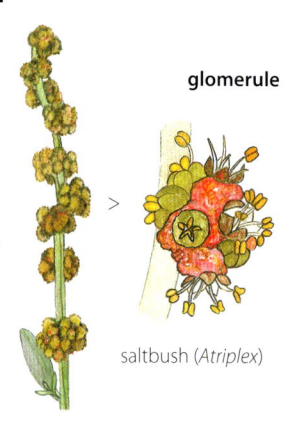

glomerule

saltbush (*Atriplex*)

glossy Having a smooth shiny surface, as cherries (*Prunus avium*).

glossy

glucose A simple sugar (monosaccharide) that is the main source of energy for cellular metabolism. In plants, a product of photosynthesis.
see also **glycolysis**

glume One of usually two bracts at the base of a spikelet in grasses (Poaceae).
The single bract subtending the flower in sedges (Cyperaceae).
glumaceous Resembling a glume. Bearing glumes.

glume

grass spikelet

glutinous Resembling glue in texture, sticky. Viscid, as the glutinous pericarp of the drupe of the dodder laurel (*Cassytha*).

glutinous

glycolysis A series of reactions in cellular respiration that breaks down glucose so that it can be stored as the readily available energy source adenosine triphosphate (ATP).
see also **Krebs cycle**

gnetales A group of non-flowering seed plants restricted to three genera today (*Gnetum, Ephedra* and *Welwitschia*).
Members of Gnetophyta, one of the four divisions of gymnosperms.

Golgi apparatus, Golgi body, Golgi complex
An organelle in the cytoplasm of a cell consisting of a system of flattened sacs and transport vesicles.
It produces, modifies and packages proteins and other products for use within the cell or outside the cell.

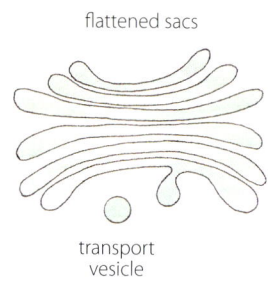

flattened sacs

transport vesicle

Golgi apparatus

Gondwana, Gondwanaland One of two vast continents believed to have existed in the southern hemisphere resulting from the break up of Pangea.
see **continental drift, Laurasia**

Gondwana

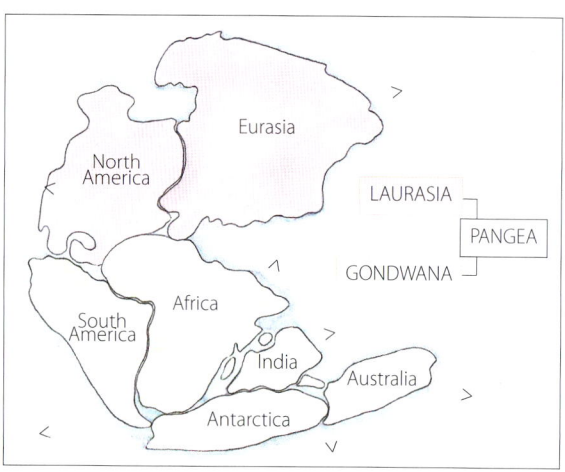

gonophore
An elongated stalk (stipe), inserted on the receptacle, that bears the stamens and pistil of a flower above the corolla and calyx.
= **androgynophore, gynandrophore**

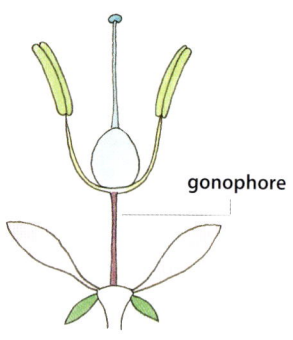

gonophore

gourd Sometimes applied generally to the fruits of the pumpkin family (Curcurbitaceae),

but more specifically to
the two genera *Lagenaria*
and *Curcurbita* that have
a hard, often irregular,
rind.
The plant that bears this
fruit.

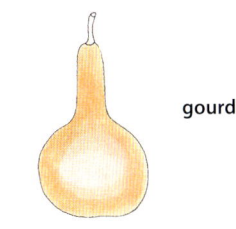

gourd

bottle gourd (*Lagenaria*)

graft A shoot or bud
(scion) to be implanted
into a growing plant.
The point of union of
the scion with the stock.
A plant produced from
the union of a scion
with a stock .
see **grafting**

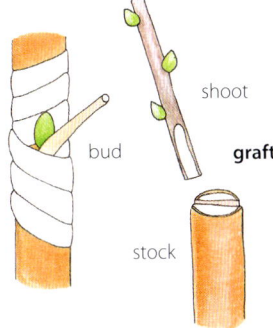

shoot

bud

graft

stock

graft chimaera A bud at the junction of the scion
and stock that has tissues with the characteristics of
both plants.
= **graft hybrid**

graft hybrid A bud at the junction of the scion
and stock that has tissues with the characteristics of
both plants.
= **graft chimaera**

graftage A method of vegetative reproduction
that consists of implanting a bud (budding), or
a shoot (scion), usually with two or three buds
(grafting), from one plant into the tissue of another
plant (stock).
It is used for plants that do not grow true from seed,
as most fruit trees, or plants with cuttings that do
not root easily.

grafting A method of
vegetative reproduction
that consists of
implanting a shoot
(scion), usually with two
or three buds, from one
plant into the tissue of
another plant (stock).
Used for plants that do
not grow true from seed,
as most fruit trees, or
plants with cuttings
that do not root easily.
see **budding, graftage,
stock**

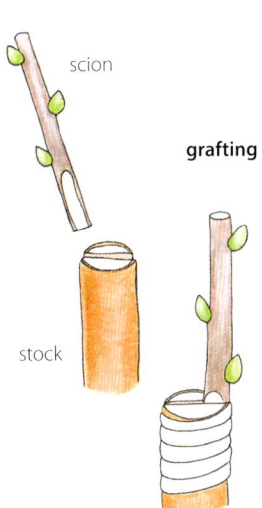

scion

grafting

stock

grain A dry indehiscent
fruit with one seed
fused to the fruit wall
(pericarp).
Derived from a one-
carpelled superior ovary.
Characteristic of grasses
(Poaceae), as wheat
(*Triticum*).
cf. **achene, cypsela,
diclesium**
= **caryopsis**

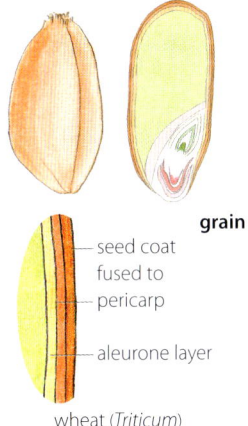

grain

seed coat
fused to
pericarp

aleurone layer

wheat (*Triticum*)

graminaceous, gramineous Pertaining to
plants of the grass family (Poaceae).

graminoid Resembling grasses.
A grass (Poaceae) or grass-like plants, such as sedges
(Cyperaceae) and rushes (Juncaceae).

granulate To form into small grains.
To become granular in texture, to become rough
and grainy in texture or appearance.

granule A small grain or particle, as the
microscopic starch grains in the cytoplasm of a
plant cell.
granular Consisting of or covered with small
grains or particles.
Having a roughened surface due to small rounded
protuberances.

granum, *pl.* **grana**
A stack of disc-
shaped thylakoids
in a chloroplast.

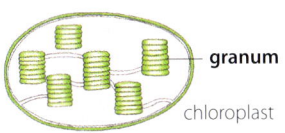

granum

chloroplast

grapnel A hook with
three or more recurved
prongs. usually found in
a ring on a stem. They are
used for climbing and
support.
see **rattan**

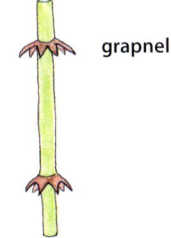

grapnel

grass   Any member of the family Poaceae that has some 9500 species that include cereal grasses, bamboos, natural and cultivated grasses (turf), and pastures.

A typical inflorescence of spikelets starting at the uppermost node. Florets are 2-ranked. Leaves are narrow, with parallel veins, and mostly arranged alternately in two opposite rows or at the base. A sheath attaches the blade to a node and at the junction of the blade and sheath there is usually a ligule. Stems (culms) are circular in cross-section and pithy or hollow with solid nodes. Roots are fibrous. The fruit is a grain (caryopsis).

see **grassland**

cf. **forb**

Grass

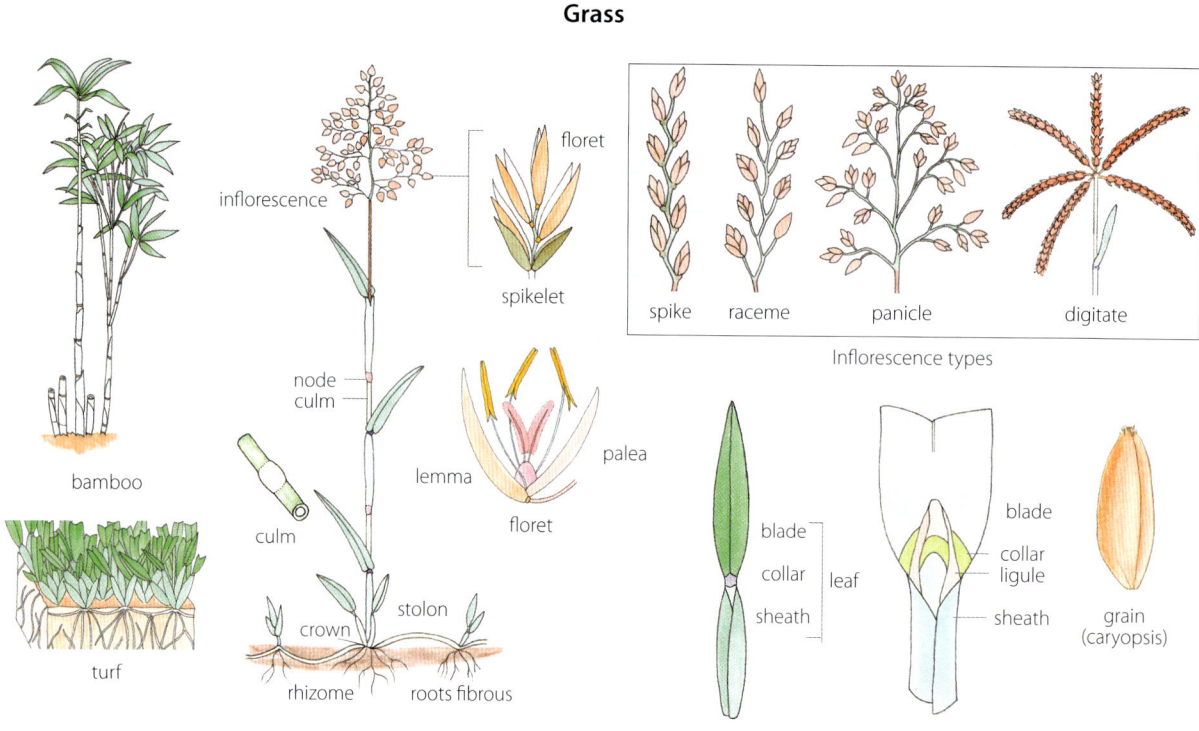

Inflorescence types

grass flower

Typically reduced to little more than reproductive organs and comprising two tiny lodicules, three stamens and a unilocular ovary bearing two stigmas. The flower, together with the two enclosing bracts, the lemma and palea, is a grass floret.

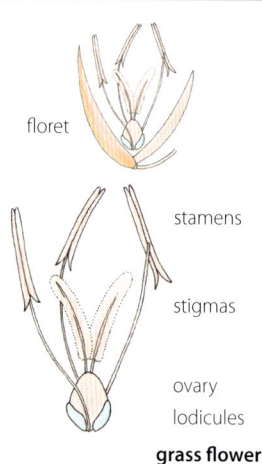

grass flower

grass inflorescence   Typically an inflorescence of spikelets arranged in spikes, racemes or panicles.

Grass inflorescence

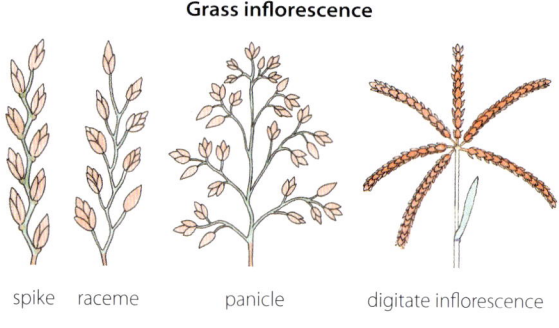

spike   raceme   panicle   digitate inflorescence

grassland   A grazing ecosystem with vegetation that is herbaceous and grasses being predominant. The rainfall is usually insufficient for trees.

Temperate grasslands include the velds of South Africa, the puszta of Hungary, the pampas of Argentina and Uruguay, the steppes in Russia and the prairies of North America.

Tropical grasslands, as the Serengeti in eastern Africa, and subtropical grasslands, have tall grasses and scattered trees.

see also **savanna**

gravitropism   Growth of a plant in response to gravity, either downwards (positive gravitropism) or upwards (negatative geotropism), as taproots that grow vertically downwards and shoots that grow vertically away from gravity.
= **geotropism**
see **tropism**
gravitropic   Of or relating to gravitropism.

gregarious flowering   Of plants within a species, and having the same seed origin, that flower at the same time, often after a long interval, although located at places distant from one another.
see also **masting**
cf. **plietesial**

grex, *pl.* **greges**   In orchid nomenclature, the hybrid progeny of an artificial cross between two parent plants.

groove   A long narrow cut or depression, as that on the side of a plum stone.

groove

ground meristem   One of three regions of primary meristematic tissue that develops behind the apical meristem.
It differentiates into ground tissue (parenchyma, collenchyma and sclerenchyma).
see **primary meristem**
see also **procambium, protoderm**

ground plan
A diagrammatic cross-section to show the organisation of the floral organs (floral ground plan) or the vascular system of a plant.
cf. **numerical plan**

Floral **ground plan**

sepals petals stamens carpels

ground substance   A clear substance that makes up most of the volume of the cytoplasm.
= **cytosol, hyaloplasm**

ground tissue   Plant tissue other than that of the vascular system and the dermal system.
Comprises the majority of the plant body and is composed of three tissues, parenchyma, collenchyma and sclerenchyma, which are distinguished by their cell wall structures.

growth ring   A distinct, usually single, annual band of secondary thickening that forms in the wood of some trees.
see also **early wood, late wood**

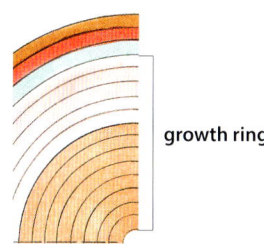

growth rings

guard cell   One of two cells located either side of a stoma in the epidermis of a leaf.
Changes in turgidity of the guard cells cause the stoma to open or close, regulating the flow of gases and water vapour into and out of the plant tissue.

Guard cell

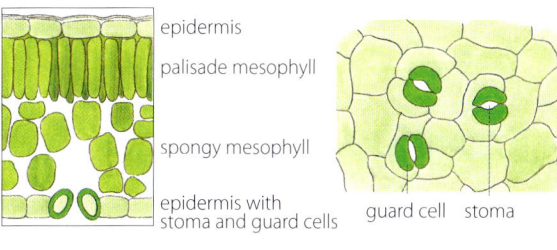

epidermis
palisade mesophyll

spongy mesophyll

epidermis with
stoma and guard cells   guard cell   stoma

gum   A viscid plant exudate that hardens. It is soluble or swells in water.
cf. **resin**

guttation   Exudation of excess water in plants when transpiration is negligible and soil moisture is high.
It occurs on leaf margins, as strawberries and roses, and at the tips of leaves, as grasses and some members of the purslane family (Portulacaceae).
see **hydathode**

guttation

gymnosperm   A seed-bearing plant that lacks flowers.

Gymnosperms have a life cycle alternating between a haploid sexual gametophyte generation and a diploid asexual sporophyte generation.

The sporophyte generation is the larger familiar green plant.

The gametophyte generation, (eggs and pollen), is microscopic and lives on the sporophyte.

The reproductive structure is usually a unisexual cone. Female cones bear ovules and seeds exposed on scales rather than in an ovary that becomes a fruit, as angiosperms. Pollen is borne in sacs on scales of the male cones.

Characteristics include simple fertilisation and haploid nutritional material (megagametophyte). Gymnosperm seeds have more than two cotyledons.

There are four divisions: conifers (Coniferophyta), cycads (Cycadophyta), ginkgo (Ginkgophyta) and gnetales (Gnetophyta).

see also **seed**

cf. **angiosperm**

Alternation of generations

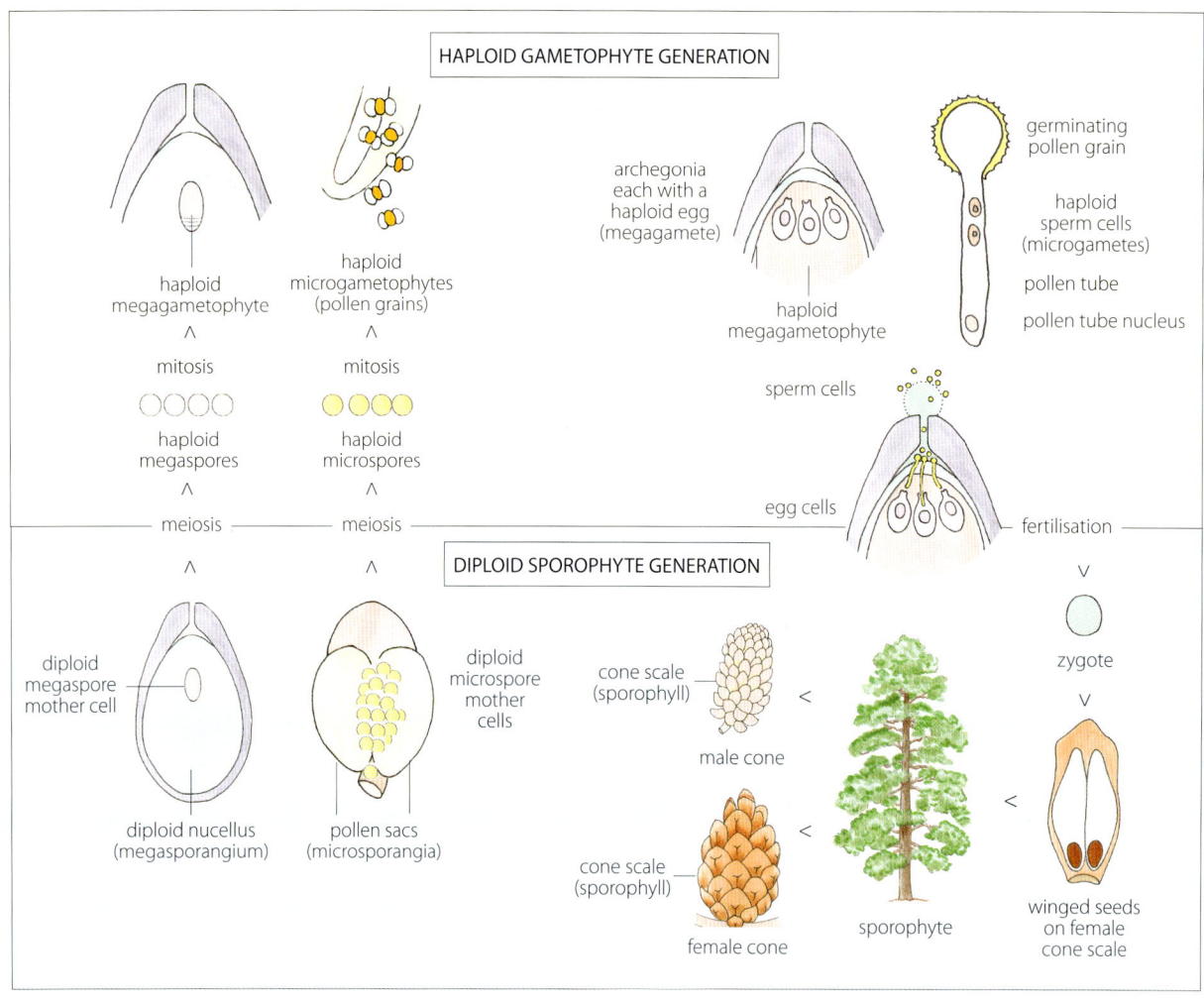

gynandrium, *pl.* **gynandria**   Of a flower, the central structure made up of the male stamen and the female stigma and style fused together, as in an orchid flower (Orchidaceae) and trigger plant flower (*Stylidium*).

= **column, gynostemium**
gynandrous   With or relating to a gynandrium.

Gynandrium

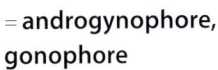
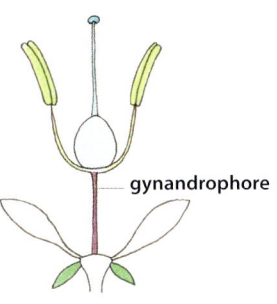

orchid gynandrium (*Caladenia*)

gynandrium trigger plant (*Stylidium*)

gynandrophore
An elongated stalk (stipe), inserted on the receptacle, that bears the stamens and pistil of a flower above the corolla and calyx.
= **androgynophore, gonophore**

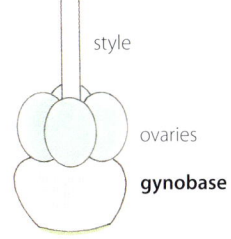

gynandrophore

gynobase
An elongation or enlargement of the receptacle bearing the gynoecium, as the swollen gynobase of the sage genus *Salvia* in the mint family (Lamiaceae).

style

ovaries

gynobase

gynobasic
Of a style arising from the gynobase and attached to the carpels.
A characteristic of the mint family (Lamiaceae).

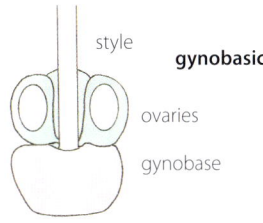

gynobasic

style

ovaries

gynobase

gynodioecious   Of a species with pistillate (female) flowers on one plant and bisexual flowers on a different plants within the same species.
see also **gynomonoecious**
cf. **androdioecious**

gynoecious   Of a plant having only female flowers.
see also **pistillate**
cf. **androecious**

gynoecium, *pl.* **gynoecia, gynecium,** *pl.* **gynecia**
The female reproductive organ of a flower (pistil). The basic unit is a carpel consisting of a stigma, style and an ovary enclosing one or more ovules each of which has an ovum (the female sex cell or gamete). A gynoecium may have more than one carpel. If there is only one carpel the terms carpel, pistil and gynoecium are synonymous.
see **apocarpous, syncarpous**

Gynoecium

stigma

style

ovary with ovules

flower

carpel

gynomonoecious   Of a species with pistillate (female) and bisexual flowers on the same plant.
see also **andromonoecious**
cf. **androdioecious**

-gynous   A suffix referring to the gynoecium. Having carpels as specified.

gynophore   A stalk-like prolongation (stipe) of the receptacle that raises the pistil above the other floral parts.
cf. **podogyne**

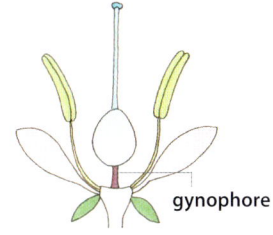

gynophore

gynostegium
Formed by the fusion of the anthers to the stigma, as in the milk-weed genus (*Asclepias*).
gynostegial
Having or relating to a gynostegium.

gynostegium

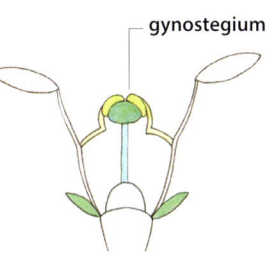

gynostemium, *pl.* **gynostemia**   Of a flower, the central structure made up of the male stamen and the female stigma and style fused together, as in an orchid flower (Orchidaceae) and trigger plant flower (*Stylidium*).

= **column, gynandrium**

gynostemial   With or relating to a gynostemium.

Gynostemium

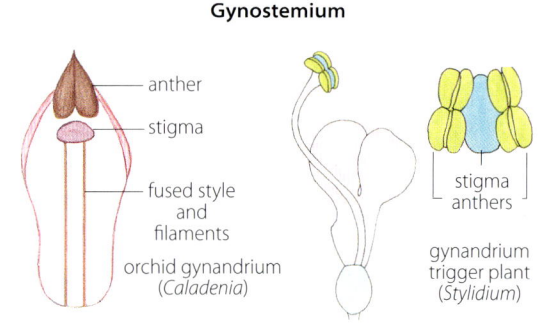

anther
stigma
fused style and filaments
orchid gynandrium (*Caladenia*)

stigma anthers
gynandrium trigger plant (*Stylidium*)

gypsophile   A plant preferring gypsum-rich soil.

gypsophilous   Flourishing in soil containing or derived from gypsum.

gypsophyte   A plant that tolerates or thrives in soil rich in gypsum.

habit   The characteristic shape and growth form of a plant, as a herb, vine, liana, shrub or tree.

Terms like erect, creeping, scandent, woody and so on further describe the habit.

cf. **habitat**

Habit

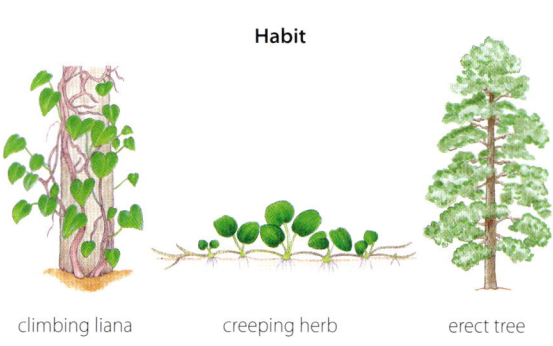

climbing liana   creeping herb   erect tree

habitat   The environment in which plants and other organisms live.

cf. **biome, community, ecosystem, population**

haft   The base of an organ when narrow or constricted, as the narrower part of the falls of an iris (*Iris*).

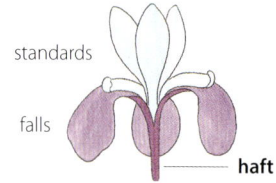

standards
falls
haft

hair   A unicellular or multicellular outgrowth of the epidermis that is branched or unbranched.

The texture varies, as a fine thread or a stiff bristle. They are variously shaped, as T-shaped or stellate and may be glandular or non-glandular.

see **indumentum**

Some hair types

silky   strigose   peltate   T-shaped / urent

stellate   pubescent   dendritic   matted   hispid

glandular   tubercle-based   septate   arachnoid

half-equitant vernation

Of young leaves in the unopened leaf bud with the margins folded to embrace one margin of another leaf.

= **obvolute vernation**

half-equitant vernation

half-inferior ovary

An ovary having the lower half embedded in the hypanthium and the upper part free.

= **semi-inferior ovary**

cf. **inferior ovary, superior ovary**

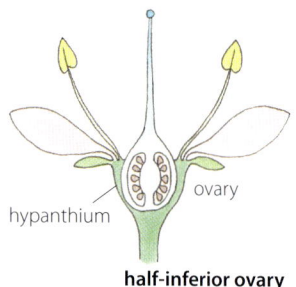

hypanthium   ovary

half-inferior ovary

halophile   A plant living and thriving in a saline environment.

halophilous   Growing in soils with high concentrations of salt.

halophobe   A plant that does not tolerate a saline envronment.
halophobic   Not tolerant of soils with high concentrations of salt.

halophyte   A salt-resistant or salt-tolerant plant that completes its life cycle in soils or waters containing high salt concentrations. Species like saltbushes (*Atriplex*) and glassworts (*Salicornia*) absorb and store salt or excrete it.
halophytic   Of or relating to a halophyte.

halosere   An ecological succession that starts in saline soil or water.

hamulus, *pl.* **hamuli**   A small hook.
Of orchids, a stipe formed from the apex of the rostellum.
cf. **tegula**
hamulate   With a small hook at the tip; with little hooks.

hamus, *pl.* **hami**   A hook.
hamate   Bent at the end like a hook.
Hooked at the apex, as the achene of some buttercups (*Ranunculus*).

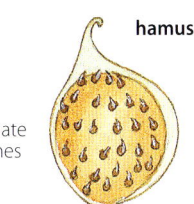

hamate spines

hamus

buttercup achene

hapaxanthic   Of stems (rather than whole plants) that flower and and bear fruit once in their lifetime then die, as palms that form stem clusters.
Most commonly used to describe palms (Arecaceae) and bamboos.
cf. **iteroparous, monocarpic, pleonanthic, polycarpic, semelparous**

haploid   The number (n) of chromosome sets in a gamete that is half the number of chromosome sets (2n) in a somatic cell.
One chromosome set occurs in gametes of diploids, two chromosome sets occur in gametes of tetraploids, and so on.
see **diploid, ploidy**

haplomorphic   With floral whorls spirally arranged and appearing hemispheric to conoidal overall, as flowers of the genus *Illicium*.

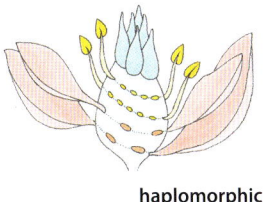

haplomorphic

haplostemonous
With a single whorl of stamens.
With as many stamens as petals.
= **isostemonous**

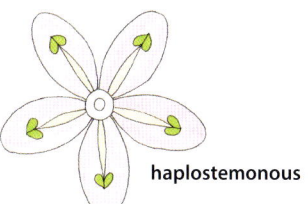

haplostemonous

hardwood   The wood of angiosperms, as eucalypts, beech and blackwood, though the wood is not always hard.
cf. **softwood**

harmemogathy   A mechanism facilitating changes in shape and size by varying the degree of hydration, as occurs in pollen grains.

Hartig net   The network of hyphae that ramify through the intercellular spaces of the roots of some plants.
see **ectomycorrhiza**

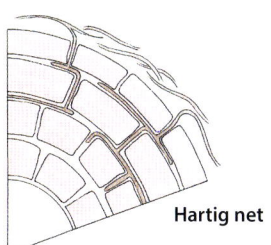

Hartig net

hastate   Shaped like the head of a halberd, with two spreading, somewhat triangular lobes at the base, as some leaves.
cf. **saggitate**

hastate

hastula   Of palms (Arecaceae), a flap of tissue at the junction of the petiole and the lamina in most palmate and costapalmate leaves.

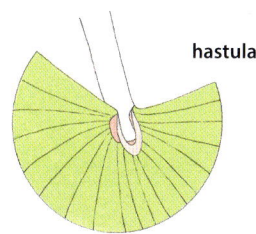

hastula

haustorium, *pl.* **haustoria**
A root-like projection from the stem of a parasitic plant that penetrates the cell walls of the host plant in order to absorb nutrients and water, as the haustoria on creeping mistletoe (*Muellerina eucalyptoides*).

haustorium

head   A compact mass of leaves, flowers or fruit, as a head of lettuce, a head of grain or a dense cluster of flowers.
Of a racemose inflorescence, as a daisy capitulum (Asteraceae).
Of a cymose inflorescence, a glomerule, as saltbushes (*Atriplex*).

heading   Of grasses (Poaceae), emergence of the inflorescence from the uppermost sheath on a culm.
see **boot stage, booting**

heading

heartwood
The central usually darker inactive central wood of a trunk or branch made up of non-functioning sapwood cells infiltrated with other substances like lignin.
= **duramen**

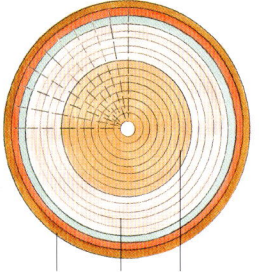

bark   sapwood   **heartwood**

heathland   A plant community, typically with low-nutrient soils, dominated by small-leaved shrubs, often of the Ericaceae family.

helicoid   Having the form of a flattened coil or spiral.

helicoid cyme
A flattened, spirally coiled cymose inflorescence with a single new stem developing repeatedly on the same side of the axis.
= **bostryx**
see also **monochasium**
cf. **scorpioid cyme**

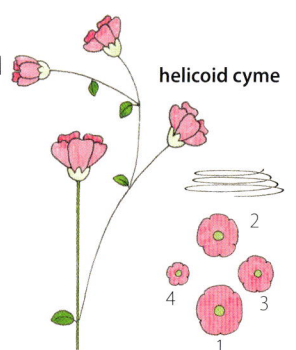

helicoid cyme

heliophile   A plant that thrives in sunlight.
heliophilic, heliophilous   Thriving in sunlight, as the sunflower (*Helianthus*).
cf. **heliophobic**

heliophobe   A shade-loving plant that has an aversion to sunlight.
heliophobic   Thriving in shade.
cf. **heliophilic**

heliophyte   A plant that tolerates or thrives in sunlight.
heliophytic   Tolerating or thriving in sunlight.

heliotropism   Originally used when movement of some plants was thought to be a response to the sun and now used to describe a plant's response to light.
see **tropism**
heliotropic   Of or relating to heliotropism.

helix   A spiral line that follows the path of a cylinder or cone at a constant angle.
helical   Of or having the shape of a helix. Spiralled.
cf. **helicoid**

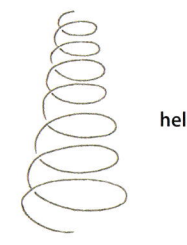

helix

helophyte   Plants with perennating buds on rhizomes in water-saturated soil like marshes and pond or lake edges, as running marsh flower (*Villarsia reniformis*).
see also **cryptophyte**

helophyte

marsh flower
(*Villarsia reniformis*)

hemi-   A prefix meaning half.

hemianatropous   Of ovule orientation, with the ovule turned at 90° so that it is at a right angle to the funicle and has the chalaza and micropyle on each side.
see **ovule orientation**

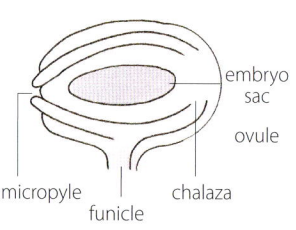

embryo sac
ovule
micropyle   chalaza
funicle
hemianatropous

hemicryptophyte   Herbaceous plants having perennating buds on the surface of the soil and aerial shoots dying down with the onset of adverse conditions.
see also **chamaephyte, cryptophyte, phanerophyte, therophyte**

hemiepiphyte   A plant with a life cycle that includes both terrestrial and epiphytic phases. Primary hemiepiphytes, like strangler figs (*Ficus*), have seeds germinating in a tree canopy and roots that grow downwards and become rooted in the ground where the plant continues to grow. Secondary hemiepiphytes, like the vanilla orchid (*Vanilla planifolia*), have seeds that germinate and grow in the soil but also roots that climb a host tree where the orchid eventually becomes epiphytic.

hemiepiphytic   Of or relating to a hemiepiphyte.

hemiparasite   A plant that can photosynthesise but also parasitises another plant, as mistletoe that gets nourishment from its host through haustoria inserted into branches.

= **semi-parasite**

hemiparasitic
Of or relating to a hemiparasite.

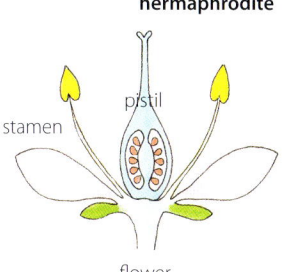

photosynthesising leaves

hemiparasite

haustoria
host branch

mistletoe (*Viscum album*)

hepta-   A prefix meaning seven.

herb   A plant that does not develop a woody stem. It may be annual, biennial or perennial. Includes both forbs and graminoids.

herbaceous   Not woody.
Usually green and soft in texture with soft stems.

herbfield   A vegetation type dominated by forbs and lacking trees and dense shrubs.

= **herbland**

herbivore   An animal that eats only plants.

herbivorous   Feeding on plants, as many animals: includes leaf chewing, sap sucking, seed predation and gall induction by insects.

= **phytophagous**

cf. **carnivorous**

herbland   A vegetation type dominated by forbs and lacking trees and dense shrubs.

= **herbfield**

heredity   The passing on of genes that determine traits from parents to offspring, by means of sexual or asexual reproduction.

= **inheritance**

herkogamy
Positioning of the anthers and stigma in a flower so that pollen is unable to reach the stigma, as the cowslip (*Primula vulgaris*).

see **pin, thrum**

cf. **dichogamy**

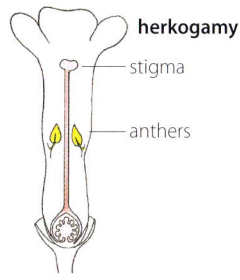

herkogamy
stigma
anthers

hermaphrodite
Of a flower with stamens and a pistil or pistils that are fertile. A bisexual or perfect flower.

hermaphroditic
Flowers with stamens and a pistil or pistils that are fertile. Bisexual.

cf. **unisexual**

hermaphrodite
pistil
stamen
flower

hesperidium   A fleshy fruit, with the pulp in segments, surrounded by a separable pithy rind: derived from a usually five-carpelled septate superior ovary.

The exocarp (outer layer of the rind) contains volatile oils in pits and the endocarp (fleshy interior) is composed of fluid-filled specialised hair cells. Typical fruit of the citrus genus (*Citrus*), as orange, lemon and grapefruit.

Hesperidium

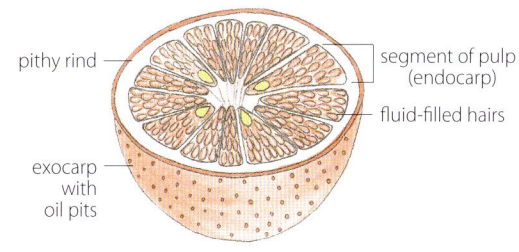

pithy rind
segment of pulp (endocarp)
fluid-filled hairs
exocarp with oil pits

hetero-   A prefix meaning different.

cf. **homo-**

heteroblastic pseudobulb
Of orchids having a pseudobulb consisting of a single internode that bears the leaves, as nodding bulbophyllum, (*Bulbophyllum nutans*).

cf. **homoblastic pseudobulb**

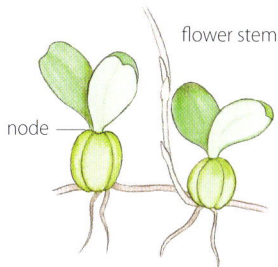

flower stem
node

heteroblastic pseudobulb

heteroblasty Having leaves that show a striking difference in appearance between the juvenile and the adult condition, as leaves of mountain clematis (*Clematis aristata*).
cf. **homoblasty**
heteroblastic Relating to heteroblasty.

adult leaf

heteroblasty

juvenile leaves

heterocarpy
Producing two kinds of fruits from the one inflorescence, as slender cotula (*Cotula vulgaris*) that has cypselas from the outer florets differing from those of the inner florets.
heterocarpous Of or relating to heterocarpy.
cf. **amphicarpic**

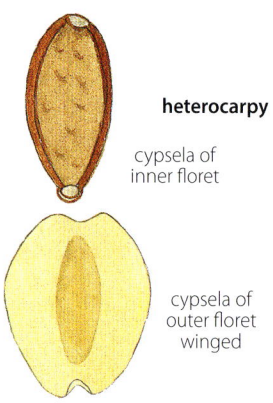

heterocarpy

cypsela of inner floret

cypsela of outer floret winged

Cotula vulgaris

heterochlamydeous, heterochlamydous
Having a perianth clearly differentiated into a calyx and and a corolla.
cf. **chlamydeous**

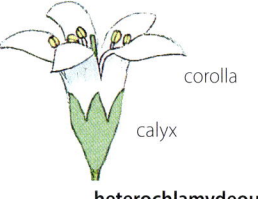

corolla

calyx

heterochlamydeous

heterochromous
Having different colours, as a daisy capitulum with the disc florets one colour and the ray florets another.
cf. **homochromous**

heterochromous

heterogamy Producing flowers of two or more kinds: bisexual, unisexual and/or neuter.
cf. **homogamy**
heterogamous Of an inflorescence with two or more kinds of flowers, as some in the daisy family (Asteraceae) that have bisexual disc florets and unisexual ray florets.
cf. **homogamous**

heterogeneous Not uniform in composition. Of ecology, having a rich variety of species, habitats, soil types etc.
cf. **homogeneous**

heterogonous Having flowers of a species with differing relative lengths of stamens and pistils.
cf. **heterostyly, homogonous**

heteromerous Having whorls with a different number of parts, as a flower with sepals, differing in number from the petals, carpels and/or stamens.
= **anisomerous**
cf. **isomerous**

heteromorphy
The condition of existing in different forms.
heteromorphic, heteromorphous
Having two or more distinct forms, as the stamens of cassia (*Cassia*).
cf. **homomorphic**

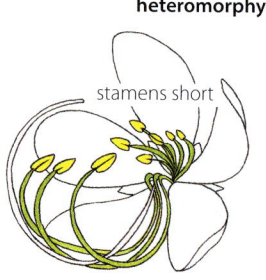

heteromorphy

stamens short

stamens long *Cassia*

heteromycotroph A plant that obtains nutrients partly as a mycotroph. It is usually not green due to inadequate photosynthesis.
heteromycotrophic Of or relating to a heteromycotroph.
see **trophic**

heterophylly Leaf form alteration in response to environmental conditions, as aquatic plants with underwater leaves differing from aerial or floating leaves.
cf. **heteroblasty**
heterophyllous Relating to heterophylly.
cf. **heteroblastic, homoblastic**

aerial leaf

heterophylly

Ranunculus

submerged leaf

heteropolar Of pollen grains in a tetrad having different proximal (facing towards the centre) and distal halves (facing away from the centre).
cf. **isopolar**

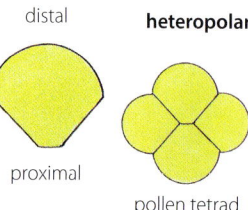

distal

heteropolar

proximal

pollen tetrad

heterosis   The tendency of cross-breeding to produce a plant with a greater hardiness than its parents.
= **hybrid vigour**

heterospory   Having male and female spores (microspores and megaspores) that develop in separate sporangia (microsporangia and megasporangia), as seed plants (gymnosperms and angiosperms) and some ferns and fern allies. Gametophytes from these spores are either male or female and produce either male or female gametes.
cf. **homospory**
heterosporous   Having different sporanagia that produce morphologically different types of spores.
cf. **homosporous**

heterostyly   Having styles of different lengths in flowers on the same plant, as the cowslip (*Primula vulgaris*).
see **distyly, pin, tristyly, thrum**
cf. **herkogamy, homostyly**
heterostylous
Exhibiting heterostyly.
cf. **homostylous**

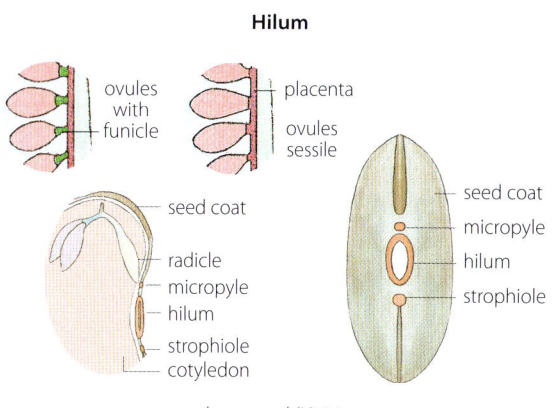
heterostyly
style

heterotroph   A living organism that gets its food from other plants or animals. Includes all animals, some fungi and most bacteria.
cf. **autotroph**
heterotrophic   Of or relating to a heterotroph.
see **trophic**

heterotypic synonym   In nomenclature, a name used with two or more names and two or more types for the same taxon.

heterozygous   Having two different alleles at a locus on a chromosome.
cf. **homozygous**

hexa-   A prefix meaning six.

hexamerous   Having flower parts, such as petals, sepals and stamens, in whorls of six or multiples of six. 6-merous.
see **-merous**

hibernal   Of or appearing in winter.
cf. **aestival, autumnal, vernal**

hierarchy   The successive levels at which taxa are classified.
see **rank, taxonomic hierarchy**

hilum, *pl.* **hila**   The scar on a seed coat that marks its point of detachment from the fruit wall.
It represents the point of attachment of the ovule to the placenta, either directly when it is sessile or by a stalk, the funicle, that is an extension of the ovule wall.
The central point in a starch grain around which starch is deposited.
hilar   Relating to a hilum.

Hilum

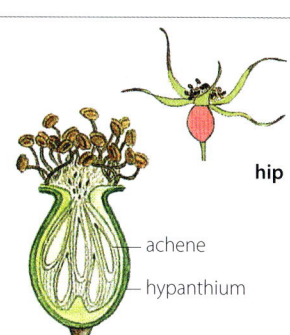
ovules with funicle
placenta
ovules sessile
seed coat
radicle
micropyle
hilum
strophiole
cotyledon
seed coat
micropyle
hilum
strophiole

bean seed (*Vicia*)

hip   The common name for the fruit of the rose genus (*Rosa*).
The hollow hypanthium contains the achenes that are the true fruit.
= **cynarrhodium**
see **accessory fruit**

hip
achene
hypanthium

hippocrepiform   Shaped like a horse-shoe.

hirsute   Rough due to a covering of coarse, longish hairs.
cf. **hispid, villous**

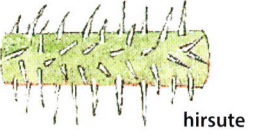
hirsute

hirsutulous   Slightly hirsute.

hirtellous   Minutely hirsute. With a covering of minute somewhat rigid hairs.

hispid   Rough due to a covering of erect stiff hairs or bristles.
cf. **hirsute**

hispid

hispidulous   Minutely hispid.

histology   Study of the cellular details of tissues.
histological   Of or relating to histology.

hoary   Pale silvery-grey.
Covered with a greyish to whitish layer of very short, closely interwoven fine hairs.
= **canescent**

holomycotroph   A plant that get its nutrients entirely from fungi and never carries out photosynthesis, as the underground orchid (*Rhizanthella*).
holomycotrophic   Of or relating to a holomycotroph.
see **trophic**

holophyte   An organism that synthesises its own food through photosynthesis, as green plants and some bacteria.
see also **autotroph**
holophytic   Of or relating to a holophyte.

holotype   The single specimen used to describe a new taxon.

homo-   A prefix meaning the same.
cf. **hetero-**

homoblasty   Having leaves that show a small and gradual change in appearance from the juvenile to adult condition.
cf. **heteroblastic**
homoblastic   Relating to homoblasty.

homoblastic pseudobulb
Of orchids having a pseudobulb with several internodes, as the pigeon orchid (*Dendrobium crumenatum*).
cf. **heteroblastic pseudobulb**

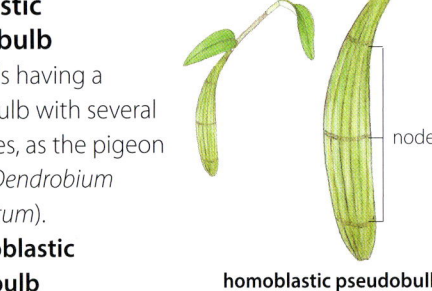
nodes
homoblastic pseudobulb

homocarpy   Having all flowers producing fruits that are of the one kind.
cf. **heterocarpy**
homocarpous   Of or relating to homocarpy.

homochlamydeous
Having a perianth composed of similar segments (tepals) and not clearly divided into a calyx and and a corolla.
cf. **chlamydeous**

homochlamydeous

homochromous
Alike in colour, as a daisy capitulum having all florets the same colour.
cf. **heterochromous**

homochromous

homogamy   Producing flowers of one kind.
Having the anthers and the stigma of a bisexual flower maturing at the same time.
Fertilisation of a flower by its own pollen or that of another flower on the same plant.
cf. **heterogamy**
homogamous   Of a plant or inflorescence having flowers of one kind, either bisexual, male or female.
Of a flower, having stamens and pistils that mature simultaneously, thus ensuring self-pollination.
cf. **heterogamous**

homogeneous   Of the same kind.
Of the ecology of an area, having a lack of variety in species, soil types and uniform habitats.
Applies mainly to man-made plant communities.
Few if any natural landscapes are composed of homogenous communities.
cf. **heterogeneous**

homogonous   Having all flowers of a plant alike in respect of the stamens and pistils.
cf. **heterogonous, homostyly**

homologous   Similar in form or function.
In genetics, of or relating to a homologue.
In phylogenetics, of similar characters in two different taxa that can be attributed to their presence in a common ancestor, as the pentadactyl limb in humans and whales.
cf. **analogous**

homologue   One of the corresponding chromosomes from the sperm and the egg that are similar because they carry the same genes in the same locations but the versions of the genes differ. Homologous chromosomes in pairs are diploid (2n), in threes are triploid (3n), in fours are tetraploid (4n) and so on.

homomorphy
Similarity of form.
homomorphic, homomorphous
Having the same kind or form, as the stamens of many different flowers.
cf. **heteromorphic**

stamen

homomorphy

homonym   In nomenclature, two or more taxa that have been given the same scientific name. The accepted name is usually the first name published and an alternative name is given to the other taxon.

homophylly   Having leaves all of the same kind.
cf. **heterophylly, isophylly**
homophyllous   Relating to homophylly.

homoplasy   In phylogenetics, similar characters that have evolved from different ancestral sources, as the evolution of eyes in very different groups like vertebrates and octopuses.
= **convergent evolution, parallel evolution**
cf. **plesiomorphy**
homoplastic   Of or related to homoplasy.

homospory   Having morphologically similar, equal-sized spores within the same sporangium, as most bryopytes and pteridophytes.
All spores germinate and develop into free-living gametophytes that are bisexual, as a thallus with male gametes (in antheridia) and female gametes (in archegonia).
cf. **heterospory**
homosporous   Having a sporanagium with morphologically similar spores that are not differentiated by sex.
cf. **heterosporous**

homostyly   With styles of a constant length and shape in flowers of the same species.
cf. **heterostyly**

homostylous   Exhibiting homostyly.
cf. **heterostylous**

homotypic synonym   In nomenclature, a name used when there are two or more names but only one type for a taxon.

homozygous   Having two identical alleles at a locus on a chromosome.
cf. **heterozygous**

hood   A hollow arched covering, as the upper sepal of monkshood (*Aconitum napellus*).
= **cucullus**

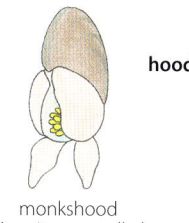

hood

monkshood
(*Aconitum napellus*)

hormone   Of plants, a chemical substance, released in small quantities, that regulates growth and development.
see **phytohormone**
cf. **pheromone**

horn   A curved, pointed appendage shaped like an animal's horn, as the appendages on the lobes of the corona of a milkweed (*Asclepias*).
horned   Having a horn.
see **corniculate**

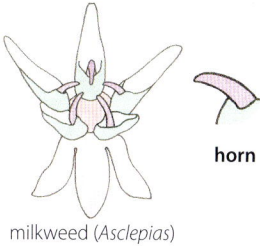

horn

milkweed (*Asclepias*)

hornworts   *see* page 138

horny   With a hard, smooth texture, as the leaf margin of some agaves.
Having an incurved, tapering appendage like the horns of cattle.
= **corneous**

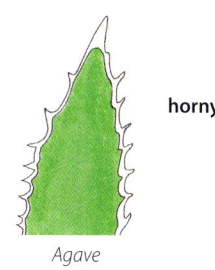

horny

Agave

hornworts   One of three major groups of nonvascular land plants (bryophytes) that are photosynthetic, reproduce by spores and lack flowers, true leaves and roots.

A hornwort has two alternating generations, the larger haploid gamete-producing gametophyte generation and the smaller diploid spore-producing sporophyte generation that has a horn-like capsule that remains attached to the female gametangium (archegonium).

The haploid gametophyte is a thallus, with gamete producing sex organs (the female archegonia and/or the male antheridia) that are partially or completely embedded in the upper surface.

The diploid sporophyte lacks a seta and comprises a horn-like capsule and a foot embedded in the thallus.

A male sperm from an antheridium fertilises a female egg in an archegonium, that then develops into a diploid spore-producing sporophyte.

The diploid spores in the capsule undergo meiosis to form haploid spores that are dispersed and develop into gametophytes, thus beginning the life cycle anew.

see **bryophyte**

Hornworts

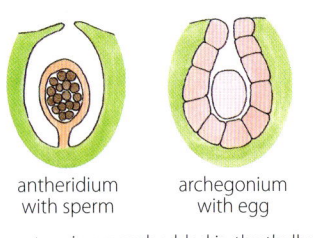

capsule splits to release spores

horn-like capsule

sporophyte

seta absent

foot embedded in thallus

gametophyte

thallus

root-like rhizoids

antheridium with sperm

archegonium with egg

gametangia are embedded in the thallus

Alternation of generations

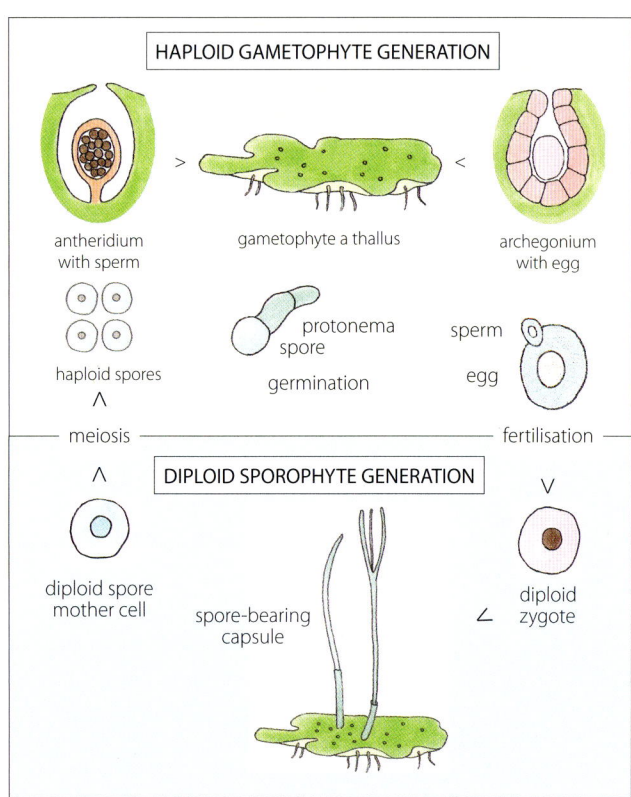

HAPLOID GAMETOPHYTE GENERATION

antheridium with sperm

gametophyte a thallus

archegonium with egg

haploid spores

protonema spore

germination

sperm

egg

meiosis

fertilisation

DIPLOID SPOROPHYTE GENERATION

diploid spore mother cell

spore-bearing capsule

diploid zygote

---

horsetail   The horsetail family (Equisetaceae) comprises one genus (*Equisetum*) of vascular plants that reproduce by spores rather than seeds. Stems are erect, jointed and grooved with whorls of branches and leaves. Spores are homosporous and borne in cone-like strobili.

see **fern allies**

horsetail

horsetail (*Equisetum*)

hort., horti   When placed after a taxon name, it indicates the use of that name in horticulture.

horticulture   The cultivation of commercial or domestic garden crops, usually fruits, vegetables and ornamental plants.

hose-in-hose Of a flower with a petaloid calyx and the tube of the corolla inserted in the tube of the calyx so that the flower appears to be double, as primrose (*Primula*).
see **double**

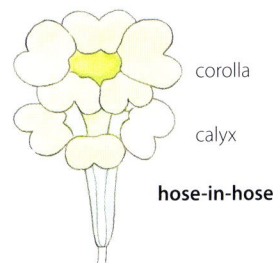

corolla

calyx

hose-in-hose

host A plant that sustains another plant or animal species. It can involve mutualism, commensalism or parasitism, as species of eucalypt (*Eucalyptus*) and wattle (*Acacia*) that host the parasitic harlequin mistletoe (*Lysiana exocarpi*).

host

mistletoe

hull The outer covering of some fruits and seeds, as the pod of peas or the husk of a grain. The persistent calyx of a strawberry.

hummock grass Tussock-forming grass with spiny leaves in the genus *Triodia* that are native to dry inland habitats in Australia.

humus That part of the soil formed by the decomposition of animal or vegetable matter.

husk The dry outer covering of some fruits and seeds, as the foliaceous leaves on a corn cob or the fibrous covering of a coconut.

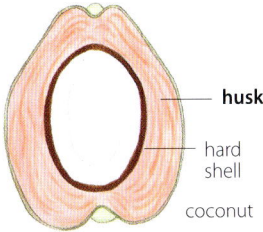

husk

hard shell

coconut

hyalescent Somewhat hyaline.

hyaline Thin, colourless and translucent, almost like clear glass, as the margins of some bracts and leaves.

hyaloplasm A clear substance that makes up most of the volume of the cytoplasm.
= **cytosol, ground substance**

hybrid The offspring from a cross between parent plants of different varieties, subspecies, species or genera.
= **cross**
cf. **purebred**

hybrid formula The names of the parents of a hybrid linked by a multiplication sign, as *Geranium macrorrhizum* x *G. dalmaticum*.

hybrid vigour The tendency of cross-breeding to produce a better plant than its parents.
= **heterosis**

hybridisation The crossing of parent plants from different varieties, subspecies, species or genera to produce offspring called a hybrid.

hydathode A secretory structure, as glands or pores, that removes excess water in plants when transpiration is negligible and moisture in the roots is high. Hydathodes occur in some herbaceous land plants like grasses and are widespread in aquatic plants.
They are usually located on the margins, surfaces and tips of leaves.
see **guttation**

hydathode

water purslane
(*Neopaxia australasica*)

hydric Of, relating to or adapted to an environment that receives abundant amounts of water, as a wetland.
cf. **mesic, xeric**

hydro- A prefix meaning water or containing hydrogen.

hydrochasy Movement caused by the absorption of water in plant parts that are mostly dead. Results in the opening of capsules after rain in most of the succulent plant family Aizoaceae.
hydrochasy page 140 (cont.)
= **hygrochasy**
cf. **xerochasy**
hydrochastic Of or related to plant movement caused by the absorption of water.
= **hygrochastic**

Hydrochasy

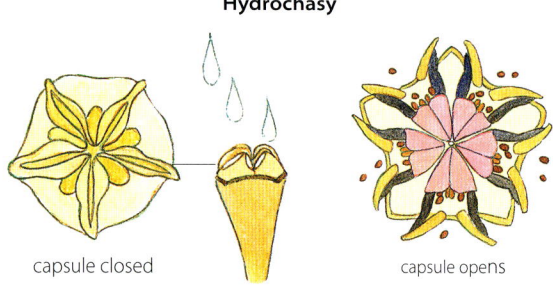

capsule closed   capsule opens

rounded noon-flower (*Disphyma crassifolium*)

hydrochore A plant that depends on water for the dispersal of its pollen, spores, seeds or fruit.

hydrochory Dispersal of pollen, spores, seeds or fruit by water.
cf. **hydrogamy, hydrophily**
hydrochorous Of or relating to hydrochory.

hydrogamy Dispersal of pollen and pollination by water.
= **hydrophily**
cf. **hydrochory**
hydrogamous Of or relating to hydrogamy.

hydrolysis A chemical reaction requiring water in which an enzyme breaks a larger molecule into smaller subunits, as starch, a polysaccharide, is broken down into maltose, a disaccharide.

hydrophile A plant species that is pollinated by water.

hydrophily Dispersal of pollen and pollination by water.
= **hydrogamy**
hydrophilous Pollinated by water-borne pollen.

hydrophyte
An aquatic plant that grows in water or waterlogged soil. It may be anchored or floating and has perennating buds below water level.
see also **cryptophyte**
cf. **hygrophyte, mesophyte, xerophyte**

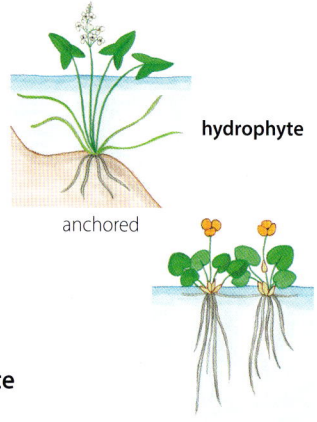

hydrophyte

anchored

floating

hydrosere An ecological succession that starts in an aquatic habitat.

hydrotropism Growth or movement of a plant towards or away from water.
see **tropism**
hydrotropic Of or relating to hydrotropism.

hygrochasy Movement caused by the absorption of water in plant parts that are mostly dead.
= **hydrochasy**
cf. **xerochasy**
hygrochastic Of or related to plant movement caused by the absorption of water.
= **hydrochastic**

hygrophile A plant living and thriving in moist places.
hygrophilous Thriving in moist places.

hygrophyte A plant adapted to waterlogged soil and a damp atmosphere, as marsh plants.
cf. **hydrophyte**
hygrophytic Needing wet or waterlogged soil to grow in.

hygroscopic Gaining or losing moisture depending on the surrounding humidity. Expanding when water is present and contracting in its absence causing changes in movement, as the opening and closing of some grass leaves or the teeth at the top of a moss capsule that open to release spores.
see **bulliform cell**

hypanthium
A tubular or cup-shaped enlargement of the receptacle and/or the bases of the floral parts. It bears the petals, sepals and stamens. It is either above the ovary, below the ovary and free, or variously united with it.
hypanthial Relating to the hypanthium.
= **floral tube**
cf. **receptacle**

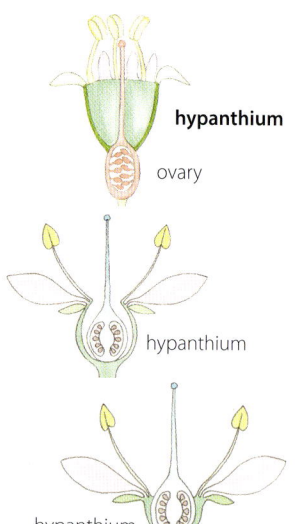

hypanthium

ovary

hypanthium

hypanthium

hypanthodium   Characteristic inflorescence of the fig genus (*Ficus*).
The fleshy, hollow receptacle is lined with numerous sessile flowers, the flowers being male, female, or female and sterile (gall flowers).
The female flowers develop into numerous fruitlets.
see **syconium**

Hypanthodium

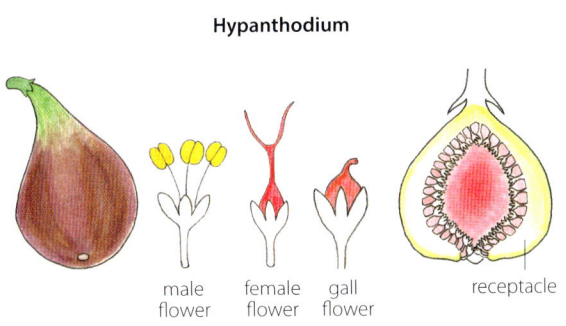

male flower   female flower   gall flower   receptacle

hyper-resupinate   In some orchids, of flowers twisted at more than 180º.

hypereutrophic   Of a nutrient rich body of water, with problems arising from lack of oxygen due to excessive plant and algal growth.
see **trophic**
cf. **eutrophic**

hyperphyll   Of palms with remote germination, and some other monocotyledons, the first structure to emerge from the seed.
It grows down into the soil, forming a swelling from which the first seedling root (radicle) and the plumular leaves emerge.
= **cotyledonary petiole**

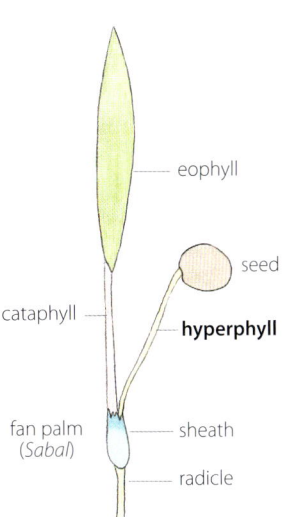

eophyll

seed

cataphyll

hyperphyll

fan palm (*Sabal*)

sheath

radicle

hyperstomatous   Of leaves with stomata only on the upper surface.
= **epistomatous**
cf. **amphistomatous, hyperstomatous**

hypertonic   Having the higher osmotic pressure of two fluids.
cf. **hypotonic, isotonic**

hypha, *pl.* **hyphae**   Long branched thread-like filaments that make up a fungus.
Some give rise to erect stalks with fruiting bodies like mushrooms or toadstools.
The network of hyphae (a mycelium) may be above ground, as bread moulds, or underground.
It extends long distances underground and can carry nutrients to roots that would otherwise be inaccessible.
see also **mycorrhiza, osmotrophy**

hyphal   Relating to or like hyphae.

Hyphae

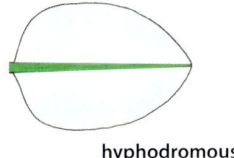

fruiting body   hyphae

mushroom

hyphodromous
Of leaves with a distinct midrib but other veins absent or concealed within the fleshy leaf.

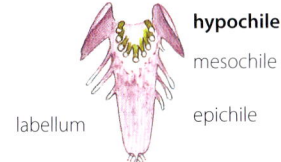

hyphodromous

hypo-   A prefix meaning beneath or under.

hypochile   Of orchids, the basal portion of the labellum, as pink fairies (*Caladenia latifolia*).

hypochile
mesochile
labellum
epichile

hypocotyl   Of a seed, in eudicots and most monocotyledons, the part of the embryo axis below the cotyledonary node and above the radicle.
In grasses, (Poaceae) that part of the embryo axis below the mesocotyl that is the embryonic root (radicle) enclosed in the protective sheathing coleorrhiza.
Of a seedling, the part of the axis below the cotyledon(s) and above the roots.
hypocotyl page 142 (cont.)
cf. **epicotyl, hypocotyl-root axis, mesocotyl**

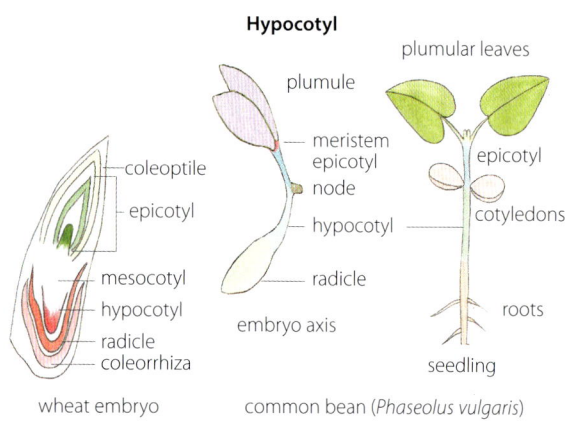

Hypocotyl

wheat embryo

- coleoptile
- epicotyl
- mesocotyl
- hypocotyl
- radicle
- coleorrhiza

embryo axis

- plumular leaves
- plumule
- meristem epicotyl node
- hypocotyl
- radicle

common bean (*Phaseolus vulgaris*)

seedling

- epicotyl
- cotyledons
- roots

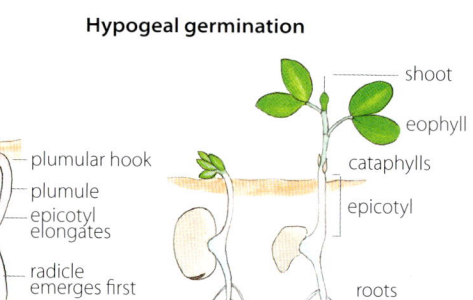

Hypogeal germination

seed

- plumular hook
- plumule
- epicotyl elongates
- radicle emerges first

- shoot
- eophyll
- cataphylls
- epicotyl
- roots

bean (*Vicia*) hypogeal germination

hypocotyl-root axis

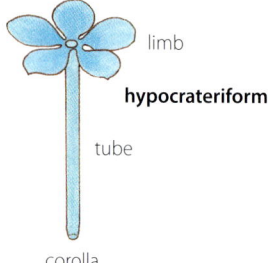

Of the embryo of a seed, the hypocotyl and radicle if they are not distinguishable from one another.

rock cress (*Arabidopsis*)

- cotyledons
- plumule
- **hypocotyl-root axis**
- root meristem

hypocrateriform

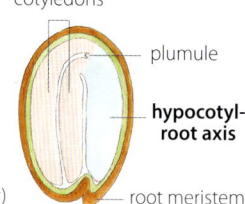

Of a corolla with a long slim tube and an abruptly expanded limb spreading at 90° to the tube, as some primroses (*Primula*).

= **salverform**

- limb
- **hypocrateriform**
- tube
- corolla

hypodermis
One or more layers of cells lying immediately below the epidermis in plants. Unlike the epidermis, it is derived from a ground tissue (collenchyma) and is sometimes modified to give structural support.

hypogeal, hypogeous
Below the ground.
see **epigeal**

hypogeal germination, hypogeous germination
Of seed germination, the cotyledons in the seed remain underground while the epicotyl together with the plumule grows upward out of the ground. During this process the plumule is protected by a plumular hook. Common in monocotyledons and found in some eudicots. Characteristic of grasses (Poaceae).
cf. **epigeal germination, viviparous germination**

hypogenous
Produced or growing on the undersurface, as fern spores growing on the undersurface of a leaf.

hypogyny
The position of the ovary when it is superior relative to the whorls of stamens, petals and sepals.
hypogynous Below the ovary.
cf. **epigynous, perigynous**

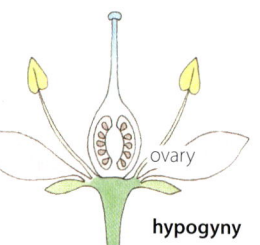

- ovary
- **hypogyny**

hyponasty
Increased growth along the lower surface of a plant part causing it to bend upward.
cf. **epinasty**
hyponastic Of or relating to hyponasty.

hypophysis,
pl. **hypophyses** The enlarged neck between the base of a moss capsule and the top of the seta .
= **apophysis**

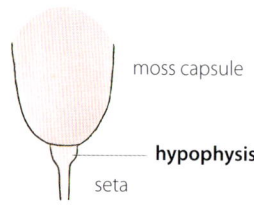

- moss capsule
- **hypophysis**
- seta

hypotonic
Having the lower osmotic pressure of two fluids.
cf. **hypertonic, isotonic**

hypotony
Development of lateral growth on the upper side of the main shoot.
cf. **amphitony, epitony**

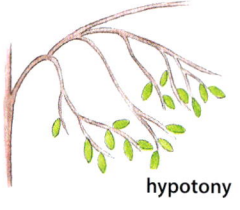

hypotony

hypotropous
Of ovule orientation, with the micropyle distal with reference to the funicle, as orthotropous (atropous) and circinotropous.
cf. **epitropous, pleurotropous**

hysteranthous
Of leaves appearing after flowers.
cf. **precocious, proteranthous, synanthous**

ICN The International Code of Nomenclature for algae, fungi and plants formerly, the International Code of Botanical Nomenclature (ICBN), is the set of rules for naming plants.

ICNP The International Code of Nomenclature for Cultivated Plants is the set of rules for naming cultivated plants.

identification Description and naming of an unknown species by comparing its characteristics with a known one, or recognising that it is new and warrants formal description and naming.
Identifying a species, usually by using a taxonomic key.
see **determination, dichotomous key**

idioblast An isolated plant cell that differs greatly from the surrounding cells or tissue, as a sclereid in pears or raphides in a cell.

illegitimate name A published name that does not accord with the rules of the International Code of Nomenclature.
= **nomen illegitimum**

imbricate Overlapping like tiles or shingles on a roof, as leaves along a stem. Having edges overlapping, as in an unopened leaf or flower bud.

imbricate

imbricate aestivation
The arrangement of petals, tepals or sepals in a bud with the margins overlapping but not involute.

imbricate aestivation

imbricate vernation
The arrangement of young leaves in an unopened leaf bud with the margins overlapping but not involute.

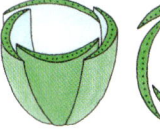 

imbricate vernation

imbricate bulb
A true bulb that consists of a compressed stem with nodes (basal plate), bearing fleshy overlapping leaves that lack a tunic, as the lily genus (*Lilium*).
= **naked bulb, scaly bulb**
see also **tunicate bulb**

fleshy leaves

imbricate bulb

immature Not yet fully developed, not ripe.
cf. **mature, senescent**

immersed Partially or wholly sunken into the surrounding parts, as seeds of pittosporum (*Pittosporum*) in a sticky pulp.
Growing under water, entirely submerged, as the aquatic plant dwarf eelgrass (*Zostera noltii*).
cf. **submerged**

seeds **immersed** in pulp

immersed in water

imparipinnate
Of a pinnate leaf with leaflets opposite or alternate, and terminating with a single leaflet.
= **odd-pinnate**

imparipinnate

imperfect Of leaf venation when lateral veins extend for less than two-thirds of the leaf surface.
Of a flower with either stamens or pistils fertile but not both.
Of a flower lacking either stamens or pistils.
= **unisexual**
cf. **perfect**

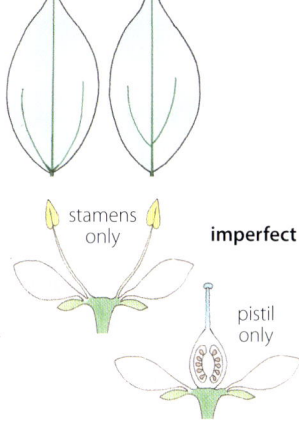

stamens only

imperfect

pistil only

impressed Sunken into the surface, as the veins of some leaves.

impressed

in- A prefix meaning absent.

inaperturate Of a pollen grain or a spore without apertures.
cf. **alete**

inbreeding The production of seeds between plants that are closely related genetically.
see also **autogamy**
cf. **outbreeding, outcrossing**

incertae sedis   A taxon or group of taxa whose relationship with others is unknown or uncertain.

incised   Cut deeply, sharply and unevenly. Of a margin, intermediate between toothed and lobed, as the margins of some leaves.

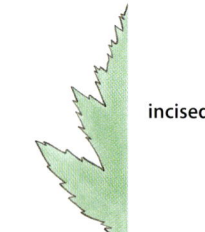

incised

included   Enclosed. Not protruding, as stamens within the corolla.
cf. **exserted**

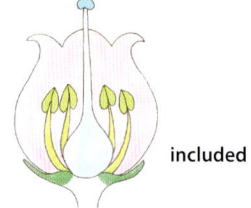

included

incompatible   Incapable of self-fertilisation or cross-fertilisation. Incapable of forming a successful graft.
cf. **compatible**

incomplete   Of a flower with one or more of the four whorls (sepals, petals, stamens and/or pistils) lacking.
cf. **complete**

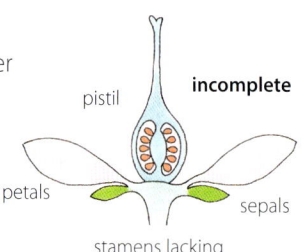

incomplete
pistil
petals
sepals
stamens lacking

incomplete dominance   The partial expression of a dominant allele that allows some expression of the recessive allele and results in an alternative phenotype with traits of both alleles.

incrassate   Thickened or swollen. Becoming thicker by degrees, as the peduncle of the egg plant (*Solanum melongena*).

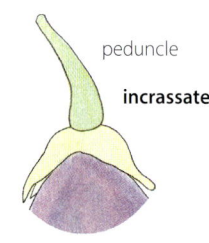

peduncle
incrassate

incubous   Of leafy liverworts, having leaves attached to the stem obliquely so that the upper margin of each leaf overlaps the base of the leaf above it. The new leaf begins under the older one.
cf. **succubous**

incubous

incumbent   Of cotyledons that are folded with the back of one lying against the hypocotyl-shoot axis, as mustard (*Erysimum cheiranthoides*).

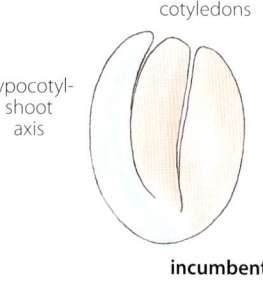

cotyledons
hypocotyl-shoot axis
incumbent

incurved   Curved or bent upward or inward. Of leaf margins curved towards the adaxial side.
cf. **involute, recurved, revolute**

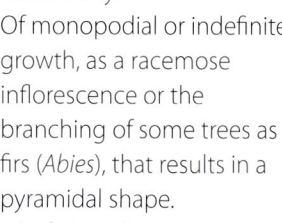

incurved
leaves   leaf margin

indefinite   Having a persistent terminal growing point and with growth similar on lateral branches. The pattern of growth in which the apex of the main stem and branches continues to grow indefinitely. Of monopodial or indefinite growth, as a racemose inflorescence or the branching of some trees as firs (*Abies*), that results in a pyramidal shape.
= **indeterminate**
cf. **definite**

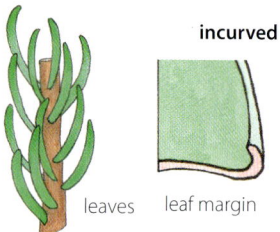

indefinite
racemose inflorescence

fir (*Abies*)

indehiscence   Of a plant part not opening at maturity to release its contents, as an indehiscent fruit, like a berry or stone fruit.
indehiscent   Not opening when ripe.
cf. **dehiscent**

indeterminate   Having a persistent terminal growing point and with growth similar on lateral branches. The pattern of growth in which the apex of the main stem and branches continues to grow indefinitely. Of monopodial or indefinite growth, as a racemose inflorescence or the branching of some trees as firs (*Abies*), that results in a pyramidal shape.
= **indefinite**
cf. **determinate**

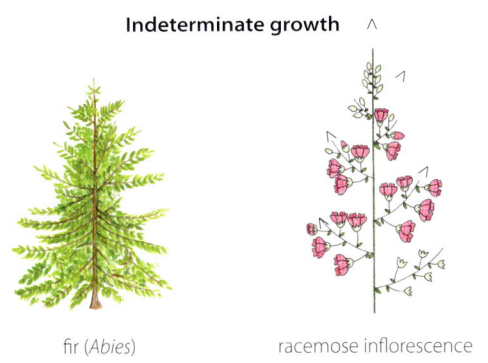

Indeterminate growth

fir (*Abies*)   racemose inflorescence

indigenous   Growing naturally in a particular place, not introduced.
= **native**

indument, indumentum   Specifically, the nature of the hairs on the surface of a plant or plant part. Generally, any type of hairiness, scaliness or other covering on the surface of a plant.
see also **vestiture**

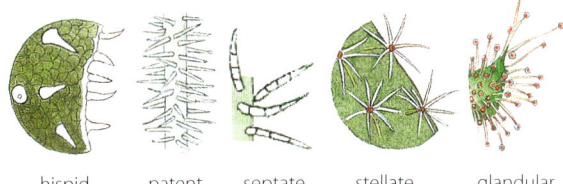

Some indumentum hairs

hispid   patent   septate   stellate   glandular

induplicate   Of palm leaflets with margins bent inwards, V-shaped in cross-section.
cf. **reduplicate**

induplicate

induplicate vernation
Of young leaves in the unopened leaf bud folded inward and arranged in a circle with the margins touching but not overlapping.

induplicate vernation

indurate   Hardened, as a thorn or the lemma and palea of some grass flowers.

thorn
indurate

indusium, *pl.* **indusia**   Of flowers, a cup-like structure at the tip of the style that collects pollen as it passes through the anther tube. A characteristic of the fan-flower family (Goodeniaceae).
Of ferns, a membranous or scale-like shield covering the sorus on some fronds. It is variously shaped, as round (*Rumohra adiantiformis*), kidney-shaped (*Dryposis apiciflora*) or linear (some *Blechnum* species).
cf. **false indusium**
indusiate   Having an indusium.
cf. **exindusiate**

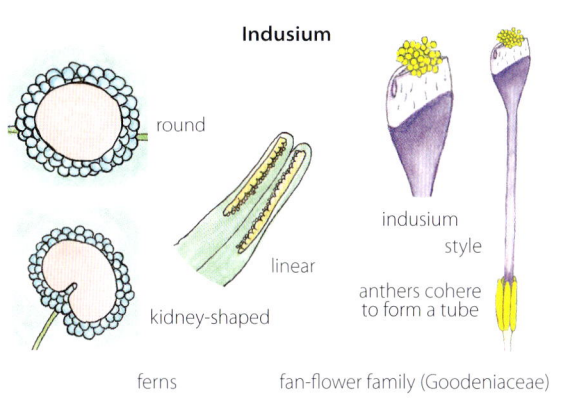

Indusium

round

kidney-shaped
linear

indusium
style

anthers cohere
to form a tube

ferns   fan-flower family (Goodeniaceae)

ined., ineditus   Unpublished.

ineditus, *abbr.* **ined.**   Unpublished.

inermous   Lacking thorns spines or prickles.
= **unarmed**
see **armature**

inferior ovary   An ovary that is embedded in the hypanthium. An ovary that is below the level of insertion of the floral parts on the hypanthium.
cf. **semi-inferior ovary, superior ovary**

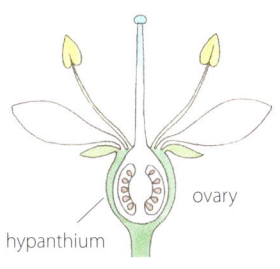

ovary

hypanthium

inferior ovary

infertile   Unable to reproduce, sterile.
Of land that is unproductive.
cf. **fertile**

inflated   Filled or expanded by, or as if by, gas or air, as the enlarged calyx of Chinese lantern (*Physalis alkekengi*).
cf. **saccate**

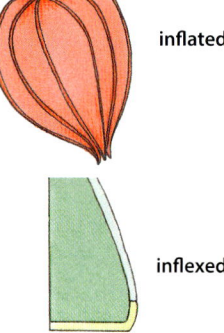

inflated

inflexed   Bent abruptly inwards, as margins of some fern fronds.

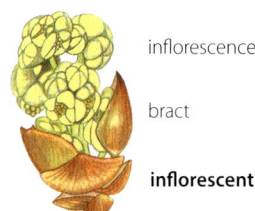

inflexed

inflorescence   *see* pages 147–148

inflorescent   Of or relating to blooming or flowering, as inflorescent shoots or the inflorescent bracts of sweet wattle (*Acacia suaveolens*).
cf. **inflorescence**

inflorescence

bract

inflorescent

infra-   A prefix meaning below or beneath.
cf. **supra-**

infrafoliar   Of an inflorescence produced below the crown of leaves, as some palms (Arecaceae).
cf. **interfoliar, suprafoliar**

infrafoliar inflorescence

infrageneric   Of a division within a genus, such as subgenus, section or series.

infraspecific   Of a division within a species, such as subspecies, variety, form or cultivar.

infratectum   A layer between the tectum and the foot layer (or endexine if the foot layer is missing).
see **pollen wall**

infravaginal
Of branching that occurs below the point of attachment of the leaf sheath, as in some bamboos (Poaceae).
cf. **extravaginal, intravaginal**

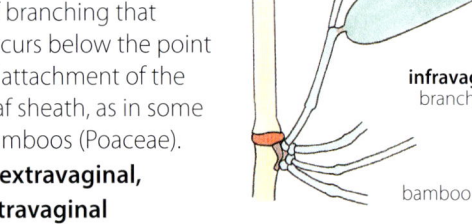

infravaginal branching

bamboo

infructescence   The mass of fruits derived from an entire inflorescence, as a pineapple (*Ananas comosus*).
see **compound fruit**

infundibular, infundibuliform
Funnel-shaped, as the corolla of bindweed (*Convolvulus*).

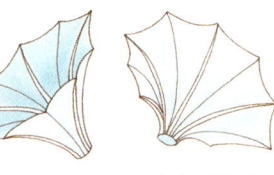

infundibular

ingroup   In phylogenetics, the group of taxa being studied.
cf. **outgroup**

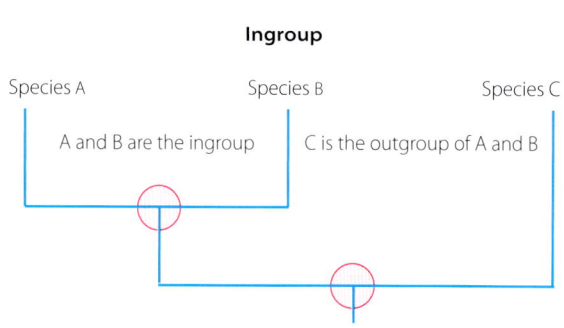

Ingroup

Species A   Species B   Species C

A and B are the ingroup   C is the outgroup of A and B

inheritance   The passing on of genes that determine traits, from parents to offspring, by means of sexual or asexual reproduction.
= **heredity**

inherited character   A feature that is passed on from the chromosomes of parents to the chromosomes of their offspring.
cf. **acquired character**

inhibitor   A chemical substance that prevents a growth process, as the hormone ethylene that causes lateral buds to remain dormant while the apex of the stem is growing.
cf. **activator**

initial cell   An actively dividing plant cell in a meristem. At each division one daughter cell remains in the meristem as an initial cell and the other will continue to divide and differentiate to become one of the three primary tissues: epidermis, ground tissue or vascular tissue.
see also **ground meristem, procambium, protoderm**

inflorescence   The arrangement of flowers on a plant.

The flowering part of a plant including stems, stalks, bracts and flowers.

The process of flowering.

There are two main types of inflorescence. In a racemose or indeterminate inflorescence the main axis and lateral branches continue to grow indefinitely, with flowers arising along the axis and branches. In a cymose or determinate inflorescence the main axis and lateral branches end in a flower.

An inflorescence may be simple or compound, terminal, axillary or intercalary.

see **anthotaxy**

see also **cauliflory, ramiflory**

cf. **inflorescent**

Inflorescence

Racemose Indefinite/Indeterminate

| spike | compound spike | raceme | compound raceme | panicle | corymb | compound corymb |

| umbel | compound umbel | head | compound head | ament | spadix |

Cymose Definite/Determinate

| solitary flower | cyme/dichasium | compound cyme/dichasium | pleiochasial cyme/pleiochasium |

| scorpioid cyme/cincinnus | helicoid cyme/bostryx | drepanium | rhiphidium | verticillaster | fascicle | glomerule |

Mixed inflorescence

mixed panicle

thyrse

Special types of inflorescence

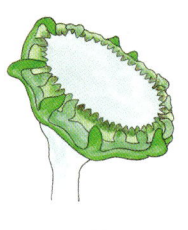

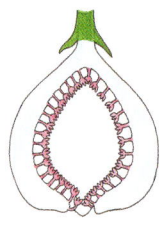

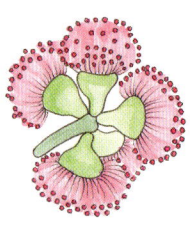

coenanthium

syconium

cyathium

umbellaster

Inflorescence of grasses, rushes and sedges

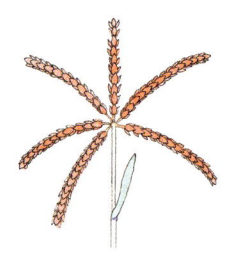

spikelet/locusta

spike

compound digitate spike

raceme

panicle

tassel

ear

anthelodium

anthela

innate   Attached at or by the base, as a stamen filament attached to the base of the anther.
= **basifixed**
see **anther attachment**

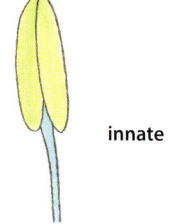

innate

innocuous   Of a plant structure or substance, harmless. Lacking the capacity to injure, as a stem without spines, thorns, prickles and the like.
cf. **inermous**

innovation   A new shoot, especially in plants with intermittent growth.
Of mosses, a new shoot that becomes independent by dying off where it joins the parent plant.

inorganic   Of chemical compounds that do not contain carbon.
Not composed of living or once living matter.
cf. **organic**

inrolled   Of a margin with the edges rolled inwards towards the upper surface.
= **involute**

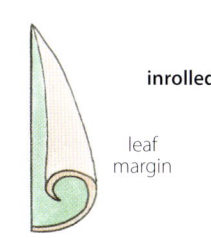

inrolled

leaf margin

insectiform   Having the appearance of an insect, as the labellum, a modified petal, of the elbow orchid (*Thynninorchis huntianus*), that mimics the female of the pollinating wasp species (*Arthrothynnus huntianus*).

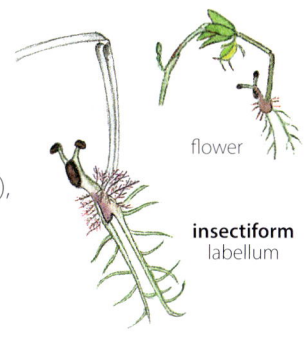

flower

insectiform labellum

insectivorous
Of plants, with adaptations to capture and digest insects, as the leaves of sundews (*Drosera*). The sticky stalked glands capture insects and drown them in digestive secretions.
see also **carnivorous**

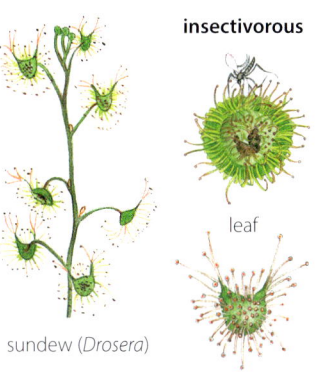

insectivorous

leaf

sundew (*Drosera*)

insertion   The point of attachment of one part to another.
inserted   Attached to or arising from, as flower parts inserted on a receptacle.

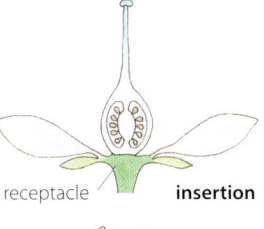

receptacle

insertion

flower

integument   Of an ovule, one or two protective layers surrounding the nucellus and having an opening at the micropyle.
Monocotyledons and most eudicots are bitegmic and have two integuments, and gymnosperms are unitegmic and have one integument.
The integuments usually give rise to the seed coat.
see **tegmen, testa**

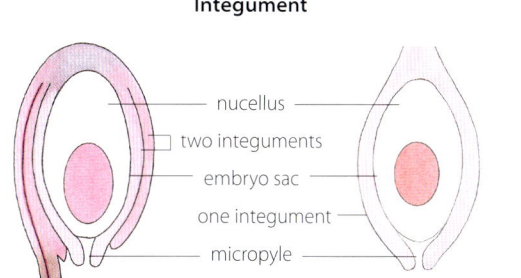

Integument

nucellus

two integuments

embryo sac

one integument

micropyle

bitegmic angiosperm ovule   unitegmic gymnosperm ovule

inter-   A prefix meaning between or among.
cf. **intra-**

intercalary   Inserted between two parts, as the intercalary meristem between the leaf blade and the sheath in grasses.
see **intercalary inflorescence**
cf. **axillary, terminal**

intercalary inflorescence
Of an inflorescence appearing to be in the middle of a leafy stem due to the axis continuing to grow vegetatively after producing flowers, as the bottlebrush genus (*Melaleuca*). Inflorescences may also be terminal or axillary.
see **auxotelic**

bottlebrush (*Melaleuca*)

intercalary inflorescence

intercalary meristem
A band of meristem interposed between two mature tissues enabling new tissue to be inserted in a longitudinal direction, as between the blade and the sheath of a grass leaf.
In grasses, it has two zones, the upper that is responsible for growth of the blade and the lower for growth of the sheath, with the upper zone eventually differentiating into the leaf collar.
It is present in many stems and at the base of a leaf, where it functions for leaf development, and in the abscission layer of deciduous leaves.
see also **apical meristem, collar, lateral meristem**

intercellular spaces

Air spaces between adjacent cells, as in parenchyma.
see also **lacuna**

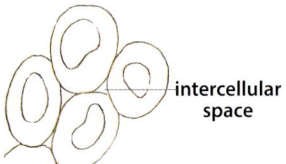

intercellular space

intercostal

Between the ribs.
The surface between the ribs of a leaf.

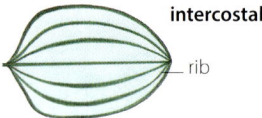

intercostal

rib

interfascicular Between the vascular bundles.

interfascicular cambium Vascular cambium that forms in secondary growth between the vascular bundles.
It forms from the interfascicular parenchyma in primary growth.
Vascular cambium and interfascicular cambium eventually join to form a continuous ring.

Interfascicular cambium

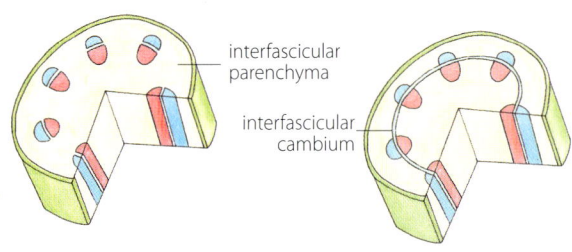

interfascicular parenchyma

interfascicular cambium

eudicot primary stem

eudicot secondary stem

interfoliar Of an inflorescence produced within the crown of leaves, as some palms (Arecaceae).
cf. **infrafoliar, suprafoliar**

interfoliar inflorescence

intergeneric Existing or occurring between different genera.

intergeneric hybrid A hybrid between two different genera.
The orchid genus x *Brassocattleya* is a hybrid between the genera *Brassovola* and *Cattleya,* with *Brassocattleya Mendelosa* being a hybrid between *Brassavola nodosa* and *Cattleya mendelii.*
= **bigeneric hybrid**

International Code of Nomenclature

The International Code of Nomenclature for algae, fungi and plants (ICN), formerly the International Code of Botanical Nomenclature (ICBN), is the set of rules for naming plants.
There is an additional set of rules for naming cultivated plants, the International Code of Nomenclature for Cultivated Plants.

International Code of Nomenclature for Cultivated Plants The set of rules for naming cultivated plants.

International Code of Phylogenetic Nomenclature A developing draft for a formal set of rules governing phylogenetic nomenclature.
see **PhyloCode**

International Union for Conservation of Nature, IUCN The chief organisation for standardising threat status of plants and animals throughout the world.

internode The part of the stem between two nodes (joints), as that on the stem of a bamboo or between one leaf base and the next.
see also **node**

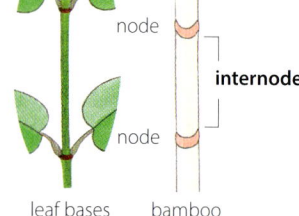

node

node

internode

leaf bases

bamboo

interpetiolar Between the petioles of opposite leaves.
Of stipules of opposite leaves fused to form one stipule on each side of the stem.

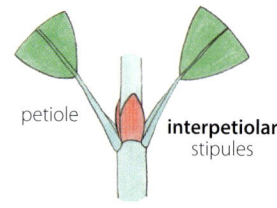

petiole

interpetiolar stipules

interrupted Not continuous in arrangement, unevenly distributed.
cf. **continuous**

interruptedly pinnate

Of a pinnate leaf with spaces along the rachis between the leaflets and the leaflets unevenly distributed, with smaller leaflets among larger leaflets.

interruptedly pinnate

interspecific   Existing or occurring between different species.

interspecific hybrid   A hybrid between two different species belonging to the same genus. *Geranium* x *cantabrigiense* is a hybrid between *G. macrorrhizom* and *G. dalmaticum*.
= **bispecific hybrid**

interstaminal   Situated between the stamens, as the staminodes of some flowers.

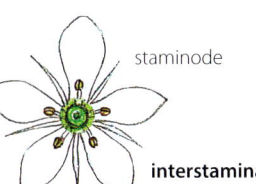
staminode
interstaminal

interstice   A small space between parts.
  interstitial   Relating to or located in one or more interstices.

intertidal   The zone of a seashore that is submerged during high tide and exposed during low tide.

intervallecular
Situated in a furrow, as the oil glands in furrows between the ribs of some fruit of the carrot family (Apiaceae).
see **vallecula**
cf. **intrajugal**

oil ducts **intervallecular**

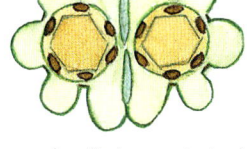
sea celery (*Apium prostratum*)

intine   The innermost layer of the wall of a pollen grain that is next to the cytoplasm.
see **pollen wall**

intra-   A prefix meaning within.
cf. **inter-**

intrajugal   Within the ribs.
cf. **intervallecular**

intramarginal
Situated inside, but close to, the margin, as an intramarginal vein on a leaf .
cf. **marginal venation**

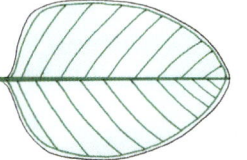
intramarginal

intrapetiolar   Between the petiole and the stem. Of a pair of stipules that fuse to form one stipule that is between the petiole and the stem.

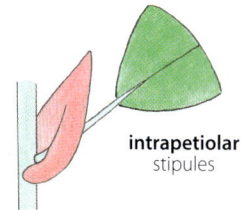
intrapetiolar
stipules

intrastaminal   Situated inside the whorl of stamens, as the nectary disc of some boronia (*Boronia*).

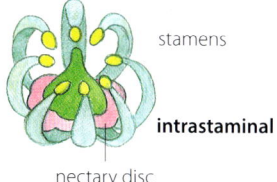
stamens
intrastaminal
nectary disc

intravaginal
Of branching, wih the young shoot growing up inside the leaf sheath, as some grasses (Poaceae).
cf. **extravaginal,**
infravaginal

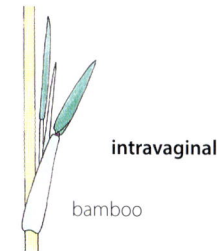
intravaginal
bamboo

intricate   Tangled, as some roots and hairs.

introduced   A non-native plant that has usually been purposefully introduced, as European blackberry (*Rubus fruticosus*) that is native to much of Europe but is now a noxious weed in Australia, New Zealand and the USA.

introgression   The incorporation of alleles from one entity (population, species etc.) into the gene pool of a second entity by hybridisation and repeated backcrossing.

introrse   Facing towards the axis.
= **anticous**
cf. **extrorse, latrorse**

introrse dehiscence
Of anthers, facing inwards and opening longitudinally to release pollen towards the centre of the flower, as the daisy family (Asteraceae).
= **anticous dehiscence**
see also **anther**
dehiscence

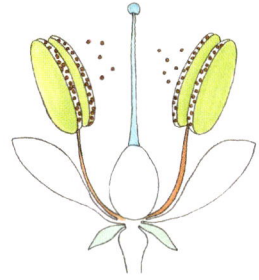
introrse dehiscence

invalid name   Publication of a new species that does not accord with the provisions of the International Code of Nomenclature for valid publication.
= **nomen invalidum**

invasive   A plant that has the ability to thrive and spread aggressively outside its natural range.

involucel   Of the carrot family (Apiaceae), a secondary involucre at the base of an umbellule within a compound umbel.
cf. **involucre**

involucre | **involucel**

involucre   A whorl of bracts around a head (Asteraceae), at the base of an umbel (Apiaceae), or around a single flower, as hibiscus (*Hibiscus*).
see also **epicalyx**
cf. **involucel**
involucral   Relating to the involucre.
involucrate   Having an involucre.

Involucre

> Asteraceae   > Apiaceae   > hibiscus (*Hibiscus*)

involute
With margins rolled inward towards the the upper surface, as the margins of some leaves.
= **inrolled**
cf. **revolute**
involute aestivation   Of young petals, tepals or sepals in the unopened bud with margins rolled inwards towards the upper surface.
involute ptyxis   Of a single leaf in bud with margins rolled inwards towards the upper surface.
involute vernation   Of young leaves in the unopened leaf bud with margins rolled inwards and arranged in a circle.

Involute

leaf margin   aestivation   ptyxis   vernation

iridescent   With shifting rainbow colours like those on a soap bubble.

irregular   Of a flower that is divisible into halves along one plane only.
= **zygomorphic**
cf. **asymmetric, regular**

irregular

pea flower

irregularly dentate
Irregularly toothed. Of a margin that appears jagged or gnawed, as the margins of some leaves.
= **erose**

irregularly dentate

irritable   Responding actively to stimuli, as the column of a trigger plant (*Stylidium*) when it is touched.

irritable

column

iso-   A prefix meaning identical.

isobilateral   Having two similar sides, where there is no evident distinction internally or externally between the upper and lower surfaces, as the leaves of an iris (*Iris*).
Typical of leaves that orient themselves parallel to the main axis, as most monocotyledons.
= **equifacial, isolateral**
cf. **dorsiventral**

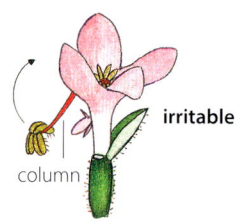

vascular bundle   isobilateral

Iris

leaves

isocotyly   Having two equally sized cotyledons.
cf. **anisocotyly**

isodiametric   Of an object with the polar axis and the equatorial diameter more or less equal.
= **spheroidal**

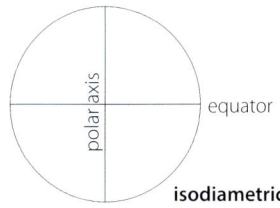
isodiametric

isodromous, isodromic   Of ferns, with the basal pinnae or pinnules opposite and having the first set of veins in a segment opposite.
cf. **anadromous, catadromous**

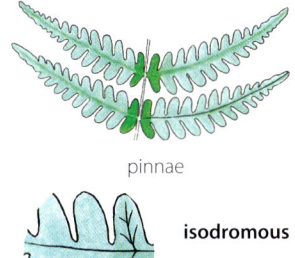
pinnae

isodromous

veins opposite

isolateral   Having similar upper and lower surfaces.
= **isobilateral**

isolectotype   A duplicate of a lectotype.

isomerous   Having whorls with the same number of parts, as a flower with equal numbers of sepals, petals, carpels and/or stamens.

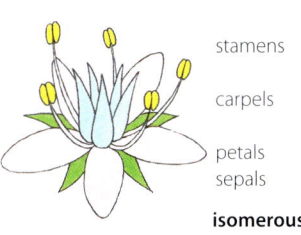
stamens

carpels

petals
sepals

isomerous

isophylly   Having leaves equal in size and shape at any point on the stem.
cf. **anisophyllous**
isophyllous   Relating to isophylly.

isophylly

isopolar   Of a pollen grain having identical proximal and distal halves.
cf. **heteropolar**

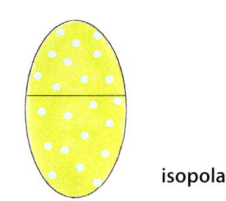
isopolar

isostemonous   With a single whorl of stamens. With as many stamens as petals.
= **haplostemonous**

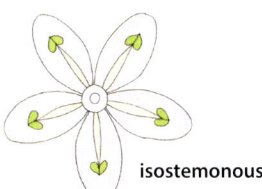
isostemonous

isostylous   Of flowers having styles all of the same length.
= **homostylous**
cf. **brachystylous, dolichostylous**

isosyntype   A duplicate of a syntype.

isotonic   Of two fluids having the same osmotic pressure.
cf. **hypertonic, hypotonic**

isotype   Any duplicate that is part of the original collection of the holotype.

isthmus   A connecting part or organ, especially when narrow and joining two larger parts.

iteroparous   Of a plant that flowers and bears fruit more than once in its lifetime, as perennial plants like roses.
= **pleonanthic, polycarpic**
cf. **hapaxanthic, monocarpic, semelparous**

IUCN   The International Code for Conservation of Nature is the chief organisation for standardising threat status of plants and animals.

jaculator   A small hooked stalk that ejects seeds from a capsule, as *Ruellia* in the acanthus family (Acanthaceae).

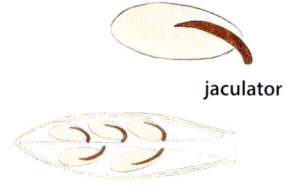
jaculator

joint   A place of union or separation of two parts. a node or articulation on a stem, as the stems of glassworts (*Salicornia*).
jointed   With joints.
cf. **continuous**

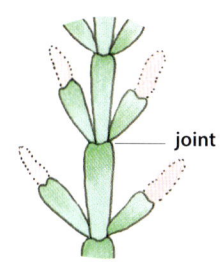
joint

jointing   Of grasses (Poaceae), the stage during which stem internodes lengthen and the culm grows taller ready for flowering and the reproductive phase.
see **vegetative phase**

jointing

153

jugate   Forming a pair.
Of leaflets arranged in pairs.
see also **conjugate, multijugate, quinquejugate**

Jugate

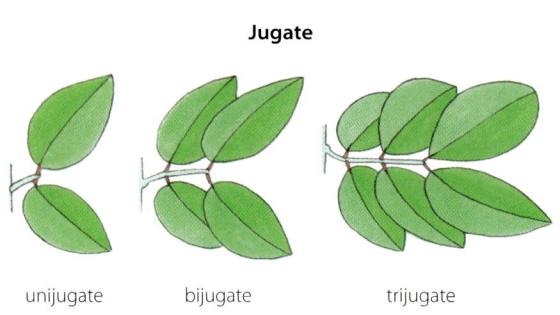

unijugate   bijugate   trijugate

juvenile   The early, non-adult phases of a plant's life or a plant organ.
Of leaves that show a striking difference in appearance between the young and the adult condition, as leaves of mountain clematis (*Clematis aristata*).
cf. **homoblastic**

adult leaf

juvenile leaves

karyokinesis   The process of mitosis.

karyotype   An image of a stained and treated set of chromosomes as they appear under the microscope.
Usually arranged in a particular order according to length, centromere location and so on.

keel   A ridge along the centre of the lower surface like that on the bottom of a boat.
The two lower petals of a pea flower united along their lower margin to form a keel enclosing the stamens and pistil.
= **carina**
keeled   Having a keel.
= **carinate**

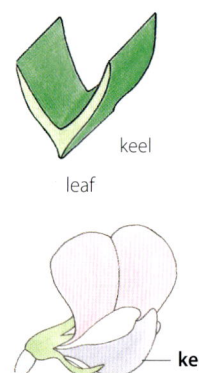

keel

leaf

keel

pea flower

keiki   A plantlet or offshoot that grows from the base of the plant or from one of the nodes on the stem of some orchids, as *Dendrobium*. It is a clone of the mother plant and can be detached and propagated once the rootlets have developed.

keiki

kernel   The softer usually edible part of a nut, seed or fruit stone that is contained within a hard shell. The grain of a cereal enclosed in the hard husk.

key   Of taxonomy, an aid for rapid identification of unknown plants, as a dichotomous key.

kingdom   One of the divisions into which natural organisms are classified.
Various numbers of kingdoms have been proposed, most recently seven: Bacteria, Archaea, Protozoa, Chromista, Plantae, Fungi and Animalia.
see **taxonomic hierarchy**

knee   In most climbing palms (rattans), a swelling on the leaf sheath at the base of the petiole.

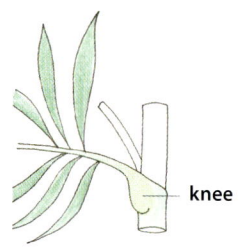

knee

knee roots   Horizontal lateral roots with emergent loops that allow for gas exchange with the atmosphere in oxygen-poor waterlogged soil. Swamp cypress (*Taxodium distichum*) and some mangroves as (*Bruguiera*) have conspicuous knee roots.
see also **lenticel**

knee roots

Krebs cycle   The stage of cellular respiration in which food molecules are broken down in the presence of oxygen to release energy.
It occurs in the mitochondria.
Also called the citric acid cycle and the tricarboxylic acid cycle (TCA).
see also **glycolysis**

kwongan   An Aboriginal term for plains that are sandy and open with scrubby vegetation and mild wet winters and warm dry summers in south-western Australia.
It covers an area the size of England, with some 3000 of the 5710 plant species being endemic.
It is a Mediterranean-type ecosystem, together with the maquis and garrigue in the Mediterranean Basin, chaparral in California, matorral in Chile and fynbos in South Africa.

labellum   A lip.
Of trigger plant flowers (Stylidiaceae), one of five petals that is usually distinctly different from the other petals.
Of orchid flowers (Orchidaceae), one of three petals that is usually distinctly different from the other petals.

Labellum

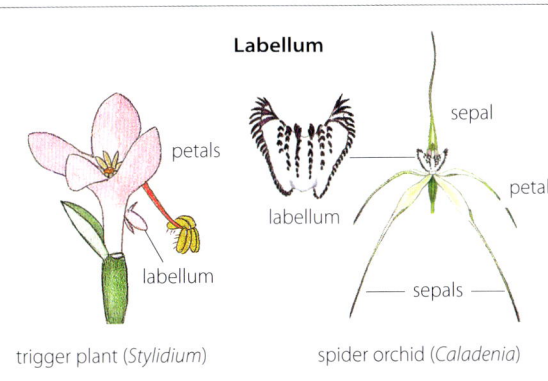

trigger plant (*Stylidium*)   spider orchid (*Caladenia*)

labium   A lip.
One of the lip-like divisions of a bilabiate corolla.
labiate   Of flowers with a tubular corolla and the limb divided into two lip-like parts that may or may not be lobed.
Of or belonging to the mint family (Lamiaceae, formerly Labiatae).

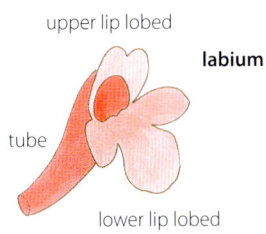

lacerate   Of a margin with irregular narrow segments that appear to have been torn, as the margins of some leaves.
cf. **fimbriate, laciniate**

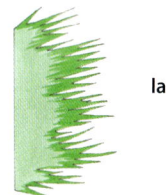

lacerate

laciniate   Irregularly and finely cut into long thin strips as if slashed, as the margins of some leaves.
cf. **fimbriate, lacerate**

laciniate

lacuna, *pl.* **lacunae**   A space, gap, cavity or depression.
An air space between adjacent cells, as in parenchyma.
Of pollen grains, a depressed area on the outer surface that is surrounded by ridges.
lacunose   Of, like or having a lacuna or lacunae.

Lacuna

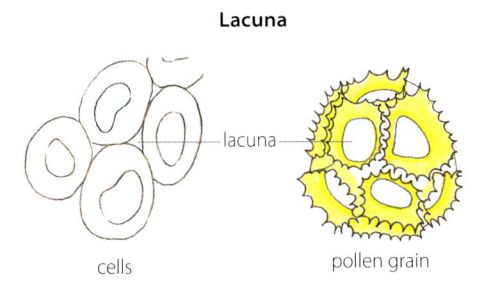

cells   pollen grain

lacunar collenchyma   Collenchyma with intercellular spaces. The cell wall thickenings are adjacent to these intercellular spaces.

laesura, *pl.* **laesurae**   A single groove (monolete) or three-rayed, Y-shaped groove (trilete) on the face of a spore.
Also sometimes, one branch of a trilete laesura.
It marks the way in which the four spores of a tetrad were in contact with each other after meiosis.
It is the area of weakness in the wall through which a spore germinates.
= **tetrad mark**

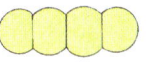

pollen tetrad linear

laesura

pollen grain
laesura monolete

pollen tetrad tetrahedral

pollen grain
laesura trilete

laevigate   Smooth as if polished, as the surface of some leaves and the seeds of some wattles (*Acacia*).

laevigate

lageniform   Much dilated and subglobose at the base with a slender neck, as a bottle gourd (*Lagenaria*).

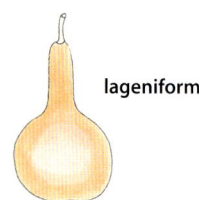

lageniform

lamella, *pl.* **lamellae**   A septum in an ovary. A thin, plate-like or scale-like structure.
   lamellar, lamellate, lamellose   Composed of a lamella or lamellae.
   lamelliform   With the shape of a thin plate or scale.

lamellar collenchyma   Collenchyma having the two sides of the cell wall that are parallel with the organ surface (the tangential cell walls) thicker than the other two (the radial cell walls).

lamellar placentation
Having carpels fused but the internal walls (septa) lacking, creating a unilocular ovary, with ovules attached to septa-like placentas that project from the wall of the ovary, as water lilies (*Nymphea*).
= **superficial placentation**
see **placentation**

Lamellar placentation

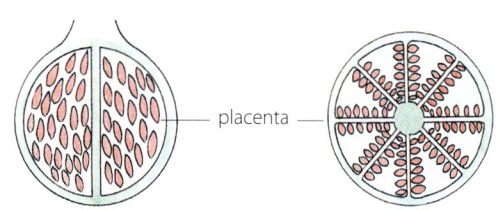

placenta

lamina, *pl.* **laminae**
A thin flat organ or part. The flat expanded part of a leaf or petal. A blade.
   laminate   Having the form of a lamina.

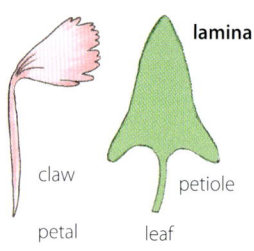

lamina

claw
petiole
petal
leaf

laminar   Blade-like, as the stamens of masiratu (*Degeneria*).

laminar

lanate, lanose   Woolly. Densely covered with long tangled fine soft curly or wavy hair, as the white-woolly bracts of blanket leaf (*Bedfordia arborescens*).

lanate

lanceolate   Shaped like the head of a lance. Distinctly longer than wide, tapering each end and widest below the middle.
cf. **oblanceolate**

lanceolate

lanceoloid   A three-dimensional shape that is much longer than wide, tapering at each end and widest below the middle, as the bulbs of chives (*Allium schoenoprasum*).
cf. **lanceolate**

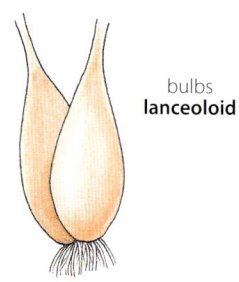

bulbs
lanceoloid

landing platform
The lower petal of the corolla of some flowers, as that of bugle flowers (*Ajuga*), adapted to form a perch for visiting insects.

landing platform

lanuginose, lanuginous
Covered with short fine soft cottony or woolly hairs. Downy.
cf. **lanate**

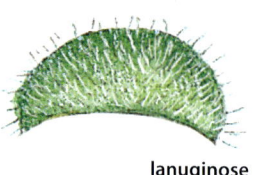

lanuginose

lanulose   Minutely woolly.

late wood   Wood in a growth ring with small thick-walled cells produced later in the growing season. It is more dense than early wood that is produced in spring when growth is faster. Early wood and late wood usually appear as two distinct bands.

latent Having the potential to develop, as a dormant bud.
cf. **dormant, quiescent**

latent buds Usually applied to buds that do not grow the following year.
Also sometimes applied to resting buds.
= **dormant buds**

lateral At, of or on the side.

lateral lobe Of an orchid with a three-lobed labellum, one of two lobes at the base of the labellum, as pink fairies, (*Caladenia latifolia*).

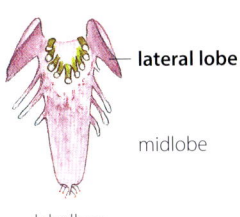

lateral lobe

midlobe

labellum

lateral meristem There are two lateral meristem tissues, vascular cambium and cork cambium. Both are found in woody plants and both give rise to secondary growth.
Vascular cambium produces secondary xylem (wood) and secondary phloem (bark) that increase the girth of the plant.
Cork cambium (phellogen) produces periderm (outer bark) that replaces the epidermis.
= **secondary meristem**
see also **intercalary meristem, primary meristem**

Lateral meristem

cork cambium
lateral meristems
vascular cambium

periderm with cork cambium
cortex
primary phloem
secondary phloem
vascular cambium

secondary xylem

primary xylem
pith

Secondary growth

latex A milky liquid in certain plants that coagulates on exposure to air, as the dune thistle stem (*Actites megalocarpus*).
see **laticifer**

latex

lati- A prefix meaning broad or wide.

laticifer An elongated cell, or series of cells joined together, that synthesises and stores latex.
laticiferous Containing, bearing or secreting latex.

latitude A line on the surface of a sphere measuring the distance north or south of the equator.
cf. **longitude**
latitudinal Across rather than lengthwise. Of or relating to latitude.
cf. **longitude**

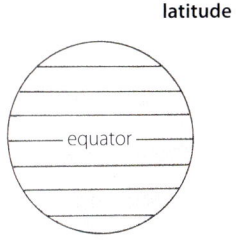

latitude

equator

latrorse Facing towards the side.

latrorse dehiscence Of anthers, opening laterally rather than facing towards or away from the centre of the flower. The anther splits longitudinally to release pollen.
see also **anther dehiscence**

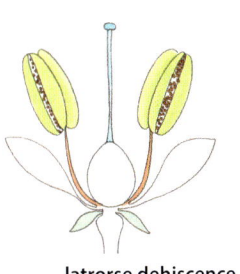

latrorse dehiscence

Laurasia One of two vast continents believed to have existed mostly in the northern hemisphere and to have resulted from the break up of Pangea.
see **continental drift, Gondwana**

Laurasia
PANGEA
Gondwana

lax Loose or open, not dense or congested, as some leaves and the inflorescence of wild radish (*Raphanus raphanistrum*).
cf. **congested**

leaves

lax

inflorescence

layering   A form of propagation that encourages root development on a stem while it is still attached to the parent plant.
This results in a new plant that can exist separately from the parent plant.
Examples of layering are simple layering, either underground or aerial, and mound layering (stool layering).
see also **stool**

Layering

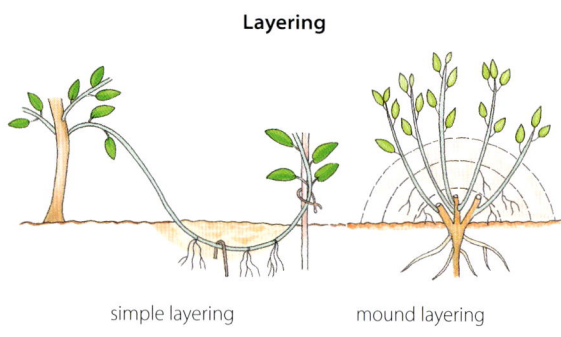

simple layering   mound layering

leader   The dominant shoot of a tree or shrub usually at the tip of the whole plant.
If the central axis dies or is damaged a lateral leader may form at the tip of a branch.
Some plants have no dominant leader.

Leader

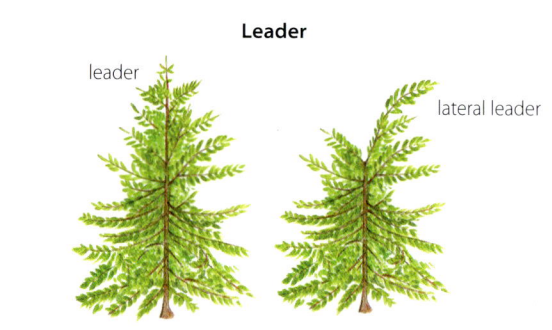

leader

lateral leader

leaf   *see* pages 159–165

leaf anatomy   A leaf blade is organised to collect sunlight and carry out photosynthesis.
A typical eudicot leaf blade has many layers.
The waxy non-cellular layer of cuticle on the top of the leaf prevents water escaping.
The epidermis covers and protects the leaf.
Palisade mesophyll is responsible for most of the photosynthesis in the leaf.
Spongy mesophyll has air spaces that work with the stomata to allow gas exchange.
Vascular tissue transports products of photosynthesis and nutrients to and from the leaf.

Leaf anatomy

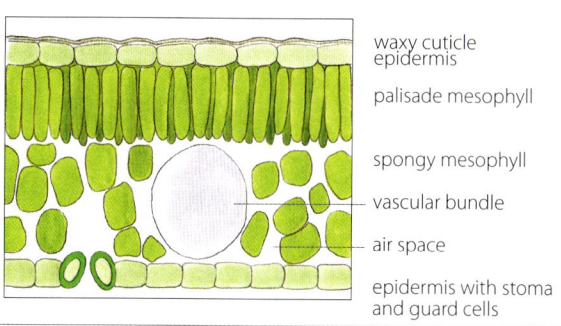

waxy cuticle
epidermis

palisade mesophyll

spongy mesophyll

vascular bundle

air space

epidermis with stoma and guard cells

leaf arrangement   *see* pages 159–160

leaf base   *see* page 160

leaf lobing   *see* pages 160–161

leaf margins   *see* pages 161–162

leaf mosaic   The arrangement of foliage in a pattern that exposes the maximum number of leaves to the direct rays of the sun.

leaf scar   The corky healing layer that forms on a stem after a leaf falls.
see also **abscission**
cf. **bundle scar**

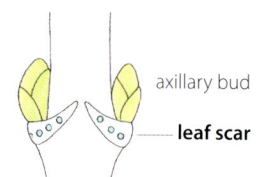

axillary bud

leaf scar

leaf shape   *see* pages 162–163

leaf sheath   A tubular or rolled structure at the base of a leaf that, at least partly, surrounds the stem, as grasses.
The expanded embracing base of the petiole, as the young *Howea* palm.
see **crownshaft**

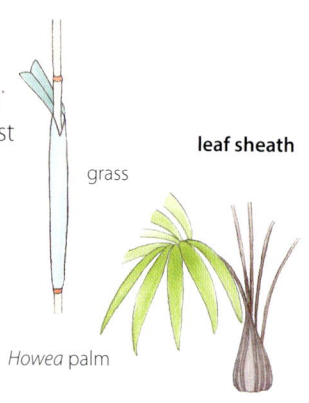

grass

leaf sheath

Howea palm

leaf tip   *see* pages 163–164

leaf venation   *see* pages 164–165

leaf vernation   *see* **vernation**

leaf   In vascular plants, a vegetative organ that is responsible for most of the plant's photosynthesis and transpiration.

A foliage leaf is a mostly flat, green outgrowth from a stem, usually consisting of a blade (lamina) and a petiole.

It may be simple and undivided or compound with a number of individual leaflets on a common stalk (rachis).

Many eudicots produce compound leaves but in monocots they are found only in palms.

A grass leaf usually consists of a blade, collar and sheath.

A seed leaf is a cotyledon that originates directly from the tissue of the embryo of seed-bearing plants.

A cataphyll is a leaf that is not photosynthetic and functions as protection or storage, as bulb scales.

Modified leaves include bracts, tendrils and spines.

Stamens, carpels, petals and sepals are considered to be modified leaves.

see also **foliage, leaf anatomy, prophyll**

Leaf

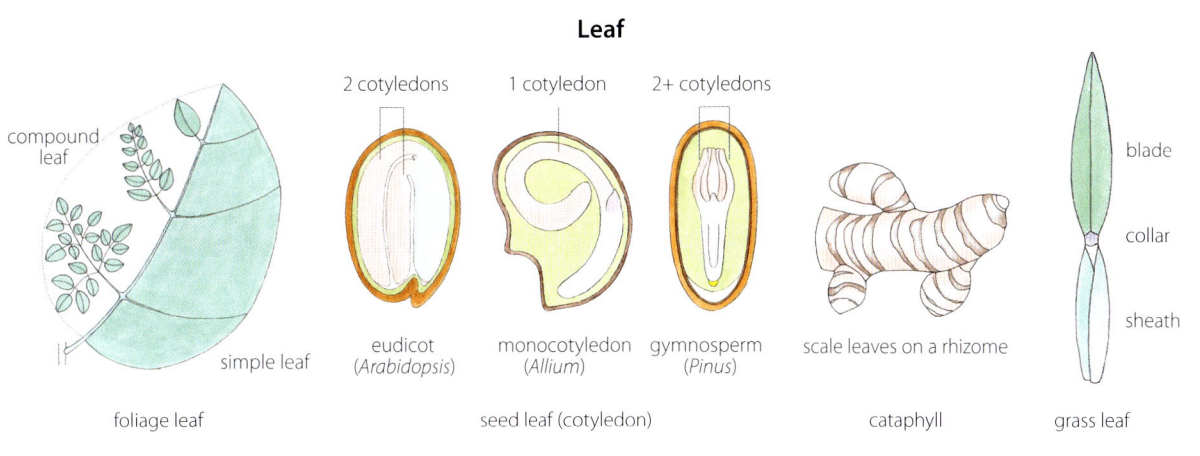

foliage leaf	seed leaf (cotyledon)		cataphyll	grass leaf

Leaf arrangement

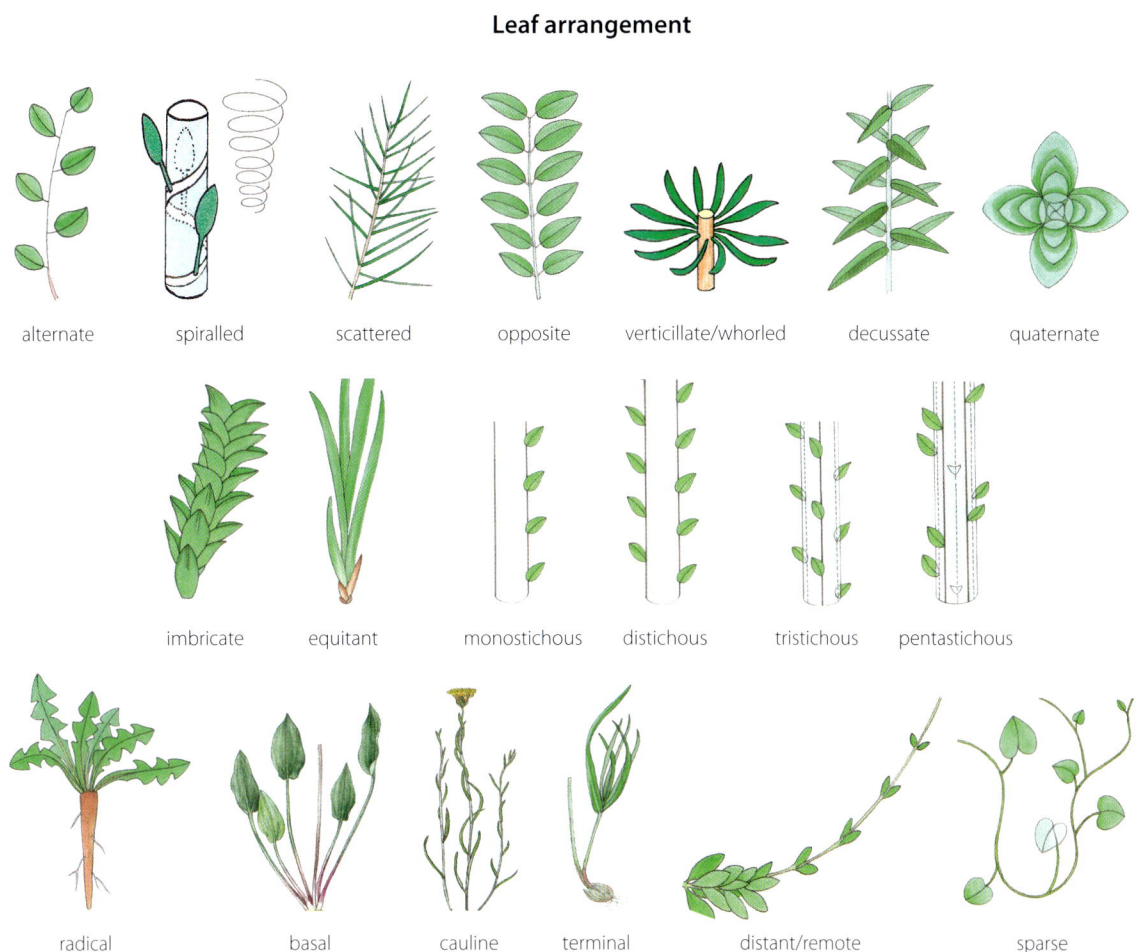

| rosulate/rosetted | caespitose/tufted | fasciculate/fascicled | clustered | marcescent |

Leaf base

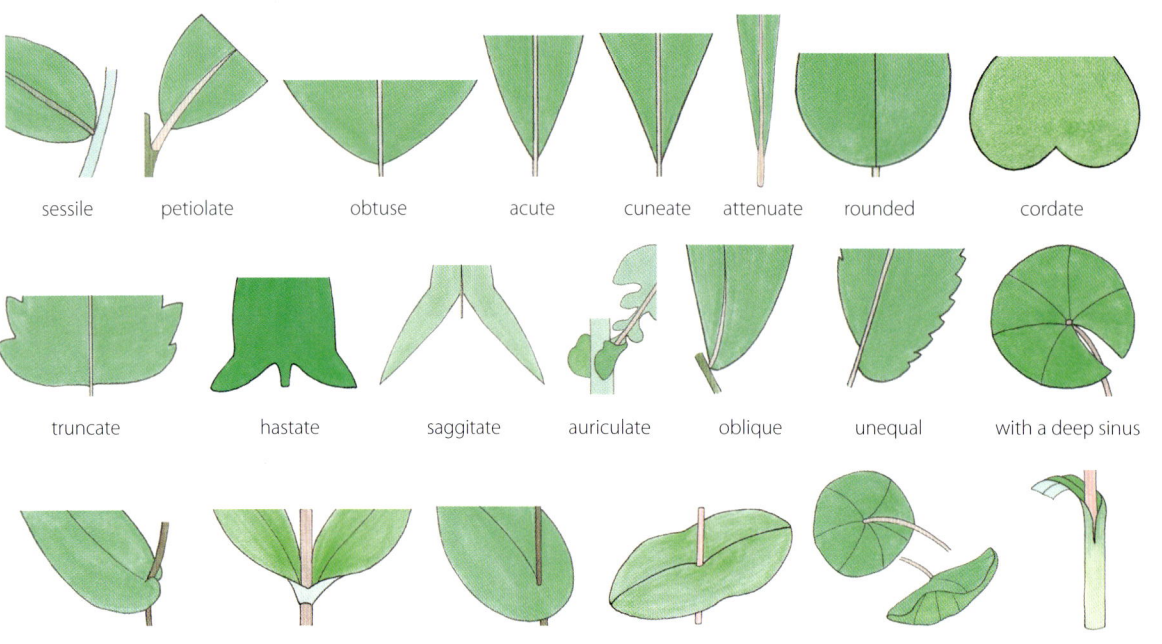

| sessile | petiolate | obtuse | acute | cuneate | attenuate | rounded | cordate |

| truncate | hastate | saggitate | auriculate | oblique | unequal | with a deep sinus |

| amplexicaul | connate | perfoliate | connate-perfoliate | peltate | sheathing/vaginate |

Leaf lobing

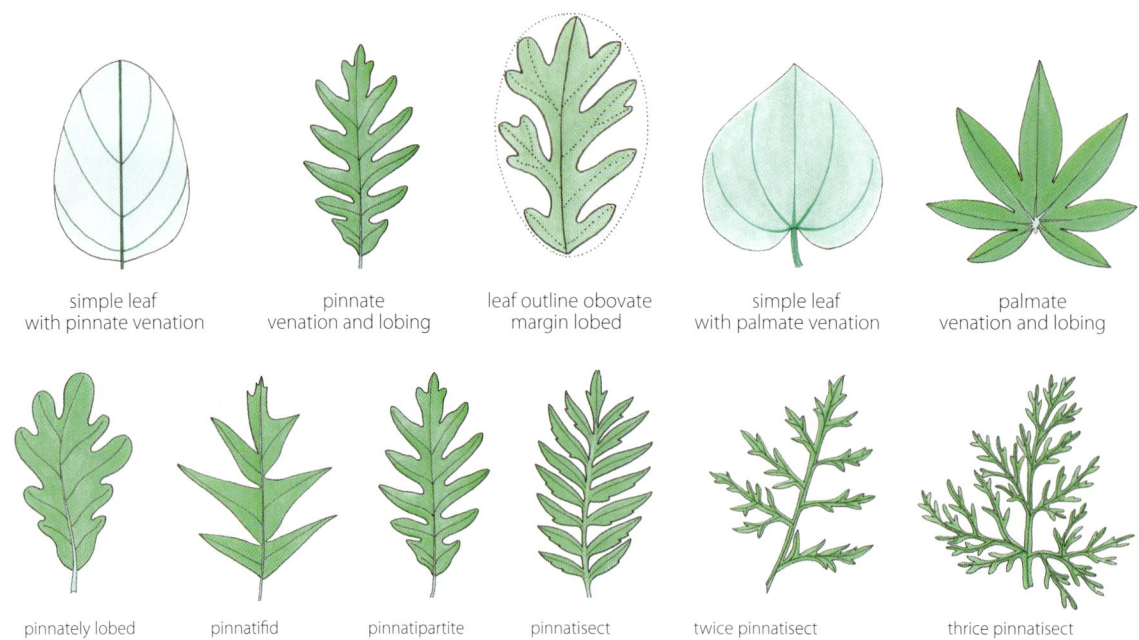

| simple leaf with pinnate venation | pinnate venation and lobing | leaf outline obovate margin lobed | simple leaf with palmate venation | palmate venation and lobing |

| pinnately lobed | pinnatifid | pinnatipartite | pinnatisect | twice pinnatisect | thrice pinnatisect |

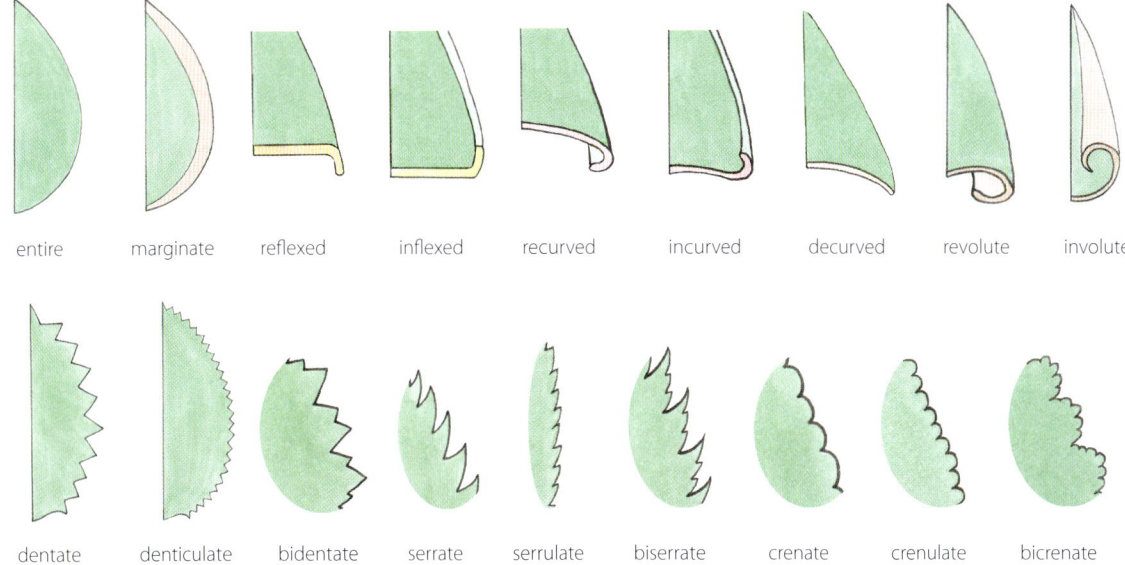

lyrate-pinnatifid   pandurate   runcinate   pectinate

palmately lobed   palmatifid   palmatipartite   palmatisect

pedately lobed   pedatifid   pedatipartite   pedatisect

fissum   bilobate   trilobate   bifid   trifid   bipartite   tripartite

Leaf margins

entire   marginate   reflexed   inflexed   recurved   incurved   decurved   revolute   involute

dentate   denticulate   bidentate   serrate   serrulate   biserrate   crenate   crenulate   bicrenate

fimbriate   lacerate   laciniate   incised   erose   dissected pinnate   dissected palmate

margin flat in cross-section

margin flat in cross-section

margin flat in cross-section

margin undulate in cross-section

sinuolate/repand   sinuate   undulate   crispate

ciliate   ciliolate   glandular-ciliate   filiferous   aculeate   bristled   spinose

Leaf shape

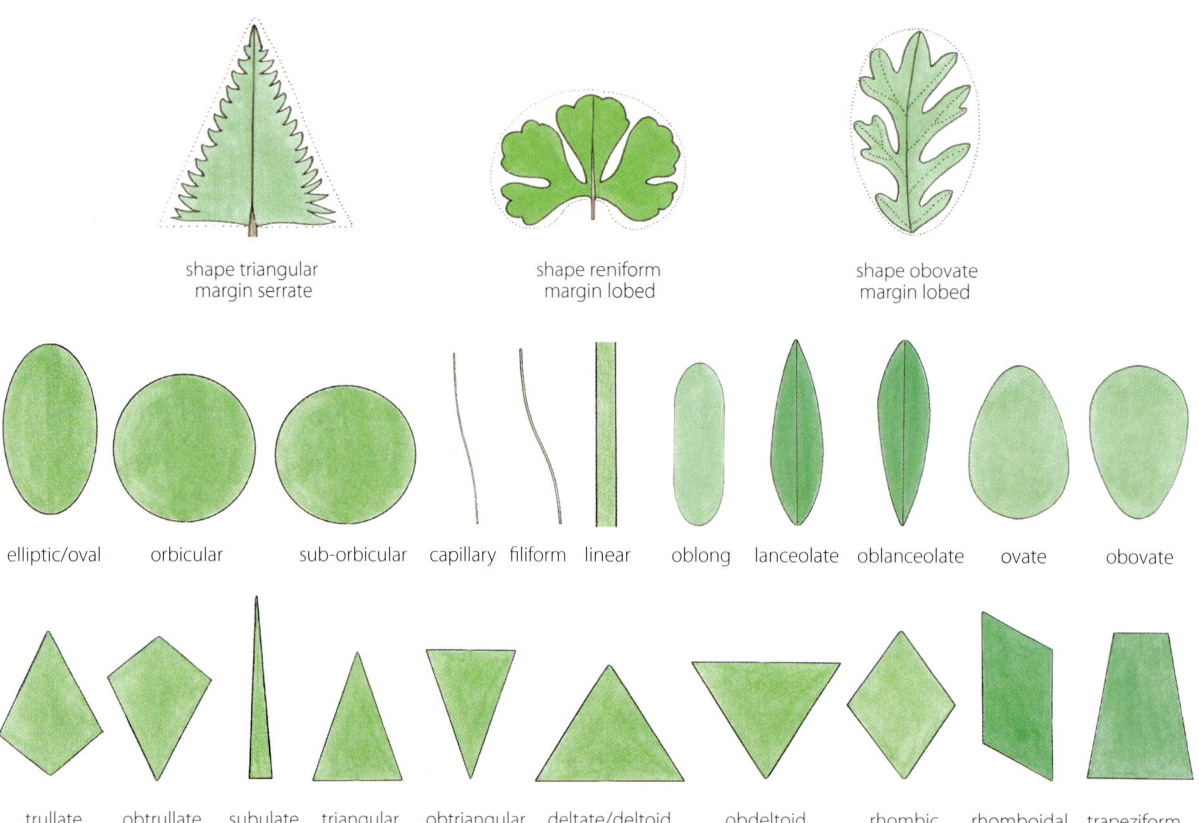

shape triangular
margin serrate

shape reniform
margin lobed

shape obovate
margin lobed

elliptic/oval   orbicular   sub-orbicular   capillary   filiform   linear   oblong   lanceolate   oblanceolate   ovate   obovate

trullate   obtrullate   subulate   triangular   obtriangular   deltate/deltoid   obdeltoid   rhombic   rhomboidal   trapeziform cordate   obcordate   reniform   flabellate   spatulate   ensiform/gladiate   falcate   acinaciform   sigmoid

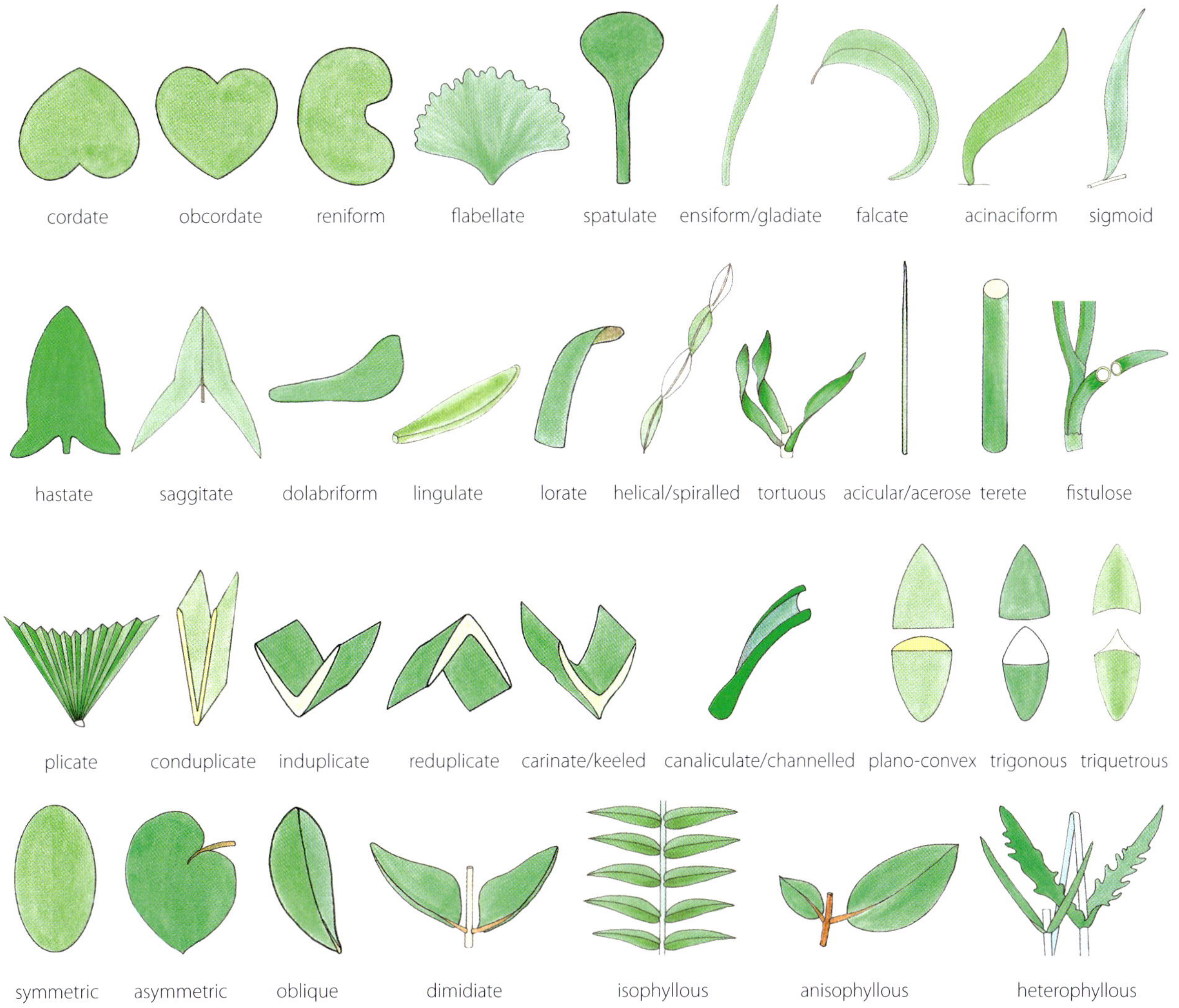

hastate   saggitate   dolabriform   lingulate   lorate   helical/spiralled   tortuous   acicular/acerose   terete   fistulose

plicate   conduplicate   induplicate   reduplicate   carinate/keeled   canaliculate/channelled   plano-convex   trigonous   triquetrous

symmetric   asymmetric   oblique   dimidiate   isophyllous   anisophyllous   heterophyllous

Leaf tip

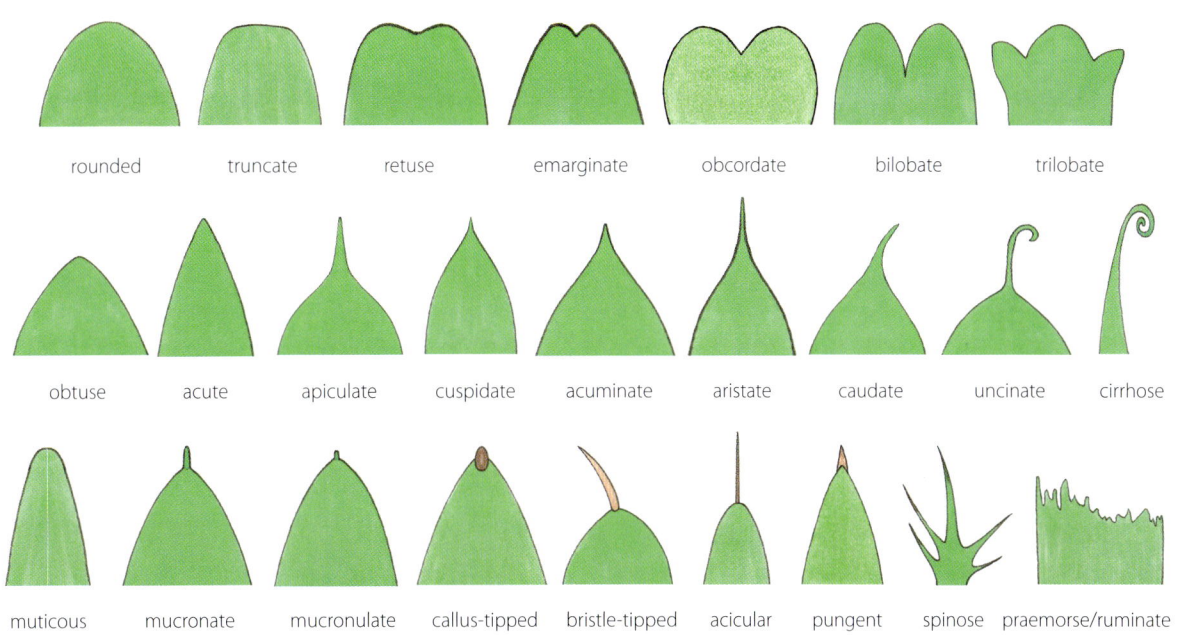

rounded   truncate   retuse   emarginate   obcordate   bilobate   trilobate

obtuse   acute   apiculate   cuspidate   acuminate   aristate   caudate   uncinate   cirrhose

muticous   mucronate   mucronulate   callus-tipped   bristle-tipped   acicular   pungent   spinose   praemorse/ruminate

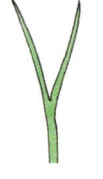  

forked/furcate   cleft   bifid

Leaf venation

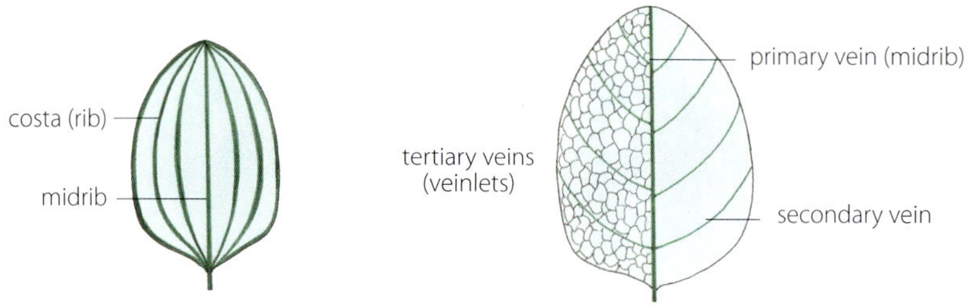

costa (rib)

midrib

primary vein (midrib)

tertiary veins
(veinlets)

secondary vein

open   closed   pinnate (open)   palmate (open)   parallel (closed)

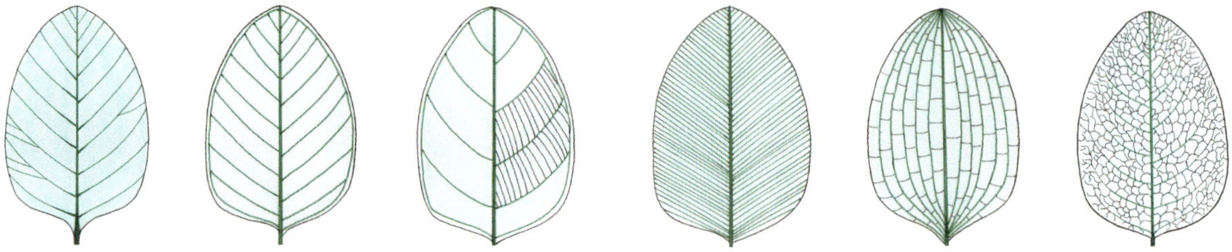

marginal   intramarginal   scalariform/percurrent   paxillate   tesselate   reticulate

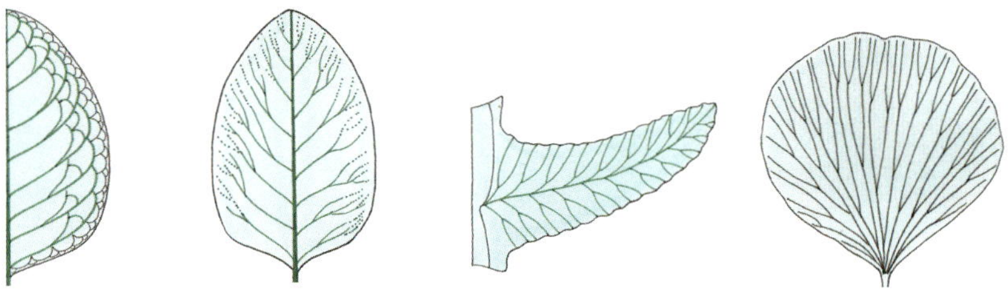

anastomosing   dendritic   furcate/forked   bifurcate/dichotomous

Leaf venation of eudicots

(Hickey LJ (1973) Classification of the architecture of dicotyledonous leaves. *American Journal of Botany* **60**, 17–33)

Parallelodromous

With two or more
primary veins
originating at the base
and converging
towards the apex.

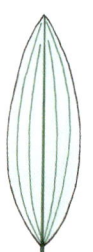

Campylodromous

With primary veins
or branches
running in recurved arches
from the base
and converging at the top.

Actinodromous

With three or more primary veins diverging radially from a single point at the base.

Palinactinodromous

With three or more primary veins diverging radially from a single point at the base and these veins branching dichotomously.

Acrodromous

With two or more primary or strong secondary veins running in converging arches towards the apex.

Perfect
With lateral veins covering at least two-thirds of the surface.

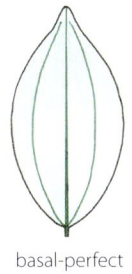

basal-perfect   suprabasal-perfect

Imperfect
With lateral veins covering less than two-thirds of the surface.

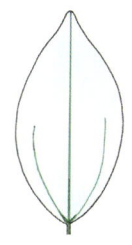 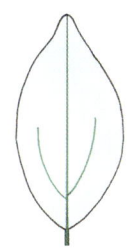

basal-imperfect   suprabasal-imperfect

Pinnate

With a single primary vein (midrib) serving as the origin for the higher order veins.

Hyphodromous
All but primary veins absent or not visible.

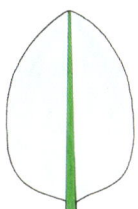

hyphodromous

Craspedodromous
Secondary veins terminating at the margins.

   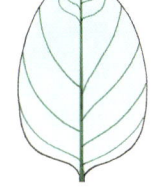

simple craspedodromous   semicraspedodromous   mixed craspedodromous

Camptodromous
Secondary veins not terminating at the margins.

   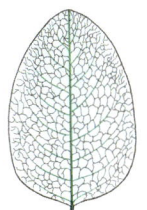

brochidodromous   eucamptodromous   cladodromous   reticulodromous

leaflet   The small leaf--like parts of a compound leaf.
see also **compound leaf, pinna, pinnule**

leaflet

leathery   Tough but somewhat flexible, as the phyllodes of some wattles (*Acacia*).
= **coriaceous**

leathery

lectotype   A specimen chosen, from the specimens available to the original author, when a holotype was never designated.

legit, *abbr.* **leg.**   From the latin *legit*, meaning he/she collected.
Used on the label of a herbarium specimen and followed by the name of the collector.

legit., legitimate   Legal. Conforming to the rules.

legitimate, *abbr.* **legit.**   Legal. Conforming to the rules.

legitimate name   In nomenclature, a name conforming to the rules of the International Code of Nomenclature.

legume   A dry dehiscent fruit, with one or more seeds, that splits into halves lengthwise along two sutures. Derived from a one-carpelled superior ovary. Characteristic fruit of wattles and peas (Fabaceae).

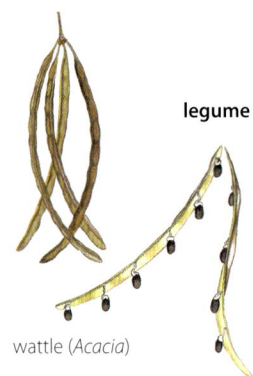

legume

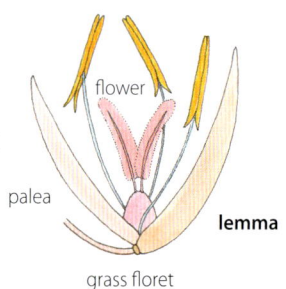

wattle (*Acacia*)

leio-   A prefix meaning smooth.

lemma *pl.* **lemmata, lemmas**   Of grasses (Poaceae), the lower and usually larger outer bract that together with the palea encloses the flower of a grass floret.
see **locusta, spikelet**

flower

palea

lemma

grass floret

lens   A strophiole.

lenticel   A small raised pore, usually elliptical in shape, for gas exchange between a plant and the atmosphere.
Found on the surface of woody stems, some roots, as the aerial roots of mangroves and on the skin of apples and pears.
cf. **stoma**

lenticellate   Having or producing lenticels.

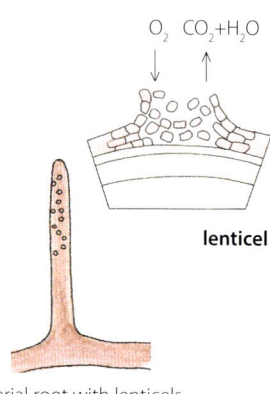

O_2 CO_2+H_2O

lenticel

aerial root with lenticels

lenticular   More or less disc-shaped but convex on both sides, as the seeds of lentils (*Lens culinaris*).

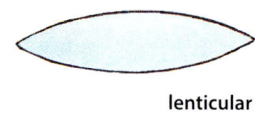

lenticular

lentiginose, lentiginous
Covered with minute dots like dust scurf or freckles, as the pod of freckled milk vetch (*Astragalus lentiginosus*).

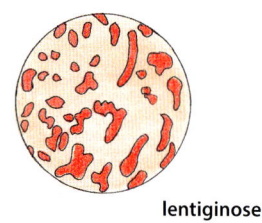

lentiginose

lepidote, leprous
Covered with scurfy scales, as the leaves of some species of *Croton*.

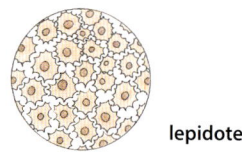

lepidote

lept-, lepto-   A prefix meaning slender, thin or narrow.

leptocaul   A growth form of plants that are deemed to be advanced.
Characterised by being much-branched with small leaves, as most angiosperms.
cf. **pachycaul**

leptoma   A thinning of the pollen wall, as found in conifers and on the spore wall of some mosses, that is presumed to function as a germination site.

leptomorph　Of woody bamboos, a long thin running and branching rhizome that spreads horizontally over a considerable distance and can develop a new plant at each node.
Such bamboos are indeterminate and invasive.
see **amphipodium, monopodium**
cf. **pachymorph, sympodium**

Leptomorph

leptophyllous　With slender leaves.

leptosporangiate　Of true ferns (leptosporangiate ferns) that arise from a single epidermal cell.
see **fern**
cf. **eusporangiate**

-lete　A suffix denoting the presence or absence of laesurae on a spore.
see **alete, monolete, trilete**

leuc-, leuco-　A prefix meaning white.

leucoplast　A colourless non-pigmented plastid in the cytoplasm of a cell, as amyloplasts that store starch, elaioplasts that store fat, or proteinoplasts that store proteins.

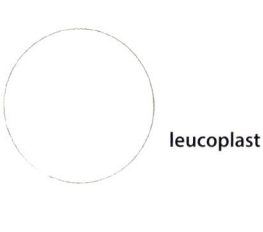

leucoplast

liana, liane　A thick-stemmed woody climber of tropical rainforest that grows to the top of the tree canopy in order to reach sunlight, as philodendron (*Philodendron eximium*). It begins life as a seedling on the forest floor and is rooted in the ground.
lianoid　Having a liana-like habit.

liana

lichen　A plant consisting of a symbiotic relationship between fungi and photosynthetic algae and/or cyanobacteria. The fungi take nutrients from the soil and the algae absorb carbon dioxide from the air and carry outh photosynthesis.
There are three main growth forms: fruticose with upright or pendulous branches, crustose that is usually flat with unlobed edges and foliose that is lobed and often more or less circular.

Lichen

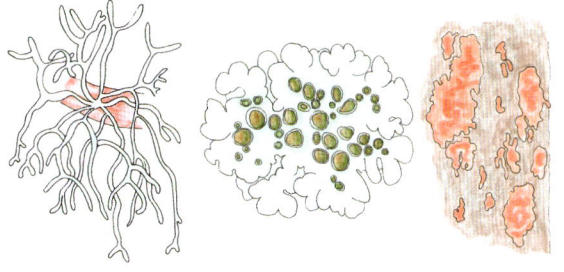

algae chains with fungal hyphae

fruticose (*Ramalina*)　　foliose (*Parmelia*)　crustose (*Caloplaca*)

life cycle　The successive stages in the life of a plant, broadly divided into a sexual phase (gametophyte generation) and a vegetative phase (sporophyte generation).
see **alternation of generations**

lignification　The deposition of lignin in the cell walls of woody plants, thereby increasing their strength and hardness.
ligneous　Woody or resembling wood.
lignified　Woody due to lignification.

lignin　A complex organic substance that adds strength and rigidity to cell walls and, together with cellulose, forms the main component of wood.
see **lignification**

lignotuber   A large rounded woody outgrowth at the base of a tree or shrub that can resist fire and drought. It consists of food reserves and a mass of vegetative buds that grow after the plant has been cut or damaged by fire.

lignotuber

ligule   A strap-shaped structure.
Of many grasses (Poaceae), the membranous outgrowth at the junction of the blade and the sheath that grows upward from the inner surface of the collar region.
Of the daisy family (Asteraceae), the strap-shaped fused petals of the corolla of a ray floret.
Of spike mosses (*Selaginella*) and quillworts (*Isoetes*), the minute appendage near the base of a leaf.
ligulate   Having a ligule. Strap-shaped.

Ligule

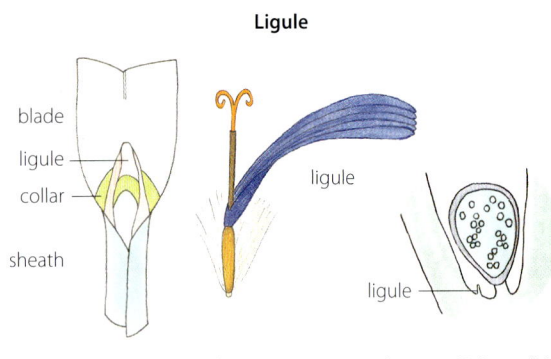

blade
ligule
collar
sheath
ligule
ligule

grass (Poaceae)   daisy (Asteraceae) spike moss (*Selaginella*)

liliaceous   Of, relating to or resembling lilies (Liliaceae).
Having a regular perianth of six tepals, similar to that of the lily genus (*Lilium*).

liliaceous

limb   The broad upper part of a petal as distinct from the claw.
The spreading, often lobed, rim of the united petals of a corolla, as distinct from the tube.
The broad upper part of a petal as distinct from the claw.

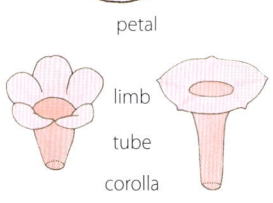
claw
limb
petal
limb
tube
corolla

limbate   Bordered, with one colour surrounded by an edging of another, as some leaves.

limbate

limicolous   Growing in mud.

limnophile   Of a plant growing in marshes or quiet shallow water.
limnophilous   Growing in quiet shallow water.

limnophyte   Of a plant growing in marshes or quiet shallow water.
It may be rooted or not, submerged or floating.
cf. **rheophyte**

lineage   A line of descent through time connecting an ancestor to its descendants.
see **clade**

linear   Long and narrow with parallel margins.

linear

ligulate floret   Of the inflorescence of a daisy (Asteraceae), a small tubular flower with the lobes united on one side into a strap-like blade.
= **ray floret**
see also **capitulum**
cf. **disc floret, tubular floret**

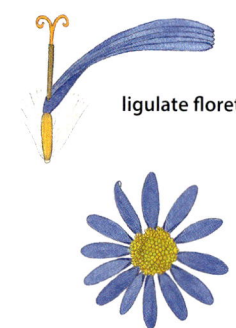
ligulate floret

lineolate   Marked with minute parallel lines, finely lineate.

linear tetrad
A uniplanar tetrad arranged with all four members cohering in a row.
see also **pollen tetrad**
see **uniplanar, viscin thread**

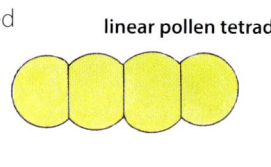
linear pollen tetrad

lineate   Striped with longitudinal lines, grooves or ridges, as the leaves of *Aloe lineata*.

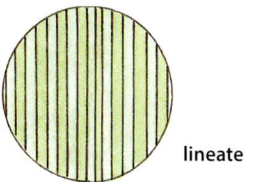
lineate

lingula, *pl.* **lingulae**   A tongue-shaped structure.
linguiform, lingulate Tongue-shaped, as some leaves.

lingula

linkage   The tendency for a group of genes located nearby on the same chromosome to be inherited together.

Linnaean System   A system of plant classification developed in the 18th century by the Swedish botanist Carl Linnaeus.
It introduced the concepts of a taxonomic hierarchy and binomial nomenclature.

lip   A labellum.
One of the lip-like divisions of a bilabiate corolla, as flowers that have a tubular corolla and the limb divided into two lip-like parts that may or may not be lobed.
Of orchid flowers (Orchidaceae), one of three petals that is usually distinctly different from the other petals.

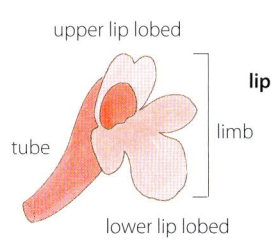

upper lip lobed

lip

limb

tube

lower lip lobed

mint family (Lamiaceae)

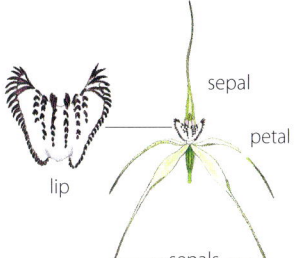

sepal

petal

lip

sepals

spider orchid (*Caladenia*)

lip cells   Cells on the sporangium of some ferns that are distinct from the annulus and part of the stomium. They are the first to separate at dehiscence.

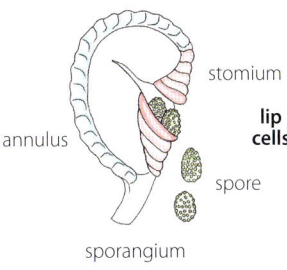

stomium

lip cells

annulus

spore

sporangium

lipid   A fatty or waxy organic compound that is insoluble in water.
It stores energy and is a structural component of the cell membrane.

lithocyst   A specialised epidermal cell in which cystoliths form, as in the leaves of species of fig (*Ficus*).

lithophyte   A plant that grows on the surface of rocks.
lithophytic   Of or relating to a lithophyte.
cf. **chasmophyte, chomophyte**

lithosere   An ecological succession that starts on a bare rock surface.

litmus test   A test for acidity or alkalinity (pH) using litmus paper.
It has a colour scale running for 0 to 14, with 0 at the red end representing the most acidic, 14, at its opposite end, a dark blue representng the most alkaline, and in the middle the pH becomes neutral.

Litmus test for pH

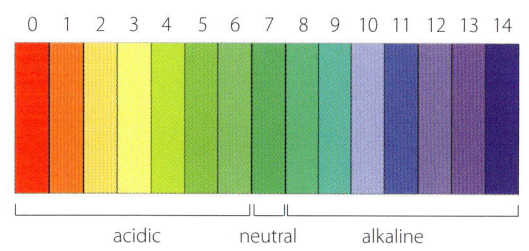

0 1 2 3 4 5 6 7 8 9 10 11 12 13 14

acidic   neutral   alkaline

litoral, littoral   A region lying along a shoreline that is periodically immersed, as a lakeside, but mostly referring to the area of the seashore between high and low tides.

liverworts   *see* page 170

loam   Soil composed of a mixture of sand, clay and organic matter.

lobe
A usually rounded, or angular, part of a margin, as the margin of a leaf or a corolla tube.
cf. **sinus**
lobed   With a lobe or lobes, as some pinnatifid or palmatifid leaves.
lobate   Having lobes or resembling a lobe.

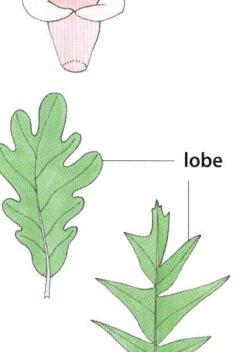

corolla lobe

lobe

liverworts   One of three major groups of nonvascular land plants that are photosynthetic, reproduce by spores and lack flowers, true leaves and roots.

A liverwort has two alternating generations, the larger haploid gamete-producing gametophyte generation and the smaller diploid spore-producing sporophyte generation that remains attached to the female gametangium (archegonium).

The haploid gametophyte, that is either a thallus or a 'leafy stem', bears the gamete producing sex organs (the female archegonia and/or the male antheridia).

A male sperm from an antheridium fertilises a female egg in an archegonium, that then develops into a diploid spore-producing sporophyte, comprising a stalk (seta), capsule and a foot.

The diploid spores in the capsule undergo meiosis to form haploid spores that are dispersed and develop into gametophytes, thus beginning the life cycle anew.

see **bryophyte**

Liverworts

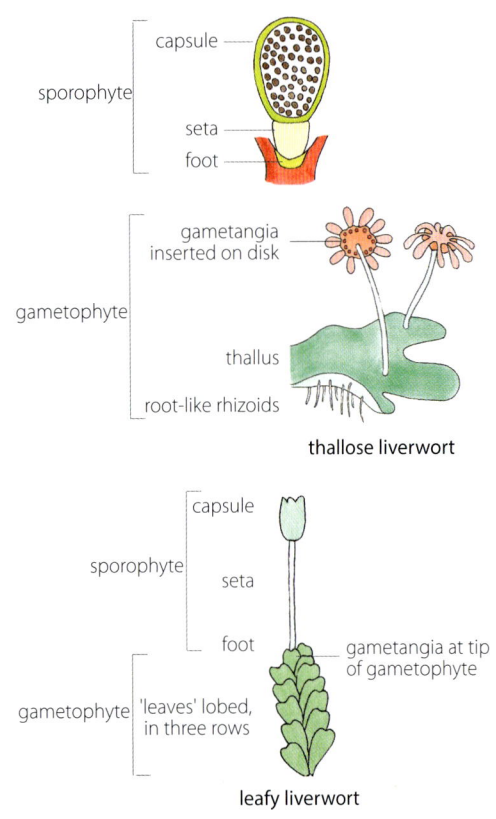

thallose liverwort

leafy liverwort

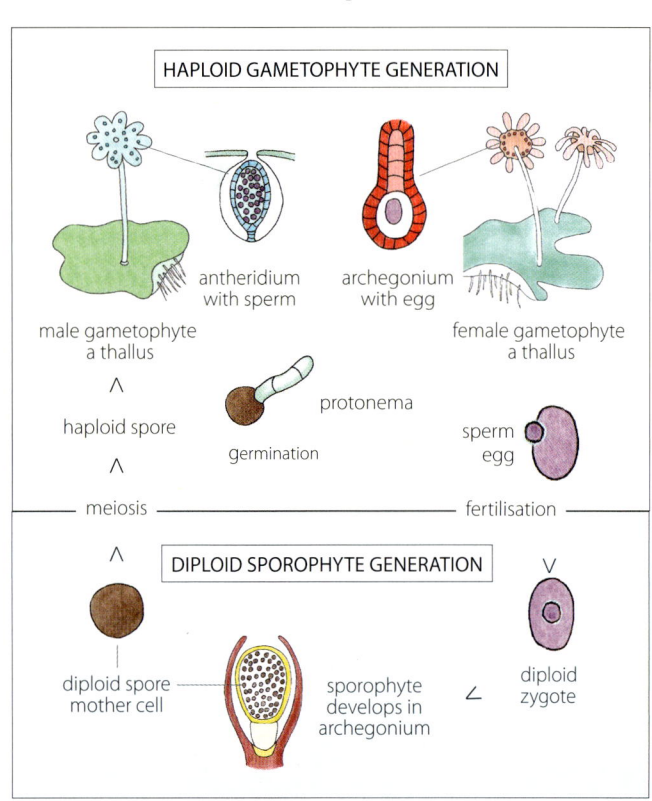

Alternation of generations

lobule   A small lobe, a subdivision of a lobe.
  lobulate   Having small lobes or resembling a lobule.

lobe

lobule

locellus, *pl.* **locelli**   A subdivision of a loculus when it is divided into even smaller cavities.
  locellate   Having small secondary cavities.

locule, loculus, *pl.* **loculi**   A small, more or less closed cavity or compartment within an organ that acts as a container.

In an ovary a locule contains the ovules, in an anther the locule (pollen sac) contains pollen and in a capsule a locule contains the seeds.

Unilocular, bilocular, trilocular and multilocular refer to the number of locules present.

locular, loculate   Having locules, divided into separate cavities or compartments.

Locule

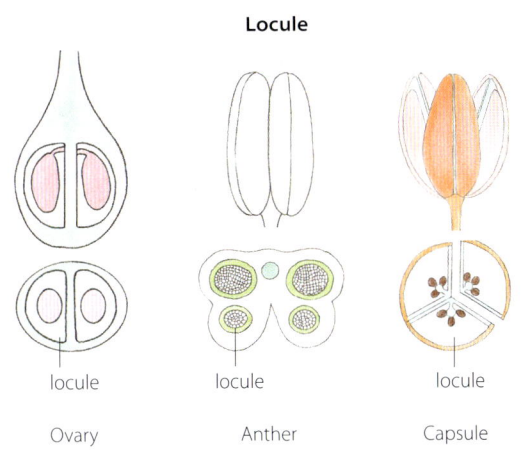

locule

Ovary

locule

Anther

locule

Capsule

loculicidal capsule

A capsule that splits lengthwise into the cavity of the locules, as a violet (*Viola*).

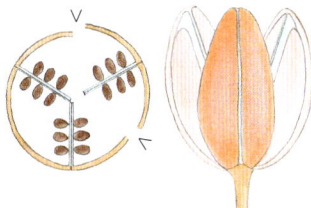

loculicidal capsule

locus, *pl.* **loci**   The position of a particular gene on a chromosome.

locusta

In grasses (Poaceae), the basic unit of an inflorescence. Typically composed of two bracts (glumes) at the base of an axis (rachilla), with one or more florets arranged alternately in two ranks.
= **spikelet**

glume

glume

locusta

lodging

Bending or collapse of the stem of a plant.
The collapse of a cereal stem when it can no longer support its own weight.

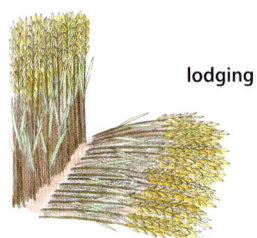

lodging

lodicule

One of usually two minute scales at the base of the pistil in a grass floret.
It swells to push the bracts (lemma and palea) apart during flowering.
= **squamula**

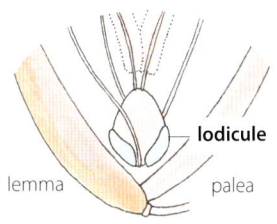

lemma

lodicule

palea

loment, lomentum, *pl.* lomenta

A legume-like pod that is contracted between the seeds and separates at maturity into one-seeded segments that don't split open. Derived from a one-carpelled superior ovary, as tick-trefoil (*Desmodium*).
lomentaceous   Having the appearance of or bearing of a loment.

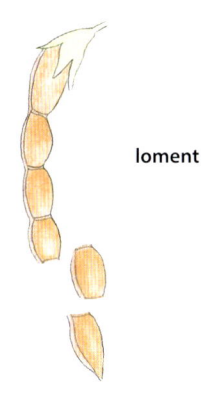

loment

long-creeping

Of ferns, having a rhizome that elongates quickly so that the fronds are some distance apart, as lance-leaf tongue fern (*Pyrrosia lanceolata*).
cf. **short-creeping**

long-creeping

longi-   A prefix meaning long.

longitude

A line on the surface of a sphere connecting the poles.
cf. **latitude**

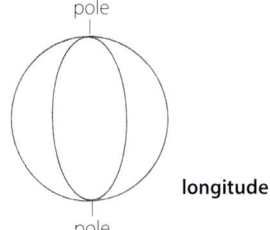

pole

pole

longitude

longitudinal dehiscence

Of anthers, opening lengthwise to release pollen, as those of devil's trumpet (*Datura*).
see also **anther dehiscence**

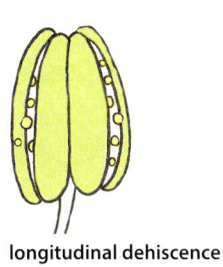

longitudinal dehiscence

lopha, *pl.* lophae

A network-like pattern of ridges surrounding spaces or depressions.
lophate   Having lophae.
cf. **lacuna**

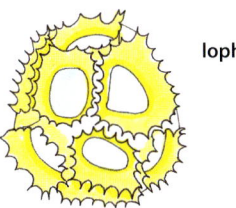

lopha

lorate Strap-shaped, as some leaves.

lumen, *pl.* **lumina** A cavity enclosed by a cell wall, as that in the centre of a fibre cell.
A space enclosed by ridges.
cf. **murus**

Lumen

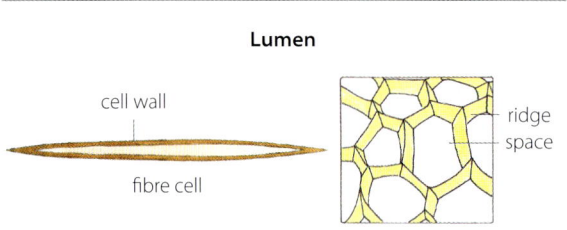

lumper A taxonomist who concentrates on similarities rather than subtle differences to sub-divide species, resulting in a smaller number of taxa.

lunate Shaped like a crescent.

lunulate Shaped like a small crescent.
Having crescent-shaped markings.

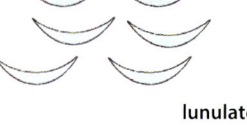

lustrous Having a smooth shining surface, glowing.

Lycophyta, lycophytes One of two groups of vascular plants that bear spores rather than seeds, the other group being Monilophyta (ferns).They possess small leaves with one vein (microphylls). Lycophyta comprises quillworts (*Isoetes*), clubmosses (*Lycopodium*) and spike mosses (*Selaginella*).
see also **fern allies**
cf. **euphyllophytes**

lyrate Lyre-shaped.
Of a leaf, lobed to resemble a lyre, with the terminal lobe or leaflet largest and the lower lobes or leaflets smaller. The leaf may be pinnate or lobed
see **lyrate-pinnate, lyrate pinnatifid**

lyrate-pinnate Of a pinnate leaf where the terminal leaflet is largest and the others successively smaller towards the base.
cf. **lyrate-pinnatifid**

lyrate-pinnatifid Of a pinnately lobed leaf with the terminal lobe largest and the lower lobes smaller.
cf. **lyrate-pinnate**

lysigenous Of intercellular spaces formed due to the breakdown of entire cells by their own enzymes.
cf. **schizogenous**

lysosome In the cytoplasm of an animal cell, one of the sacs with enzymes that break down cellular waste and debris.
Sacs in plants with similar functions are considered by some to be lysosomes.

macro- A prefix meaning large or long.
cf. **micro-**

macrocotyledon The larger of two unequally developed cotyledons that grows to resemble a foliage leaf, as nodding violet (*Streptocarpus caulescens*).
A petiolode with normal leaves arises from the mesocotyl situated between the microcotyledon and the macrocotyledon.
see **anisocotyly**
cf. **microcotyledon**

Macrocotyledon

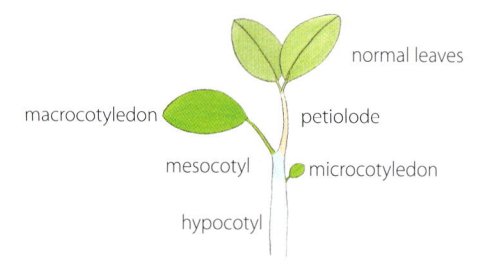

macroevolution Evolution of large groups (above species level) over the history of life, as mammals and flowering plants.
cf. **microevolution**

macronutrient   Nutrients needed by a plant in higher amounts than other nutrients, as nitrogen, potassium and phosphorus.
cf. **micronutrient, trace element**

macrophyll   A leaf with a complex system of veins. Typical of seed plants and ferns.
= **megaphyll**
cf. **microphyll**
macrophyllous   With large leaves.
= **megaphyllous**

macropodal, macropodous
Of an embryo having an enlarged hypocotyl and lacking well-developed cotyledons, as souari trees (*Caryocar*).

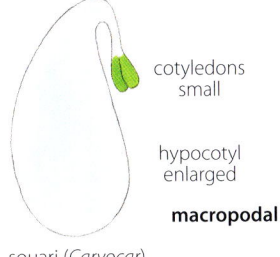

cotyledons small

hypocotyl enlarged

macropodal

souari (*Caryocar*)

macrospore   Another term for megaspore.

macula, *pl.* **maculae**   A spot or blotch.
maculate   Spotted or blotched.

maidenhair fern
Maidenhair ferns (*Adiantum*) belong to one of about 50 genera of mostly terrestrial ferns in the family (Pteridaceae), that have variously divided fronds and erect or creeping scaly or woolly rhizomes.
see **fern**

maidenhair fern
(*Adiantum*)

male flower   A flower with functional stamens but no functional carpels.

malodorous, malodourous   Smelling unpleasant or offensive.

maltose   A sugar fromed by the action of diastase on starch.

mammilla,
pl. **mammilae**
A nipple-shaped protuberance.
mammiform   Shaped like a breast or teat.
mammillate   With nipple-like projections, as the cactus genus *Mammillaria*.

mammila

mangal   One of the mangrove forests along much of the coastline in tropical areas of the world.

mangrove   A shrub or small tree variously adapted to growing in brackish water and the muddy, oxygen-poor soils of tidally inundated areas along the coast and in river estuaries.
Adaptations include prop roots for support and breathing roots that project above the mud and have small openings (lenticels) for gas exchange with the atmosphere.
Found in many families, including Rhizophoraceae, Acanthaceae, Combretaceae and Arecaceae.
The name given to the habitat in which these plants grow.
see also **mangal, vivipary**

Mangrove

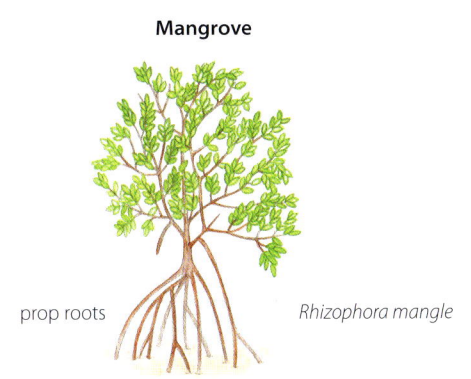

prop roots

Rhizophora mangle

mantle   An outer covering.
Of fungi, a network of hyphae that form a sheath, as that around the outside surface of some roots.
see **ectomycorrhiza**

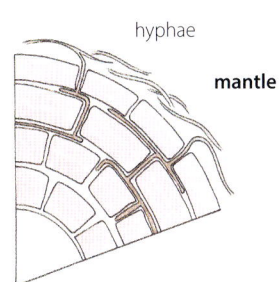

hyphae

mantle

maquis   A dense shrubby vegetation more than three metres high, growing on siliceous soils in low rainfall areas in the Mediterranean Basin.
It is a Mediterranean-type ecosystem together with the garrigue, also in the Mediterranean Basin, the chaparral in California, matorral in Chile, kwongan in southwestern Australia and fynbos in South Africa.

marcescent   Withered but remaining attached, as the leaves of some plants and the fronds of some ferns or palms.

marcescent

margin   An edge or border.
marginal   On the edge or border, relating to the margin.

marginal placentation
With ovules arranged along the fused margins of a single free carpel (simple ovary), as peas (*Pisum*) that have one free carpel, and buttercups (*Ranunculus*) that have several free carpels.
see **placentation**

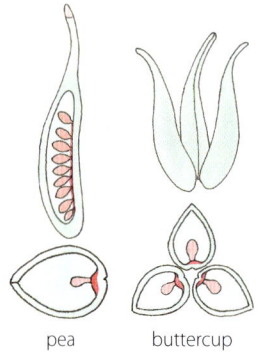
pea   buttercup
marginal placentation

marginal venation
Of leaf venation with the veins reaching the margin of the leaf.
cf. **intramarginal**

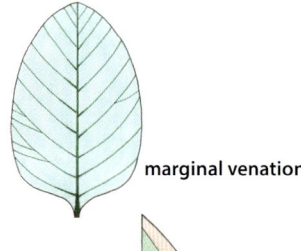
marginal venation

marginate
With a distinct margin or border, as the margins of some leaves.
cf. **emarginate**

marginate

margo   An ornamented margin surrounding a colpus on the wall of a pollen grain.
cf. **annulus**

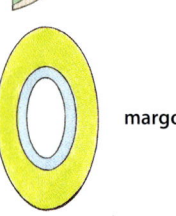
margo

marine   Of or relating to the sea, as marine habitat rather than a freshwater habitat.

marsh   A freshwater or saltwater wetland that is dominated by herbaceous plants, as grasses, sedges rushes, low shrubs and forbs rather than trees.
see also **halophyte**
cf. **bog, swamp**

massula, *pl.* **massulae**
Of aquatic ferns in the Salviniaceae family, a group of microspores enclosed in a hardened mucilage, often with hooked bristles.
Of orchids, a cohering mass of pollen grains produced from a single pollen mother cell.
Many of these make up a pollinium.

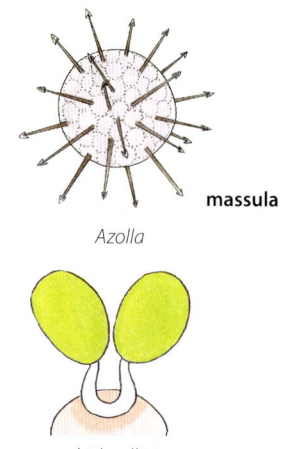
massula
Azolla

orchid pollinia

mast   The fallen fruit of trees and shrubs that are eaten by wildlife, as acorns (*Quercus*) and chestnuts (*Castanaea*).

masting   Within a species, mass synchronised flowering and fruiting, at intervals, across a large area.
see also **gregarious flowering, plietesial, semelparous**

mat grass   Any of various stoloniferous or rhizomatous grasses, with tillers on long, often branched lateral stems that spread across the ground.
cf. **bunch grass**

mat grass

mat-forming   Prostrate and dense, as some foxtails (*Ptilotus*), or a plant spreading by sending out roots from the nodes on the horizontal stems, as creeping thyme (*Thymus serpyllum*).

matorral   A vegetation type in Chile, with drought-deciduous and evergreen sclerophyllous shrubs, trees and diverse herbaceous plants having winter rain and summer droughts.
It is a Mediterranean-type ecosystem together with the garrigue and maquis in the Mediterranean Basin, the chaparral in California, kwongan in southwestern Australia and fynbos in South Africa.

matrix   The material or tissue in which more specialised structures are embedded.

matt, matte   Having a dull lustreless surface that reflects little if any light.
cf. **glossy**

matted   With hairs tangled and adhering closely together.
Tomentose, as the soft cottony or woolly hairs of cudweed (*Gnaphalium*).

matted

mature   Fully developed, ripe.
cf. **immature, senescent**

matutinal   Occurring or active early in the morning, as the flowers of morning glory (*Ipomoea purpurea*) that open at dawn.
cf. **crepuscular, diurnal, nocturnal, vespertine**

mealy
Covered with a crumbly, scurfy flour-like powder, as young saltbushes (*Atriplex*).

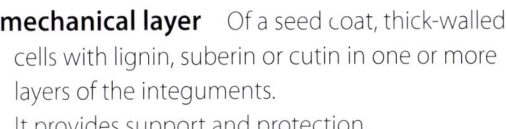

mealy

mechanical layer   Of a seed coat, thick-walled cells with lignin, suberin or cutin in one or more layers of the integuments.
It provides support and protection.

medial, median   Of or situated at or towards the middle, as the midvein of a leaf.
Midway between the base and the apex.
cf. **admedial, distal, exmedial, proximal**

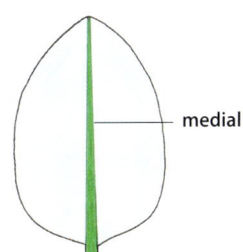

medial

medifixed   Attached at or by the middle, as a stamen filament attached to the middle of the connective at the back of an anther.
cf. **dorsifixed, ventrifixed**
see **anther attachment**

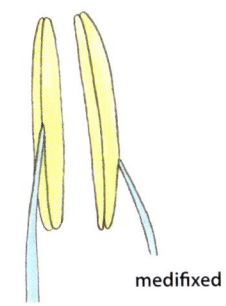

medifixed

Mediterranean-type ecosystems
Characterised by mild wet winters and warm dry summers and dominated by evergreen sclerophyllous shrublands.
Termed maquis and garrigue in the Mediterranean Basin, chaparral in California, matorral in Chile, fynbos in South Africa and kwongan in southwestern Australia.

medulla   Pith.

medullary ray   Of primary growth, one of the bands of mostly parenchyma tissue in young eudicot stems and gymnosperms extending radially from the pith (medulla) between the vascular bundles to the cortex.
Medullary rays store nutrients and transport them radially compared with the vascular bundles that transport nutrients vertically.
see **vascular ray**
see also **ray initials**

Medullary ray

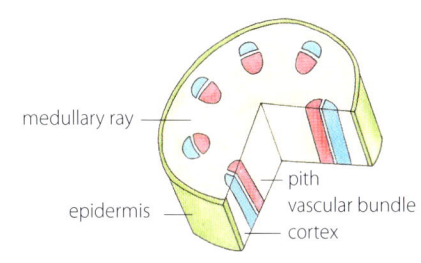

medullary ray

epidermis

pith
vascular bundle
cortex

mega-   A prefix meaning large in size or amount: one million units of the base word.

megagamete   The larger female gamete formed by plants that have unequal-sized male and female gametes.
= **egg cell, ovum**
see **megagamophyte**
cf. **microgamete**

megagametogenesis
Formation of a megagamete from a megaspore mother cell.
After meiosis three of the microspores degenerate and the one remaining microspore is the megamete (egg cell).
megagametogenesis page 176 (cont.)
cf. **microsporogenesis**

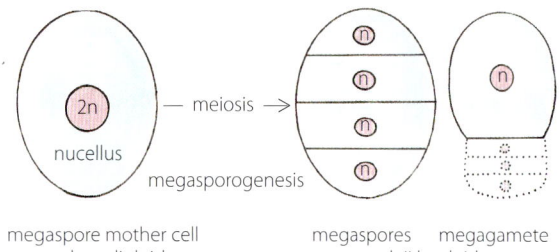

Megagametogenesis

megaspore mother cell
nucleus diploid

meiosis →

megasporogenesis

megaspores
nucleii haploid

megagamete

megagametophyte

Of seed plants, the female gametophyte that develops within each ovule.

In angiosperms it commonly consists of seven cells: three antipodal cells, two synergid cells, one egg cell and one central cell with two polar nuclei.
Found in the ovule that is enclosed in the ovary.

= **embryo sac**

In gymnosperms, the haploid nutritional tissue, formed from the megaspore, in which the archegonia develop.
Found in the exposed ovule on the upper surface of a cone scale on the female cone.

cf. **microgametophyte**

Megagametophyte

Angiosperms

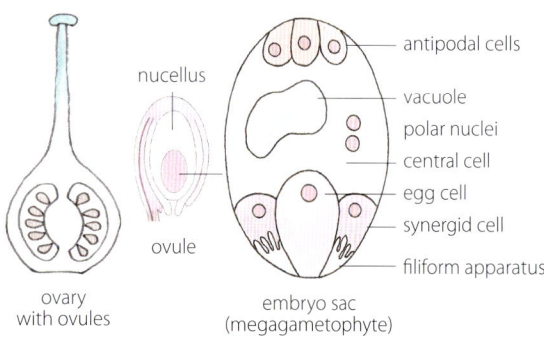

nucellus

ovule

ovary
with ovules

antipodal cells
vacuole
polar nuclei
central cell
egg cell
synergid cell
filiform apparatus

embryo sac
(megagametophyte)

Gymnosperms

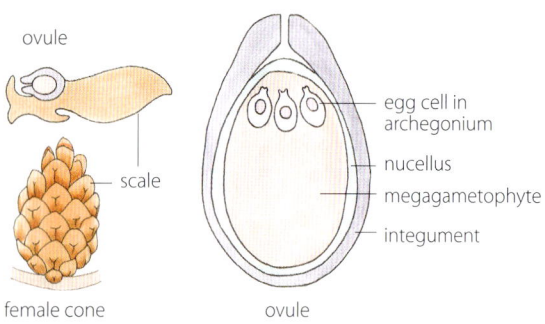

ovule

scale

female cone

egg cell in
archegonium
nucellus
megagametophyte
integument

ovule

megalo- A prefix meaning exceptionally large.

megaphyll A leaf with a complex system of veins; typical of seed plants and ferns.
= **macrophyll**
cf. **microphyll**
megaphyllous With large leaves.
= **macrophyllous**

megasporangium, *pl.* megasporangia

The structure in which megaspores are formed in a heterosporous plant.

In seed plants (gymnosperms and angiosperms), the tissue in the immature ovule in which megaspores are formed.

= **nucellus**

Some fern allies, like the spike moss (*Selaginella*) are heterosporous. Spike moss has megasporangia in the axils of the leaves (sporophylls) and each consists of a short stalk and a wall enclosing the sporogenous tissue from which the megaspores are formed.

see **megasporogenesis**
cf. **microsporangium**

Megasporangium

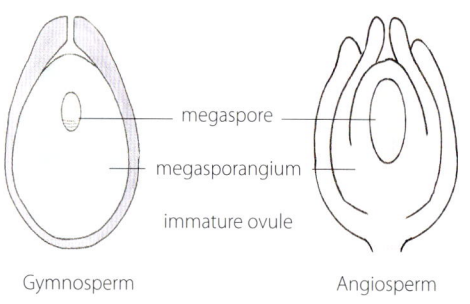

megaspore
megasporangium
immature ovule

Gymnosperm

Angiosperm

megaspore One of two kinds of spores, produced by meiosis in a heterosporous plant. It gives rise to the female gametophyte (megagametophyte).
All seed-bearing plants, as well as some ferns and other seedless plants, are heterosporous.
In angiosperms and gymnosperms, it is one of usually four haploid cells produced by the diploid megaspore mother cell. Three of these cells degenerate and the remaining megaspore develops into the the embryo sac (megagametophyte).
The fern ally, spike moss *(Selaginella)* is a seedless heterosporous plant that produces megaspores in megasporangia located in the leaf axils on the upper part of the stem.
= **macrospore**
see **macrosporangium, macrosporogenesis**
cf. **microspore**

Megaspore

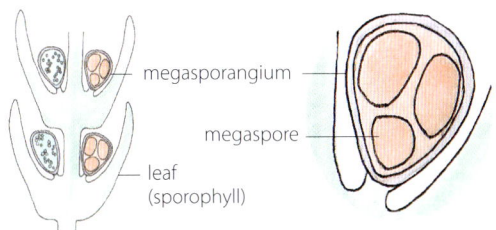

megaspore

megasporangium

immature ovule

Gymnosperm   Angiosperm

megasporangium

megaspore

leaf
(sporophyll)

Spike moss (*Selaginella*)

megaspore mother cell

In seed plants, a diploid cell in the immature ovule that gives rise, by meiosis, to four female haploid spores (megaspores).

There are usually four megaspores and three of these disintegrate.

The one remaining megaspore develops into the megagametophyte.

= **megasporocyte**

see **megasporogenesis, megasporangium, mother cell**

cf. **microspore mother cell**

Megaspore mother cell

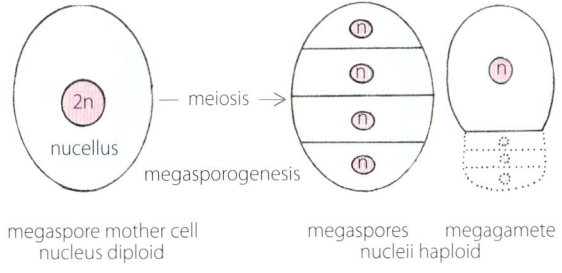

2n

nucellus

meiosis →

megasporogenesis

n
n
n
n

n

megaspore mother cell
nucleus diploid

megaspores
nucleii haploid

megagamete

megasporocarp

The reproductive structure in some heterosporous fern families consisting of a single megasporangium with a single megaspore.

see **sporocarp**

cf. **microsporocarp**

Megasporocarp

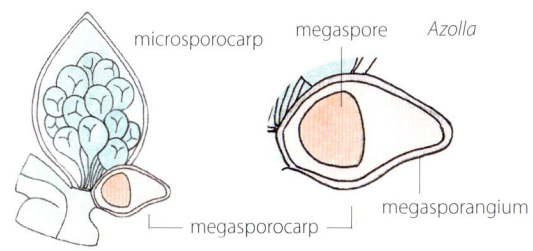

microsporocarp   megaspore   *Azolla*

megasporangium

megasporocarp

megasporocyte

In seed plants, a diploid cell in the immature ovule that gives rise, by meiosis, to four female haploid spores (megaspores).

There are usually four megaspores and three of these disintegrate, with the one remaining megaspore developing into the megagametophyte.

= **megaspore mother cell**

see **mother cell**

cf. **microsporocyte**

megasporogenesis

In heterosporous plants, the process of forming, by meiosis, haploid megaspores from the diploid megaspore mother cell.

Macrosporogenesis takes place in the megasporangium.

There are usually four megaspores and three of these disintegrate.

The one remaining megaspore develops into the megagametophyte.

cf. **megagametogenesis, microsporogenesis**

Megasporogenesis

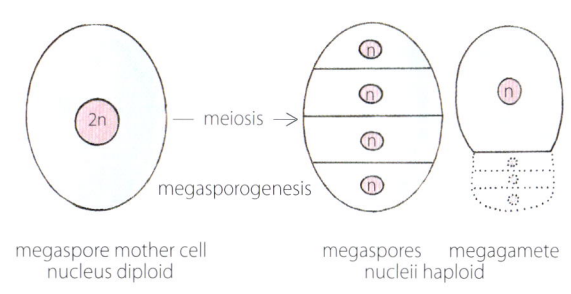

2n

meiosis →

megasporogenesis

n
n
n
n

n

megaspore mother cell
nucleus diploid

megaspores
nucleii haploid

megagamete

megasporophyll

A modified leaf that bears the megasporangia in a heterosporous plant.

In angiosperms, it is the ovule-bearing carpel.

In gymnosperms, it is the ovule-bearing scale, as on a female pine cone.

megasporophyll page 178 (cont.)

In some heterosporous ferns allies, like spike moss (*Selaginella*), it is a scale-like leaf on the cone at the top of the stem.
see **megasporangium**
cf. **microsporophyll**

Megasporophyll

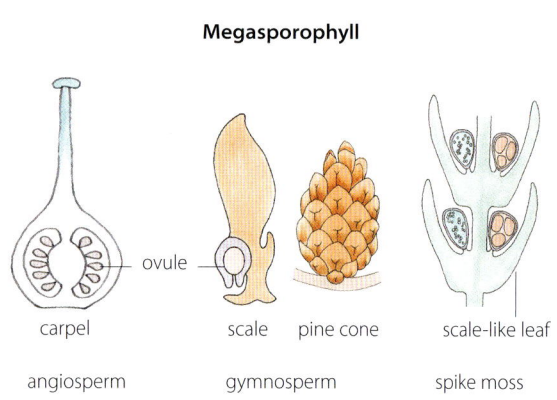

ovule

| carpel | scale | pine cone | scale-like leaf |
| angiosperm | gymnosperm | | spike moss |

megastrobilus,
pl. **megastrobili**
The ovule-producing cone of gymnosperms, as the female pine cone. Rarely, it can be fleshy, as the cones of juniper (*Juniperus communis*).
cf. **microstrobilus**

megastrobilus

pine

juniper

meiosis *see* page 179

melittophile One of thousands of plant species pollinated by bees.

melittophily Dispersal of pollen and pollination of bees.
melittophilous Pollinated by bees.

membrane A thin, rather soft, pliable and semi-transparent tissue.
membranaceous, membranous, membraneous Like a membrane, as the stipules of some leaves.

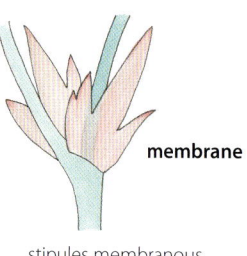

membrane

stipules membranous

Mendelian inheritance The key ideas of this model of inheritance are that traits are determined by genes, that genes come in pairs and in different versions called alleles, that gametes receive one of each gene from each parent, and that traits are inherited independently of each other.

mentum In some orchids, a chin-like projection formed by the bases of the column and the lateral sepals, as the genus *Dendrobium*.

mericarp
A single-seeded dry fruitlet of a schizocarp. It may be indehiscent, as the carrot family (Apiaceae) or dehiscent and releasing a seed, as geranium (*Geranium*).
see **carcerule, regma**
= **coccus**

indehiscent **mericarp**

carrot family (Apiaceae)

geranium (*Geranium*)

merocarp dehiscent

meristem A region of primary undifferentiated cells (initial cells) that divide and grow continuously. It is classified according to its location.
Apical meristem occurs at the tips of shoots and roots, lateral meristem occurs in vascular tissue (vascular cambium and cork cambium) and intercalary meristem between two mature tissues. Meristematic cells differentiate to form plant tissues.
meristematic Of or relating to a meristem.

merosity The number of parts in each whorl of a plant structure, as the number of sepals in a calyx, the number of petals in a corolla, the number of stamens in an androecium and the number of carpels in a gynoecium.
see **numerical plan**

Merosity

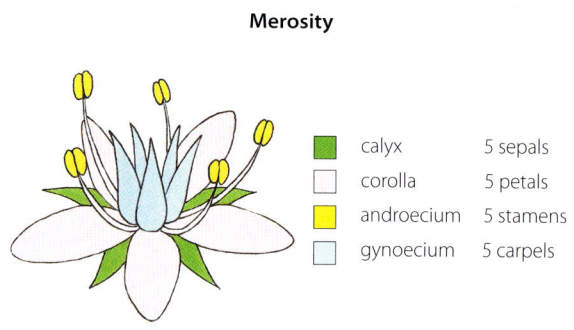

calyx	5 sepals
corolla	5 petals
androecium	5 stamens
gynoecium	5 carpels

-merous, -merus Suffix indicating number of parts.
see **dimerous, monomerous, pentamerous, polymerous, tetramerous, trimerous**

meiosis   The process of division of a diploid reproductive cell.
The cell divides twice to produce four haploid daughter cells that are genetically unique due to crossing over.
At the end of meiosis the cell is haploid.
cf. **mitosis**
meiotic   Of or pertaining to meiosis.

Meiosis

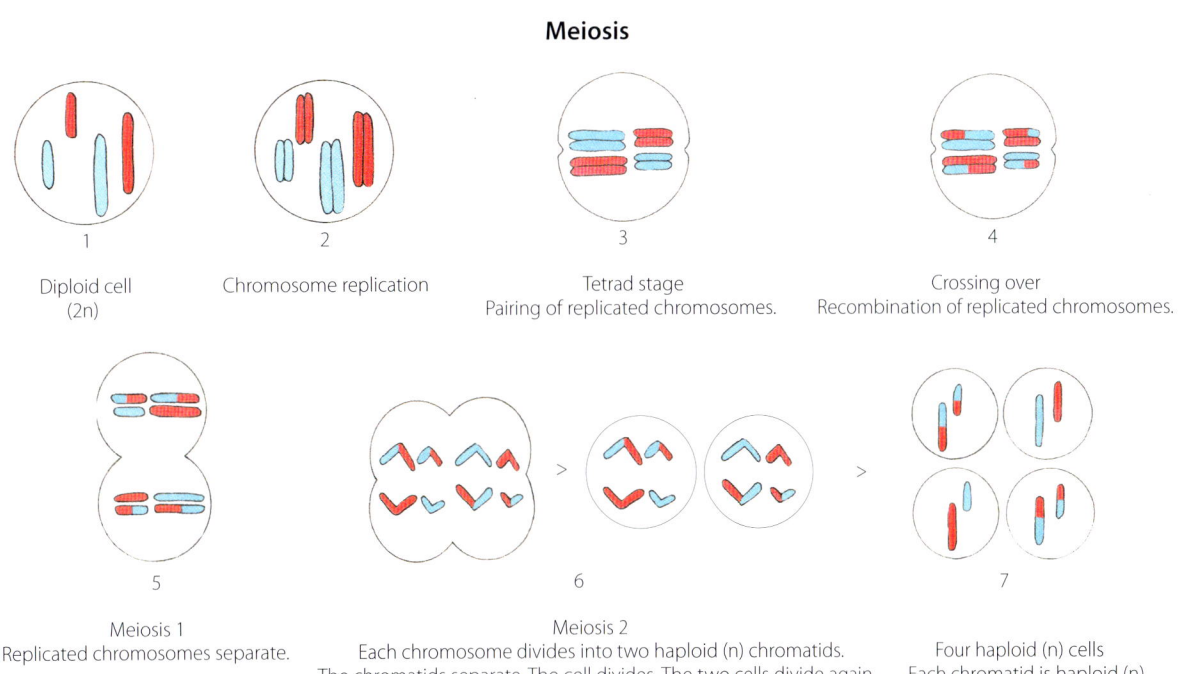

1	2	3	4
Diploid cell (2n)	Chromosome replication	Tetrad stage Pairing of replicated chromosomes.	Crossing over Recombination of replicated chromosomes.

5
Meiosis 1
Replicated chromosomes separate.

6
Meiosis 2
Each chromosome divides into two haploid (n) chromatids.
The chromatids separate. The cell divides. The two cells divide again.

7
Four haploid (n) cells
Each chromatid is haploid (n).

mesarch   Describes radial differentiation of xylem according to the relative position of protoxylem and metaxylem in which protoxylem is positioned in the centre of the stem and is surrounded by metaxylem. Found in many species of ferns.
cf. **endarch, exarch**

mesic   Of, related to or adapted to an environment that receives a moderate or well-balanced supply of moisture, as a temperate hardwood forest or prairies that have good drainage and high moisture available during the growing season.
cf. **hydric, xeric**

meso-   A prefix meaning middle.

mesocarp   The middle layer of the fruit wall that is usually fleshy, as the succulent layer of a stone fruit.
see **pericarp**

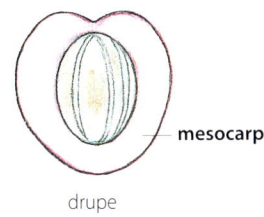

drupe

mesochile   Of orchids, the middle portion of the labellum, as pink fairies (*Caladenia latifolia*).

hypochile

mesochile

epichile

labellum

mesocotyl   Of grasses (Poaceae), part of the embryo axis, or of a seedling, below the coleoptile. Elongation of the mesocotyl during germination pushes the coleoptile above the soil surface.
cf. **epicotyl, hypocotyl**

Mesocotyl

coleoptile
shoot primordia
meristem
mesocotyl
hypocotyl
radicle
coleorrhiza

embryo

true leaf
eophyll
coleoptile
nodal roots
mesocotyl
caryopsis
radicle

seedling
grass (*Poaceae*)

mesogamy   Entrance of the pollen tube through the middle part, the integuments or funicle, of the ovule.
cf. **chalazogamy, porogamy**

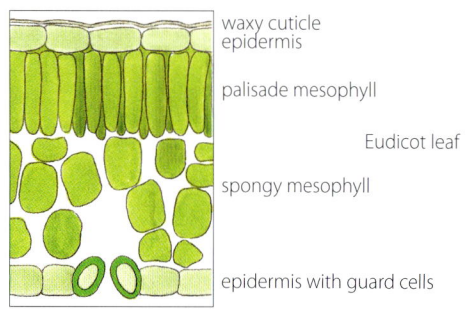

mesogamy

mesophyll
Photosynthetic cells that make up the bulk of the internal tissue of a leaf.
Eudicots have two types of mesophyll: palisade mesophyll and spongy mesophyll.
Mesophyll cells in monocotyledonous leaves are not differentiated into two types.
see **chlorenchyma**

Mesophyll

waxy cuticle
epidermis
palisade mesophyll
Eudicot leaf
spongy mesophyll
epidermis with guard cells

mesophyte
A plant that requires adequate water during its growing season but does not tolerate environmental extremes like drought.
Most angiosperms, grasses, herbs and woody plants are mesophytes.
They may be evergreen or deciduous.
cf. **tropophyte, xerophyte**

mesotegmen
Of the bitegmic seed coat of angiosperms, the middle layer of the inner integument (tegmen).
cf. **endotegmen, exotegmen**
see also **endotesta, exotesta, mesotesta**

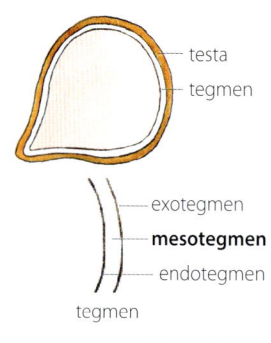

testa
tegmen

exotegmen
mesotegmen
endotegmen
tegmen

chick pea seed (*Cicer arietinum*)

mesotegmic seed   Of angiosperms, having the mechanical layer in the mesotegmen.
see **tegmic seed**

mesotesta,
pl. **mesotestae**
Of the bitegmic seed coat of angiosperms, the middle layer of the outer integument (testa).
cf. **endotesta, exotesta**
see also **endotegmen, exotegmen, mesotegmen**

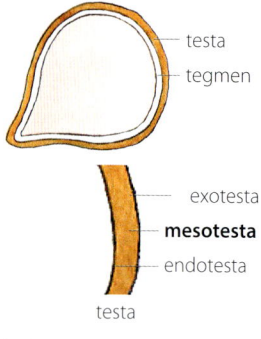

testa
tegmen

exotesta
mesotesta
endotesta

testa

chick pea seed (*Cicer arietinum*)

mesotestal seed   Of angiosperms, having the mechanical layer in the mesotesta.
see **testal seed**

mesotonic   Having growth strongest in the median part of the plant.
cf. **acrotonic, basitonic**

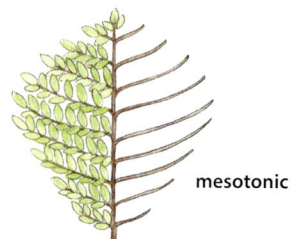

mesotonic

mesotrophic   Of a body of water having a moderate amount of dissolved nutrients, as a lake with clear water and submerged aquatic plants.
see **trophic**
cf. **dystrophic, eutrophic, oligotrophic**

messenger RNA, mRNA   A molecule in a cell that carries information encoded in DNA to a ribosome, a specialised organelle in the cytoplasm, for protein synthesis.

meta-   A prefix meaning changed or substituted for.

metabolism   The complex physical and chemical processes that together allow a plant to live and survive (as photosynthesis and respiration).
It promotes growth and development.
see **anabolism, catabolism**

metamer

The basic structural unit of a plant that is repeated.

It derives from the meristem of a root or shoot apex.

Of stem shoots, it comprises a node with its leaf or leaves, its axillary bud(s) and the subtending internode. Metamers also occur on lateral growth.

= **phytomer**

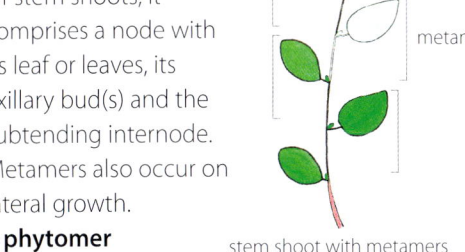

metamer

stem shoot with metamers

metaphloem

Phloem tissue that differentiates from procambium in the apical meristem of root and shoot tips during primary growth of a vascular plant. It replaces protophloem.

It is conducting tissue that occurs in regions that are no longer actively elongating and functions indefinitely.

see **primary phloem**

metaxylem

Found in primary growth, the primary xylem in a vascular bundle that develops from the procambium after protoxylem.

It is distinguished from protoxylem by having more tracheids and cells that are not capable of elongating.

see **primary xylem**

micro-

A prefix meaning small.

cf. **macro-**

microbe

A mostly unicellular living organism that is too small to be seen with the naked eye.

Includes bacteria and some fungi.

= **microorganism**

microclimate

A smaller area within a climate zone that has its own unique climate.

microcotyledon

The smaller of two unequally developed cotyledons that grows to resemble a tiny foliage leaf, as nodding violet (*Streptocarpus caulescens*).

A petiolode with normal leaves arises from the mesocotyl situated between the microcotyledon and the macrocotyledon.

see **anisocotyly**

cf. **macrocotyledon**

Microcotyledon

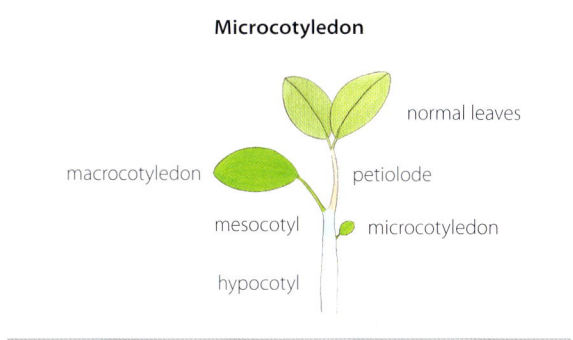

microevolution

Change in allele frequency within the population of a species.

Observable over a short period of time.

cf. **macroevolution**

microgamete

One of two male sperm cells in the pollen grain that are derived from the generative cell.

see **microgametogenesis, pollen tube**

cf. **megagamete**

microgametogenesis

Formation of pollen grains from microspores and development of microgametes in the pollen grain.

cf. **megagametogenesis, microsporogenesis**

Microgametogenesis

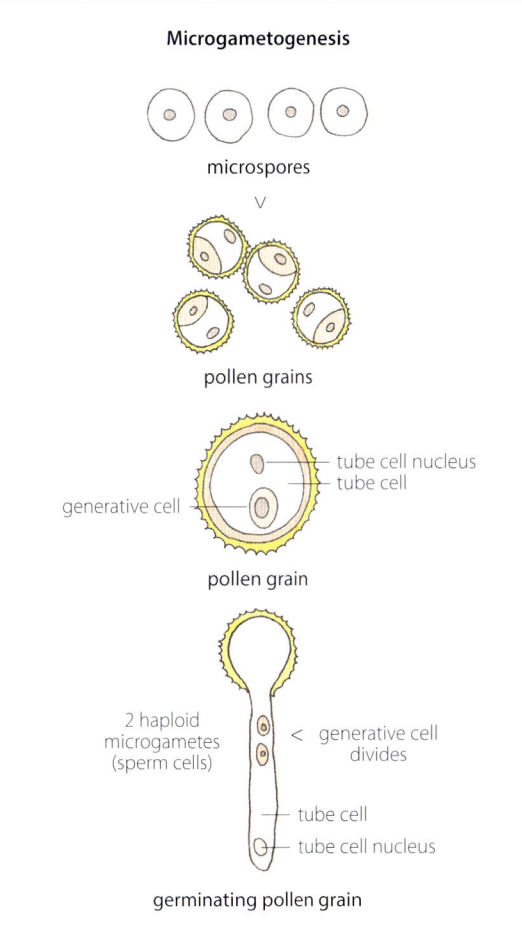

microgametophyte   Of seed plants, a pollen grain, the male gamete-bearing entity produced from a microspore.

In angiosperms, pollen grains develop in the pollen sacs in the anthers. Each grain has two cells. The reproductive generative cell divides to form two male gametes (sperm cells). The vegetative cell, the tube cell, elongates during germination and carries the gametes to the egg.

In gymnosperms, the pollen grains develop in the pollen sacs, typically exposed on the lower surface of cone scales on the male cone. The generative cell produces two gametes but only one survives.

Grains may be solitary or cohere in units of two (diads), four (tetrads), eight or sixteen and so on, for dispersal.

= **pollen grain**

see also **massula, pollen tetrad, pollen wall, pollinia, polyad**

cf. **megagametophyte**

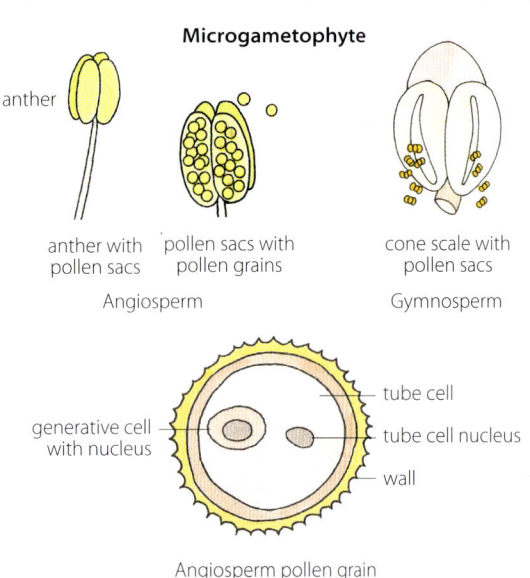

Microgametophyte

anther

anther with pollen sacs | pollen sacs with pollen grains

Angiosperm

cone scale with pollen sacs

Gymnosperm

generative cell with nucleus — tube cell — tube cell nucleus — wall

Angiosperm pollen grain

micronutrient

Nutrients essential to plant growth and health that are only needed in very small quantities, including iron, manganese, boron, zinc, molybdenum, chlorine and copper.

= **trace element**

cf. **macronutrient**

microorganism

A mostly unicellular living organism that is too small to be seen with the naked eye.

Includes bacteria and some fungi.

= **microbe**

microphyll

A usually small, sessile leaf with only one vein. Unique to lycophytes like spike mosses (*Selaginella*), quillworts (*Isoetes*) and clubmosses (*Lycopodium*).

cf. **megaphyll**

microphyllous   With small leaves.

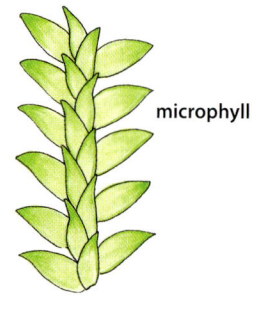

microphyll

Selaginella

micropropagation

The production of a new plant or plants from a small piece of plant tissue grown in a nutrient medium.

cf. **propagation**

micropyle

The opening in the integuments of an ovule through which the pollen tube grows to reach the egg cell in the embryo sac.

At germination the radicle usually emerges through the micropyle.

micropylar   Relating to a micropyle.

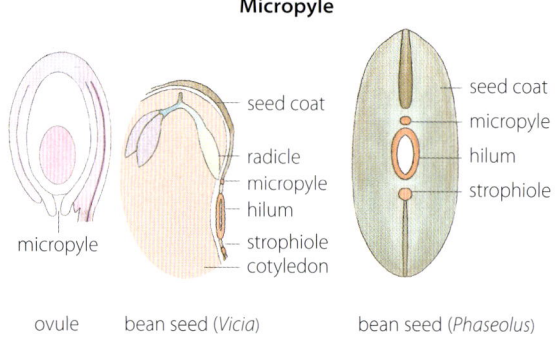

Micropyle

seed coat — radicle — micropyle — hilum — strophiole — cotyledon

micropyle

ovule | bean seed (*Vicia*)

seed coat — micropyle — hilum — strophiole

bean seed (*Phaseolus*)

microsporangial wall

The wall of an immature pollen sac (microsporangium) that surrounds a mass of sporogenous tissue.

In angiosperms it has four layers: the epidermis, endothecium, middle layers and tapetum. Gymnosperms, conifers for example, have three layers: the epidermis, a single middle layer and a tapetum.

Microsporangial wall

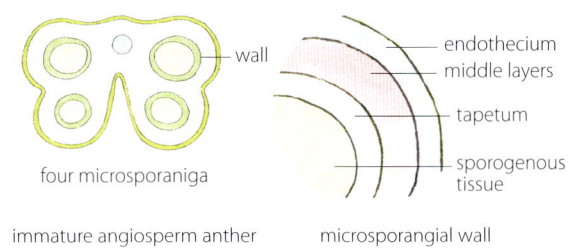

four microsporaniga

immature angiosperm anther

wall

endothecium
middle layers
tapetum
sporogenous tissue

microsporangial wall

microsporangium, *pl.* microsporangia

The structure in which microspores are formed in a heterosporous plant.

In angiosperms the microsporangium is the pollen sac in the anther and in gymnosperms it is the pollen sac borne on the lower surface of scales (microsporophylls) on a male cone.

Some fern allies, like the spike moss (*Selaginella*) are heterosporous. Spike moss has microsporangia in the axils of the upper leaves (sporophylls).

cf. **megasporangium**

microsporangial   Of or relating to microsprangia.

Microsporangium

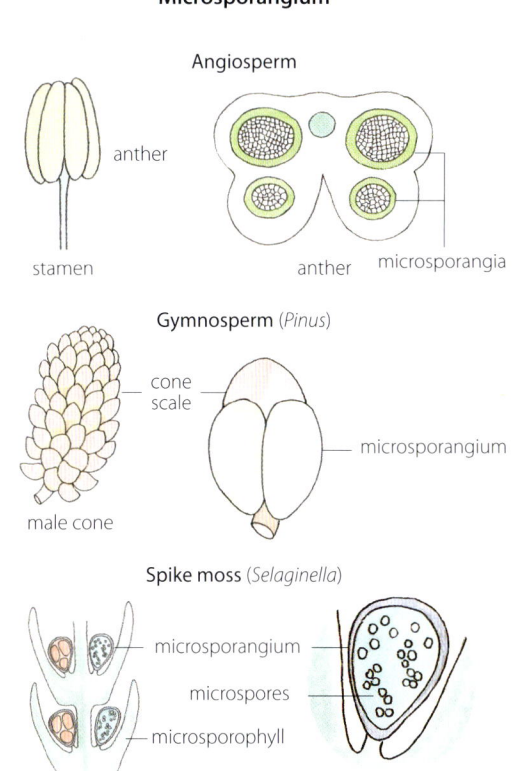

Angiosperm

anther

stamen

anther   microsporangia

Gymnosperm (*Pinus*)

cone scale

microsporangium

male cone

Spike moss (*Selaginella*)

microsporangium

microspores

microsporophyll

microspore   One of two kinds of spores, produced by meiosis in a heterosporous plant, that gives rise to the male gametophyte (microgametophyte). All seed-bearing plants, as well as some ferns and other seedless plants, are heterosporous.

In angiosperms and gymnosperms, one of four haploid cells produced by a diploid microspore mother cell that develops into a pollen grain (microgametophyte).

The fern ally, spike moss (*Selaginella*) is a seedless heterosporous plant that produces microspores in microsporangia located in the leaf axils on the upper part of the stem.

see **microsporogenesis, microsporangium**

cf. **megaspore**

microspore mother cell   A diploid cell in the immature pollen sac that gives rise, by meiosis, to four haploid male spores.

In seed plants (angiosperms and gymnosperms) the mother cells are in the pollens sacs.

The four microspores are usually viable and will develop into pollen grains.

see **microsporogenesis, microsporangium, mother cell**

cf. **megaspore mother cell**

microsporocarp

The male reproductive structure in some heterosporous ferns, as *Azola*.

It encloses many microsporangia. The microsporangia contain the microspores.

see **sporocarp**

cf. **megasporocarp**

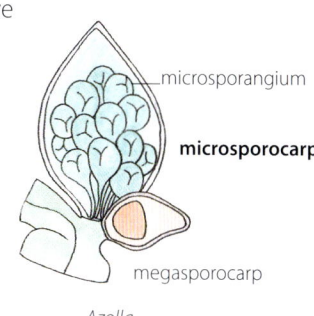

microsporangium

microsporocarp

megasporocarp

Azolla

microsporocyte   In seed plants, a diploid cell in the immature pollen sac that that gives rise, by meiosis, to male haploid spores (microspores). These will develop into the microgametophytes.

= **microspore mother cell**

see **mother cell**

cf. **megasporocyte**

microsporogenesis   In heterosporous plants, the division, by meiosis, of a diploid microspore mother cell to form four haploid microspores.
These will develop into microgametophytes.
Microsporogenesis takes place in the microsporangium.
cf. **megasporogenesis, microgametogenesis**

Microsporogenesis

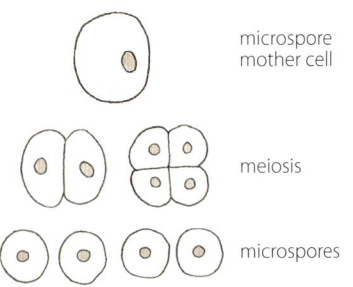

microspore mother cell

meiosis

microspores

microsporophyll   A modified leaf that bears the microsporangia in a heterosporous plant.
In angiosperms, it is the anther that bears the pollen sac.
In gymnosperms, it is the pollen sac-bearing scale, as on a male pine cone.
In some heterosporous ferns allies, like spike moss *(Selaginella)*, it is a leaf-like scale on the cone at the top of the stem.
see **microsporangium**
cf. **megasporophyll**

microstrobilus,
   pl. **microstrobili**   The male, pollen-producing cone of gymnosperms.
   cf. **megastrobilus**

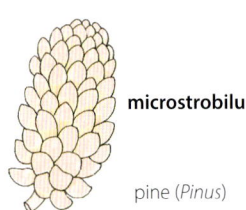

microstrobilus

pine *(Pinus)*

middle lamella   The adhesive layer between two fully grown adjacent cell walls that binds them to one another and is mainly composed of pectin.
see **primary cell wall, secondary cell wall**

Middle lamella

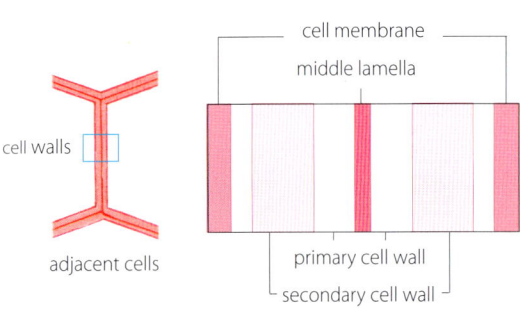

cell membrane
middle lamella
cell walls
adjacent cells
primary cell wall
secondary cell wall

middle layer   Of angiosperms, one to three layers of the wall of an immature pollen sac (microsporangium) that is made up of parenchyma cells.
Gymnosperms have a single middle layer in the wall of the immature pollen sac (microsporangium).
see **microsporangial wall**

midlobe
Of an orchid with a three-lobed labellum, the middle lobe, as pink fairies, *(Caladenia latifolia)*.

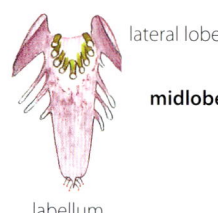

lateral lobe   lateral lobe

midlobe

labellum

midrib   The main vein of a leaf or leaflet, usually running up the centre as a continuation of the petiole or petiolule.
= **midvein, primary vein**

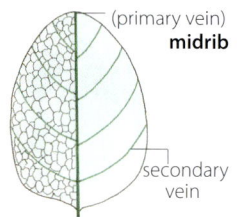

(primary vein)
midrib
tertiary veins (veinlets)
secondary vein

midvein   The main vein of a leaf or leaflet, usually running up the centre as a continuation of the petiole or petiolule.
= **midrib, primary vein**

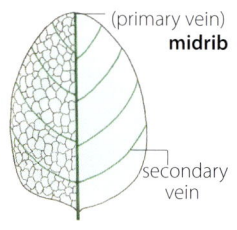

(primary vein)
midrib
tertiary veins (veinlets)
secondary vein

migration   The exchange of genes between individuals within a population or between different populations of the same species.

mimic   To copy or imitate closely, as orchid flowers that mimic a female wasp to attract a male wasp pollinator.
see **pseudocopulation**

mimicry   A form of deception whereby one organism adopts the colour, smell or structure of another organism.
see **Bakerian mimicry, Dodsonian mimicry, Pouyannian mimicry, Vavilovian mimicry**

mitochondrial DNA, mtDNA   The small amount of DNA unique to mitochondria.
Unlike nuclear DNA, mtDNA is arranged in rings.

mitochondrion,

pl. **mitochondria**   An organelle in the cytoplasm of a cell enclosed by a double membrane. It produces energy in the form of adenosine triphosphate (ATP).

see also **aerobic respiration, glycolosis, Krebs cycle**

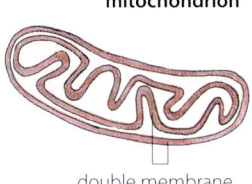

mitochondrion

double membrane

mixed inflorescence

An inflorescence in which some of the branching is racemose and some is cymose, as a thyrse.

branches cymose

main axis branching racemose

mixed inflorescence

---

mitosis   The process of division in a diploid somatic cell in which the chromosomes replicate exactly. The cell divides once into two identical daughter cells, each with a complete copy of the parent chromosomes. At the end of mitosis the cell is diploid.

cf. **meiosis**

mitotic   Of or pertaining to mitosis.

Mitosis

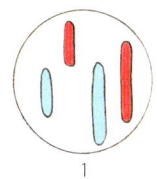 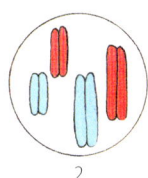 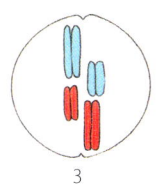 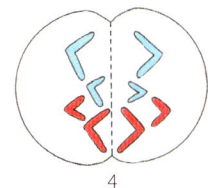 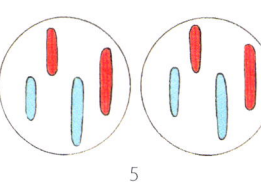

| 1 | 2 | 3 | 4 | 5 |

One diploid (2n) cell.   Chromosome replication.   Replicated chromosomes line up at equator.   Replicated chromosomes separate. The cell divides.   There are now two diploid (2n) cells.

---

mitra

Of orchids, the hood on a column formed by the fusion of the staminodia and the filament of the fertile stamen, as the midlobe in some sun orchids (*Thelymitra*) that forms a hood over the anther.

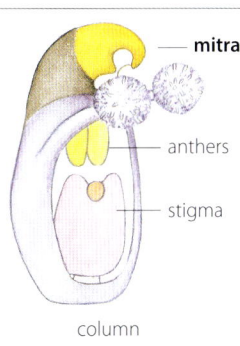

mitra

anthers

stigma

column

sun orchid (*Thelymitra*)

mitriform

Shaped like a peaked cap, as the calyptra of a moss.

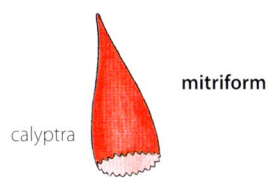

mitriform

calyptra

mixed craspedodromous

Of leaves with about half of the secondary veins terminating at the margin and the remainder not.

see **craspedodromous**

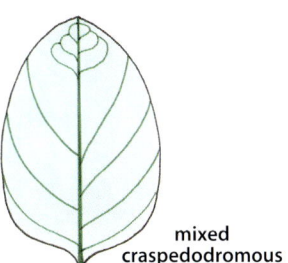

mixed craspedodromous

modification

Of evolution, a change in the genetic material that is transferred from parents to offspring. A result of mutation or recombination.

modified

Changed, altered. The petals, sepals, pistils and stamens of a flower are modified leaves.

monad

A single individual. A solitary, sessile or pedicellate flower in an inflorescence, as some mistletoes (*Amyema*). A pollen grain or spore that is not united with others.

cf. **diad, polyad, tetrad, triad**

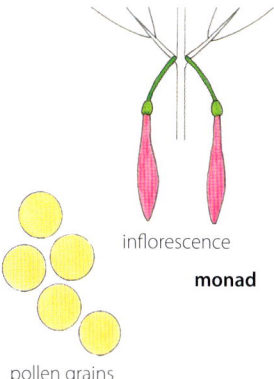

inflorescence

monad

pollen grains

185

monadelphous

Of stamens united by
their filaments into one
bundle, as wood sorrel
(*Oxalis*).

see **adelphous**

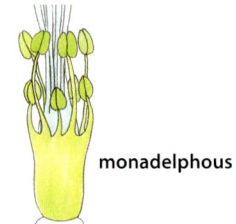

monadelphous

monandrous   Having
one stamen, as the
flowers of red valerian
(*Centranthus ruber*).

cf. **diandrous,
pentandrous,
polyandrous,
tetrandrous, triandrous**

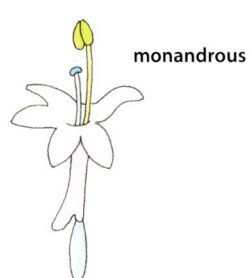

monandrous

monembryony   The condition of having or
producing one embryo.

cf. **polyembryony**

moniliform   Like a
string of beads, as the
pods of river cooba
(*Acacia stenophylla*).

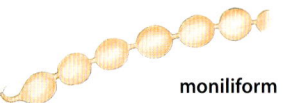

moniliform

Monilophyta, monilophytes   One of two
groups of vascular plants that bear spores rather
than seeds, the other group being Lycophyta.
They have large leaves (megaphylls) usually with
more than one vein. It comprises ferns, horsetails
(*Equisetum*) and whisk ferns (*Psilotum*).

see **euphyllophytes**

mono-   A prefix meaning one.

monocarp   An annual or other plant that flowers
and fruits only once then dies.

see **monocarpic**

cf. **hapaxanthic, iteroparous, pleonanthic,
polycarpic**

monocarpellary,
monocarpellate,
monocarpous

Of a flower having a
carpel with one locule,
as peas.

= **stylodious,
unicarpellate**

cf. **apocarpous,
syncarpous**

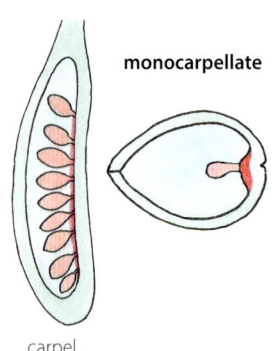

monocarpellate

carpel

monocarpic   Of a plant that reproduces once in
its lifetime then dies.

An annual is monocarpic, as rice (*Oryza sativa*).
Some monocarpic plants are long-lived, as century
plant (*Agave americana*), the talipot palm
(*Corypha*) and some species of bamboo.

= **semelparous**

see **monocarp, monocarpy**

cf. **hapaxanthic, iteroparous, pleonanthic,
polycarpic**

monocarpy   The condition of being monocarpic.
The process of bearing fruit once then dying.

monocephalous

Having a solitary flower
head, as some plants
in the daisy family
(Asteraceae).

monocephalous

monochasium, *pl.* monochasia, monochasial
cyme   A cymose inflorescence with the main axis
bearing a terminal flower and a lateral branch that
bears a terminal flower developing in one of the
subtending bracts or bracteoles.

There are four kinds of monochasium: the helicoid
cyme (bostryx) and the scorpioid cyme (cincinnus),
that are spirally coiled, and the drepanium and the
rhipidium, that are coiled and flattened.

see also **uniparous**

cf. **dichasium, pleiochasium, polychasium**

monochasial   Of a monochasium.

Monochasial cymes

spirally coiled

helicoid cyme/bostryx   scorpioid cyme/cincinnus

flattened, coiled

drepanium

flattened,
fan-shaped

rhiphidium

monochlamydeous Having a perianth of a single whorl, that is, with either a calyx or a corolla.
  cf. **chlamydeous**

monoclinous Of a flower having both stamens and pistils.
  = **hermaphrodite**
  see **perfect**

monocolpate Of a pollen grain, having a single colpus.

monofoliolate Of a compound leaf having a single leaflet, with the leaflet on a petiolule attached to the top of the petiole, as some bossiaea (*Bossiaea*) and lemon (*Citrus*).
  cf. **unifoliate**
  = **unifoliolate**

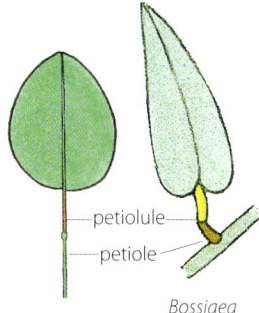
Bossiaea
monofoliolate

monocotyledon One of the largest divisions of flowering plants (22% of all angiosperms).
  Characterised by a seed with one cotyledon (seed leaf) in the embryo, endosperm that persists, flower parts that are usually in multiples of three, pollen with a single furrow or pore, leaves with main veins parallel, a stem in cross-section with vascular bundles scattered and roots that are usually fibrous.
  cf. **dicotyledon, eudicot**

Monocotyledons

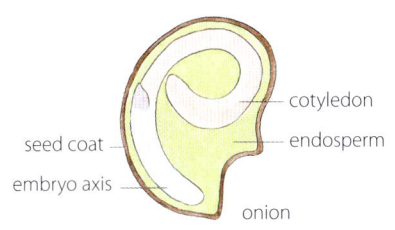
seed coat
embryo axis
cotyledon
endosperm
onion
Seed with one seed leaf (cotyledon). Endosperm persists.

water plantain

iris

lily
Flower parts usually in multiples of three. Sepals often same colour and shape as petals.

Pollen with a single pore or furrow.

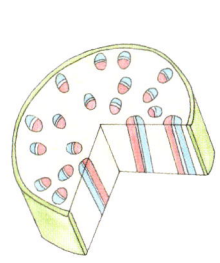
Vascular bundles in stem scattered.

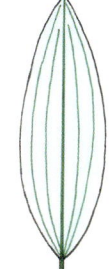
Leaves with main veins parallel.

garlic   roots   grass   bamboo

lily
Usually herbaceous plants that lack side shoots. Root system usually fibrous, never a taproot.

monocotyledonous Of a plant embryo having one cotyledon, as most monocotyledons.
  Of a plant producing such embryos.
  cf. **dicotyledonous, polycotyledonous**

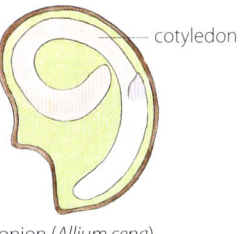
cotyledon
onion (*Allium cepa*)
monocotyledonous

monogeneric Of a family having only one genus.

monoecious Of a species with unisexual flowers on the same plant.
  see **diclinous, dioecious, trioecious**

187

monolete Of a spore with a single laesura that is a result of the way the four spores of the uniplanar tetrad are in contact with each other after meiosis. It is the area of weakness in the wall through which the spore germinates.
cf. **alete, trilete**

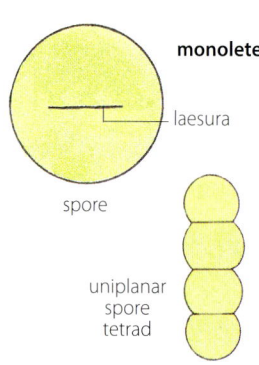

monolete

laesura

spore

uniplanar spore tetrad

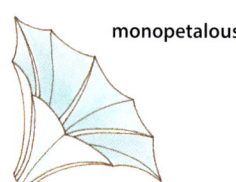

monomerous Having flower parts, such as petals, sepals and stamens, each solitary. 1-merous.
see **-merous**

monomorphic Having only one form that is invariable across a taxon, as the stamens and pistils of most species.
cf. **dimorphic, polymorphic, trimorphic**

monopetalous Having only one petal, or the petals united by their edges into one piece, as bindweed (*Convolvulus*).

monopetalous

monophyletic Of a group of organisms that includes a common ancestor and all of its descendants.
cf. **paraphyletic, polyphyletic**

Descendents
A   B   C

ancestor

monophyletic

monoploid, monoploidy Having one chromosome set (x) in each somatic cell.
see **ploidy**

monopodium A pattern of growth in which the apex of the main stem continues to grow indefinitely and lateral stems grow similarly.
A rhizome with a persistent growing point producing buds at nodes along its length.
The growth pattern of an invasive running bamboo.
see **pachymorph**
cf. **amphipodium, leptomorph**
monopodial Having a persistent terminal growing point.
Of indeterminate or indefinite growth, as a racemose inflorescence or the branching of some trees, as firs (*Abies*), that results in a pyramidal shape.

Monopodium

rhizome of a running bamboo

racemose inflorescence

fir (*Abies*)

monosepalous Having only one sepal, or the sepals united by their edges into one piece, as some species of native fuchsia (*Correa*).

monosepalous

sepals

native fuchsia (*Correa*)

monoseriate Arranged in one row or whorl.
= **uniseriate**
see also **seriate**

monospecific Of a genus having only one species.

monosporangiate Of an anther with only one pollen sac, as the dwarf mistletoes (*Arceuthobium*).
= **unisporangiate**
cf. **bisporangiate, tetrasporangiate**
Of conifers, having cones that are either male or female but not both.

monostichous Arranged in a single row on one side of a stem, as some leaves, with any leaf above the one below it.
see **orthostichy**

monostichous

4

3

2

1

monosulcate   Of a
pollen grain having
a single elongated
aperture (sulcus) situated
at one of the poles.
see also **monocotyledon**

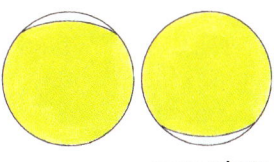

monosulcate

monosymmetric
Divisible through the
centre, on one plane
only, into exactly similar
halves, as flowers in the
pea family (Faboideae).
Bilaterally symmetrical.
= **zygomorphic**
cf. **bisymmetric,
polysymmetric**

monosymmetric

pea flower

monotelic   Of an inflorescence axis that ends
with a flower and thus stops growing, the oldest
flower being at the tip, as a determinate or definite
inflorescence.
see **cymose inflorescence**
cf. **polytelic**

**monothecal,
monothecous**
Of a stamen having
a single anther lobe
(theca) with two pollen
sacs (microsporangia).
= **bisporangiate**
cf. **dithecal**

monothecal

one-lobed anther

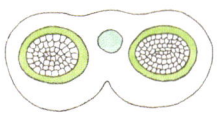

two pollen sacs

monotypic   A taxon with only a single
representative, as a genus with only one species or a
family with only one genus.

montane   Of mountainous areas, but usually
excluding the true alpine zone.

morphology   The external form or structure of a
plant or plant part.
The study of plant structure that compares features
and observes patterns of development, as in the
reproductive system or the root system.
= **phytomorphology**
cf. **anatomy**
morphological   Of or relating to morphology.

morphotaxon, *pl.* **morphotaxa**   A taxon
classified according to its morphology alone.

mosaic   An organism having a mixture of
genetically different tissues that originate from one
zygote.
A plant disease cause by various strains of viruses,
as tobacco mosaic disease that stunts the growth of
tobacco plants.
see **graft chimaera**

mosses   *see* page 190

mother cell   In mitosis, a cell that divides to
produce two new daughter cells that are genetically
identical to itself.
In meiosis, a cell that divides to produce four new
haploid daughter cells that are genetically different
from itself.
A cell that divides to form another type of cell, as
a diploid microspore mother cell that gives rise to
haploid microspores.

motile   Capable of independent movement.
Self-propelled, as the sperm of mosses and ferns
that swim through water to the non-motile egg.

mound layering
A form of propagation
whereby a plant is cut
back to near ground
level and covered with
layers of soil as new
shoots develop. Rooted
shoots are later separated
and grown as new plants.
= **stool layering**

mound layering

mRNA, messenger, RNA   A molecule in a cell
that carries information encoded in DNA to a
ribosome, a specialised organelle in the cytoplasm,
for protein synthesis.

mtDNA, mitochondrial DNA   The small amount
of DNA unique to mitochondria.
Unlike nuclear DNA, mtDNA is arranged in rings.

mosses   One of three major groups of nonvascular land plants that are photosynthetic, reproduce by spores and lack flowers, true leaves and roots.

A moss has two alternating generations, the larger perennial haploid gamete-producing gametophyte generation and the smaller short-lived diploid spore-producing sporophyte generation.

The haploid gametophyte is a leafy stem that bears the gamete-producing sex organs, the female archegonia and/or the male antheridia at the tip, and the diploid sporophyte comprises a capsule, a seta and a foot.

A male sperm from an antheridium fertilises a female egg in an archegonium, that then develops into a diploid spore-producing sporophyte, comprising a long stalk (seta) and a capsule capped by a calyptra and an operculum. The diploid spores developing in the capsule undergo meiosis to form haploid spores that are dispersed and develop into gametophytes, thus beginning the life cycle anew.

see **bryophyte**

Mosses

Alternation of generations

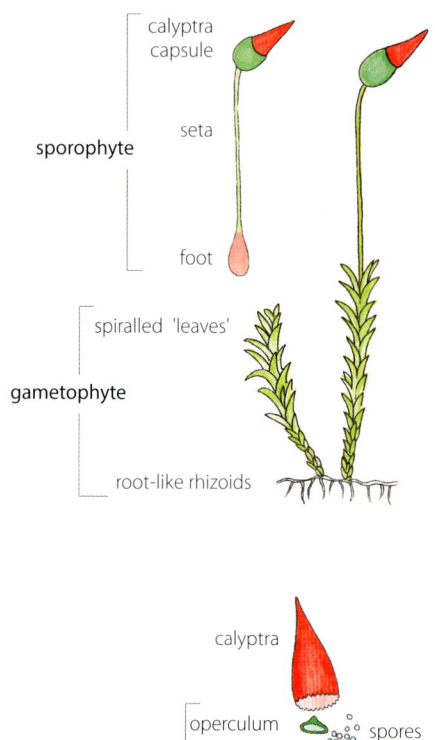

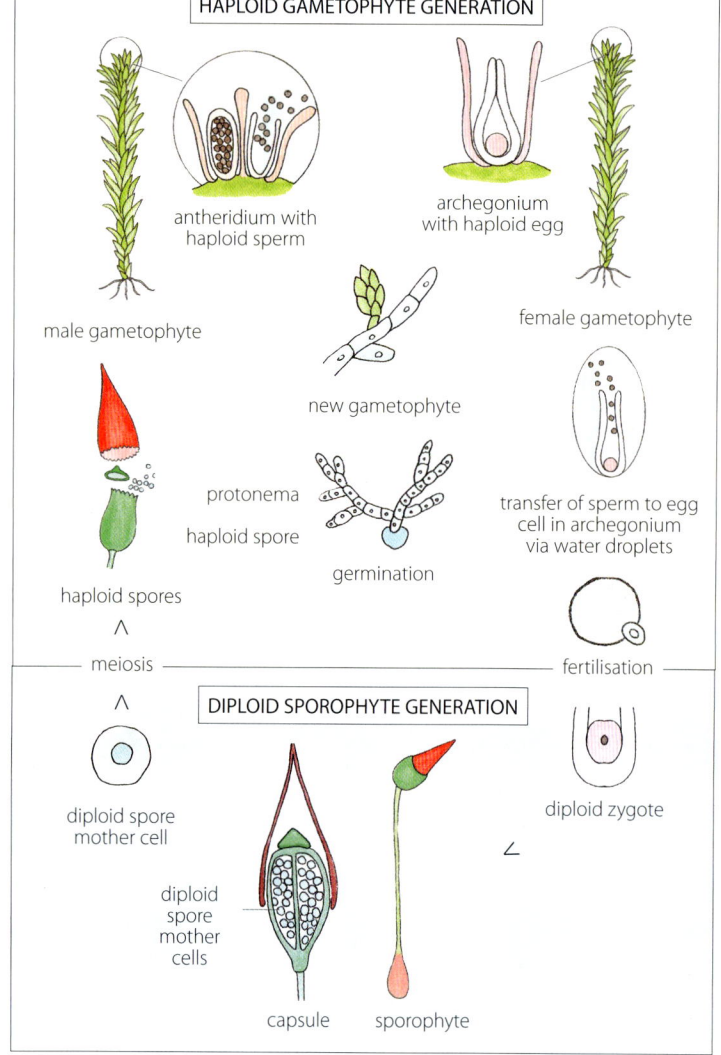

mucilage   A slippery or jelly-like substance, secreted by specialised plant cells when they are exposed to water, as that surrounding the seeds of plantain (*Plantago*).

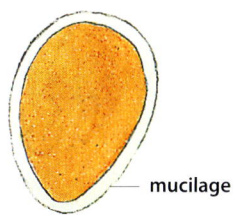

mucilage

mucilaginous   Moist, soft and sticky.

mucous   Slimy. Of a thick and slippery texture.

mucro   A short, sharp, hard point, as at the tips of some leaves.

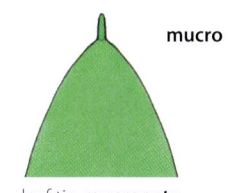

mucro

leaf tip **mucronate**

mucronate   With a mucro.
cf. **mucronulate**

mucronulate   Ending with a very short, sharp, hard point, such as a mucronulate leaf tip.
cf. **mucronate**

mucronulate

multi-   A prefix meaning many.

multicarpellary, multicarpellate
Of a flower having a gynoecium with more than one carpel, the carpels being either free or variously fused.
= **polycarpellate, polygynous**
see also **apocarpous, syncarpous**

Multicarpellate

 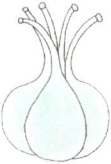  

gynoecium with free carpels   gynoecium with fused carpels

multicipital   Of a rhizome or root, producing many crowns.

multijugate   Of a pinnate leaf having more than one pair of leaflets. Having many pairs of leaflets.
see **jugate**

multiparous   Of a cymose inflorescence forming more than two stems at each branching point as a multiparous cyme.
see also **polychasium, pleiochasium**
cf. **uniparous, biparous**

multiparous

multiplanar   Having or situated on more than one plane, as pollen grains in a decussate tetrad or a tetrahedral tetrad.
see **pollen tetrad**
cf. **uniplanar**

multiple fruit   A fruit derived from an entire inflorescence with more than one flower.
It may incorporate parts of the flower other than the carpels, as a sorosis (mulberry and pineapple), a bibacca (honeysuckle) or a syconium (fig).
= **composite fruit**
cf. **accessory fruit, aggregate fruit**

Multiple fruit

sorosis
pineapple   white mulberry   honeysuckle   fig
(*Ananas comosus*)   (*Morus alba*)   (*Lonicera*)   (*Ficus*)

bibacca   syconium

multiseriate   Arranged in two or more rows or whorls.
see also **seriate**

muricate   Covered with minute short hard-pointed protuberances, as the surface of the fruit of soursop (*Annona muricata*).

muricate

muriculate   Slightly muricate.

murus, *pl.* **muri**
A ridge that forms part of
a mesh-like reticulum.
cf. **lumen**

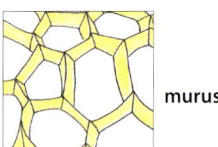

murus

mutant   A new entity or individual that has arisen
as a result of a mutation.

mutation   A permanent change in DNA resulting
from a new allele on a chromosome.
It may occur spontaneously or it may be induced
artificially.

muticous   Blunt.
Of an apex lacking a
projection, as a mucro or
an awn.
cf. **aristate, cuspidage,
mucronate,
mucronulate**

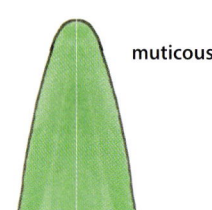

muticous

mutualism   A relationship between two
organisms where both benefit, as the association
between the nitrogen-fixing bacteria *Rhizobium*
that live in the nodules on the roots of legumes
like clover. The bacteria fixes nitrogen from the
atmosphere for the plant to use and the plant
supplies shelter and organic acids that can be used
as a food source for the bacteria.
see **symbiosis**

mycelium, *pl.* **mycelia**
A network of fungal
hyphae that reproduce
by spores. It may be
above ground, as bread
mould, or underground
where it can spread for
kilometres.
During favourable
conditions a fruiting
body that produces
spores will grow. These
include mushrooms,
toadstools, bracket fungi
and truffles.
see **fungus**

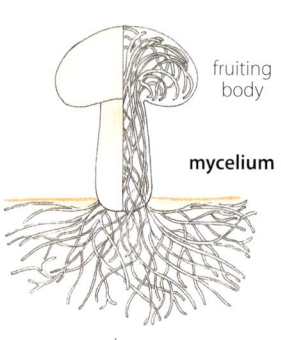

fruiting
body

mycelium

mushroom

mycology   The study of fungi.
mycologist   A biologist specialising in the study
of fungi.

mycorrhiza, *pl.* **mycorrhizae, mycorrhizas**
mycorrhizal   *see* page 193

mycotroph   A plant that obtains part or all of
its nutrients through a symbiotic association
with fungi. A three-way system that includes
the mycotroph, a mycorrhizal fungus and a
photosynthesising (green) plant.
The mycotroph parasitises the fungus which in turn
gets its nutrients from the roots of the green plant,
as some heath species that indirectly parasitise the
roots of a conifer via a fungus.
mycotrophic   Of plants that obtain nutrients
through an intermediary mycorrhizal fungi.
see also **mycorrhiza, trophic**
cf. **saprophyte**

myophile   Flies, as bee flies and hoverflies, that
feed on nectar and pollen.
cf. **sapromyophile**

myophily   Dispersal of pollen and pollination by
flies.
myophilous   Pollinated by flies.

myrmechory   Dispersal of seeds by ants.
myrmechorous   Having seeds that are dispersed
by ants.
see **elaiosome**

myrmecophile   A plant or other organism in a
symbiotic association with ants.
myrmecophily   Ant love. A symbiotic relationship
between an organism and ants.
myrmecophilous   Of or relating to myrmecophily
or a myrmecophile.
see **myrmecophyte**

myrmecophyte
A plant adapted to
maintaining a symbiotic
relationship with ants.
Adaptations include
swollen stems with
cavities for ants, and
domatia, as those
in the thorns of the
bullhorn acacia (*Vachellia
cornigera*).

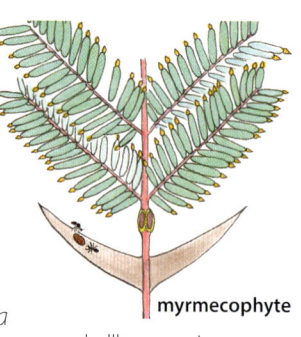

myrmecophyte

bullhorn acacia

myxogenic   Producing mucus or slime, as some
hairs and cells.

mycorrhiza, *pl.* **mycorrhizae, mycorrhizas**   A mutually beneficial association between a fungus (or several different kinds of fungi) and the roots of a particular plant.

The fungus supplies mineral nutrients like phosphorus and in turn receives nutrients from the plant like carbon, that results from photosynthesis.

Mycorrhiza differ in how they enter the root structure. Fungal hypae either enter the spaces between the cells (ectomycorrhizas) or penetrate the cells (arbuscular mycorrhizas, endomycorrhizas, ericoid mycorrhizas and orchidaceous mychorrhizas) or both (ectendomycorrhizas). Hyphae reproduce asexually but from time to time reproduce by sending up fruiting bodies, like sac fungi (Ascomycota), that produce spores.

mycorrhizal   Of or relating to a mycorrhiza.

Mycorrhiza

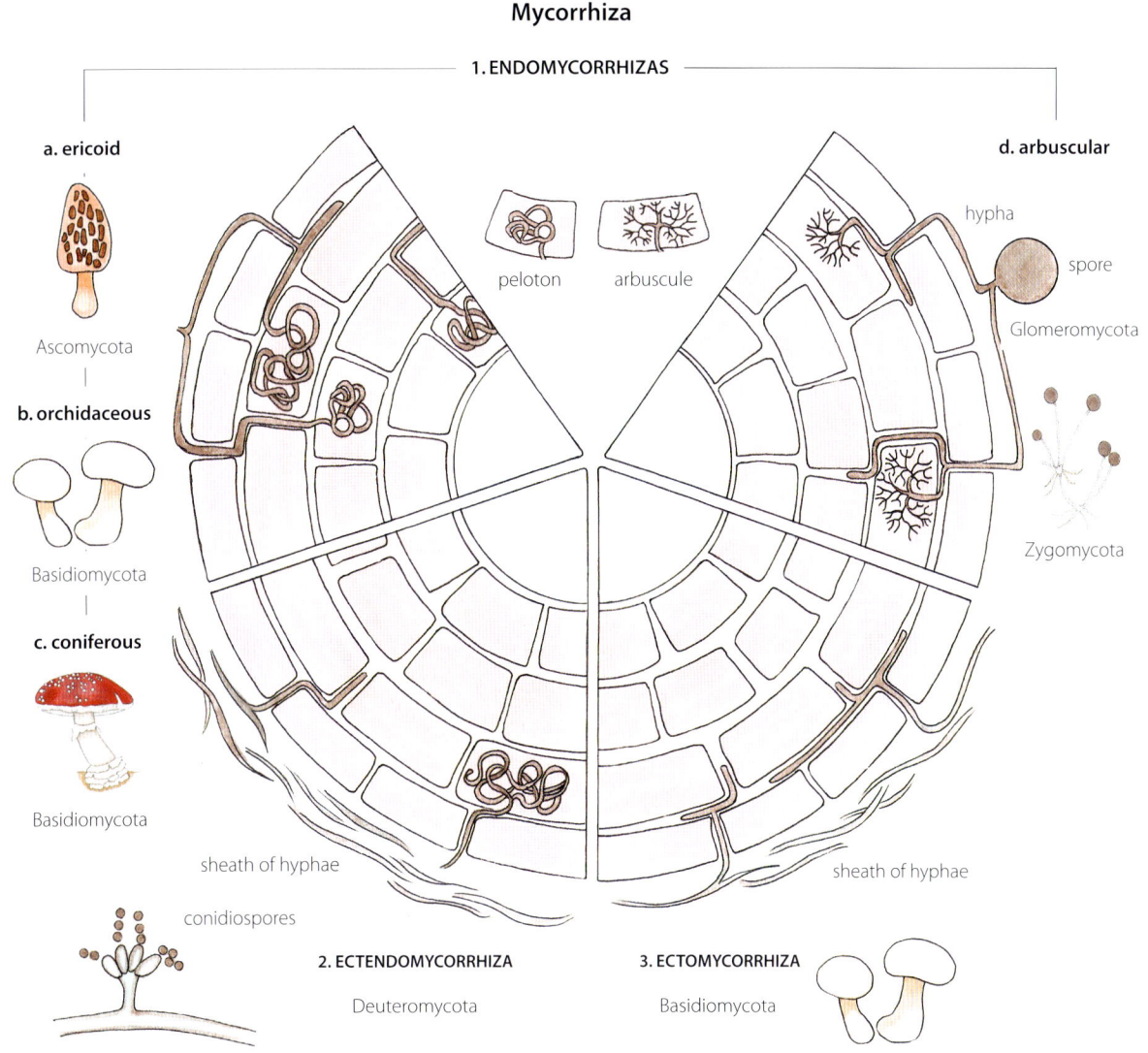

1. ENDOMYCORRHIZAS

a. ericoid

Ascomycota

b. orchidaceous

Basidiomycota

c. coniferous

Basidiomycota

peloton   arbuscule

d. arbuscular

hypha

spore

Glomeromycota

Zygomycota

sheath of hyphae

sheath of hyphae

conidiospores

2. ECTENDOMYCORRHIZA

Deuteromycota

3. ECTOMYCORRHIZA

Basidiomycota

---

n   The haploid number of chromosomes that occurs in a reproductive cell. It represents half the number (2n) of chromosomes in a somatic cell.

nacreous   Having the lustrous rainbow-like colours of mother-of-pearl.

naked   Lacking a covering.

Of a bud that lacks protective scales.

see **bud scale**

Of a flower, without a perianth, as the female flower of willows (*Salix*).

see **achlamydeous**

Of gymnosperms, having ovules and seeds that are exposed on the surface of a cone scale.

naked bud   A bud lacking protective bud scales.

naked bulb   A true bulb that consists of a compressed stem with nodes (basal plate), bearing fleshy overlapping leaves that lack a tunic, as the lily genus (*Lilium*).
= **imbricate bulb, scaly bulb**
see also **tunicate bulb**

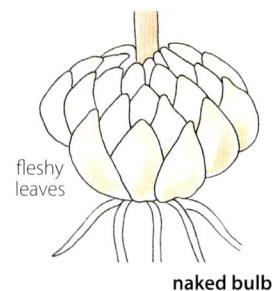

fleshy leaves

naked bulb

naked flower   Of a flower having no tepals. Without a perianth, as the female flower of willows (*Salix*).
= **achlamydeous, atepalous**

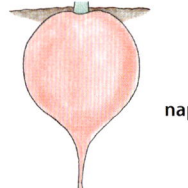

stigma

naked flower

ovary

female flower of *Salix*

nan-, nann-, nano-, nanno-   A prefix meaning dwarf.

napiform   Turnip-shaped, large and round in the upper part and slender below.

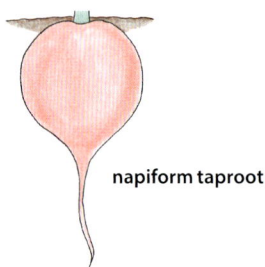

napiform

napiform taproot   A main, descending root large and round in the upper part and slender below, as a turnip.
see **conical taproot, fusiform taproot**

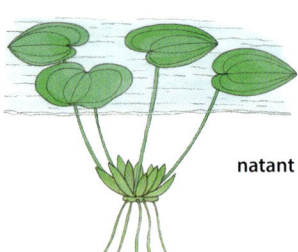

napiform taproot

nastism   Plant movement, not related to the direction of growth, caused by an external stimulus, as the opening and closing of flowers in response to alteration in temperature or light.
cf. **tropism**

nastic   Of or relating to nastism.

natant   Floating on water, as the leaves of water lily (*Nymphaea*) and some other aquatic plants.
see also **floating**

natant

native   Growing naturally in a particular place, not introduced.
= **indigenous**

natural selection   A natural process that favours the survival of organisms best adapted to an environment at the expense of those that are not. One of the basic mechanisms of evolution together with genetic drift, mutation and gene flow (gene migration).
cf. **artificial selection**

naturalised   Of a non-native plant that has become adapted to its new environment.

navicular   Shaped like the bow of a boat, as the united lower petals (keel) of pea flowers (Fabaceae).
= **cymbiform**

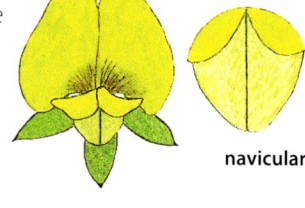

navicular

neck   A narrow supporting part in a plant. The swelling between the base of the capsule and the top of the seta in mosses.
= **apophysis**
The slender tube-like extension of an archegonium through which the sperm swim to reach the egg. The junction between the base of the stem and the roots.

Neck

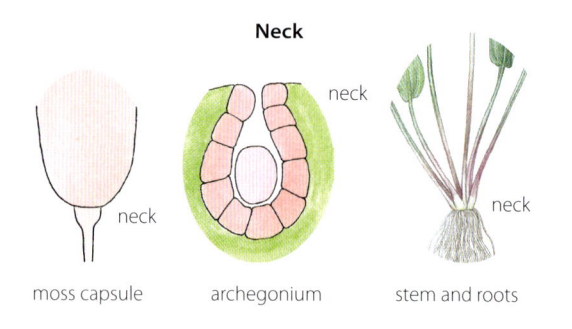

neck

neck

neck

moss capsule   archegonium   stem and roots

necrosis   Localised death of cells occurring in living tissue.

necrotic   Of necrosis.

nectar   A sugar substance secreted by glands (nectaries) that attract pollinators.

nectar guide   Patterns on petals of some flowers, as those on pelargoniums (*Pelargonium*), that guide pollinators to a nectar reward.

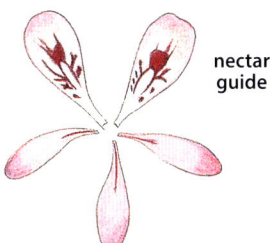

nectar guide

nectariferous   Producing nectar. Bearing nectar producing glands.

nectary
A gland that secretes nectar.
Floral nectaries are located on flower parts, as those at the base of the petals of buttercups (*Ranunculus*), and extrafloral nectaries are located on plant parts other than the flower, as on the phyllodes of wattles (*Acacia*).

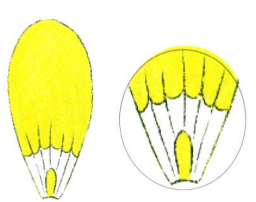

buttercup petal nectary

nectary

phyllode nectary

needle   A long narrow leaf, characteristic of many conifers.

needle

neo-   A prefix meaning new or recent.

neoteny   Reaching sexual maturity while retaining juvenile characteristics.

neotropics   The New World tropical regions that extend south of the Mexican desert and includes Central America and all of South America to the sub-Antarctic zone.
Vegetation ranges from tropical rainforest to savanna.
cf. **palaeotropics**
neotropical   Of or relating to the neotropics.

neotype   A specimen used instead of the holotype if the specimens available to the original author have been lost or destroyed.

nervation   The arrangement of veins in a leaf.

nerve   A vascular bundle in a leaf or other plant part that circulates water, minerals and other substances.
It typically divides or branches and provides support and strength.
= **vein**

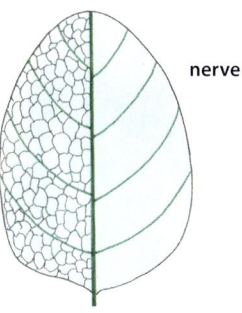

nerve

nerved, nervose   With ribs or veins.

nested   Embedded, as a taxon or clade within another clade in a phylogenetic tree.

Nested

Clade 3
Clade 1
Clade 2

Clade 1 and clade 2 are nested in clade 3.

net-veined   Leaf venation with smaller veins joined in a net-like or mesh-like pattern.

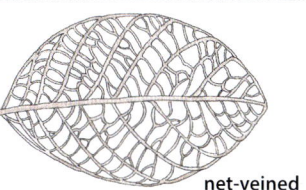

net-veined

neuter   Of a flower lacking stamens and pistils, or stamens and pistils present but not functional.
cf. **bisexual, unisexual**

new clade name   Of phylogeny, a new clade name.
= **nomen cladi novum**

new combination   A new name for a taxon that has the specific or infraspecific epithet used with a new genus or species name respectively.
= **combinatio nova**

new name   A replacement name.
= **avowed substitute, nomen novum**

nexine   The inner non-sculptured layer of the exine that lies below the sexine in the pollen grain wall.
see **pollen wall**

niche   The position a species or population has within an ecosystem. It is related to how it finds food and shelter, reproduces and survives. Different plants may coexist in the same niche, or compete, as native and introduced species.

nidulent   Of seeds that are immersed in pulp, as those of a berry.

nigrescent   Blackish, turning black.

nitid   Having a smooth, shiny, polished surface.

nitrogen, (N), (N$_2$)   A common, non-metallic element that is normally a colourless, odourless and tasteless gas. It makes up 78 per cent of the earth's atmosphere and is a constituent of all living tissues. Plants do not absorb nitrogen directly from the air but through their roots, from nitrate (NO_3^-) and ammonium (NH_4^+) ions in the soil. It is a major component of chlorophyll and amino acids, the building blocks of proteins. It is also a component of adenosine triphosphate that is involved in energy conservation and release in cells
see **nitrogen fixation**

nitrogen fixation   Nitrogen diffuses into the soil from the atmosphere and needs to be changed or 'fixed' before plants can use it.
Microorganisms, like rhizobia in the root nodules of many plants, but primarily legumes such as clover, bind nitrogen and water to form ammonium that can be used by plants.
Lightning converts atmospheric nitrogen into ammonia and nitrates that enter the soil with rainfall.

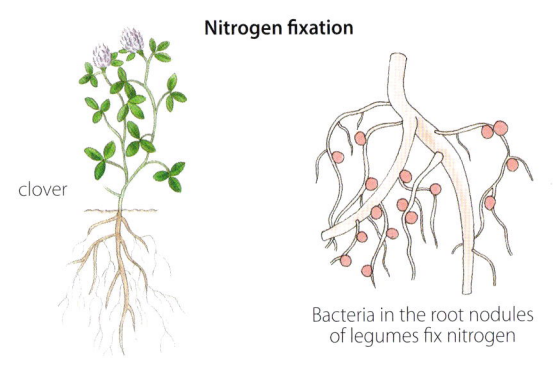

Nitrogen fixation

clover

Bacteria in the root nodules
of legumes fix nitrogen

niveous   Snow white.

nocturnal   Of or appearing at night, as the flowers of the cactus queen of the night (*Epiphyllum oxypetalum*) that open only during the night.
cf. **diurnal**

nodding   Having the tip hanging downwards, as the flower of the nodding greenhood (*Pterostylis nutans*).
= **nutant**

nodding

node   A joint.
That part of a stem, sometimes swollen or knob-like, from which leaves, branches or flowers arise.
Horizontal stems, (rhizomes and stolons) have nodes that give rise to scale leaves, roots and shoots.
see also **internode**
nodal   Of or like a node.
nodose   With closely packed nodes. Knotty.

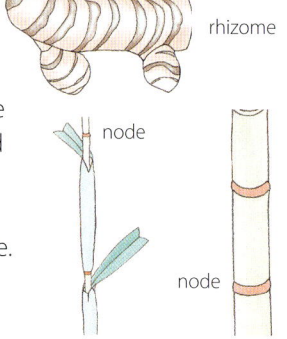

node

node

rhizome

node

node

grass   bamboo

node
Of phylogenetics, the point at which two new branches occur that represents a speciation event in a phylogenetic tree.
A hypothetical ancestor.
see also **sister group, sister taxa**

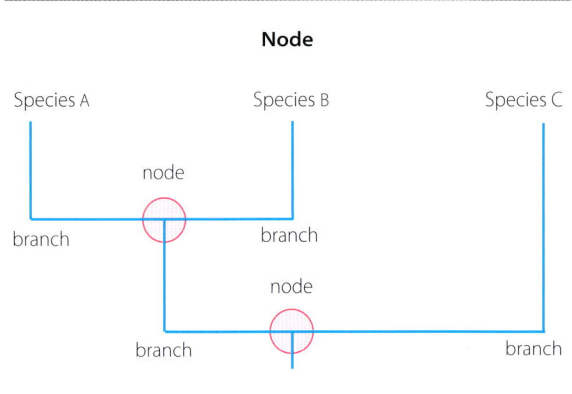

Node

Species A   Species B   Species C

node

branch   branch

node

branch   branch

node diaphragm
Of bamboos, a rigid membrane inside a node that forms a partition between adjacent internodes.

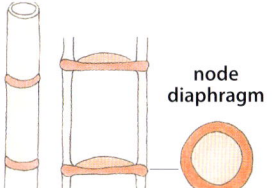

node diaphragm

nodule   A small rounded protruberance. A small swelling on the roots of legumes, as clover, and some other plants, resulting from infection by nitrogen-fixing bacteria.
see also **nitrogen fixation, rhizobium**
nodular   Relating to or characterised by having nodules.
nodulose   Bearing nodules.

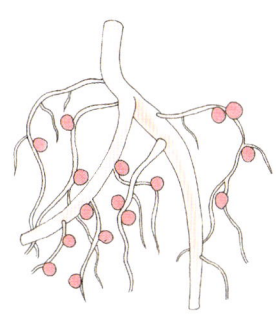

roots with **nodules**

nom., nomen   Name.

nomen, *abbr.* **nom.**   Name.

nomen cladi conversum   Of phylogeny, a clade name converted from a pre-existing name.
= **converted clade name**

nomen cladi novum   Of phylogeny, a new clade name.
= **new clade name**

nom. cons., nomen conservandum
Conserved name. A name that is kept even though it may formerly have been invalid.

nomen conservandum, *abbr.* **nom. cons.**
Conserved name. A name that is kept even though it may formerly have been invalid.

nom. illeg., nom. illegit., nomen illegitimum
Illegitimate name.
A published name that does not accord with the rules of the International Code of Nomenclature.

nomen illegitimum, *abbr.* **nom. illeg., nom. illegit.**   Illegitimate name.
A published name that does not accord with the rules of the International Code of Nomenclature.

nom. inval., nomen invalidum   Publication of a new species that does not accord with the provisions of the International Code of Nomenclature for valid publication.
cf. **nomen nudum**

nomen invalidum, *abbr.* **nom. inval.**
Publication of a new species that does not accord with the provisions of the International Code of Nomenclature for valid publication.
cf. **nomen nudum**

nom. leg., nom. legit., nomen legitimum
A validly published name that is in accordance with the rules of the International Code of Nomenclature.

nomen legitimum, *abbr.* **nom. leg., nom. legit.**
A validly published name that is in accordance with the rules of the International Code of Nomenclature.

nom. nov., nomen novum   New name.
A replacement name.
= **avowed substitute**

nomen novum, *abbr.* **nom. nov.**   New name.
A replacement name.
= **avowed substitute**

nom. nud., nomen nudum   A name published without a description or diagnosis and is therefore invalid.
cf. **nomen invalidum**

nomen nudum, *abbr.* **nom. nud.**   A name published without a description or diagnosis and is therefore invalid.
cf. **nomen invalidum**

nom. rejic., nomen rejiciendum   A name rejected in favour of a conserved name.

nomen rejiciendum, *abbr.* **nom. rejic.**   A name rejected in favour of a conserved name.

nomenclatural type   The specimen or illustration with which the name of a taxon is permanently associated, not necessarily the most typical or representative of the taxon.

nomenclature The procedure of naming plants according to the International Code of Nomenclature and the PhyloCode.

non-endospermic seed One having endosperm absorbed by the growing embryo, with the embryo itself having cotyledons that store food for germination. Found in most eudicots, as the bean family (Fabaceae).
= **exalbuminous seed**
cf. **endospermic seed**

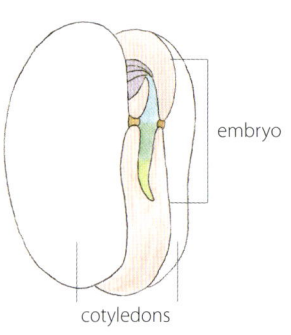

broad bean (*Vicia faba*)

non-endospermic seed

non-resupinate Not twisting through 180°. Of orchids, flowers that have the dorsal sepal below the lateral sepals, as leek orchids (*Prasophyllum*).
cf. **hyper-resupinate, resupinate**

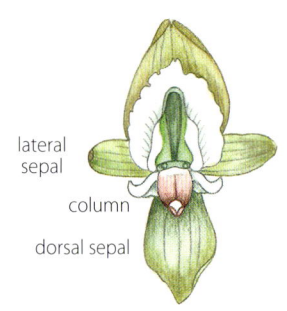

lateral sepal

column

dorsal sepal

non-resupinate

nonvascular plants Mosses, liverworts, hornworts and some algae that lack specialised tissues (xylem and phloem) for conducting water and nutrients.
see also **cryptogams**
cf. **vascular plants**

Nonvascular plants

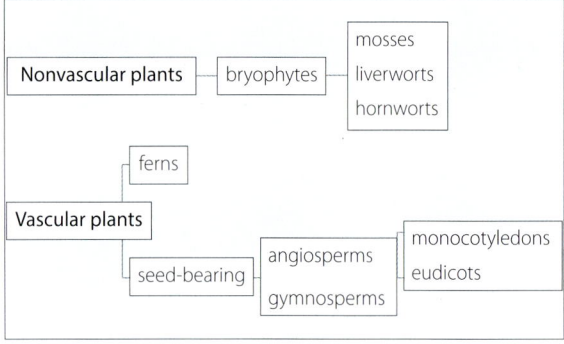

notch A V-shaped cut or indentation. A nick.
notched With a notch.

nothogenus A hybrid genus produced by crossing plants from different genera and indicated

by x before the genus name, as the orchid *xBrassocattleya* that is a hybrid between *Brassavola* and *Cattleya*.

nothomorph A now obsolete term denoting any hybrid variant derived from the same parent species and denoting a subordinate taxon within a nothospecies. Names published at the rank of nothomorph are now treated as varieties.

nothospecies A hybrid from a cross beween two species belonging to the same genus, as *Geranium xcantabrigiense* that is a hybrid between *Geranium macrorrhizum* and *G. dalmaticum*.

nothotaxon A hybrid taxon.
see **nothogenus, nothospecies**

nucellus, *pl.* **nucelli** In seed plants, the tissue in the immature ovule in which megaspores are formed.
The megaspore grows to become the gametophyte which is nourished by the nucellus in the mature ovule.
see **megasporangium**

Nucellus

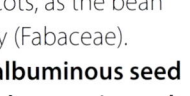

megaspore

nucellus

immature ovule

archegonia

gametophyte

nucellus

gametophyte (embryo sac)

nucellus

mature ovule

Gymnosperm   Angiosperm

nuciferous Bearing nuts, nut-like.

nuclear envelope A membrane barrier that separates the nucleus from the cytoplasm in eukaryotic cells.
It is composed of an inner and outer membrane, with the perinuclear space in between.

nuclear pores   Numerous small holes in the nuclear envelope through which material passes into and out of the nucleus.

nuclear sap   The gelatinous substance inside the nucleus, enclosed by the nuclear envelope.
The nucleolus and chromatin are suspended in the nucleoplasm.
Processes that occur in the nucleoplasm include transcription and replication.
= **nucleoplasm**

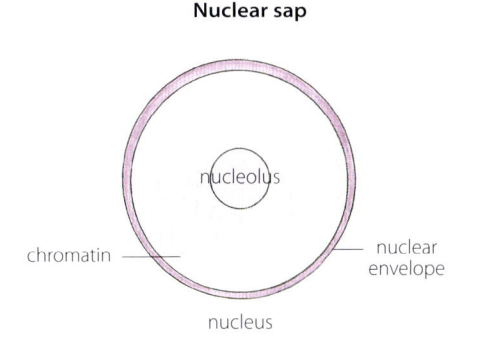

Nuclear sap

nucleolus, *pl.* **nucleoli**
A body within the cell nucleus that consists of RNA and protein.
It assembles ribosomes.

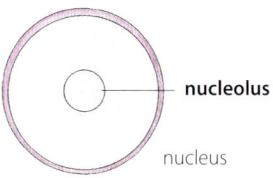

nucleoplasm   The gelatinous substance inside the nucleus, enclosed by the nuclear envelope. The nucleolus and chromatin are suspended in the nucleoplasm.
Processes that occur in the nucleoplasm include transcription and replication.
= **nuclear sap**

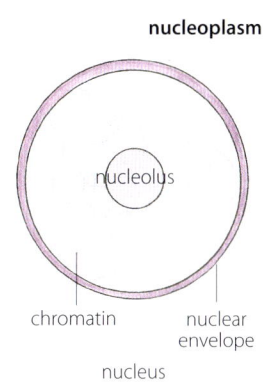

nucleotide   A basic building block of DNA and RNA.
In DNA it is composed of a nitrogen base (adenine, guanine, cytosine or thymine), a five-carbons sugar base (desoxyribose) and a phosphate group.
In RNA, thymine is replaced by uracil.
see also **DNA sequencing**

nucleus, *pl.* **nuclei**   A membrane-bound organelle that contains the genetic material of a cell, as DNA (deoxyribose nucleic acid).
cf. **nucleolus**
nuclear   Of or relating to the nucleus of a cell.

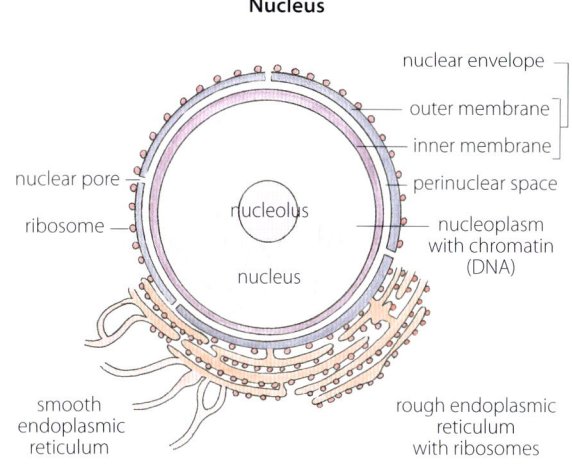

Nucleus

nucule   A nutlet.

numerical plan   The numerical order discernible in the petals, sepals, gynoecium and androecium of a flower.
Commonly arranged in multiples of three, four or five and then termed 3-merous, 4-merous and 5-merous. Monocots are typically 3-merous and eudicots usually 4- or 5-merous.
see also **ground plan, merosity**

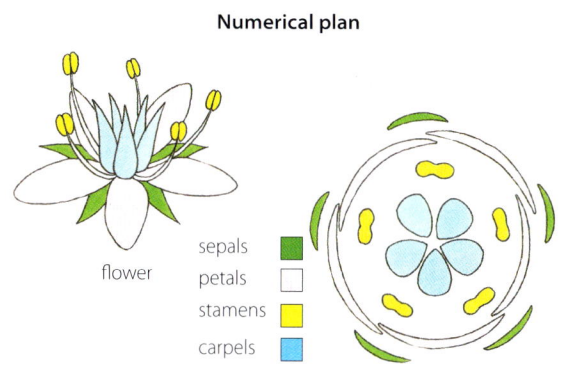

Numerical plan

numerous   Many, often indefinite in number. More than ten.

nut A dry indehiscent fruit with a hard pericarp (shell) that encloses one seed, as hazelnut (*Corylus*).
Also used loosely to describe any hard dry indehiscent fruit with a single seed, as an achene or mericarp.

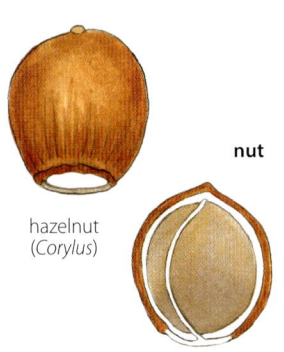
nut

hazelnut (*Corylus*)

nutant Having the tip hanging downwards, as the flower of the nodding greenhood (*Pterostylis nutans*).
= **nodding**

nutant

nutlet A small indehiscent nut.
A unit of a compound fruit that splits at maturity, as a mericarp of a schizocarp in the borage family (Boraginaceae) and mint family (Lamiaceae).

mericarp
nutlet
schizocarp

nutrient One of the substances that provides nourishment essential for plant growth and development.
cf. **macronutrient, micronutrient, trace element**

nyctanthous Flowering at night.

nyctinasty Plant movement in response to nightly changes in temperature and light intensity.
see akso **phytochrome, pulvinus**
nyctinastic Of or relating to nyctinasty.

nyctitropism The tendency of some plant parts to assume a different position at night, as the folding of the leaflets of some leguminous plants.

ob- A prefix indicating an inversion or rotation of 180°.

obconic, obconical Cone-shaped with the narrow end at the base.
cf. **conical**

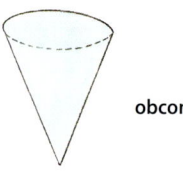
obconic

obcordate Heart-shaped in outline with the notch at the top.
cf. **cordate**

obcordate

obdeltoid A triangle with sides of about equal length and broad at the top.
cf. **deltoid**

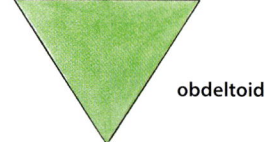
obdeltoid

obdiplostemonous With two whorls of stamens, the inner whorl alternate with the petals and the outer whorl opposite the petals.

obdiplostemonous

oblanceolate Shaped like the inverted head of a lance.
Distinctly longer than wide, tapering each end and widest above the middle.
cf. **lanceolate**

oblanceolate

oblate Of an object with the polar axis shorter than the equatorial diameter.
cf. **prolate**

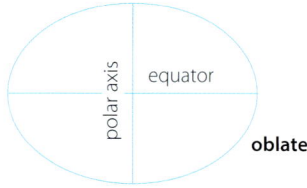
polar axis
equator
oblate

obligate Essential, restricted to a particular function, as some orchids that must cross-pollinate in order to produce viable seeds.
cf. **facultative**

obligate outcrosser Pollen from a different plant is required for successful seed set.
cf. **autogamy, self-pollination, selfing**

oblique Slanting.
Of a leaf or leaflet larger on one side of the midrib than on the other.
Of a midrib that is nearer one margin than the other.
Asymmetrical.
see also **unequal**

oblique

obloid A three-dimensional oblong shape.

oblong About two or three times longer than broad, with parallel sides and rounded ends.

oblong

obovate Egg-shaped in outline, with the broad end at the top.
cf. **obovoid**

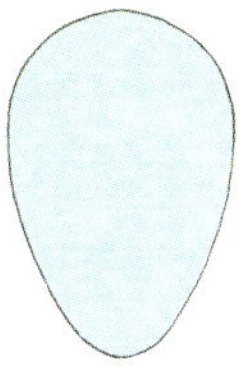

obovate

obovoid Obovate but three-dimensional with the broad end above the middle.

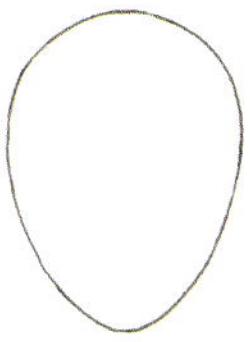

obovoid

obsolescent Having dwindled to a rudimentary state or disappeared altogether.
cf. **obsolete**

obsolete No longer in use.
cf. **obsolescent**
see also **rudimentary, vestigial**

obtrullate Shaped like a bricklayer's trowel but broadest at the top.
cf. **trullate**

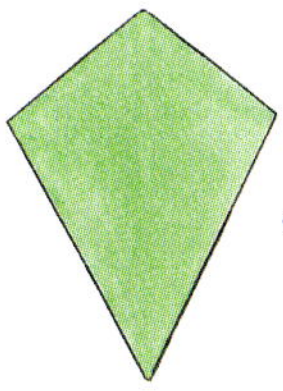

obtrullate

obturator An outgrowth, usually of the funicle, producing a secretion that promotes the passage of the pollen tube into the ovule.

obtuse Terminating gradually into a blunt or rounded tip or base, with the margins meeting at an angle of more than 90°.
cf. **acute**

leaf tip **obtuse**

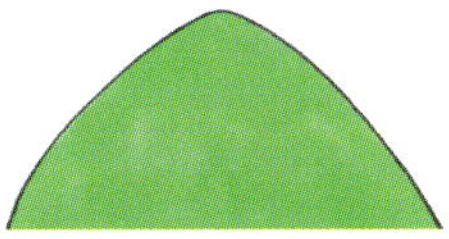

obvolute vernation Of young leaves in the unopened leaf bud with the margins folded to embrace one margin of another leaf.
= **half-equitant vernation**

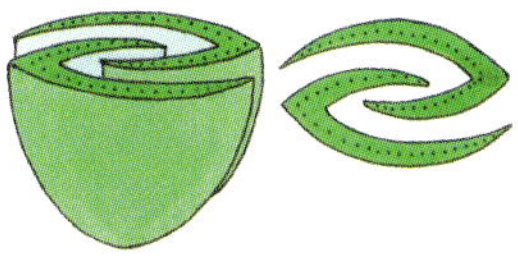

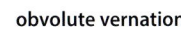

obvolute vernation

ochrea, ocrea A sheath formed by stipules of alternate leaves that fuse around a stem.
ochreate, ocreate With an ochrea.

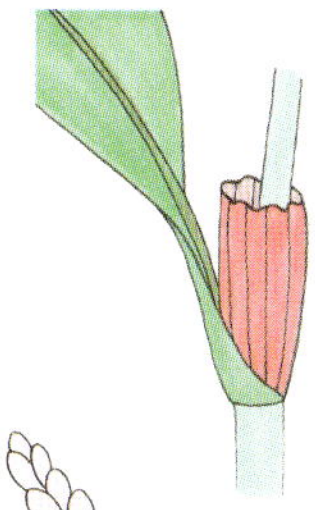

ochrea

ocreola A small ocrea subtending the flowers in the inflorescence of the knotweed family (Polygonaceae).

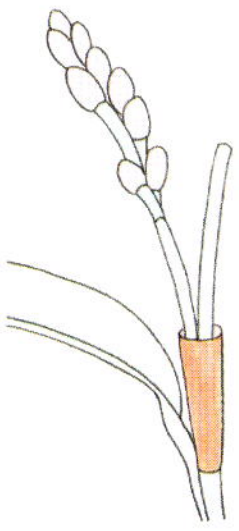

ocreola

oct-, octa-, octo- A prefix meaning eight.

odd-pinnate Of a pinnate leaf with leaflets alternate, or arranged in pairs, and terminating with a single leaflet.
= **imparipinnate**

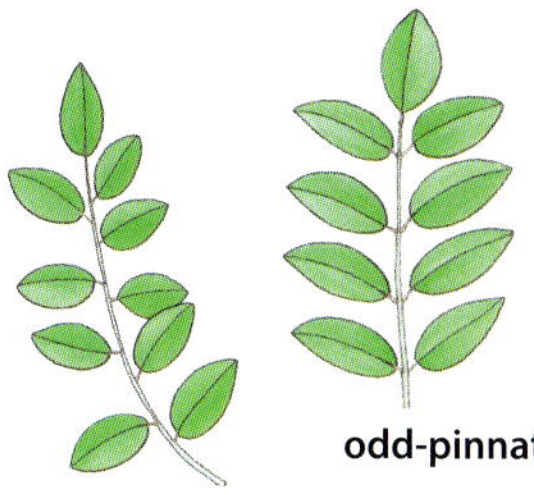

odd-pinnate

odor, odour Having a smell, whether pleasant or unpleasant.
odiferous, odoriferous, odorous Having a characteristic smell, as fragrant or malodorous.

offset A young plantlet developed at the base of the parent plant, as many of the globular cacti. It is often rooted and can be detached and propagated as a new plant.

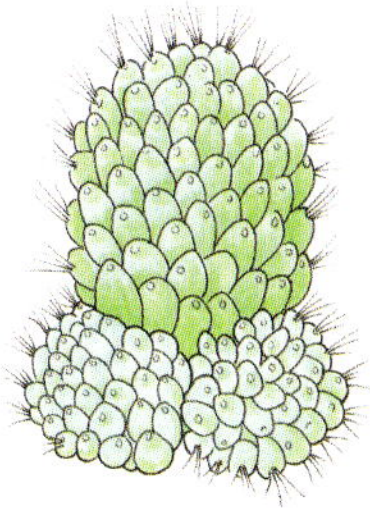

offset

oleaginous Resembling or having the properties of oil, oily.
Containing or producing oil, as the fruit of the olive tree (*Olea europaea*).

oleiferous Producing oil, as seeds of the sunflower (*Helianthus annuus*).

oleoresin A mix of volatile oil and resinous substances, as turpentine.

oligo- A prefix meaning few.

oligomery   A whorl of a flower that has a lesser number of members than the other whorls, as the whorl of two or three carpels in guinea flower (*Hibbertia*) that has five petals, five sepals and many stamens.
cf. **pleiomery**
oligomerous   Of or relating to oligomery.

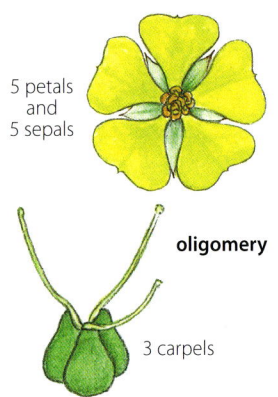

5 petals and 5 sepals

oligomery

3 carpels

guinea flower (*Hibbertia*)

oligospermous   Having few seeds.

oligotrophic   Of a body of clear water low in nutrients, like nitrogen and phosphate, that supports few plants and is rich in dissolved oxygen.
see **trophic**
cf. **dystrophic, eutrophic, mesotrophic**

olive   Dark yellow-green.
olivaceous   Relating to the olive. Dark yellow-green in colour.

omni-   A prefix meaning all.

ontogenesis, ontogeny   The sequence of stages through which an organism passes during its lifetime.
The stages of development of an anatomical or behavioural feature.
cf. **phylogeny**
ontogenic   Of or related to ontogenesis or ontogeny.

oosphere   The non-motile female gamete (egg cell) of bryophytes and ferns that is retained in the archegonium.
cf. **spermatozoid**

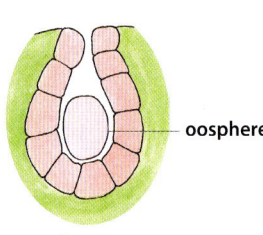

oosphere

archegonium

opaque   Having a surface that does not reflect light and is neither transparent or translucent.

open vascular bundle   With a cambium layer, that promotes secondary growth, between the phloem and xylem. Typical of most eudicots.
cf. **closed vascular bundle**

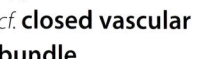

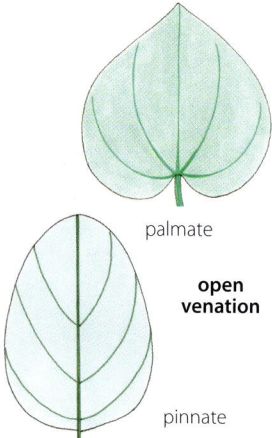

phloem

cambium

xylem

open vascular bundle

open venation   Venation that is branched with free-ending veins: leaves may be toothed, lobed or compound. The two forms of open venation are pinnate and palmate.
see **cladodromous, dendritic**
cf. **anastomosis, closed venation**

palmate

open venation

pinnate

operculate-poricidal capsule   A capsule that dehisces through pores in a lid called an operculum, as the opium poppy (*Papaver somniferum*).
cf. **poricidal capsule**

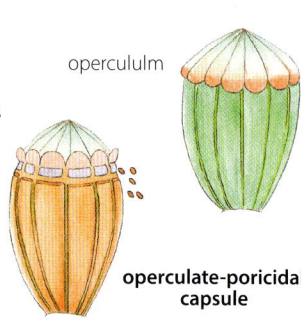

opercululm

operculate-poricidal capsule

operculum, *pl.* **operculi**   A hood or lid. Of mosses, a small cap covering the mouth of the capsule that is shed at spore dispersal.
cf. **calyptra**
operculate, opercular   Relating to, bearing or resembling an operculum.

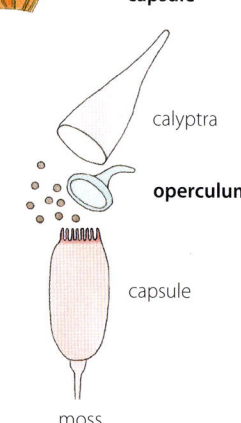

calyptra

operculum

capsule

moss

opportunistic   Of plants adapted to exploit newly available habitats, as on a clear-cut forest floor that is colonised rapidly by opportunistic plants with windborne seeds.
These plants typically have a short life cycle and establish themselves quickly.

opposite

Of plant parts, as leaves and flowers, arranged in pairs on each side of a stem.

Of plants parts, as stamens above petals, arranged with one on top of the other.

Of phyllotaxy, having two leaves or shoots occurring at a node on opposite sides of a stem. The simplest form of a whorled arrangement.

cf. **distichous, whorled**

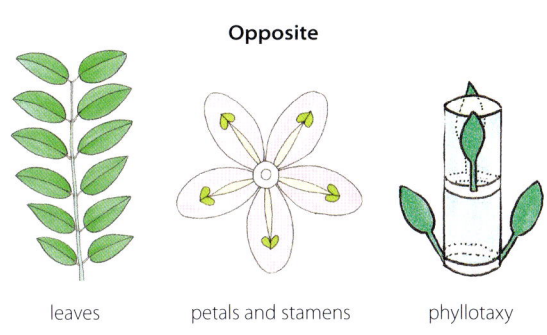

Opposite

leaves   petals and stamens   phyllotaxy

opposite vernation

Of young leaves in the unopened leaf bud arranged in alternating pairs.

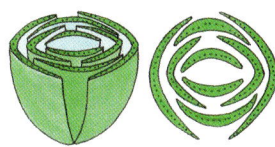

opposite vernation

opposite-bipinnate

Of a bipinnate leaf with the primary divisions arranged in pairs on opposite sides of the rachis.

see also **pinnate**

opposite-pinnate

Of a pinnate leaf with leaflets in pairs on opposite sides of the rachis.

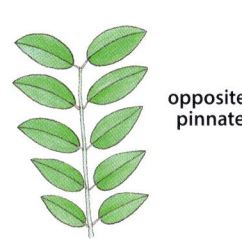

opposite-pinnate

orbicular

A two-dimensional shape, with a circular or almost circular outline.

cf. **globose**

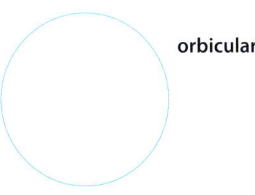

orbicular

orchidaceous   Of, relating to or resembling the orchid family (Orchidaceae).

orchidaceous mycorrhiza

Mutually beneficial symbiosis formed between a Basidiomycota fungi and the roots of orchids (Orchidaceae).

One of the endomycorrhizas with hyphae that penetrate the cells of the root cortex and form pelotons.

see **mycorrhiza**

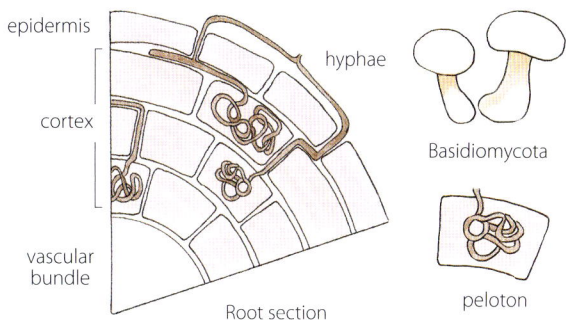

Orchidaceous mycorrhiza

epidermis   hyphae

cortex   Basidiomycota

vascular bundle

Root section   peloton

orchids   *see* page 204

organ

A structure that performs a particular function and is composed of all three types of plant tissue (dermal, ground and vascular).

Organs include: stems that support and mediate growth, leaves that produce food through photosynthesis, flowers and fruits that reproduce the plant, and roots that absorb water and nutrients. Cells are organised into tissues, tissues are organised into organs and organs function together in systems.

order   In taxonomic classification, a rank below class and above family. Names of orders end in *-ales*.

see **taxonomic hierarchy**

organelle   A structure in a eukaryotic cell (cell with a nucleus) that has a specific function, as the nucleus that holds the cell's genetic information, chloroplasts that provide energy for cell metabolism, mictochondria and the Golgi apparatus.

organelle page 204 (cont.)

orchids, Orchidaceae   The largest flowering plant family in the world.

Flowers have three sepals and three petals, with the column in the middle, and one petal (labellum) usually different from the others and often modified to to attract pollinators.

Stamens and pistils are united into a single structure, the column, with one fertile anther.

Pollen is bound together into two pollinia that sit in the anther and are easily carried away by pollinating insects. The stigmatic disc sits below the anther.

Most orchids are pollinated by insects.

Fruit is a capsule.

Seeds lack endosperm and after germination rely on a relationship with mycorrhizal fungi for nutrition.

Orchids also reproduce vegetatively by tubers.

Orchids (Orchidaceae)

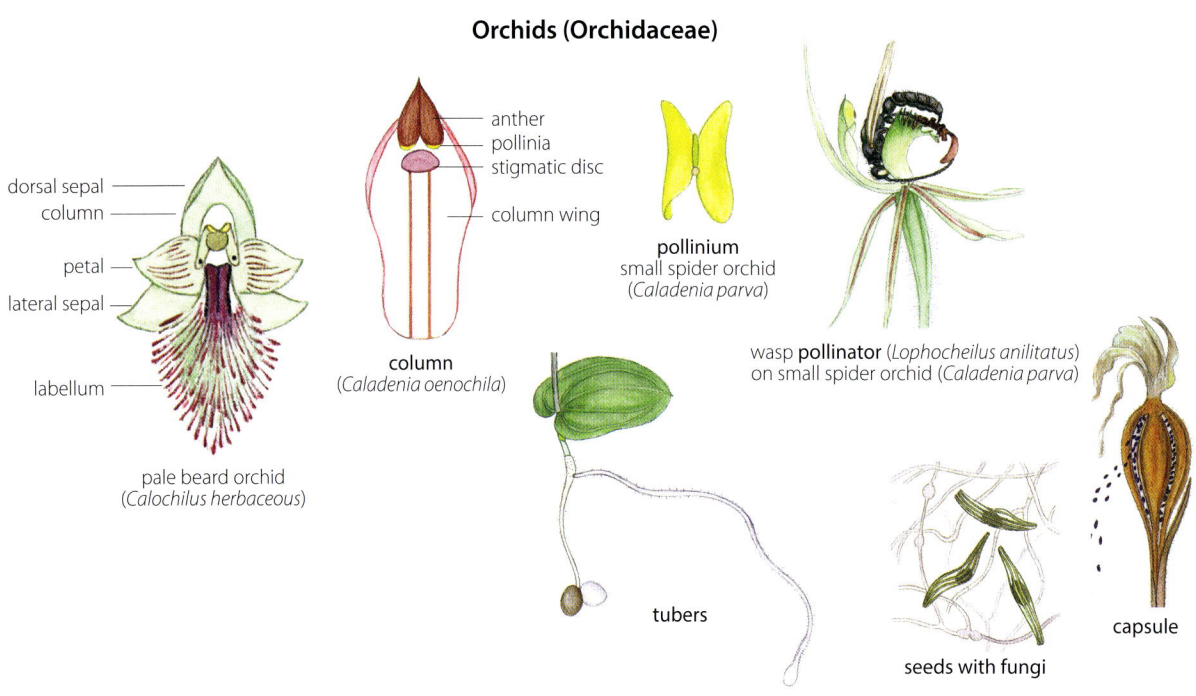

dorsal sepal
column
petal
lateral sepal
labellum

pale beard orchid
(*Calochilus herbaceous*)

anther
pollinia
stigmatic disc
column wing

column
(*Caladenia oenochila*)

pollinium
small spider orchid
(*Caladenia parva*)

wasp **pollinator** (*Lophocheilus anilitatus*)
on small spider orchid (*Caladenia parva*)

tubers

seeds with fungi

capsule

Organelle

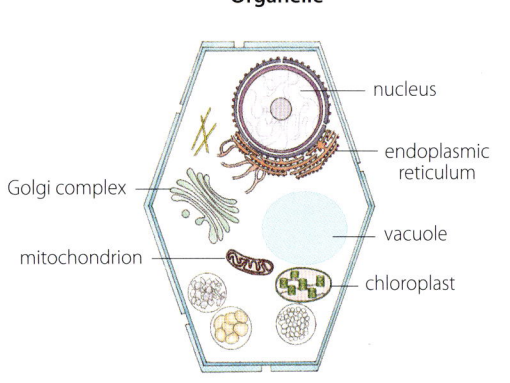

nucleus

endoplasmic
reticulum

Golgi complex

vacuole

mitochondrion

chloroplast

organic   Describes a molecule that consists mainly of carbon and hydrogen atoms.

Of, relating to, derived from or characteristic of living plants and animals.

cf. **inorganic**

organism   An individual life form, as an animal, plant, fungus or alga.

orifice   An opening or mouth, as the tubular part of a corolla where it joins the limb.

cf. **palate**

limb

orifice

tube

ornamentation   Surface features, as those on a pollen grain.

see **sculpturing**

ornamental   Of plants grown for decorative purposes.

Ornamentation of pollen grains

punctate   faveolate   areolate

papillate   verrucate   fossulate

echinate   striate   reticulate   lacunate

ornithochore   A plant whose seeds, spores or fruits are dispersed by birds.

ornithochory   Dispersal of seeds and fruit by birds.
cf. **ornithophilous**
ornithochorous   Of or relating to ornithochory.

ornithogamy   Having flowers adapted to attract bird pollinators. Pollination by birds.
ornithogamous   Of or relating to ornithogamy.

ornithophile   A plant that is pollinated by birds.

ornithophily   Pollination of flowers by birds.
cf. **ornithochory**
ornithophilous   Pollinated by birds.

orophile   Thriving in subalpine or mountainous regions.
orophilous   Thriving in subalpine habitats.

orophyte   A plant growing in subalpine habitats.
orophytic   Thriving in subalpine habitats.

orthostichy   *see* page 206

orthotropism   Tendency of a plant to grow in a vertical direction, either upwards as some stems or downwards as taproots.
see **tropism**
orthotropic   Of or relating to orthotropism.

orthotropous
Of ovule orientation, with the chalaza at the base the micropyle at the top and facing upwards and the ovule between, with all aligned with the funicle on a straight axis.
The most primitive ovule orientation.
= **atropous**
see **ovule orientation**

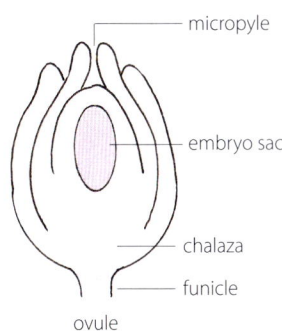

micropyle

embryo sac

chalaza

funicle

ovule

orthotropous

osmophore
Scent-producing glands, as those on the sepals of some spider orchids (*Caladenia*).

osmophores

osmosis   Movement of solvent molecules from a region of higher concentration to a region of lower concentration across a semipermeable membrane. A solvent (like water) can pass through the membrane but dissolved solids (like sugar) cannot. Movement continues across the membrance until the concentrations of solvent and solute are equal on both sides.

Osmosis

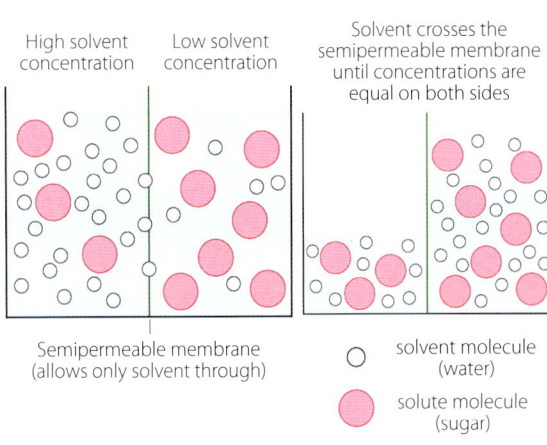

High solvent concentration

Low solvent concentration

Solvent crosses the semipermeable membrane until concentrations are equal on both sides

Semipermeable membrane (allows only solvent through)

○   solvent molecule (water)

●   solute molecule (sugar)

orthostichy   A vertical row or rank of leaves, flowers or scales along a stem.
A stem with one row is monostichous, two rows is distichous, three rows tristichous and five rows pentastichous.
see also **phyllotaxy**
cf. **parastichy**

Orthostichy

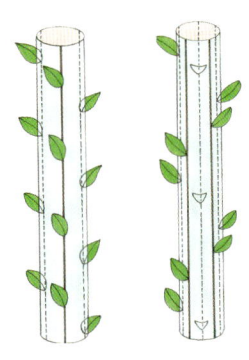

| monostichous | distichous | tristichous | pentastichous |

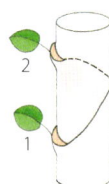
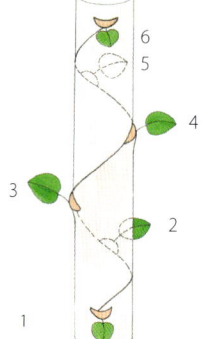

monostichous	**distichous**	**tristichous**	**pentastichous**
One vertical row of leaves on the stem	Two vertical rows of leaves on the stem	Three vertical rows of leaves on the stem	Five vertical rows of leaves on the stem
One turn around the stem	One turn around the stem	One turn around the stem	Two turns around the stem
The second leaf is above the first and begins a new series	The third leaf is above the first and begins a new series	The fourth leaf is above the first and begins a new series	The sixth leaf is above the first and begins a new series
One turn in the series Two leaves in the series	One turn in the series Two leaves in the series	One turn in the series Three leaves in the series	Two turns in the series Five leaves in the series
1/1 Turns around stem once, vertical rows one	1/2 Turns around stem once, vertical rows two	1/3 Turns around stem once, vertical rows three	2/5 Turns around stem twice, vertical rows five

osmotic pressure   Pressure of a solvent against a semipermeable membrane when moving in the direction of a low solvent concentration.
It persists until concentrations of solvent and solute are equal on both sides of the membrane.
see also **osmosis**

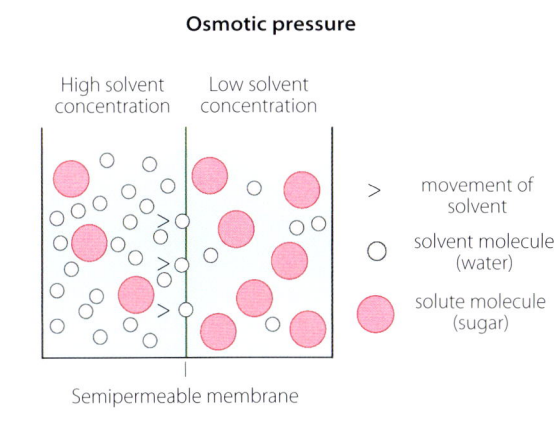

Osmotic pressure

osmotrophy   Digestion outside the cells before absorbing nutrients into the cells by osmosis, as in fungi.

osseous   Of a bone-like hardness and texture, as the stone in the fruit of olives (*Olea europaea*).

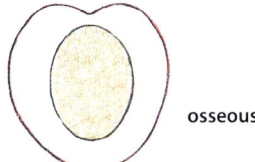

osseous

ostiole   A small opening or pore.
The opening at the apex of the fig inflorescence through which fig wasps enter and pollinate the tiny flowers.
ostiolar   Pertaining to an ostiole.

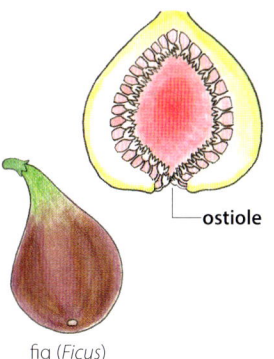

ostiole

fig (*Ficus*)

outbreeding   The production of seeds between plants that are from different populations, subspecies or species.
see also **autogamy**
cf. **outbreeding, outcrossing**

outcrossing   Pollination or fertilisation by pollen from a different plant of the same species.
cf. **self-pollination**

outgroup   In phylogenetics, taxa of interest to a study that are closely related to but not included in the ingroup.
cf. **ingroup**

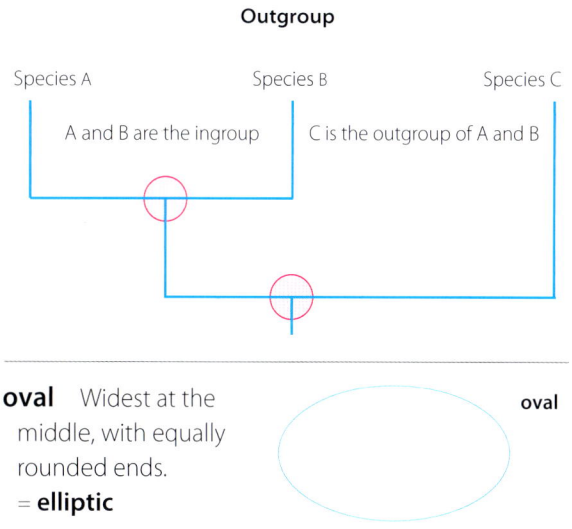

Outgroup

Species A   Species B   Species C

A and B are the ingroup   C is the outgroup of A and B

oval   Widest at the middle, with equally rounded ends.
= **elliptic**

oval

ovary
The part of the carpel below the stigma and style that encloses one or more ovules, with each ovule containing an ovum (female sex cell or gamete).
A simple ovary is a single carpel containing one ovary, as a pea (*Pisum*).
A compound ovary results from two or more fused carpels. If the carpels of a compound ovary are aseptate it has one locule.
The ovary matures into the seed-bearing fruit.
see also **paracarpous, syncarpous**

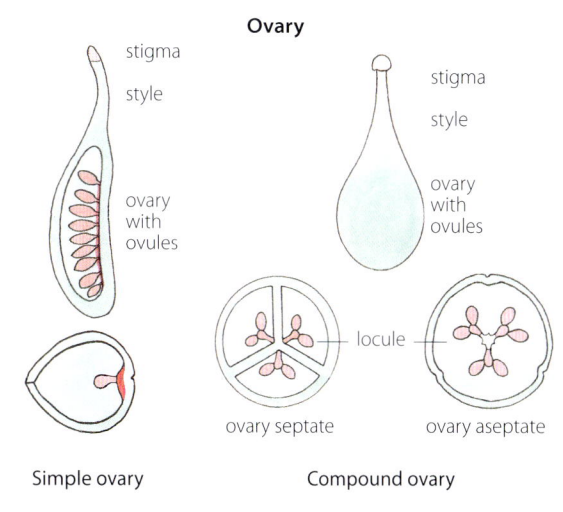

Ovary

stigma
style
ovary with ovules

stigma
style
ovary with ovules

locule

ovary septate   ovary aseptate

Simple ovary   Compound ovary

ovate Egg-shaped in outline, with the broad end at the base.
cf. **ovoid**

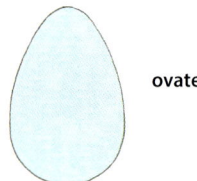

ovate

ovoid, oviform Ovate, but three-dimensional with the broad end below the middle. Egg-shaped.

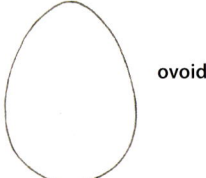

ovoid

ovulate cone The female, ovule-producing cone of gymnosperms.
= **megastrobilus**
cf. **staminate cone**

ovule In seed plants, a haploid egg (the female gamete) that develops into a seed after fertilisation. in angiosperms, it is the embryo sac (female gametophyte, megagametophyte), enclosed in the ovary, in which one egg (female gamete) develops. In gymnosperms, the ovule is exposed on the upper surface of a scale on the female cone. Typically, at maturity, it consists of an integument, a thin layer of nucellus and an ovoid female gametophyte (megagametophyte) that is undifferentiated except for several archegonia, each with an egg (female gamete).

Ovule

Angiosperms

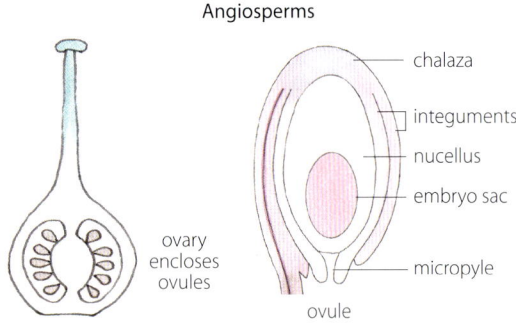

chalaza
integuments
nucellus
embryo sac
micropyle

ovary encloses ovules

ovule

Gymnosperms

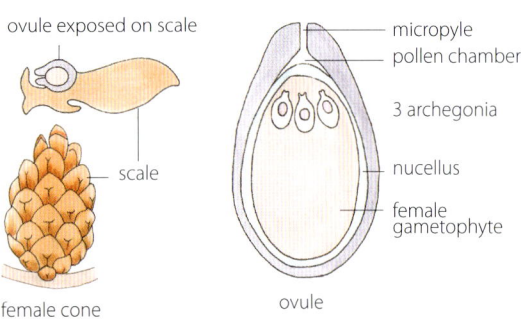

ovule exposed on scale

scale

female cone

micropyle
pollen chamber
3 archegonia
nucellus
female gametophyte

ovule

ovule orientation The positioning of the ovule in the ovary in seed plants (angiosperms and gymnosperms).
Generally, it takes into account the relative positions of the ovule and funiculus.

Examples of ovule orientation in angiosperms

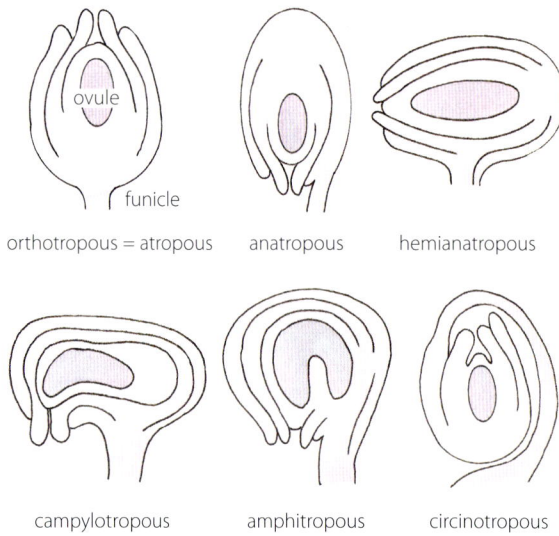

ovule

funicle

orthotropous = atropous   anatropous   hemianatropous

campylotropous   amphitropous   circinotropous

ovuliferous Bearing ovules.

ovuliferous scale Of gymnosperms, typically one of the scales on a female cone, usually bearing two ovules on its upper surface.
see **megasporophyll**

ovulode A sterile reduced ovule. Common in the myrtle family (Myrtaceae).
cf. **staminode**

ovum, *pl.* **ova** The female reproductive cell of plants and animals.
= **egg, egg cell, megagamete**

oxygen, (O), (O$_2$) Plants aquire oxygen by the breakdown of carbon dioxide (CO_2) during photosynthesis.
A small amount is used in cellular respiration and the rest is released back into the atmosphere as molecular oxygen.

P generation The parental generation.
The two true breeding varieties that are crossed to produce a hybrid (F1 generation).

P/E-ratio   The ratio of the length of the polar axis to the equatorial diameter.

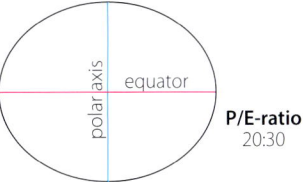

polar axis
equator
P/E-ratio
20:30

pachy-   A prefix meaning thick.

pachycaul   A growth form of plants that is deemed to be a trait of primitive angiosperms. Characterised by thick unbranched trunks and few large leaves, as cucumber tree (*Dendrosicyos socotranus*).
cf. **leptocaul**

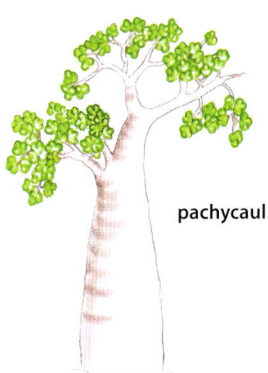

pachycaul

pachymorph   Of woody bamboos, a short thick clumping rhizome that usually curves upwards, with a culm produced at the tip of each rhizome. Such bamboos are determinate and non-invasive.
see **amphipodium, sympodium**
cf. **leptomorph, monopodium**

Pachymorph

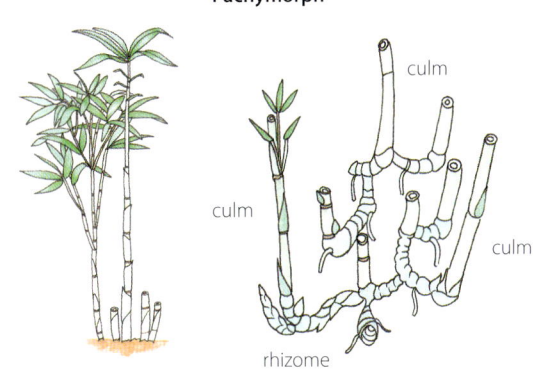

culm

culm

culm

rhizome

palaceous   With the stalk attached to the margin, as the petiole of a leaf.
cf. **peltate**

palaceous

palae-, paleo-   A prefix meaning ancient or from a time in the geological past.

palaeotropics   The Old World tropical continental regions of Africa and southwest Asia (excluding Australia) and islands of the Indian and Pacific oceans.
cf. **neotropics**

palate   A projection of the lower lip of a bilabiate corolla that partially or completely closes the throat, as that on the snapdragon flower (*Antirrhinum*).
cf. **throat**

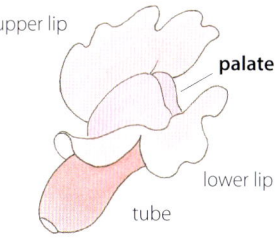

upper lip

palate

lower lip

tube

palea, *pl.* **paleae, paleas**   Of grasses (Poaceae), the inner bract that, together with the lemma, encloses the typical flower. One of the chaffy bracts or scales subtending a floret on the receptacle of the head of many Asteraceae.
= receptacular bract
paleate   Chaffy. Bearing paleae.

flower

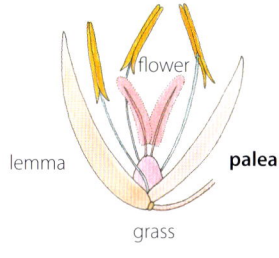

lemma

palea

grass

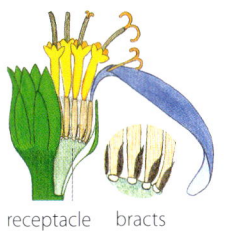

receptacle

bracts

daisy

paleaceous   Consisting of chaff, as the winnowed glumes lemmas and paleas of grains and other grasses. Bearing chaff or chaffy scales. Chaff-like. Of grasses, bearing a palea.

paleaceous

paleola   A diminutive or secondary palea.
paleolate   With a paleola.

palinactinodromous   Of leaves with primary veins diverging radially from a single point and with these main veins themselves branching dichotomously.
cf. **actinodromous**

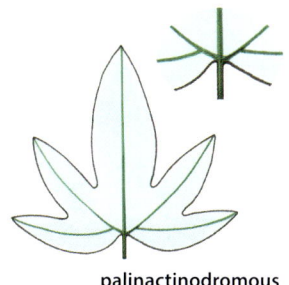

palinactinodromous

palisade mesophyll   A layer of elongated cells located beneath the upper epidermis of a eudicot leaf.
It contains most of a leaf's chlorophyll.

Palisade mesophyll

waxy cuticle
epidermis

palisade mesophyll

spongy mesophyll

epidermis with guard cells

Eudicot leaf

palm   Common name for a member of the Arecaceae family. A subtropical tree, shrub or climbing vine (rattan). Except for rattans, mostly characterised by a crown of pinnate or palmate fronds on a tall unbranched stem.

palm

palman   The undivided central part of a palmate leaf.

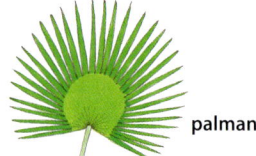

palman

palmate   Having the shape of a hand.
Of a compound leaf with sessile or petiolulate leaflets attached to the apex of the petiole.
cf. **pinnate**

Palmate

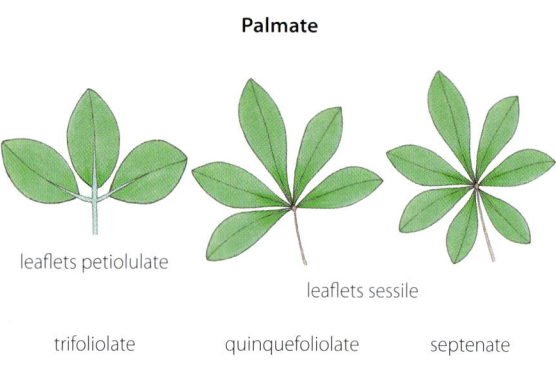

leaflets petiolulate

leaflets sessile

trifoliolate   quinquefoliolate   septenate

palmate venation, palmately veined
Leaf venation with the main veins radiating from one point at the base, usually near the tip of the petiole.
cf. **parallel venation, pinnate venation**

palmate venation

palmately lobed
Having lobes that are rounded and with the divisions not more than half way to the top of the petiole, as some palmate leaves.
= **palmatilobate, palmatilobed**

palmately lobed

palmately trifoliolate
Of a palmate leaf with leaflets arranged in groups of three at the top of the petiole.
Leaflets can be sessile or petiolulate.
see also **ternate**
cf. **pinnately trifoliolate**

leaflets sessile

petiole

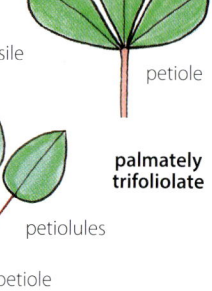

palmately trifoliolate

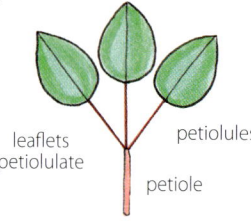

leaflets petiolulate

petiolules

petiole

palmatifid   Of a palmate leaf that is split into pointed rather than rounded lobes.

palmatifid

palmatilobate, palmatilobed   Having lobes that are rounded and with the divisions not more than half way to the top of the petiole, as some palmate leaves.
= **palmately lobed**

palmatilobate

palmatipartite
Of a palmate leaf having lobes with incisions extending from about half to three-quarters of the way towards the top of the petiole.

palmatipartite

palmatisect Of a palmate leaf having lobes with incisions that extend almost, but not quite, to the top of the petiole.

palmatisect

pan- A prefix meaning all, whole or all-inclusive.

pandurate, panduriform Fiddle-shaped. Of a pinnately lobed leaf that is obovate in outline with an indentation on each side, like a violin, as the leaf of fiddle dock (*Rumex pulcher*).

pandurate

Pangaea, Pangea A vast continent, including all the exposed landmass of the earth, believed to have existed prior to the Triassic period when it split into Laurasia and Gondwana.
see **continental drift**

Pangea

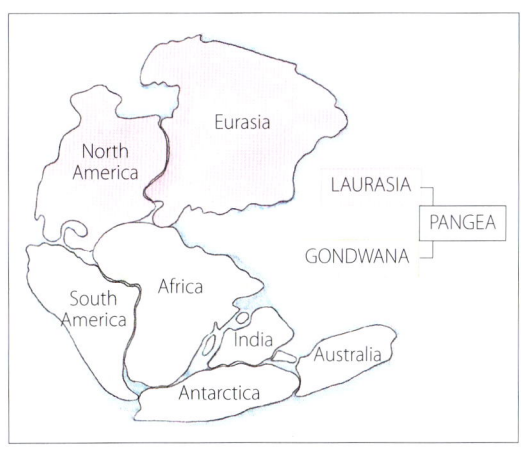

panicle Loosely, a much-branched inflorescence. Specifically, a compound racemose inflorescence in which a raceme branches into racemes, or a raceme branches in diverse ways and becomes irregularly compound.
In grasses (Poaceae), the arrangement of pedicellate spikelets, (rather than flowers), along the spreading or contracted branches of the inflorescence.
cf. **digitate inflorescence, raceme, spikelet, thyrse**
paniculate Growing or arranged in racemes.
paniculiform With the appearance, but not necessarily the structure, of a panicle.

Panicle

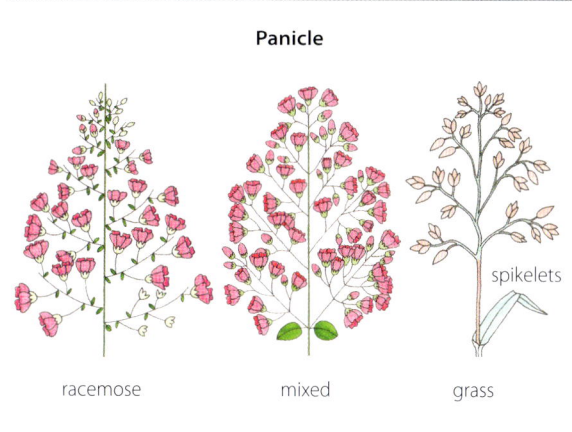

racemose   mixed   grass

spikelets

pannose Covered with felted woolly hairs.

panto- A prefix meaning global.

pantoaperturate Of a pollen grain with apertures distributed more or less evenly over its surface.

pantoporate Of a pollen grain with pori distributed more or less evenly over its surface.

pantropical Distributed throughout all of the tropical regions of the world.

papery Thin and dry like paper, as the outer bracts of blunt everlasting (*Argentipallium obtusifolium*).
= **chartaceous**

bracts **papery**

papilionaceous Butterfly-like, as the corolla of peas (family Fabaceae, subfamily Faboideae). It has five petals comprising a large upright standard, two clawed lateral wings and two lower petals united along their lower margin to form a keel enclosing the stamens and pistil.

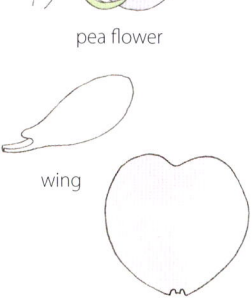

pea flower

wing

standard

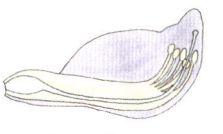

pistil and stamens

keel

papilionaceous aestivation Another term for vexillary aestivation.

papilionate Butterfly-like, as the flowers of peas (family Fabaceae, subfamily Faboideae).

papilla, *pl.* **papillae**
A small rounded protuberance.
papillate, papillose
Bearing papillae.

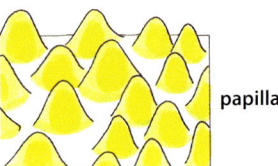
papilla

pappus, *pl.* **pappuses, pappi**
A modified calyx, composed of a ring of hairs, bristles or scales, borne at the base of the floret of many daisies (Asteraceae).
It persists and assists dispersal of the fruit, a cypsela, by wind.

Pappus

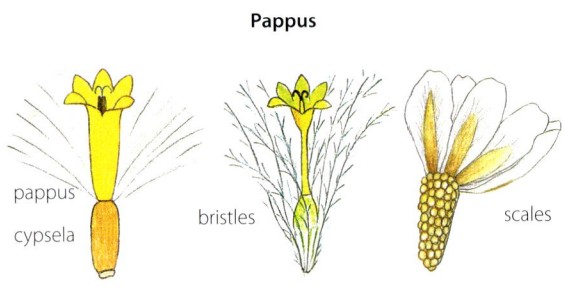

pappus
cypsela
bristles
scales

papyraceous Thin and dry like paper, as the bark of the paper birch (*Betula papyrifera*) that peels off in papery strips.

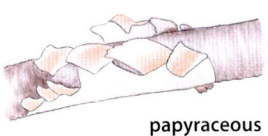
papyraceous

para- A prefix meaning by the side of, near, compared with or similar to.

paracarpous
Of a flower having a compound gynoecium of two or more carpels with the ovaries, styles and stigmas fused together to form one unit and the ovary lacking septa inside.
cf. **syncarpous**

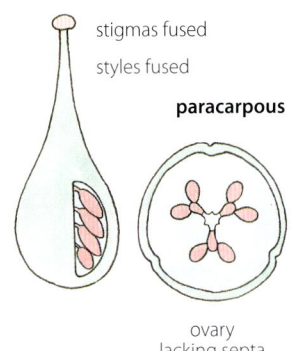
stigmas fused
styles fused
paracarpous
ovary lacking septa

paraclade, paracladium, *pl.* **paracladia**
A lateral inflorescence in a synflorescence that repeats the pattern of the terminal inflorescence.
cf. **coflorescence**

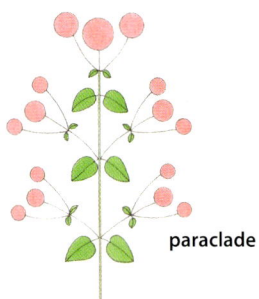
paraclade

parallel evolution The independent evolution of similar traits in two or more unrelated or distantly related organisms.
= **convergent evolution, homoplasy**
cf. **divergent evolution**

parallel venation
Leaf venation with two or more parallel main veins originating at the base and converging towards the apex and more or less parallel with the midrib or with veins at acute or right angles to the midrib.
cf. **palmate venation, pinnate venation**

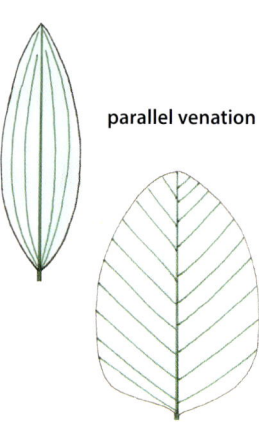
parallel venation

parallelodromous
Of leaves with two or more parallel main veins originating at the base and converging towards the apex.

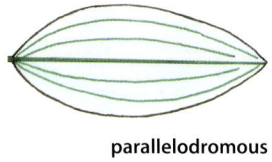
parallelodromous

parapatric Of distribution, occurring in the same geographic area but having adjacent but non-overlapping ranges.
cf. **allopatric, sympatric**

parapetalous Beside or between petals, as stamens that are beside or between the petals.

parapetalous

paraphyletic
Of a group of
organisms that
includes the most
recent common
ancestor but not
all of its descendants.
cf. **monophyletic,
polyphyletic**

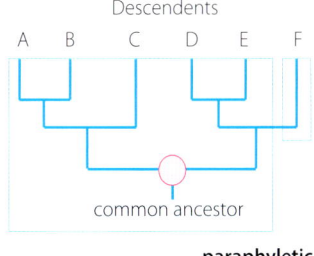

Descendents

common ancestor

paraphyletic

paraphyllidium *pl.* **paraphyllidia**   A reduced
leaflet at the base of the pinna in mimosa (*Mimosa*).

paraphysis,
pl. **paraphyses**
A sterile filament among
the sporangia of some
ferns, as tree ferns
(*Cyathea*), and some
cryptogams.

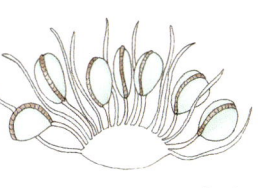

paraphysis

parasite   An organism (the parasite) that lives
on or in another organism (its host) from which
it derives some or all of its nourishment, to the
detriment of its host.
see **parasitism, symbiosis**
cf. **epiphyte, saprophyte**
parasitic   Of or relating to a parasite.

parasitism   A close and
prolonged interaction
between organisms
of different species in
which one benefits (the
parasite) and the other
(the host) is harmed, as
dodder laurel (*Cassytha*)
that parasitises and
harms other plants.

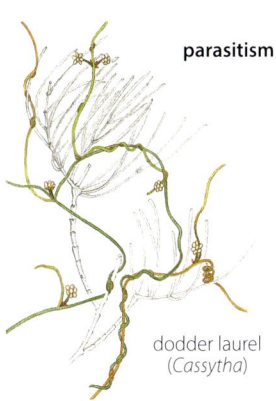

parasitism

dodder laurel
(*Cassytha*)

parastichy   An
imaginary spiral around
a stem connecting
the bases of a series of
crowded leaves, scales
or bracts.
see also **phyllotaxy**
cf. **orthostichy**

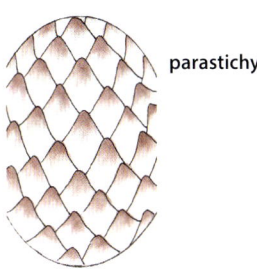

parastichy

paratype   Any specimen used, together with
the holotype, in the original description of a taxon
(protologue).

parenchyma   Composed of thin-walled
isodiametric cells that are flexible, still capable of
dividing and typically with a large central vacuole.
Functions include photosynthesis in the
parenchyma of leaves, storage (mainly starch) and
secretion of hormones and enzymes.
Found, for example, in the cortex and pith of stems,
the pulp of fruits and the endosperm of seeds.
parenchymatous   Like or consisting of
parenchyma.

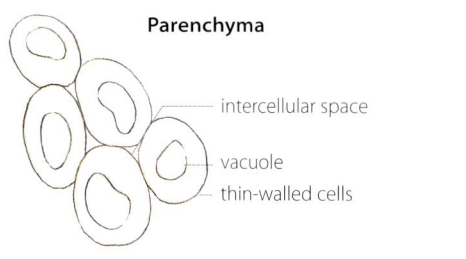

Parenchyma

intercellular space

vacuole

thin-walled cells

parietal   Attached to the wall of a structure, as
ovules attached to placentas on the wall of the
ovary.
cf. **axile**

parietal placentation
With carpels fused
but the internal walls
(septa) lacking, creating
a unilocular ovary, with
ovules arranged along
each line of fusion, as
violets (*Viola*).
see **placentation**

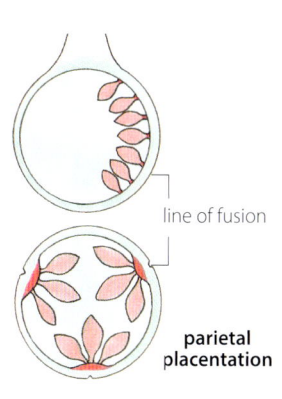

line of fusion

parietal
placentation

paripinnate
Of a pinnate leaf with
leaflets arranged in
pairs and terminating
with a pair of leaflets.
= **abruptly-pinnate,
even-pinnate**

paripinnate

parsimony   In phylogenetics, the principle that
when more than one evolutionary tree (cladogram)
can be made from available data the simplest tree is
chosen as the hypothesis to study.
= **phylogenetic hypothesis**

parthenocarpy Formation of fruit without fertilisation of ovules, resulting in seedless sterile fruits.
It may occur naturally or be artificially induced, as cultivars of bananas, pineapples and navel oranges.
cf. **parthenogenesis, stenospermocarpy**

parthenogenesis The development of an embryo from an unfertilised egg cell.
A form of agamospermy.
cf. **parthenocarpy**

-partite A suffix meaning divided into parts.

partite Divided into segments extending about three-quarters of the way towards the midrib of a pinnately lobed leaf or to the base of a palmately lobed leaf.

partite

pinnate

palmate

patent Horizontal, spreading more or less at right angles, as leaves on a stem.

patent

paxillate venation Leaf venation with very closely parallel veins running from the midrib to the margin at a slight angle.

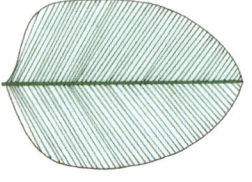

paxillate venation

pectin Any of various water-soluble polysaccharides that bind adjacent cell walls in plant tissues.
see **middle lamella**

pectinate Of a pinnately lobed leaf with the lobes numerous, close and narrow, like the teeth of a comb.
Of leaf arrangement, arranged like the teeth of a comb, as the narrow leaves of some firs (*Abies*).

pectinate

pedate Palmately divided but with the lowest leaflet on each side attached to the petiole of the leaflet above.
see also **palmate, pedately lobed**

pedate

pedately lobed Palmately lobed but with the lowest lobe on each side itself lobed.
= **pedatilobate, pedatilobed**

pedately lobed

pedatifid Of a pedate leaf that is split into pointed rather than rounded lobes.

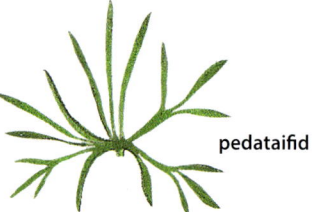

pedataifid

pedatilobate, pedatilobed Palmately lobed but with the lowest lobe on each side itself lobed.
= **pedately lobed**

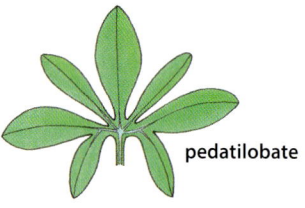

pedatilobate

pedatipartite Of a pedate leaf having the main lobes with incisions extending from about half to three-quarters of the way to the top of the petiole.

pedatipartite

pedatisect Of a pedate leaf having the main lobes with incisions that extend almost, but not quite, to the top of the petiole.

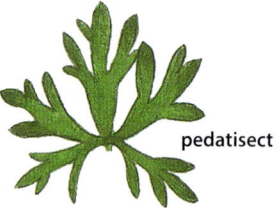

pedatisect

pedicel   The stalk of a single flower in a compound inflorescence.

Of a geniculate pedicel, the stalk beyond the bend in the stem between the peduncle and the calyx, as bent goodenia (*Goodenia geniculata*).

cf. **peduncle**

pedicellate   Borne on or having a pedicel.

Pedicel

compound inflorescence   geniculate pedicel

peduncle   The stem that supports a group of pedicels in a compound inflorescence.

The stem of a solitary flower.

cf. **pedicel, petiole, scape**

pedunculate   Borne on or having a peduncle.

peduncular   Of or relating to a peduncle.

Peduncle

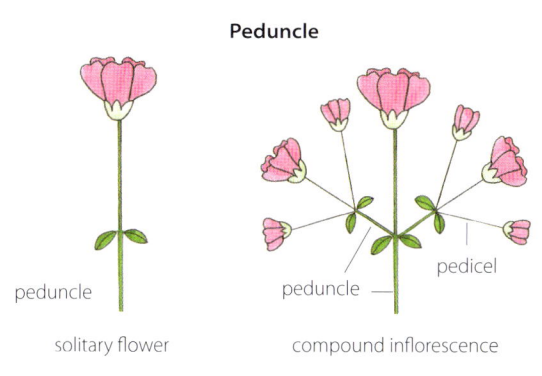

solitary flower   compound inflorescence

peg roots
A slender cylindrical pneumatophore found in mangroves.

= **pencil roots**
see also **cone roots, knee roots**

peg roots

pellicle   A thin skin-like or membranous covering, as that on the outer surface of the seed of a pomegranate (*Punica granatum*).

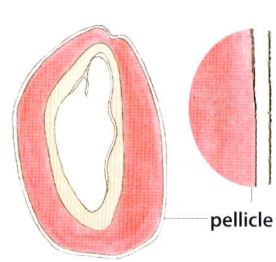

pellicle

pellucid   Letting light through, transparent or translucent.

peloria   Abnormal development, due to a mutation, of a typically zygomorphic flower so that it becomes actinomorphic.

peloric   Of or relating to peloria.

peloton   Of ericoid and orchidaceous mycorrhizas and ectendomycorrhizas, coils of hyphae found in the cortical cells of roots.

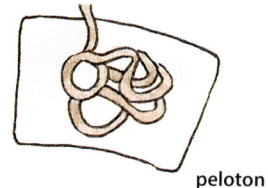

peloton

peltate
Shield-like.
More or less circular, with the stalk attached on the lower surface at or near the centre, as some leaves and scales.

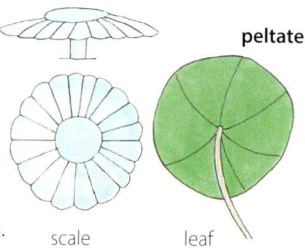

peltate

scale   leaf

pencil roots
A slender cylindrical pneumatophore found in mangroves.

= **peg roots**
see also **cone roots, knee roots**

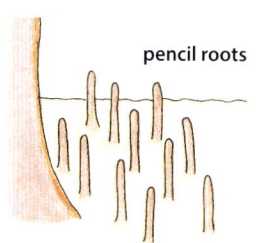

pencil roots

pendent   Hanging with the apex pointing vertically downward, as a flower or a fruit.

pendent

pendulous   Hanging loosely.

pendulous

penicillate
Resembling a tuft or brush of fine hairs, as those on the column lobes of the trim sun orchid (*Thelymitra peniculata*).

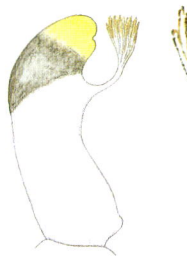

penicillate

215

penninerved   Of leaves with the secondary veins starting from a point on the midrib and running more or less parallel to each other towards the margin.
= **penniveined**
see **pinnate venation**

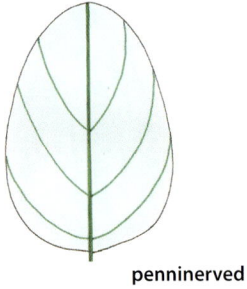
penninerved

penniveined   Of leaves with the secondary veins starting from a point on the midrib and running more or less parallel to each other towards the margin.
= **penninerved**
see **pinnate venation**

penniveined

penta-   A prefix meaning five.

pentadelphous
Of stamens united by their filaments into five bundles, as some species of *Hypericum*.
see **adelphous**

pentadelphous

pentamerous   Having flower parts, such as petals, sepals and stamens, in whorls of five or multiples of five. 5-merous.
see **-merous**

pentandrous
Having five stamens, as flowers of the carrot family (Apiaceae).
cf. **diandrous, monandrous, polyandrous, tetrandrous, triandrous**

pentandrous

pentastichous
Arranged on a stem in five vertical rows, as some leaves, with any sixth leaf above the one below it.
= **five-ranked**
see **orthostichy**

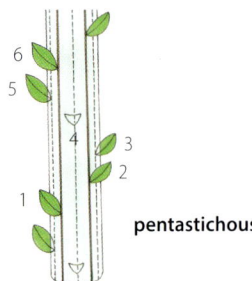
pentastichous

pepo   A many-seeded, fleshy fruit with a firm hard rind.
Derived from a three-carpelled non-septate inferior ovary.
Typical fruit of the gourd family (Curcurbitaceae), as melon and squash.
cf. **amphisarca**

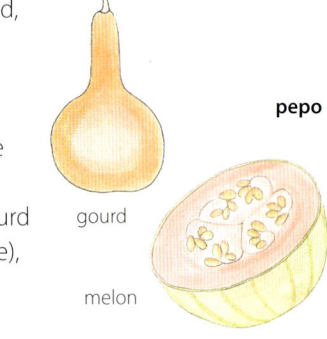
pepo
gourd
melon

per-   A prefix meaning through.

percurrent   Running through the entire length from the base to the tip, as the midrib of a leaf.

percurrent venation
Leaf venation with veins of the same order that run parallel to each like a ladder along the entire area between veins of a higher order.
see **scalariform**

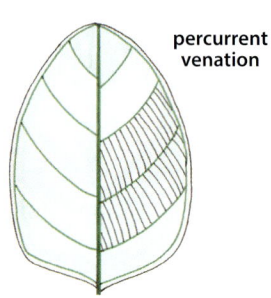
percurrent venation

perennating bud   The vegetative buds on aerial stems and underground stems (bulbs, rhizomes, tubers, corms etc.) that survive during dormancy and produce new growth for the next season.
A potato is a tuber with perennating buds.

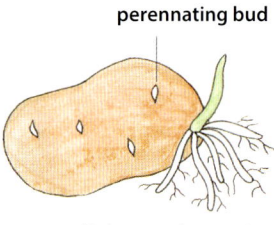
perennating bud
potato (*Solanum tuberosum*)

perennation   Survival from one season to the next, often in unfavourable conditions like drought or cold when the plant becomes dormant or dies back.
Buds, tubers, bulbs and rhizomes are examples of organs that allow a plant to perennate.
perennate, perennating   To last from one season to the next often with a period of dormancy or reduced growth between seasons.

perennial   A plant living and usually flowering for more than two years.
It may die back annually or persist above ground.
see also **herbaceous**
cf. **annual, biennial**

perfect   Of a flower with both stamens and pistils fertile.
= **bisexual, hermaphrodite**
Of leaf venation, having lateral veins extend for at least two-thirds of the leaf surface.
cf. **imperfect**

pistil   stamen

perfect

perfoliate   With the two basal lobes of an alternate leaf or bract united so that the stem appears to pass through the blade.

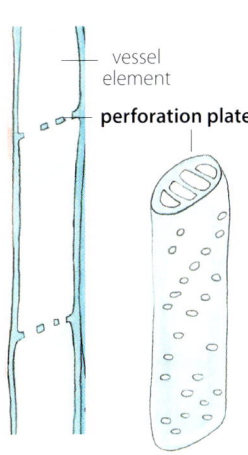

perfoliate

perforate   The presence of holes, as in a fenestrate leaf.
cf. **pit, punctate**

perforate

perforation plate
The remains of the end walls between two vessel elements in a xylem vessel.
Dying vessel element cells digest large holes in their end walls that allow free flow of water vertically in the xylem.
see **vessel**
cf. **sieve plate**

vessel element

perforation plate

peri-   A prefix meaning around or enclosing.

perianth   A collective term for the calyx and corolla of a flower, especially when both are similar, as the day lily (*Hemerocallis*).
The sterile parts of a typical flower.
= **floral envelope, perigone**

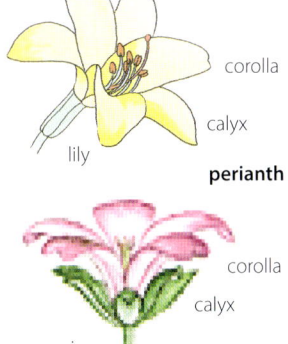

corolla

calyx

lily

perianth

corolla

calyx

geranium

pericarp   The dry or fleshy fruit wall that develops from the ovary wall or the ovary wall plus accessory parts and encloses the seed(s).
It consists of one or more of three layers: the outer layer (exocarp), the middle layer (mesocarp) and the inner layer (endocarp).
In fleshy fruits, as drupes, three layers are present.
see also **achene, caryopsis**

Pericarp

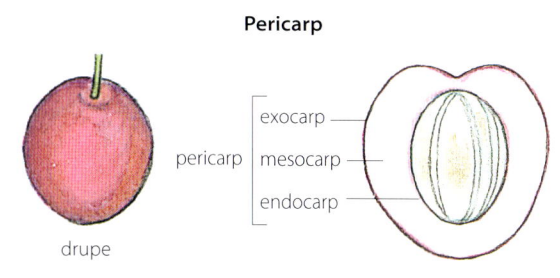

exocarp

pericarp   mesocarp

endocarp

drupe

perichaetium   A cluster of modified leaves around the reproductive organs of some bryophytes.

periclinal   Of the cell division plane, or any lines generally, parallel to the surface of the plant body.
cf. **anticlinal**

periclinium   The involucre surrounding flowers on the common receptacle of a flower head (capitulum) in the daisy family (Asteraceae).

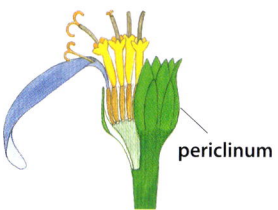

periclinum

pericycle
A layer of parenchyma cells that surrounds the vascular cylinder in roots and is in turn surrounded by the endodermis.
In eudicots it gives rise to lateral roots.

Pericycle

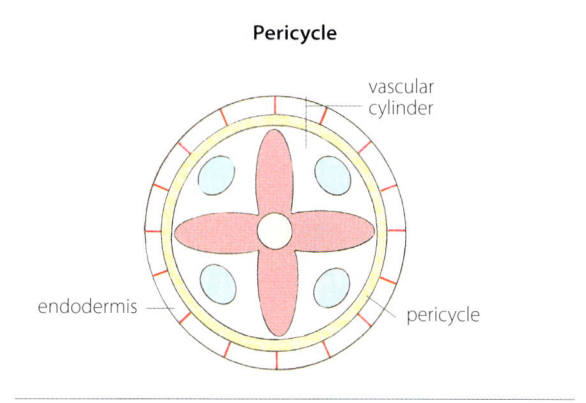

vascular cylinder

endodermis

pericycle

217

periderm A protective layer that replaces the epidermis in gymnosperms and woody eudicots and in the oldest parts of some herbaceous eudicots.

Consists of cork cells (phellem), cork cambium (phellogen) and parenchyma-like cells (phelloderm).

Periderm

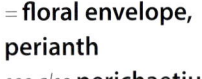

epidermis
phellem
periderm
phellogen
phelloderm
cortex

perigone, perigonium, *pl.* **perigonia**

A collective term for the calyx and corolla of a flower, especially when both are similar, as the day lily (*Hemerocallis*).

The sterile parts of a typical flower that enclose the reproductive stamens and/or pistil.

= **floral envelope, perianth**

see also **perichaetium**

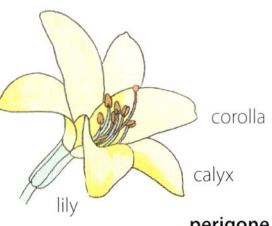

corolla
calyx
lily
perigone

corolla
calyx
geranium

perigynium The papery sac-like structure that encloses the female flower in the sedge genus *Carex*.

= **utricle**

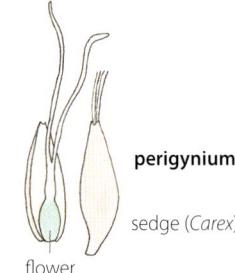

perigynium
sedge (*Carex*)
flower

perigyny The position of the ovary when it is surrounded by a hypanthium that bears the calyx, corolla and stamens.

perigynous Around the ovary.

Of the calyx, corolla and stamens inserted on the rim of a hypanthium and the ovary free within it.

cf. **epigynous, hypogynous**

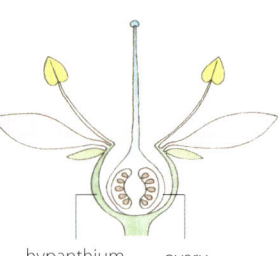

hypanthium   ovary
perigyny

perine The often highly ornamented layer of sporopollenin on the exine of a pollen grain wall.

see also **pollen wall**

cf. **perispore**

perinuclear Situated around or surrounding the nucleus of a cell.

perinuclear space

Of the nucleus, the empty space between the inner and outer nuclear membranes of the nuclear envelope.

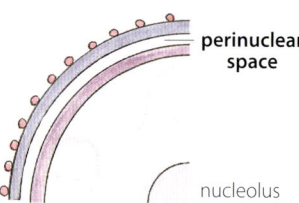

perinuclear space

nucleolus

perisperm Storage tissue in some seeds that is derived from the diploid nucellus.

It is present in some gymnosperms, as pine (*Pinus*).

Occurs in a few eudicot families, as the goosefoot family (Amaranthaceae).

see also **cotyledon, endosperm**

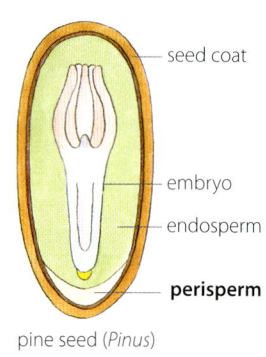

seed coat
embryo
endosperm
perisperm
pine seed (*Pinus*)

perispore The often highly ornamented sporopollenin layer that surrounds the exospore of a spore wall.

see **sporoderm**

cf. **perine**

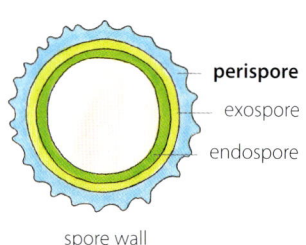

perispore
exospore
endospore
spore wall

peristome A single or double fringe of teeth around the mouth of a moss capsule.

It usually regulates spore dispersal.

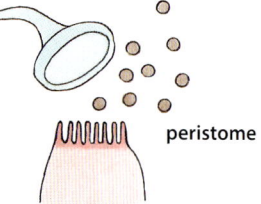

peristome

persistent

Remaining attached to the plant beyond the usual time of falling, as the calyx that remains on the capsule of bluebells (*Wahlenbergia*).

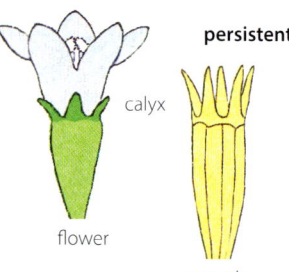

persistent
calyx
flower
capsule

pers. comm., personal communication
Personal communications from which data is not recoverable, as emails, conversations and speeches.

personal communication, *abbr.* **pers. comm.**
Personal communications from which data is not recoverable, as emails, conversations and speeches.

personate Mask-like. Of a corolla, in the form of a face, with two lips, a prominent palate closing or almost closing the throat and a gibbous tube, as snapdragons (*Antirrhinum*).
cf. **ringent**

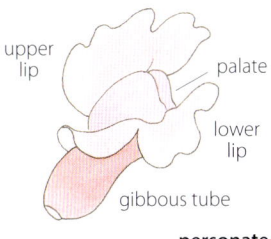

upper lip
palate
lower lip
gibbous tube

personate

perula, perule,
pl. **perulae, perules**
A modified leaf that acts as a protective covering and tightly encloses the developing flower or leaf bud of some plants, as the leaf bud of sugar maple (*Acer saccharum*).
= **bud scale**

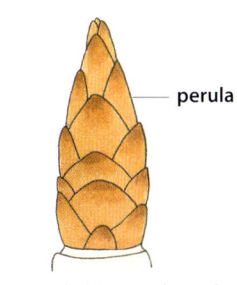

perula

sugar maple (*Acer saccharum*)

petal A segment of the corolla of a flower that is typically coloured and soft in texture. Pollinators are attracted by their colour and shape.
cf. **sepal, tepal**
petaliferous Bearing petals.
petaline Relating to, attached to or resembling a petal.

petal

petalantherous Of a stamen, with a terminal anther and a petaloid filament.

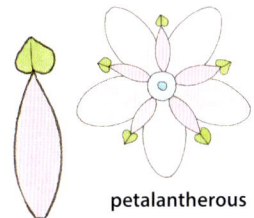

petaloid filament

petalantherous

petalody The abnormal development of floral parts into petals or petaloid organs.

petaloid Having the form or appearance of a petal, as the sepals of day lilies (*Hemerocallis*).

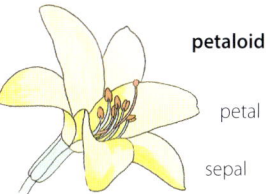

petaloid
petal
sepal

petalostemonous
With filaments of stamens fused to the petals or corolla tube and anthers free.

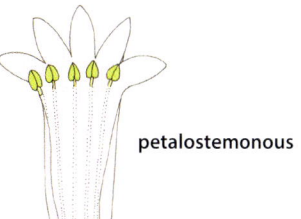

petalostemonous

petiole The stalk of a leaf that attaches the blade to a stem or branch.
petiolar Borne on or relating to a petiole.
petiolate With a petiole.

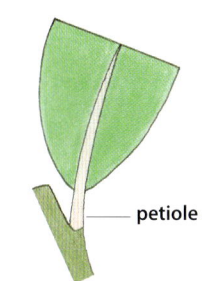

petiole

petiolode
The rooting petiole-like stalk of a leaf-like phyllomorph, as cape primrose (*Streptocarpus haygarthii*).

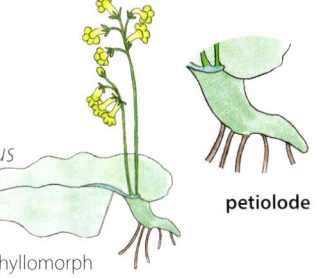

phyllomorph
petiolode

petiolule A small petiole. The stalk of a leaflet.
petiolulate Having a petiolule.
cf. **sessile**

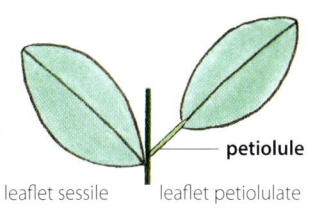

petiolule
leaflet sessile   leaflet petiolulate

pH The pH scale measures how acidic or alkaline a solution is.
The scale ranges from 0 to 14, with 7 being neutral, less than 7 having a high concentration of hydrogen ions and being acidic, and greater than 7 having a low concentration of hydrogen ions and being alkaline.
see **litmus test**

phalaenophily Dispersal of pollen and pollination by moths.
phalaenophilous Pollinated by moths.

phalange   A bundle of stamens united by their filaments, as some members of the gourd family (Curcurbitaceae).
see **adelphous**

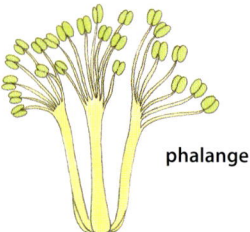

phalange

phanerocotyly   Of seed germination, having the cotyledons emerge from the seed coat at germination.
cf. **cryptocotyly**

phanerogams   Seed plants that comprise the flowering plants (angiosperms) and a group of non-flowering plants (gymnosperms) that includes conifers and cycads.
= **spermatophyte**
cf. **cryptogam**

phanerophyte   Plants with perennating buds exposed on branches and twigs as most woody trees and shrubs, woody lianas and some epiphytes. They may be deciduous or evergreen.
see also **chamaephyte, cryptophyte, hemicryptophyte, therophyte**

phellem   Cork tissue that is one of three layers of the periderm.
It is produced by the cork cambium (phellogen).

Phellem

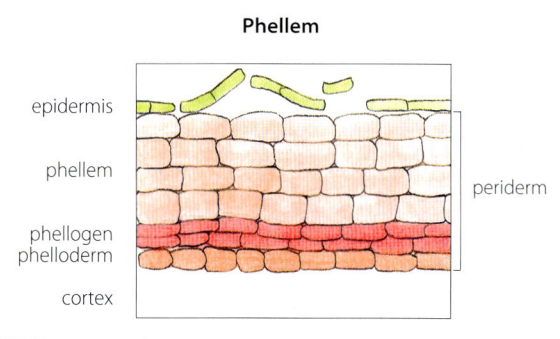

epidermis
phellem
phellogen
phelloderm
cortex
periderm

phelloderm   Parenchyma tissue that is one of three layers of the periderm.
It is produced by the cork cambium (phellogen).

Phelloderm

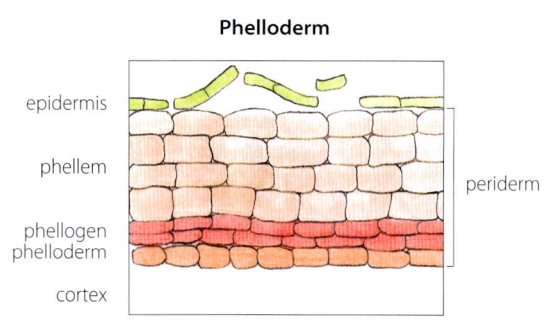

epidermis
phellem
phellogen
phelloderm
cortex
periderm

phellogen   A layer of meristem that is responsible for secondary growth and is part of the periderm in woody plants and some herbaceous plants.
It produces cork cells (phellem) on the side towards the surface of the plant and parenchymatous tissue (phelloderm) on the inner side.
= **cork cambium**
see **cambium, lateral meristem**
see also **fusiform initials, ray initials**

Phellogen

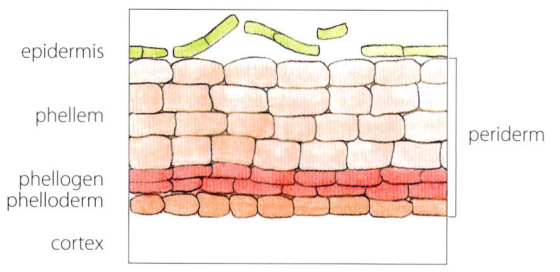

epidermis
phellem
phellogen
phelloderm
cortex
periderm

phenetics   A system of classifying plants according to their shared morphological characterstcs only.
cf. **phylogenetics, taxonomy**

phenology   The study of the timing of biological events in plants, such as flowering and reproduction, in relation to changes in season and climate.

phenotype   The observable characteristics of a cell or an organism, such as its size, shape and metabolism, that is a result of its genetic makeup and its interaction with the environment.
cf. **genotype**

pheromone   A volatile chemical compound released by plants into the environment that affects the behaviour of another species.
Notable in orchids that produce pheromones to attract a pollinating insect.
cf. **hormone**

phloem   Tissue in a vascular bundle composed of living cells that transport sap containing nutrients from the shoots to all parts of the plant.
There are two types of phloem. In primary growth, primary phloem is formed from procambium in the apical meristem of shoots and roots, and in secondary growth, secondary phloem is formed from vascular cambium and is the inner layer of the bark.

Typical phloem components are sieve elements (sieve tube elements in angiosperms and sieve cells in gymnosperms and lower vascular plants), companion cells (in angiosperms) and albuminous cells (in gymnosperms).
Fibre and/or sclereid cells and parenchyma cells can also be found in the phloem.
see **bast, metaphloem, protophloem, vascular bundle**

Phloem

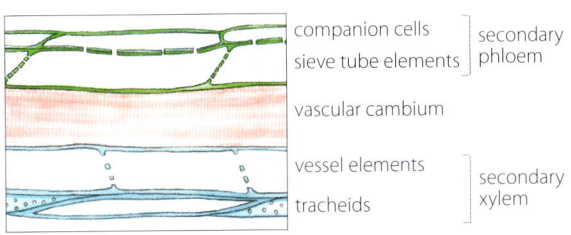

companion cells
sieve tube elements } secondary phloem

vascular cambium

vessel elements
tracheids } secondary xylem

phorophtye　Any plant, most often a tree, on which an epiphyte grows.

phosphorylation　A biochemical process that adds a phosphate group to an organic compound, as the addition of a phosphate group to adenosine diphosphate (ADP) forms adenosine triphosphate (ATP).
see **dephosphorylation**

photoautotroph, phototroph　An organism that uses sunlight, through photosysnthesis, to synthesise its own food from inorganic substances like carbon dioxide and water.
Almost all plants are autotrophs.
see **autotroph**
photoautotrophic, phototrophic　Of or relating to a photoautotroph or phototroph.
see **trophic**

photophile　Describing a phase in which light affects flowering.

photophilic, photophilous　Of a plant that seeks or thrives in light.

photophobe　Describing a phase in which darkness affects flowering.

photophobic　Of a plant that seeks or thrives in shade or indirect sunlight.

photosynthesis　The process that converts light energy from the sun into chemical energy. This energy is used to make sugars from carbon dioxide that is absorbed by plants from the atmosphere.

phototropism　Growth of a plant towards or away from light that is controlled by the flow of auxin to or away from a plant part.
see **tropism**
cf. **heliotropism**
phototropic　Of or relating to phototropism.
cf. **phototrophic**

phyllary　One of the bracts in the involucre that surrounds the capitulum in the daisy family (Asteraceae).

involucre　　phyllary

phylloclade　A type of cladode or phyllode that resembles and/or functions like a leaf, as those in the cactus family. A term with variable definitions
see **cladode**
= **cladophyll**

prickly pear
(*Opuntia*)　　Christmas cactus
(*Schlumbergera*)

phyllode, phyllodium,
pl. **phyllodia**
A modified petiole that functions as a leaf, as many wattles (*Acacia*).
cf. **cladode**
phyllodineous
Relating to or having phyllodes.

Acacia　　phyllode

phyllody　The abnormal development of floral parts into leafy organs.

phyllome　A plant part that is a leaf or derived from a leaf.
cf. **caulome**

phyllomorph   A term coined to distinguish the peculiar leaf-like structures of cape primrose (*Streptocarpus*) from true leaves and a cotyledon.
It consists of a foliose lamina that is the single enlarged cotyledon with a rooting petiole-like stalk called the petiolode. Species are unifoliate, as cape primrose (*Streptocarpus haygarthii*), or in a rosette of several unifoliate units, as African violet (*Saintpaulia*).

phyllomorph

petiolode

cape primrose
(*Streptocarpus haygarthii*)

phyllopodic
Of some sedges in the genus *Carex*, having the base of a flowering stem bearing dead leaves from the previous year's vegetative growth.

phyllopodic

phyllopodium,
pl. **phyllopodia**
The articulation at the base of the stipe of some ferns, as *Oleandra*. It remains on the rhizome once the frond has fallen off.

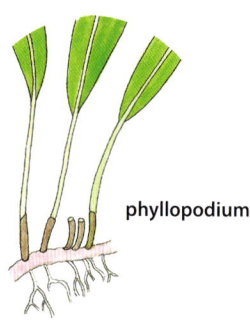

phyllopodium

phyllotaxy
The order in which leaves, scales or bracts with flowers are arranged on the stem.
There are three basic types: spiralled, with one leaf at each node on alternate sides, including alternate, coiled and scattered; opposite, with two leaves arising from each node on opposite sides of the stem; and whorled (verticillate), with at least three leaves at each node.
see also **orthostichy, parastichy**

Phyllotaxy

Arrangement spiralled, leaves one at a node

alternate   coiled   scattered

Arrangement whorled, leaves more than one at a node

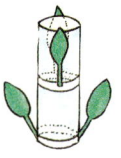

opposite   verticillate/whorled

PhyloCode   A code for classifying organisms according to ancestry and descent (phylogeny), using clades within clades, rather than a hierarchy of taxonomic ranks (species, genus, family etc.) as set out in the International Code of Nomenclature.

phylogenesis   The evolutionary history of a species or other taxonomic entity.
It is commonly represented as a cladogram (phylogenetic tree).
= **phylogeny**
cf. **ontogeny**

phylogenetic, phylogenic   Of or relating to the evolutionary development of organisms.

phylogenetic hypothesis   In phylogenetics, the principle that when more than one evolutionary tree (cladogram) can be made from available data the simplest tree is chosen as the hypothesis to study.
= **parsimony**

phylogenetic nomenclature   A rank-free system of nomencalture consisting of species and clades.

phylogenetic systematics   The study of organisms and their classification into groups based on their evolutionary descent.

phylogenetic taxon   The members of a named entity with a common evolutionary history of descent.
The node subtending a clade that may be real or, more often, hypothesised.
A phylogenetic taxon gains or loses characteristics over time.

phylogenetic taxonomy   A system of classification that shows how species may be related by descent from a common hypothetical ancestor.
It groups these species, together with their common ancestor, in a clade.

phylogenetic tree   A branching diagram or indented list depicting the lines of descent of different species, organisms or genes from a common hypothetical ancestor.
A phylogeny.
see **branch, node, root node**

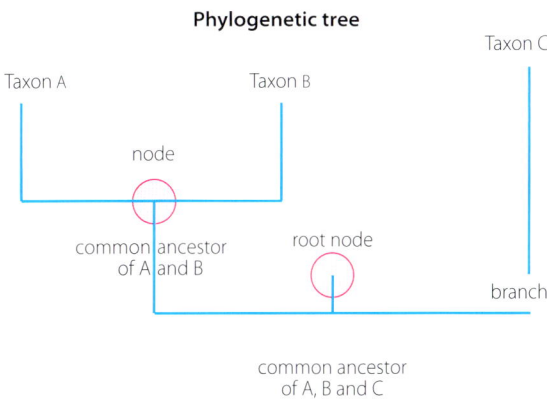

Phylogenetic tree

phylogenetics   A type of classification that expresses theoretical relationships between plants in terms of their evolutionary history.
DNA sequencing methods are used to study these relationships between organisms.
see **phylogeny, systematics, taxonomy**

phylogeny   The evolutionary history of a species or other taxonomic entity.
It is commonly represented as a cladogram (phylogenetic tree).
= **phylogenesis**
cf. **ontogeny**

phylum   A taxonomic classification between kingdom and class.
Formerly applied correctly only to the animal kingdom.
The name of a division or class in the plant kingdom ends in *-phyta*.
= **division**
see **taxonomic hierarchy**

physiology   The study of the vital processes that take place in a plant, as metabolism, growth and development, reproduction and responses to the environment.
physiological   Of or relating to physiology.

-phyte   A suffix meaning relating to plants.

phyte-   A prefix meaning relating to plants.

phytochrome   A blue-green pigment in plants that can only be seen when it is purified.
It regulates plant development, including seed germination, stem growth, leaf expansion and pigment synthesis.

phytohormone   A compound produced in small quantities that is a plant growth and development regulator. Includes: abscisic acid, auxins, cytokinins, ethylene and gibberellins.

phytomelan, phytomelanin   A black, inert organic material that forms a crust-like covering on some seeds, as hippeastrum (*Hippeastrum*) and the cypselas of some daisies (Asteraceae), as sunflowers (*Helianthus*).

phytomer, phytomere
The basic structural unit of a plant that is repeated.
It derives from the meristem of a root or shoot apex.
Of stem shoots, it comprises a node with its leaf or leaves, its axillary bud(s) and the subtending internode.
Phytomers also occur on lateral growth.
= **metamer**

axillary bud at node

leaf
internode
phytomer

shoot apex with meristem

phytomer

stem shoot with phytomers

phytomorphology   The study of the external
form or structure of a plant or plant part that
compares features and observes patterns of
development.
phytomorphological   Of or relating to
phytomorphology.

phytophage   An animal that eats only plants.
phytophagous   Feeding on plants as many
animals; includes leaf chewing, sap sucking, seed
predation and gall induction by insects.
= **herbivorous**

phytotomy   The study of the anatomy of plant
tissues and organs.
phytotomical   Of or relating to phytotomy.

pigment   A substance produced by plants that
absorbs specific wavelengths of sunlight and
reflects it back as colour.
Main plant pigment groups are chlorophylls,
carotenoids, flavonoids and betalains.

pileus, *pl.* **pilei**   The apex of the angular drupe of
pandanus (*Pandanus*).
The cap of a mushroom or toadstool.

piliferous   Tipped with
a single slender bristle or
hair, as the leaves of some
mosses.
Bearing hairs.

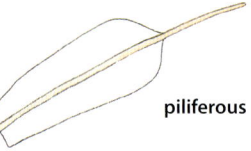
piliferous

piliferous layer   The
region near the root tip
that is covered with the
root hairs responsible for
water uptake, particularly
in young plants.

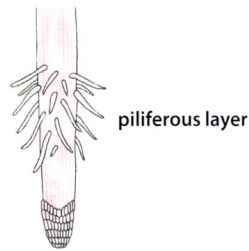
piliferous layer

piliform   Resembling a hair, hair-like.

pilose   Hairy. Covered
with soft slender hairs,
as some leaves
cf. **villous**

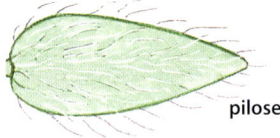
pilose

pilosulose   Finely pilose.

pilosulous   Slightly pilose.

pin, pin-eyed
Presentation of the stigma
above the level of the
anthers so that floral visitors
contact the stigma first, as
the cowslip (*Primula vulgaris*).
see **herkogamy**
cf. **thrum**

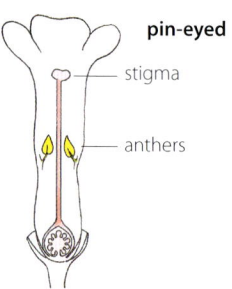
pin-eyed
stigma
anthers

pinna *pl.* **pinnae**   A leaflet of a pinnate leaf.
cf. **pinnule**

Pinna

opposite   sessile or petiolulate   alternate

pinnate   Resembling a feather.
One of two ways in which a leaf is divided into
leaflets (a compound leaf), the other being palmate.
The leaflets (pinnae) arranged on either side of a
common rachis.
Leaflets may be opposite or alternate, sessile or with
a petiolule.
see also **pinnately unifoliolate**

Pinnate

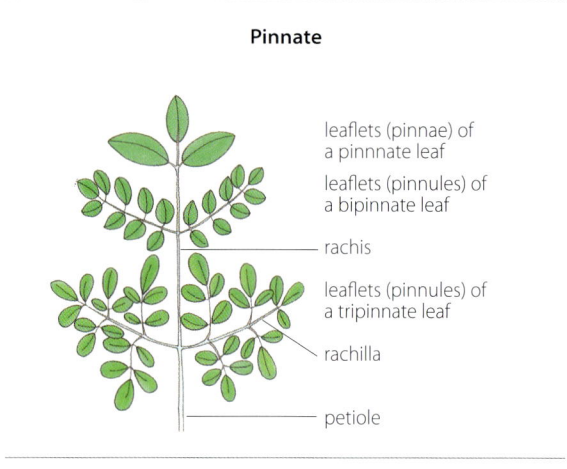
leaflets (pinnae) of
a pinnnate leaf
leaflets (pinnules) of
a bipinnate leaf
rachis
leaflets (pinnules) of
a tripinnate leaf
rachilla
petiole

pinnate venation, pinnately veined

Leaf venation with the secondary veins starting from a point on the midrib and running more or less parallel to each other towards the margin.

= **penniveined**
cf. **basinerved, palmate venation, parallel venation**

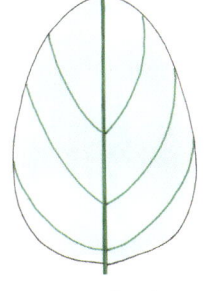
pinnate venation

pinnate-pinnatifid

Of a pinnate leaf with leaflets pinnately lobed.

pinnate-pinnatifid

pinnately decompound

Of a pinnate leaf divided more than three times or irregularly divided many times, as fennel (*Foeniculum vulgare*).
see **decompound**

pinnately decompound

pinnately lobed

Having lobes that are rounded and with the divisions not more than half way to the midrib, as some pinnate leaves.

= **pinnatilobate, pinnatilobed**

pinnately lobed

pinnately trifoliolate

Of a pinnately compound leaf with three leaflets. Leaflets can be petiolulate or sessile.
see also **ternate**
cf. **palmately trifoliolate**

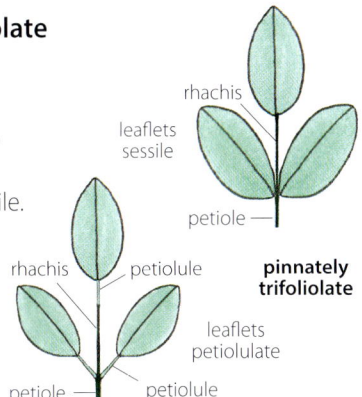
rhachis
leaflets sessile
petiole
rhachis
petiolule
leaflets petiolulate
petiole
petiolule
pinnately trifoliolate

pinnatifid

Of a pinnately lobed leaf, split almost to the midrib, by sharp sinuses, into pointed rather than rounded lobes.
see also **pinnatipartite, pinnnatisect**
cf. **bipinnatisect, tripinnatisect**

lobe
sinus
pinnatifid

pinnatilobate, pinnatilobed

Having lobes that are rounded and with the divisions not more than half way to the midrib, as some pinnate leaves.

= **pinnately lobed**

pinnatilobate

pinnatipartite

Of a pinnately lobed leaf with the incisions extending about three-quarters of the way towards the midrib.
see also **pinnatifid, pinnnatisect**

pinnatipartite

pinnatisect

Of a pinnately lobed leaf with the incisions extending almost, but not quite, to the midrib.
see also **pinnatifid, pinnatipartite**
cf. **bipinnatisect, tripinnatisect**

pinnatisect

pinnule

A leaflet that is one of the divisions of a bipinnate or tripinnate compound leaf.
see also **bipinnate, pinnately decompound, tripinnate**
cf. **pinna**

pinnule

pioneer species   Hardy species that are the first to colonise degraded ecosystems and disturbed areas, such as construction sites or bush that has been burnt.
see also **colonisation, coloniser**

pistil   The female reproductive organ of a flower (gynoecium).
A single carpel or group of free or fused carpels.
see also **compound gynoecium, simple gynoecium**

Pistil

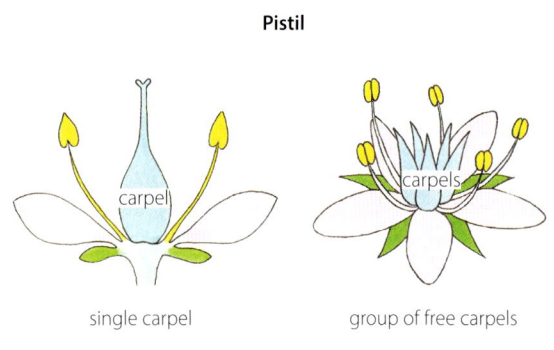

single carpel   group of free carpels

pistillate   Of a unisexual female flower having a carpel or carpels and no functional stamens.
cf. **staminate**

pistillode   A sterile or rudimentary pistil, often present in male flowers, as coconut (*Cocos nucifera*).
cf. **staminode**

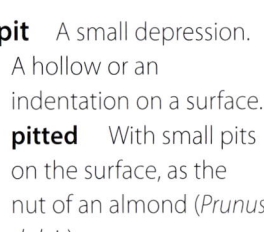

pistillode

pit   A small depression.
A hollow or an indentation on a surface.
pitted   With small pits on the surface, as the nut of an almond (*Prunus dulcis*).

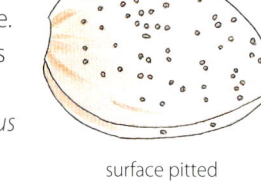

pit

surface pitted

pit   Of palms (Arecaceae), a cavity in the receptacle that bears flowers.

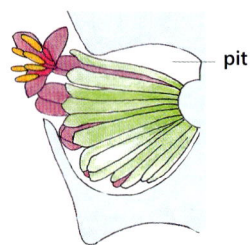

pit

pit   A cavity in a cell wall with channels (plasmodesmata) that connect adjacent cells.

Pit

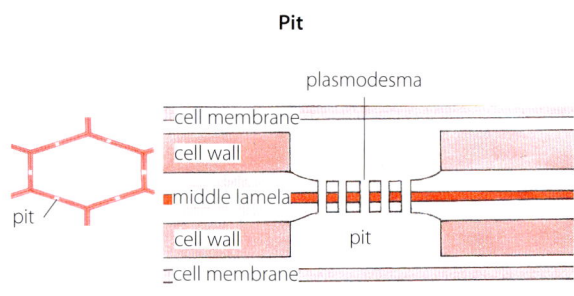

plasmodesma
cell membrane
cell wall
middle lamela
pit
cell wall
pit
cell membrane

pith   A core of spongy ground tissue in the centre of the stems of most flowering plants, gymnosperms and ferns. It functions primarily as storage.
Also the tissue inside the rind of some fruits, as oranges and lemons.
= **medulla**

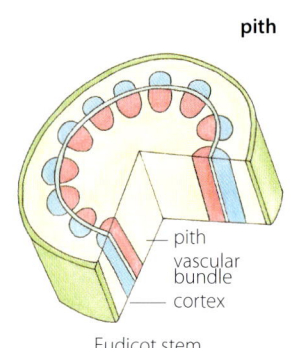

pith
pith
vascular bundle
cortex
Eudicot stem

placenta, placentae
Mostly undifferentiated tissue, in the ovary of a flower, on which the ovules are borne.
A vascular bundle that provides nutrition extends through the placenta and connects to the ovules.
The ovules may be sessile or attached to the placenta by a stalk-like funicle.

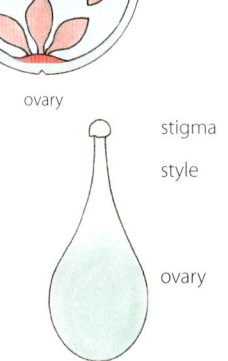

placenta
ovary
stigma
style
ovary

placentation   The arrangement of the ovules on the placenta in the ovary of a flower.
There are several types of placentation, including axile, apical, basal, free central, lamellar, marginal and superficial.

Placentation

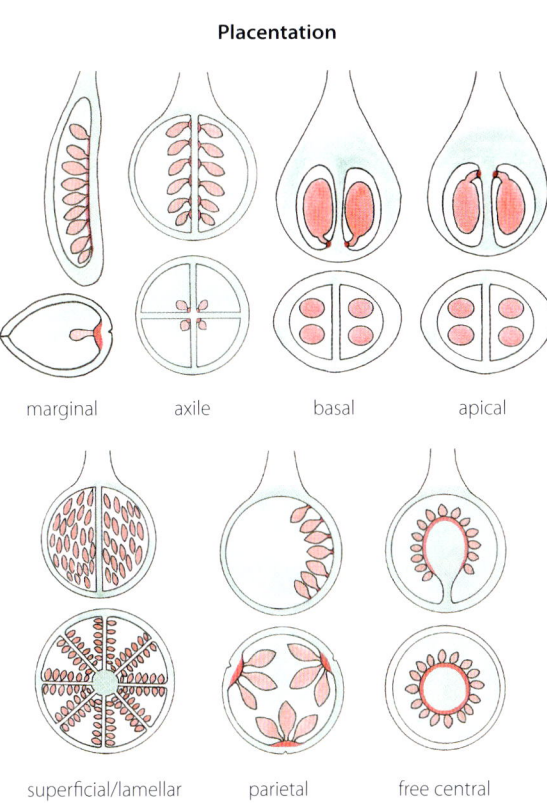

marginal   axile   basal   apical

superficial/lamellar   parietal   free central

plagiogeotropism   The tendency of a plant part to grow at an oblique angle to the direction of the orienting growth stimulus, as lateral branches that grow at an oblique angle to the stimulus of gravity.
see **diatropism, plagiotropism, tropism**
plagiogeotropic   Of or relating to plagiogeotropism.

plagiotropism   The tendency of a plant part to grow at an oblique angle or perpendicular to the direction of the orienting growth stimulus, as lateral branches and roots that grow at an oblique angle or perpendicular to the stimulus of gravity.
see **diatropism, plagiogeotropism, tropism**
plagiotropic   Of or relating to plagiotropism.

plane   A real or imaginary flat surface.
see also **equatorial plane**
cf. **axis**

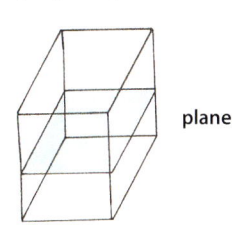

plane

plano-convex   Flat on one side and convex on the other.

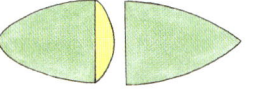

plano-convex

plantlet   A small plant, as that growing on a leaf of the piggyback plant (*Tolmiea menziesii*) and on the fronds of some ferns.

plantlet

plasma membrane   A thin semipermeable layer of tissue enclosing the cytoplasm of a cell and, in plants, surrounded by the cell wall. It allows movement of some substances into and out of the cytoplasm.
= **cell membrane, cytoplasmic membrane**

Plasma membrane

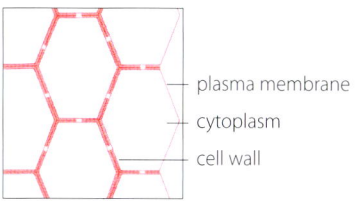

plasma membrane
cytoplasm
cell wall

plasmodesma, *pl.* **plasmodesmata**   A minute connecting channel in the pits of adjacent cell walls. exchange of molecules between cells takes place here.

Plasmodesma

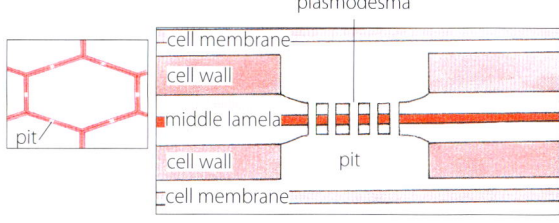

plasmodesma
cell membrane
cell wall
middle lamella
pit
cell wall
pit
cell membrane

plasmolysis   Shrinkage of the cytoplasm away from the cell wall due to loss of water through osmosis.
cf. **flaccidity, turgor**

Plasmolysis

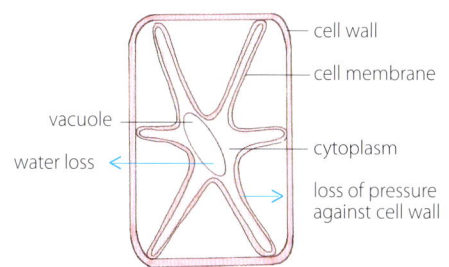

cell wall
cell membrane
vacuole
water loss
cytoplasm
loss of pressure against cell wall

plastid　An organelle found in the cytoplasm of a plant cell.

The different types of plastid have specific functions in the cell.

Plastid

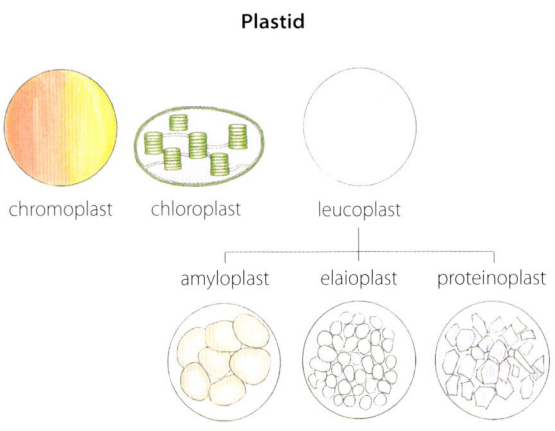

chromoplast　chloroplast　leucoplast

amyloplast　elaioplast　proteinoplast

pleated　Folded back and forth longitudinally, as some leaves.

= **plicate**

pleated

pleiochasium

A compound cymose inflorescence, in which the main axis has more than two lateral branches.

= **polychasium**
pleiochasial　Relating to a pleiochasium.

see also **multiparous cyme**

cf. **monochasium, dichasium**

pleiochasium

pleiomery　A whorl of a flower that has a greater number of members than the other whorls, as the whorl of many stamens in guinea flower (*Hibbertia*) that has five petals, five sepals and two or three carpels.

cf. **oligomery**
pleiomerous　Of or relating to pleiomery.

petals and sepals

pleiomery

stamens andcarpels

pleiomorphic, pleomorphic　Able to change shape or form.

cf. **amorphic**

pleiotropy　The phenomenon of a single gene affecting multiple traits.

see also **genotype, phenotype**

pleonanthic　Of a plant or stem that flowers and bears fruit more than once in its lifetime.

Most commonly used to describe palms, as bamboos.

Of palms that have unlimited vegetative growth with flowering shoots produced on axillary branches year after year.

= **iteroparous, polycarpic**
cf. **hapaxanthic, monocarpic, semelparous**

plesiomorph, plesiomorphy　In cladistics, an ancestral or primitive character.

cf. **apomorph, apomorphy**
plesiomorphic　In cladistics, ancestral or primitive.

pleurotropous　Of ovule orientation, with the micropyle lateral, as hemianatropous.

cf. **epitropous, hypotropous**

plicate

Folded back and forth longitudinally, as some leaves.

= **pleated**
Of carpels, folded with edges sealed together, as peas (*Pisum*).

plicate ptyxis
Of a single leaf in bud, folded like a fan.

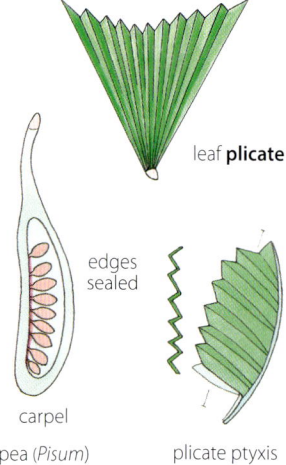

leaf **plicate**

edges sealed

carpel

pea (*Pisum*)　　plicate ptyxis

plietesial　Of plants within a species that flower at the same time, often after an interval of years and over a wide area, then set seed and die.

see **semelparous**
see also **masting**
cf. **gregarious flowering**

ploidy　The number of sets of chromosomes in a cell. Each set is designated by n.

The chromosomes in a set may be haploid (n) (occurring singly) as in gametes, or in somatic cells diploid (2n) (occurring in pairs), triploid (3n) (occurring in threes), tetraploid (4n) (occurring in fours) and so on.
see **chromosome set**

plumose　Like a feather, with fine hairs branching from a central axis, as the stigmas of many grasses (Poaceae).

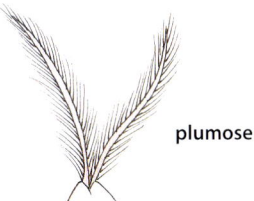

plumose

plumular hook

The hook-like curve of the epicotyl found in seedlings that germinate buried in the soil.

It protects the shoot from damage as the seedling is pushing upwards towards light.
= **apical hook**
see **hypogeal germination**

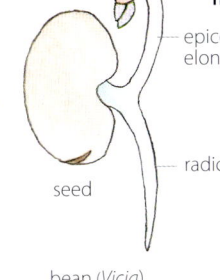

plumular hook

epicotyl elongates

radicle

seed

bean (*Vicia*)

plumular leaves　The first seedling leaves, as cataphylls and eophylls.

They are transitional leaves produced in succession to the cotyledons and are usually simpler in shape and smaller in size than the true adult leaves.

Plumular leaves

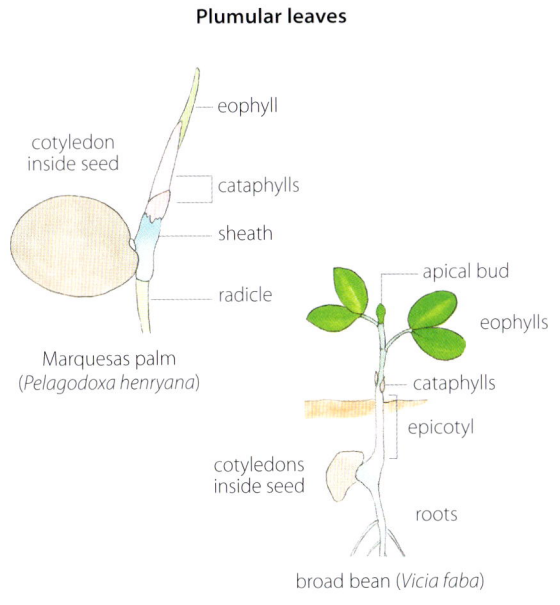

eophyll

cotyledon inside seed

cataphylls

sheath

radicle

Marquesas palm (*Pelagodoxa henryana*)

apical bud

eophylls

cataphylls

epicotyl

cotyledons inside seed

roots

broad bean (*Vicia faba*)

plumule　Of a seed embryo, the primordial shoot system initiated by the shoot apial meristem at the tip of the epicotyl.

Usually a small conical structure but sometimes leafy, as broad bean (*Vicia faba*).
cf. **radicle**
see **embryo axis**
see also **hypogeal germination**
plumular　Relating to a plumule.

Plumule

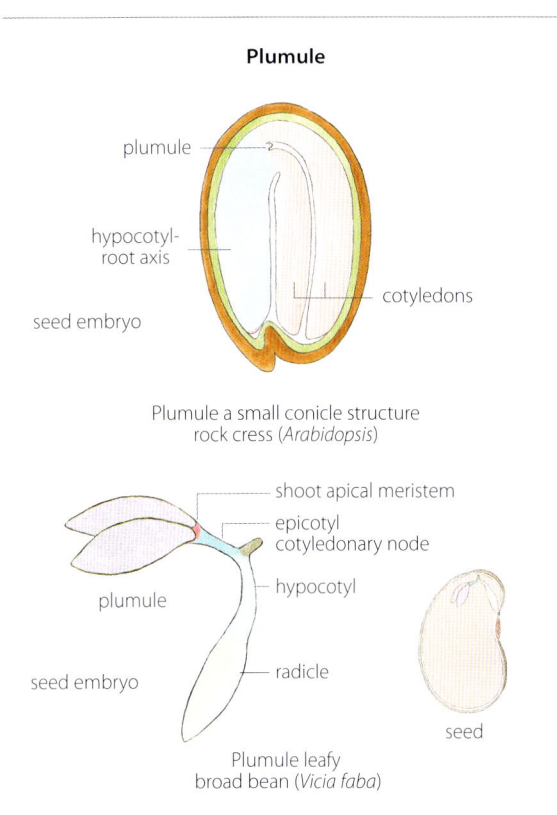

plumule

hypocotyl-root axis

cotyledons

seed embryo

Plumule a small conicle structure rock cress (*Arabidopsis*)

shoot apical meristem

epicotyl cotyledonary node

hypocotyl

plumule

radicle

seed embryo

seed

Plumule leafy broad bean (*Vicia faba*)

pluri-　A prefix meaning several.

plurilocular

Of an ovary, anther or fruit, having several or many locules or cavities for ovules, pollen or seeds.

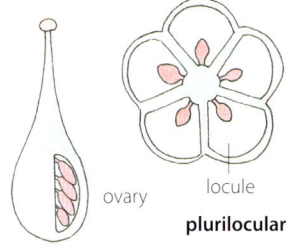

ovary

locule

plurilocular

pneumatophore　A usually vertical aerial root, specialised for gas exchange with the atmosphere, found in plants that grow in oxygen-poor soil in waterlogged conditions.

Examples of pneumatophores in mangroves are knee roots, cone roots and peg roots.
pneumatophore page 230 (cont.)
see **lenticel**

Pneumatophore

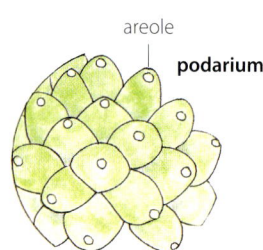

knee roots   cone roots   peg roots

Polar nucleus

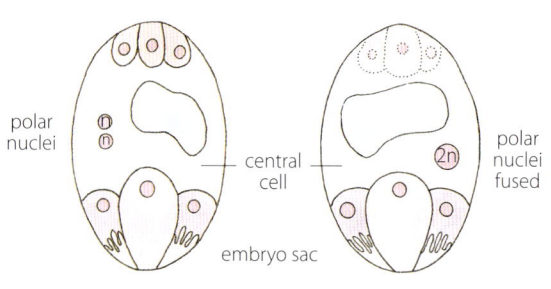

polar nuclei — central cell — polar nuclei fused

embryo sac

pod   A general term for a legume or dry legume-like fruit, as that of the vanilla orchid.
It has one or more seeds and dehisces when mature.
Derived from a one-carpelled ovary.

podarium   Of some succulents, the modified photosynthetic leaf base that is a nipple-like projection with an areole at the tip, as the cactus genus *Mammillaria*.

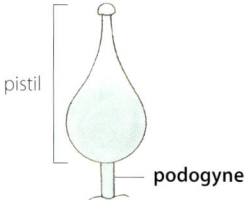

areole

podarium

podogyne   A small stalk (stipe) that is a prolongatiion of the base of the pistil, as the aquatic genus *Ruppia*.
cf. **gynophore**

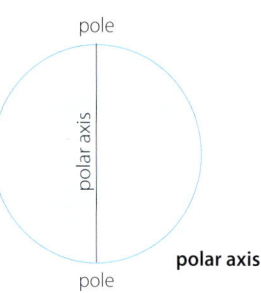

pistil

podogyne

polar   A biome, usually covered by ice most of the year, that includes the Arctic and Antarctic regions around the North and South poles.
see **biome**

polar axis
An imaginary line connecting two poles, as the line that connects the proximal and distal poles of a pollen grain.
cf. **equatorial plane**

pole

polar axis

pole

polar axis

polar nucleus   Either of two female haploid nuclei in the central cell of the embryo sac.
These fuse before fertilisation, then, at fertilisation, combine with one of the two male sperm cell.
The central cell then has a triploid nucleus and will divide to form the endosperm of the seed.
see **double fertilisation**

polar view   An object as it appears when the pole is in the line of sight.
cf. **equatorial view**

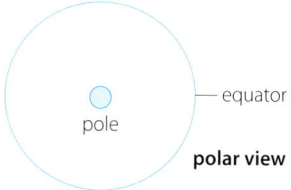

equator

pole

polar view

polarity   Having two distinct poles.
The orientation of a pollen grain in a tetrad.
see also **distal pole, polar axis, proximal pole**
cf. **apolar, heteropolar, isopolar**

Polarity

distal pole

polar axis

distal pole

distal pole

proximal poles

Pollen tetrad

pole   Either extremity of an axis through a sphere.
cf. **equator**
polar   Relating to a pole.
Of pollen grains with a distinctly recognisable polarity.
cf. **apolar**

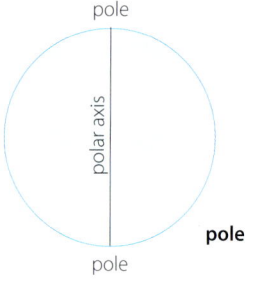

pole

polar axis

pole

pole

pollen   Of seed plants, minute grains, commonly yellow in colour, produced in a pollen sac and shed from the anther of a flower or from the exposed pollen sacs of gymnosperms.
Pollen is the result of two stages: microsporogenesis followed by microgametogenesis.
see **pollen dispersal, pollen grain**

pollen chamber

Of gymnosperms, a small space in the ovule, near the micropyle, for collecting pollen. Pollen grains germinate here.

see also **pollen tube, pollination drop, siphonogamy**

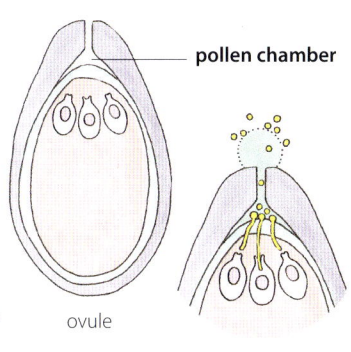

pollen chamber

ovule

germinating
pollen grains

pollen dispersal

In gymnosperms pollen is dispersed by wind (anemophily) and in angiosperms pollen it is commonly dispersed by wind (anemophily), insects (entomophily), water (hydrophily) or animals (zoophily).

Dispersal units can be solitary pollen grains (monads), grains aggregated into diads, tetrads or polyads, or a massula, pollinium or pollinarium.

see **multiplanar, uniplanar**

Pollen dispersal units

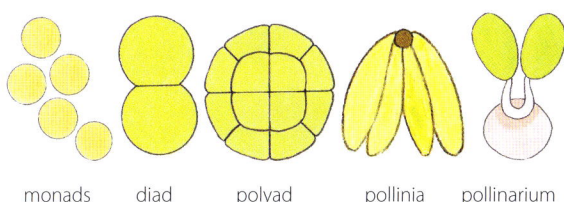

monads   diad   polyad   pollinia   pollinarium

tetrads

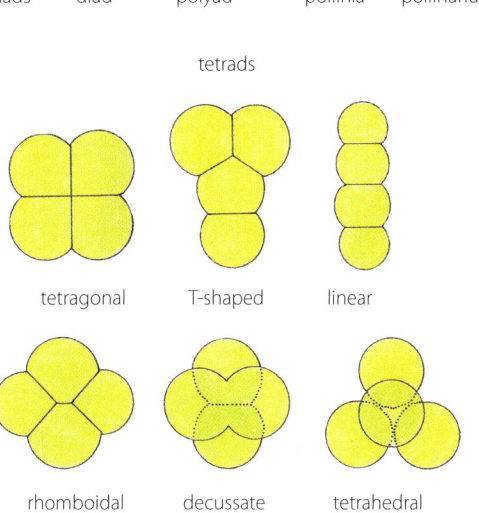

tetragonal   T-shaped   linear

rhomboidal   decussate   tetrahedral

pollen grain

In seed plants, a male gametophyte, the male gamete-bearing entity produced from a microspore.

In angiosperms pollen grains develop in the pollen sacs in the anthers. Each grain has two cells, a reproductive generative cell from which the two male gametes (sperm cells) develop and a vegetative cell, the tube cell.

In gymnosperms, the pollen grains develop in the pollen sacs, typically exposed on the lower surface of scales on the male cone. The generative cell produces two gametes (sperm cells) but only one survives.

Grains may be solitary or cohere in units of two (diads), four (tetrads), eight or sixteen and so on, for dispersal.

= **microgametophyte**

see also **massula, ornamentation, pollen dispersal, pollen wall, pollinium, polyad**

cf. **megagametophyte**

Pollen grain

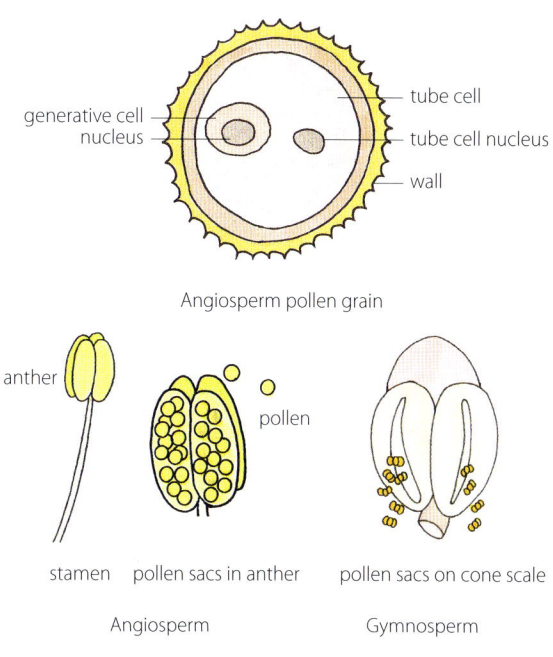

generative cell
nucleus

tube cell

tube cell nucleus

wall

Angiosperm pollen grain

anther

pollen

stamen   pollen sacs in anther   pollen sacs on cone scale

Angiosperm   Gymnosperm

pollen kit, pollenkitt

A sticky material commonly found coating a pollen grain.

see **tryphine**

pollen presenter

A floral structure other than an anther that presents pollen for cross-fertilisation. It is a strategy to prevent self-pollination.

Anthers release their pollen before the stigma is receptive. The style emerges and pollen is presented on hairs on the style and stylar arms in Asteraceae. Pollen is presented on the expanded area around the small stigma in Proteaceae.

In flowering plants, an indusium is a pollen presenter.

pollen presenter page 232 (cont.)

Pollen presenter

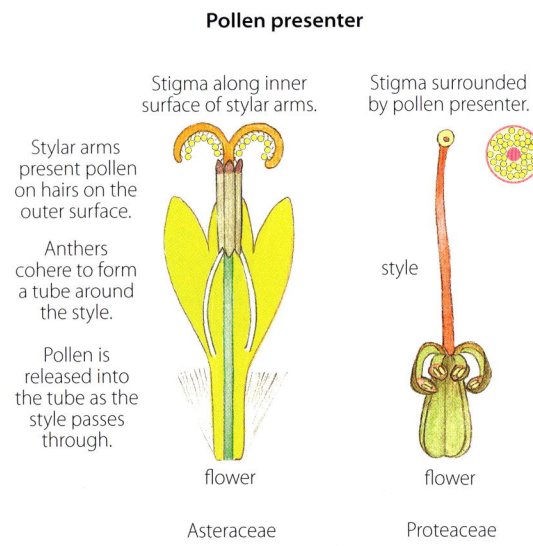

Stylar arms present pollen on hairs on the outer surface.

Anthers cohere to form a tube around the style.

Pollen is released into the tube as the style passes through.

Stigma along inner surface of stylar arms.

Stigma surrounded by pollen presenter.

style

flower — Asteraceae

flower — Proteaceae

pollen sac   The chamber (locule) in an anther of a flowering plant (angiosperm) in which pollen is produced. In gymnosperms, the structure on the lower surface of a scale on the male cone in which pollen is produced.
= **microsporangium**
see **anther sac**

Pollen sac

Angiosperm

anther

stamen

pollen sacs

Gymnosperm (*Pinus*)

cone scale

male cone

pollen sac

pollen tetrad   Four cohering pollen grains.
see **pollen dispersal**

Pollen tetrad

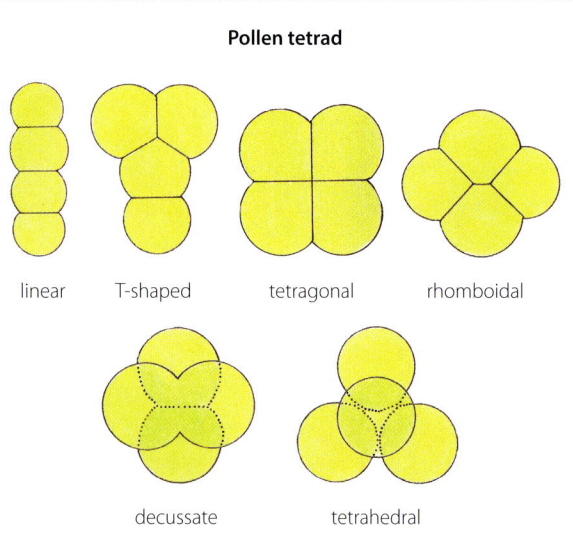

linear   T-shaped   tetragonal   rhomboidal

decussate   tetrahedral

pollen tube   In the course of germination of a pollen grain, the structure that will transport the non-motile male gametes through the style to the the egg cell in the embryo sac of flowering plants. It develops from the wall of the pollen grain.
see also **double fertilisation, siphonogamy**

Pollen tube

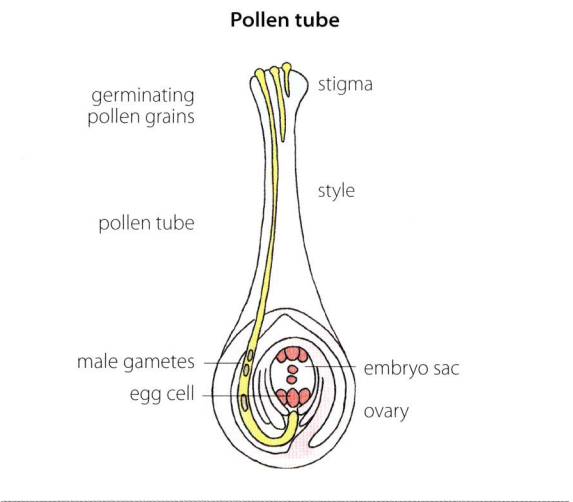

germinating pollen grains

stigma

style

pollen tube

male gametes

egg cell

embryo sac

ovary

pollen wall   The layers enclosing the cytoplasm of a pollen grain.
It is almost always composed of two major layers: the outer exine and the inner intine.
see also **sporoderm**

Pollen wall

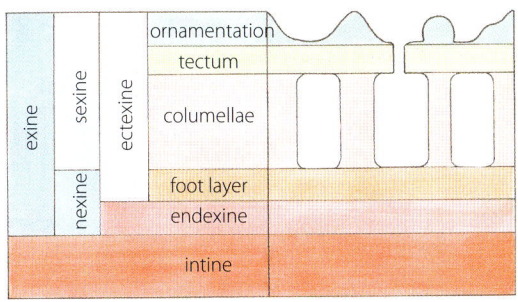

pollen grain — exine, intine, cytoplasm

ornamentation, exine, tectum, columellae, foot layer, endexine, intine

exine, sexine, ectexine, nexine, ornamentation, tectum, columellae, foot layer, endexine, intine

polleniser   A plant that provides fertile pollen.
cf. **pollinator**
pollenise   To provide pollen for pollination.
cf. **pollinate**

pollinarium,
pl. **pollinaria**
The pollen dispersal unit of orchids (Orchidaceae) and milkweeds (Asclepiadaceae) that includes the pollinia, and its accessories like the stipe, caudicle and viscidium.
cf. **pollinium**

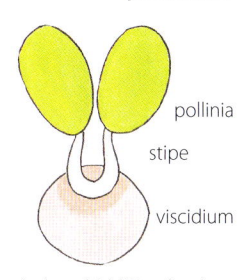

pollinarium

pollinia, stipe, viscidium

hyacinth orchid (*Dipodium*)

pollination   Of angiosperms, the transfer of pollen from the anther of a flower to the stigma of the same flower, or to the stigma of another flower, usually of the same species, commonly by animals, insects and wind.
see **abiotic, biotic, pollinator**

Of gymnosperms, the transfer of pollen from a male cone to the ovules on a female cone belonging to the same species, commonly by wind. Pollen grains are often winged to assist wind dispersal.
see **pollination drop, sporophyll**
cf. **fertilisation**
pollinate   To bring about pollination.
cf. **pollenise**

pollination drop   Of many gymnosperms, a drop of sugary fluid, exuded at the tip of each ovule, that traps pollen then retracts, transporting the pollen to the pollen chamber near the egg.
see **siphonogamy**

Pollination drop

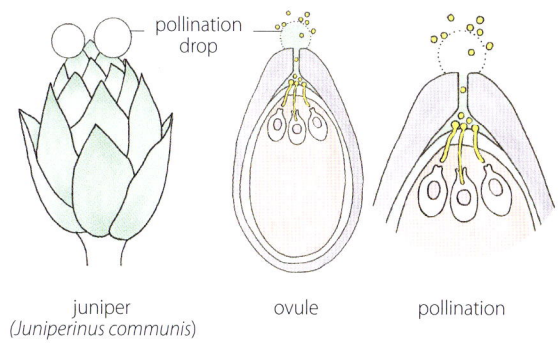

pollination drop

juniper
(*Juniperinus communis*)
ovule
pollination

pollination syndrome   A system of pollen transfer.
It may be abiotic (as wind and water) or biotic (as birds, mammals and insects).
cf. **pollinator syndrome**

pollinator   The agent that moves pollen from the anther of a flower to the stigma of the same flower, or to the stigma of another flower, or from the male cone to the ovules on a female cone, so that fertilisation and seed production can occur.
Bees, wasps, wind and water are examples of pollinators.
cf. **polleniser**

pollinator syndrome   Flower characteristics related to a certain type of pollinator, as form, colour, odour and nectar attract pollinators like birds, bees and flies.
Plants that use wind or water as pollinators produce large amounts of pollen.
cf. **pollination syndrome**

polliniferous Producing or bearing pollen.

pollinium, *pl.* **pollinia** Of plants with one anther, as orchids (Orchidaceae) and milkweeds (Asclepiadaceae), a cohering mass of pollen grains produced in the anther and transferred as a unit during pollination.

The pollen mass may be soft, hard or mealy.

see also **massula, sectile**

cf. **pollinarium**

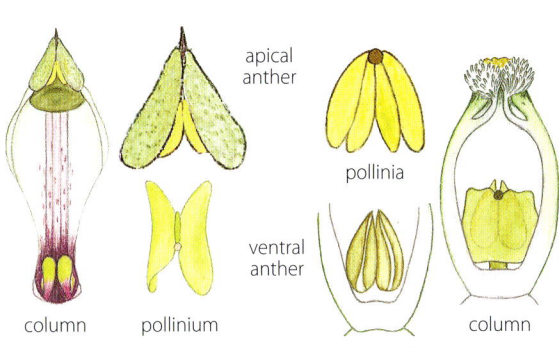

Pollinium

apical anther

pollinia

ventral anther

column   pollinium   column

spider orchid (*Caladenia*)   sun orchid (*Thelymitra*)

poly- A prefix meaning many.

polyad A group consisting of an undefined number. A unit of many cohering pollen grains.

cf. **monad, diad, tetrad, triad**

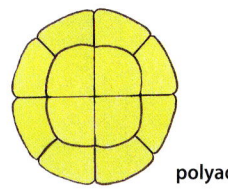

polyad

polyadelphous

Of stamens united by their filaments into several bundles, as paperbark (*Melaleuca*).

see **adelphous**

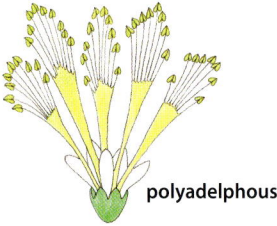

polyadelphous

polyandrous Having many stamens, as the flowers of buttercups (*Ranunculus*).

= **polystemonous**

cf. **diandrous, monandrous, pentandrous, tetrandrous, triandrous**

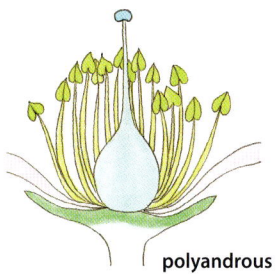

polyandrous

polycarp A plant that flowers and fruits an indefinite number of times. A perennial.

cf. **monocarp, monocarpy**

polycarpic Of or relating to a polycarp.

= **iteroparous, pleonanthic**

cf. **hapaxanthic, monocarpic, semelparous**

polycarpellary, polycarpellate, polycarpous

Of a flower having a gynoecium with more than one carpel, the carpels being either free or variously fused.

= **multicarpellate, polygynous**

see also **apocarpous, compound pistil, syncarpous**

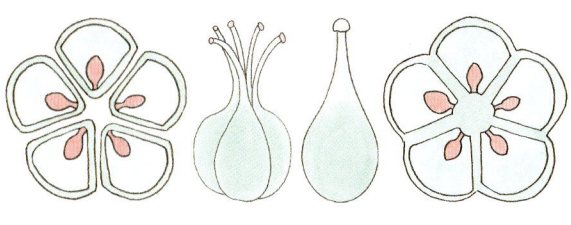

Polycarpellate

gynoecium with free carpels   gynoecium with fused carpels

polycarpic Of a plant that reproduces more than once in its lifetime, as perennial plants.

= **iteroparous**

see **pleonanthic**

cf. **hapaxanthic, monocarpic, semelparous**

polychasium

A compound cymose inflorescence. The main axis has more than two lateral branches.

= **pleiochasium**

polychasial Relating to a polychasium.

see also **multiparous**

cf. **dichasium, monochasium**

polychasium

polycotyledonous
Of a plant embryo, having more than two cotyledons, as many conifers that belong to the non-flowering seed plants (Gymnosperms).
Of a plant producing such embryos.
cf. **dicotyledonous, monocotyledonous**

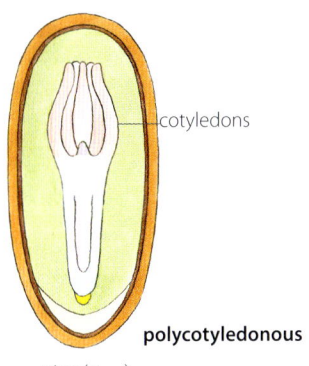

cotyledons

polycotyledonous
pine (*Pinus*)

polyembryony The formation of two or more embryos from a single fertilised ovum.
The presence of two or more embryos in a seed resulting in more than one seedling emerging from a seed.
polyembryonic Having more than one embryo.

polygamodioecious Having bisexual and male flowers on some plants and bisexual and female flowers on others.
see **polygamomonoecious, polygamous**

polygamomonoecious Of a plant with male flowers, female flowers and bisexual flowers.
see **polygamodioecious, polygamous**

polygamous Of a species with bisexual and unisexual flowers on the same plant.
see **polygamodioecious, polygamomonoecious**

polygynous Of a flower having a gynoecium with more than one carpel, the carpels being either free or variously fused.
= **multicarpellate, polycarpellate**
see also **apocarpous, syncarpous**

Polygynous

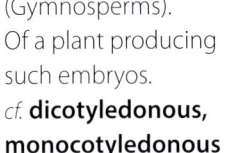

gynoecium with free carpels   gynoecium with fused carpels

polymerous Having petals, sepals and stamens, in whorls, each with many parts.
see **-merous**

polymorphic Having two or more distinct forms as the styles of purple loosestrife (*Lythrum salicaria*).
cf. **dimorphic, monomorphic, trimorphic**

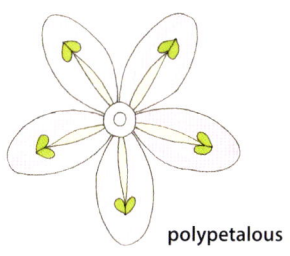

polymorphic

style

polymorphism Multiple forms (alleles) of a single gene existing in an individual or population that give rise to different traits, as species that have genes for distinctly different flower colour or style lengths (heterostyly).
Many polymorphisms are not visible and require genetic techinques to identify them.
polymorphic Of or relating to polymorphism.

polypetalous
With petals free from each other.
= **apopetalous, choripetalous, dialypetalous**
cf. **gamopetalous, sympetalous**

polypetalous

polyphyletic
Of a group of organisms that does not include a single common ancestor.
cf. **monophyletic, paraphyletic**

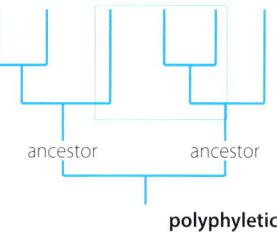

ancestor   ancestor

polyphyletic

polyploid, polyploidy Having three or more complete sets of chromosomes in each somatic cell.
The two types of polyploidy are autopolyploidy and allopolyploidy.
see **ploidy**

polysepalous
With sepals free from each other.
= **aposepalous, chorisepalous, dialysepalous**
cf. **gamosepalous, synsepalous**

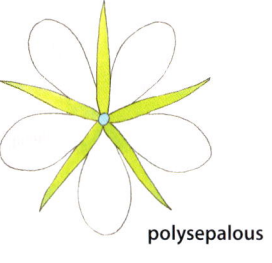

polysepalous

polystemonous
Having many stamens,
as the flowers of
buttercups
(*Ranunculus*).
= **polyandrous**

polystemonous

**polysymmetric,
polysymmetrical**
Divisible into like
halves on more
than one plane.
cf. **radial symmetry**

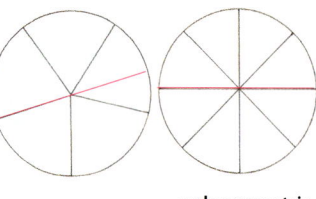

polysymmetric

polytelic   Of an inflorescence axis that grows
indefinitely, the oldest flower being at the base, as
an indeterminate or indefinite inflorescence.
see **racemose inflorescence**
cf. **monotelic**

polytepalous   With tepals free from each other.
Of flowers having more petals and/or sepals than
usual, as some daylilies with more than six tepals.
see **apotepalous, choritepalous**
cf. **gamotepalous**

pome   The fleshy
false fruit in which the
receptacle enlarges,
encloses and adheres
to the true fruit.
Characteristic of the
apple genus (*Malus*) and
the pear genus (*Pyrus*).
see **accessory fruit**

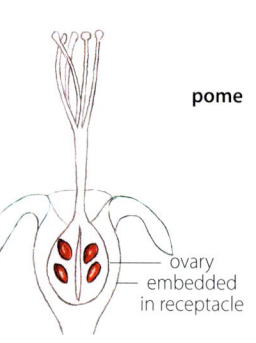

pome

ovary
embedded
in receptacle

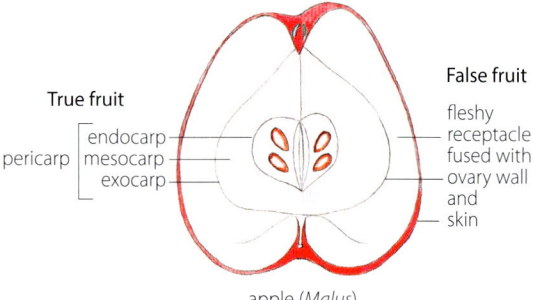

True fruit

endocarp
pericarp   mesocarp
exocarp

False fruit

fleshy
receptacle
fused with
ovary wall
and
skin

apple (*Malus*)

population   A group of organisms of a particular
species living in a given area.
cf. **biome, community, ecosystem, habitat**

pore, porus *pl.* **pori**
A small, usually rounded
opening.
A small, more or less
circular aperture on a
pollen grain located at
the equator or evenly
distributed over its
surface.
see **pantoaperturate**
cf. **colporus, colpus,
sulcus**
porate, porose   Having
a pore or pores.

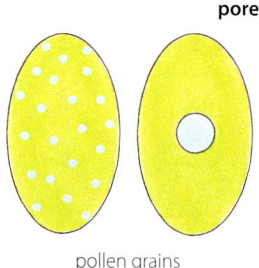

pore

pollen grains

poricidal   Opening by pores, as some capsules.
see **poricidal capsule**

poricidal capsule
A capsule that dehisces
through pores, as the
those on the sides of the
capsule of Venus' looking
glass (*Triodanis perfoliata*).
cf. **operculate-poricidal
capsule**

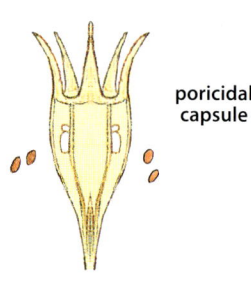

poricidal
capsule

poricidal dehiscence
Of anthers, opening by
pores to release pollen,
as the potato and tomato
genus (*Solanum*).
see also **anther
dehiscence**

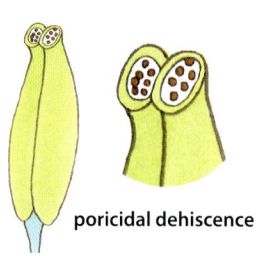

poricidal dehiscence

porogamy
Entrance of the pollen
tube through the
micropyle of the ovule.
cf. **chalazogamy,
mesogamy**

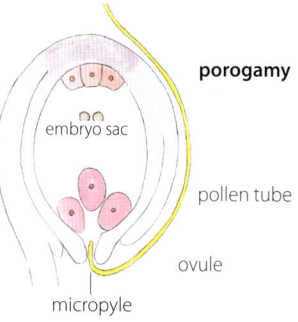

porogamy

embryo sac

pollen tube

ovule

micropyle

porrect   Extending
forward. Stretched out as
the column arms of some
sun orchids (*Thelymitra*).

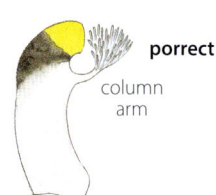

porrect

column
arm

post-   A prefix meaning after.

posterior   Nearest the
axis. The side of a flower
nearest the stem.
cf. **anterior**

posterior

anterior

posticous   Facing away from the axis.
= **extrorse**
cf. **introrse, latrorse**

posticous dehiscence
Of anthers, facing
outwards and opening
longitudinally to release
pollen away from the
centre of the flower, as
the dayflower family
(Commelinaceae).
= **extrorse dehiscence**
see also **anther
dehiscence**

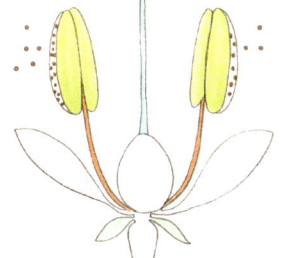

posticous dehiscence

pouch   A little sac, a bulge.
pouched   With small sac-like structures, as on the
fronds of pouched coral fern (*Gleichenia dicarpa*).
= **saccate**

Pouyannian mimicry   Evolution of plants to
appear like other organisms in order to attract a
pollinator, as the labellum of the elbow orchid
(*Thynninorchis huntianus*) that mimics the female of
the wasp of its pollinator *Arthrothynnus huntianus*.
see **pseudocopulation**
cf. **Bakerian mimicry, Dodsonian mimicry,
Vavilovian mimicry**

prae-, pre-   A prefix meaning before.

praemorse, premorse
Appearing as if bitten off
at the apex. Ruminate.
The same as truncate but
with the apex ragged
and irregular.

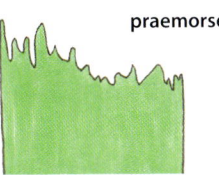

praemorse

prairie   Vast stretches of flat temperate grasslands
and herbfields in the middle of North America.
Equivalent to the steppe of Europe and Siberia.
see **biome**

precocious   Appearing or developing early.
Of flowers, appearing before leaves, as on the
deciduous peach tree (*Prunus persica*).
cf. **coetaneous, serotinous**

**prefoliation,
praefoliation**
The arrangement of
young leaves in the
unopened leaf bud.
= **vernation**

prefoliation

prickle
A sharp outgrowth on a
plant derived that is from
the epidermis, as on a
rose (*Rosa*).
cf. **spine, thorn**

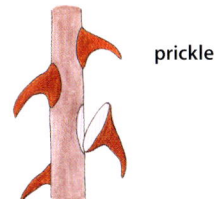

prickle

primary cell wall   The first layer of wall laid down
next to the cell membrane.
Mainly composed of cellulose.
see **middle lamella, secondary cell wall**

Primary cell wall

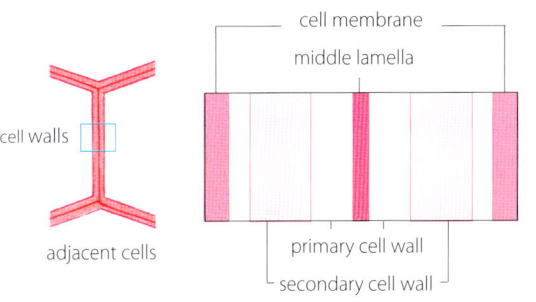

cell membrane

middle lamella

cell walls

adjacent cells

primary cell wall

secondary cell wall

primary growth   Occurs in the apical meristems
located at the tips of roots and above-ground
shoots and in buds on stems.
It increases the length of the plant.
Plants with only primary growth are herbaceous.
cf. **secondary growth**

primary meristem   Any one of three apical
meristem tissues: protoderm gives rise to
the epidermis; ground meristem gives rise to
parenchyma, collenchyma and sclerenchyma; and
procambium gives rise to primary xylem and
primary phloem and produces two secondary
meristems, the cork cambium and the vascular
cambium.
primary meristem page 238 (cont.)
see **primary tissue**
cf. **secondary meristem**

Primary meristem

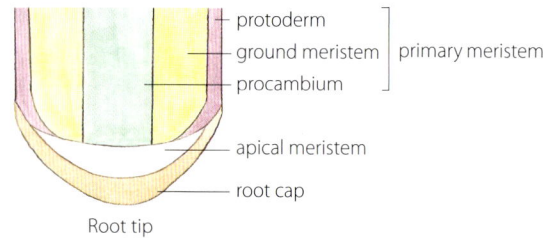

- protoderm ⎤
- ground meristem ⎬ primary meristem
- procambium ⎦
- apical meristem
- root cap

Root tip

primary phloem   Phloem tissue that differentiates from procambium in the apical meristem of root and shoot tips during primary growth of a vascular plant.

There are two types of primary phloem: protophloem and the later forming metaphloem. Together with primary xylem, it is a major component of vascular bundles that run the length of the plant.

see **secondary phloem**

Primary phloem

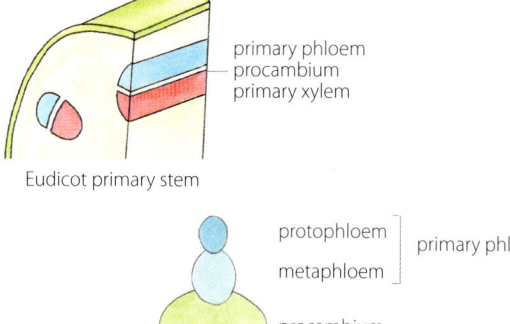

- primary phloem
- procambium
- primary xylem

Eudicot primary stem

- protophloem ⎤
- metaphloem ⎬ primary phloem
- procambium
- metaxylem ⎤
- protoxylem ⎬ primary xylem

primary tissue   Tissues derived from one of three primary meristems (protoderm, ground meristem or procambium) that generate the three primary plant tissues. Protoderm generates the epidermis. Ground meristem generates the ground tissue. Procambium generates the two primary vascular tissues, phloem and xylem.

see **apical meristem**
cf. **secondary tissue**

Primary root tissue

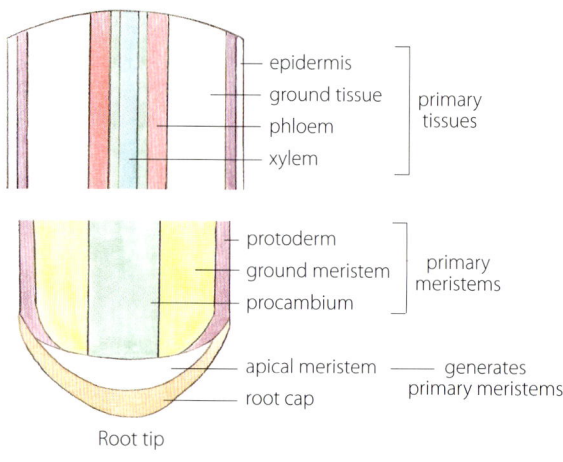

- epidermis ⎤
- ground tissue ⎬ primary
- phloem ⎬ tissues
- xylem ⎦
- protoderm ⎤
- ground meristem ⎬ primary
- procambium ⎦ meristems
- apical meristem — generates primary meristems
- root cap

Root tip

primary vein

The main vein of a leaf or leaflet, usually running up the centre as a continuation of the petiole or petiolule.

= **midrib, midvein**

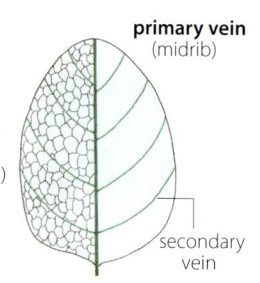

primary vein (midrib)

tertiary veins (veinlets)

secondary vein

primary xylem   Xylem tissue that differentiates from procambium in the apical meristem of root and shoot tips during primary growth of a vascular plant.

Together with primary phloem, it is a major component of vascular bundles that run the length of the plant.

There are two types of primary xylem: protoxylem and the later forming metaxylem.

see **secondary xylem**

Primary xylem

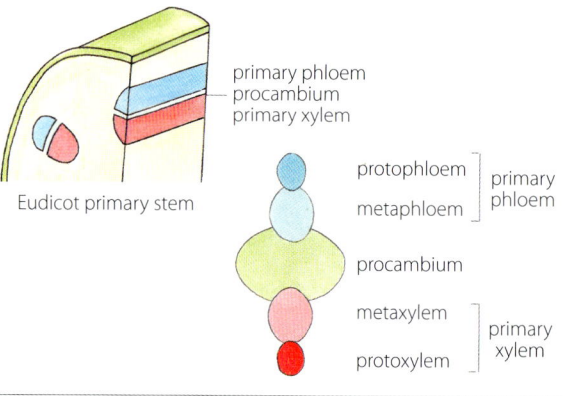

- primary phloem
- procambium
- primary xylem

Eudicot primary stem

- protophloem ⎤
- metaphloem ⎬ primary phloem
- procambium
- metaxylem ⎤
- protoxylem ⎬ primary xylem

primitive character   In phylogenetics, of a character belonging to or inherited from an ancestor.

Ancestral or plesiomorphic are preferred terms as primitive may falsely infer inferiority.
cf. **derived character**

primocane   A first year cane of a bramble that will flower and fruit the next year.
cf. **floricane**

primocane

primordium, *pl.* **primordia**   Plant tissue, in its earliest stage of development, at the very beginning of differentiation into a particular cell type.

priority   In taxonomy, when there is more than one name available for a taxon, the principle of the right of the first name validly published to take precedence, with other names becoming synonyms.

p.p., pro parte   In nomenclature, denotes that a taxon includes more than one crurrently recognised entity, and that only one of those entities is being considered.

pro parte, *abbr.* **p.p.**   In nomenclature, denotes that a taxon includes more than one crurrently recognised entity, and that only one of those entities is being considered.

procambium   One of three regions of primary meristematic tissue that develops behind the apical meristem of stems and roots.
It differentiates into the first vascular tissue (primary xylem and primary phloem).
see **primary meristem**
see also **ground meristem, protoderm**

procumbent   Trailing loosely along the ground without taking root, as the stems of some goodenias (*Goodenia*).
cf. **decumbent, prostrate**

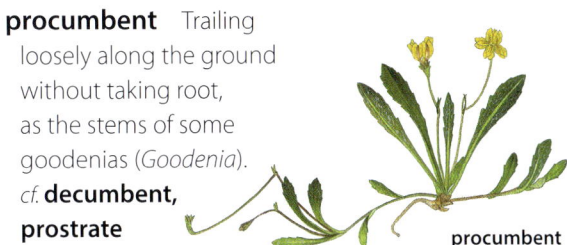
procumbent

proembryo   In eudicots, the earliest multicellular stage of the embryo before cells specialise into the tissues of the embryo.
see **embryogenesis**

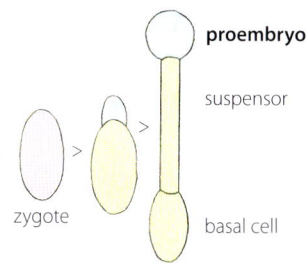
proembryo
suspensor
zygote
basal cell

progenitor   A direct ancestor of a plant.

prokaryote   A unicellular organism that does not have DNA in a well-defined nucleus that is surrounded by a membrane.
It lacks organelles due to the absence of internal membranes, as bacteria.
cf. **eukaryote**
prokaryotic   Of a cell that does not have a clearly defined nucleus and lacks organelles.

prolate   Of an object with the polar axis longer than the equatorial diameter.
cf. **oblate**

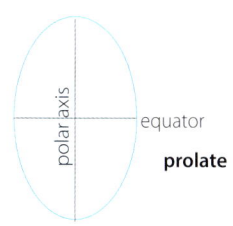
polar axis
equator
prolate

proliferation   Rapid growth or reproduction of new parts, as roots.
Producing buds or offshoots, especially from unusual organs, as bulblets from leaves.
cf. **prolification**
proliferate   To increase by proliferation.
proliferous   Of plants or parts that increase by proliferation.

prolification   The production of offspring.
A mutation or disorder of flowers having one or more buds form in an already open bloom, as can occur in roses (*Rosa*).
cf. **proliferation**

prop roots   Adventitious roots that grow out from the lower trunk of a tree and into the soil to provide support. Found in some mangroves (*Rhizophora*) and the banyan (*Ficus benghalensis*).
= **stilt roots**
cf. **buttress root**

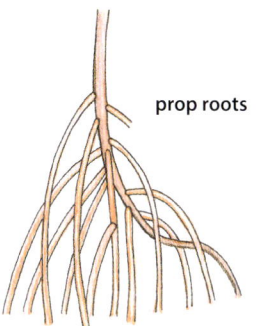
prop roots

propagation   The reproduction of plants by any number of natural means, as seeds, offsets and bulbs, or by artificial means, as cuttings, layering and grafting.
see also **micropropagation**

propagule   A structure capable of producing a new plant.
Seeds, bulbs, suckers and runners are propagules. In non-flowering plants like fungi, ferns and bryophytes, a spore is a propagule.
see also **micropropagation**

prophyll, prophyllum   A modified leaf-like structure.
The modified first leaf (monocotyledons) or two leaves (eudicots), as a bracteole, produced at the first node of a lateral shoot.
Of palms, the usually two-keeled bract that encloses the inflorescence.
Of grasses, a two-keeled modified leaf, without a blade, attached to the node within the leaf sheath on the adaxial side of a branch.

Prophyll

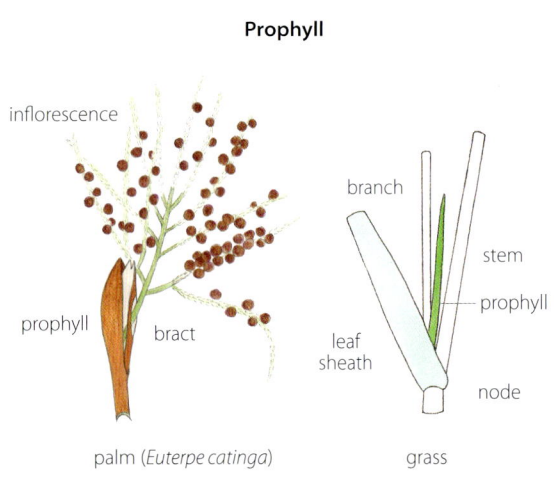

inflorescence

prophyll bract

branch

stem

prophyll

leaf sheath

node

palm (*Euterpe catinga*)   grass

proplastid   An undifferentiated plastid that can differentiate into a specific kind of plastid with a particular function in the cell, as chloroplasts that are the site of photosynthesis.

prostrate   Growing closely along the ground without taking root, as running postman (*Kennedia prostrata*).
cf. **decumbent, procumbent**

prostrate

protandry   The condition of a flower having anthers mature and shed pollen before the stigma becomes receptive.
protandrous   Of or relating to protandry.
see **dichogamous, protogynous**

protein   Any of a group of complex organic molecules that contain carbon, hydrogen, oxygen and nitrogen.
An essential component of living cells and substances like enzymes and hormones.

proteinoplast
A leucoplast that stores proteins in crystalline or amorphous bodies.

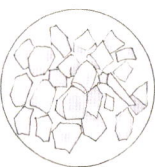

proteinoplast

proteranthous   Having flowers appearing before the leaves.
cf. **hysteranthous, synanthous**

prothallus   The gametophyte phase of ferns, most fern allies and some algae.
In ferns and most fern allies a germinating spore gives rise to a thread-like protonema that develops into a multicellular prothallus. It is usually cordate, photosynthetic and has no vascular system or differentiation into root, stem or leaf-like structures. The lower surface bears archegonia, antheridia and rhizoids.
In gymnosperms, as pines (*Pinus*), the megaspore germinates and gives rise to the female gametophyte or prothallus.

Prothallus

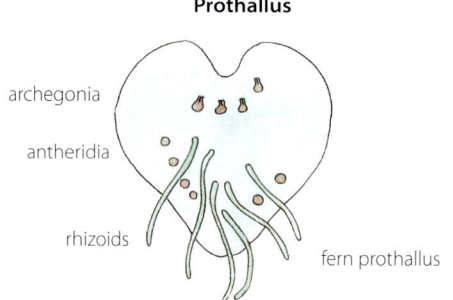

archegonia

antheridia

rhizoids

fern prothallus

Protista   The taxonomic kingdom that contains organisms that do not fit into any other category. They are a very diverse group that are not plants, animals, bacteria or fungi.

proto-   A prefix meaning first, earliest form or original.

protocorm   Of orchids, following fertilisation, the tuber-like body with rhizoids that gives rise to the first leaf and root shoots.

Protocorm

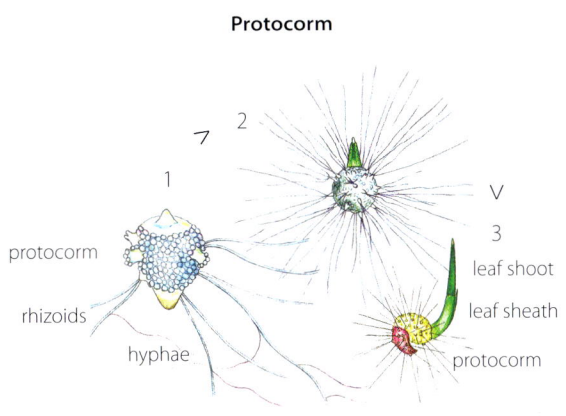

protoderm   One of three regions of primary meristematic tissue that develops behind the apical meristem.
It differentiates into epidermis.
see **primary meristem**
see also **ground meristem, procambium**

protogyny   The condition of a flower having the stigma become receptive before pollen is shed from the anthers.
protogynous   Of or relating to protogyny.
see **dichogamous, protandry**

protologue   The original description of a taxon including any other elements like illustrations, references, specimens, synonyms etc.

protonema, *pl.* **protonemata**   The thread-like chain of cells produced by a germinating bryophyte, fern or fern ally spore.
Buds on a bryophyte protonema develop into the thalloid or leafy gametophyte.
Gametophyte buds on the protonema of a fern and most fern allies develop into the prothallus.

Protonema

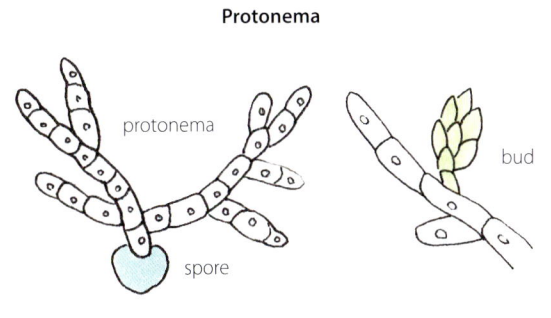

protophloem   The first formed phloem tissue that differentiates from procambium in the apical meristem of root and shoot tips during primary growth of a vascular plant.
It is conducting tissue that occurs in regions that are actively elongating and is transient, being replaced by metaphloem that functions indefinitely.
see **primary phloem, protoxylem**

protoplasm   Everything inside the plasma membrane of a cell, including the nucleus in eukaryotes.
It contains organelles, that carry out metabolic functions, and ergastic substances that are the product of the cell's metabolism.
cf. **protoplast**

protoplast   A plant cell from which the cell wall has been removed.
see also **cell membrane**
cf. **protoplasm**

protostele   Stele having a solid core of primary xylem surrounded by a cylinder of phloem.
A primitive type of stele.

protoxylem   Found in primary growth, the conducting tissue formed at the beginning of vascular differentiation.
It has mostly thin-walled cells that elongate rapidly and a few tracheids with thickened rings or spirals of lignin.
It differentiates from the procambium and is followed by the formation of metaxylem.
see **primary xylem**

provisional name   A name proposed in anticipation of the future publication and acceptance of a taxon.

proximal   Near the attached end as opposed to the free (distal) end.

Proximal

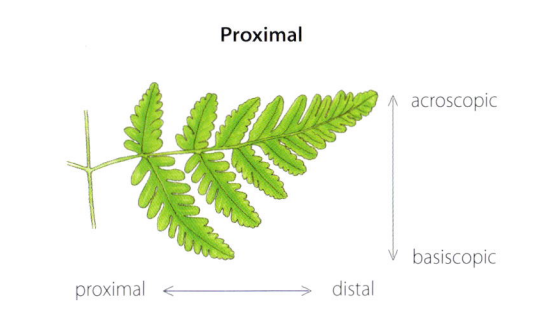

proximal pole   Of a pollen grain, that part of the polar axis orientated towards the inside.
cf. **distal pole**

Proximal pole

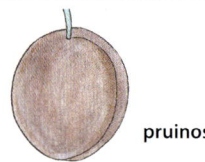

distal pole

polar axis

distal pole

distal pole

proximal poles
pollen tetrad

pruinose   Covered with a usually white powdery or waxy bloom, as that on a prune.

pruinose

psammosere   An ecological succession that starts on new sand dunes.

psamophyte   A plant thriving in shifting sands, primarily of deserts and dunes.

pseudanthium
A false flower.
A compact inflorescence of many small flowers that collectively look like a single flower.
Characteristic inflorescence of the daisy family (Asteraceae).
see also **capitulum, head**

pseudanthium

pseudo-   A prefix meaning false.

pseudobulb   Of most epiphytic orchids, an enlarged portion of the stem, that serves as a water and storage organ, from which all leaves and infloresences arise.
Regardless of shape a pseudobulb is either heteroblastic (with a single internode) and bearing leaves at the apex or homoblastic (with many internodes) and bearing leaves along its length or at the apex.
pseudobulbous   Bearing pseudobulbs.

Pseudobulb

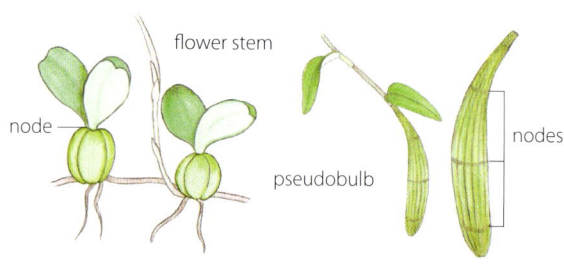

flower stem

node

nodes

pseudobulb

pseudocarp   A fruit derived from a simple ovary or compound ovary and some additional non-ovarian tissue like the receptacle. A strawberry has the true fruits (achenes derived from the ovaries) embedded in the fleshy receptacle.
Other accessory fruits include hips, pomes and pineapples.
= **accessory fruit, false fruit**

Pseudocarp

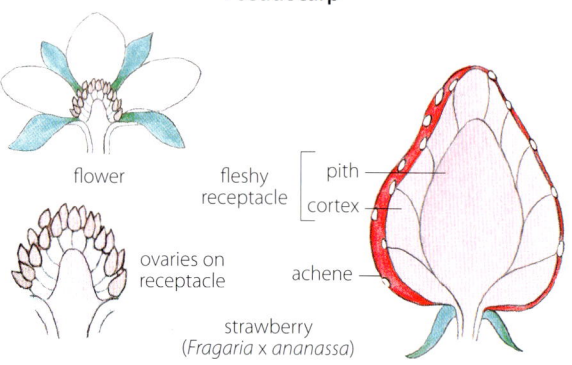

flower

fleshy receptacle

pith

cortex

ovaries on receptacle

achene

strawberry
(*Fragaria* x *ananassa*)

pseudocopulation
An attempt by a male insect to mate with a flower that resembles its female and in doing so takes pollen from the flower and carries it to another flower of the same species, thus pollinating the flower. The wasp pollinator (*Arthrothynnus huntianus*) pollinates the elbow orchid (*Thynninorchis huntianus*) when it attempts to mate with the labellum that mimics the female of its species.

labellum

1   flower

2

pseudocopulation

3

4

pollinia

5

pseudodrupe

A nut surrounded by a fleshy fused indehiscent involucre, as the walnut (*Juglans*).

see **accessory fruit**
cf. **tryma**

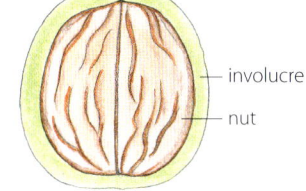

pseudodrupe

pseudopetiole

A narrow petiole-like extension of the leaf in some monocotyledons, as grasses and bamboos.

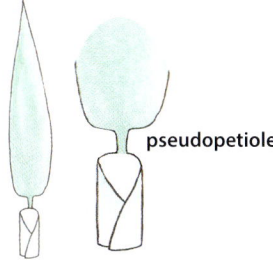

pseudopetiole

pseudopollen   Of some orchids, a mealy material resembling pollen that is collected by pollinators, as that on the labellum of the potato orchid (*Gastrodia sesamoides*).

It may or may not be a true food reward.

pseudovein

Of fern fronds, a vein-like strand with no vascular bundle.

= **false vein**

pseudovein

psilate   With a more or less smooth surface.

psychophily   Pollination of flowers by butterflies.
 psychophilous   Pollinated by butterflies.

pteridophyte   A term informally used for ferns and fern allies that are vascular spore-bearing plants.

pteridosperm   A member of an extinct group of plants that is believed to be intermediate between ferns and seed-bearing plants.
Now included in gymnosperms.

ptero-   A prefix meaning wing or feather.

pterocarpous

Having winged fruit, as the wing pennywort (*Hydrocotyle pterocarpa*).

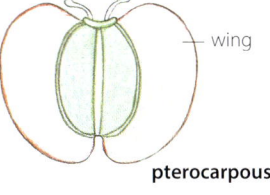

wing

pterocarpous

pterospermous

Having winged seeds, as elms (*Ulmus*).

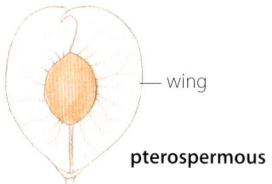

wing

pterospermous

ptyxis   *see* page 244

puberulent, puberulous   Minutely pubescent.

pubescence

A covering of short soft hairs, as on the buds, leaves and fruit of some plants. More generally, hairiness.

= **down**

pubescence

pubescent   Covered with short soft hairs or down.

pulp   Soft or fleshy and moist plant tissue, as the sticky pulp in which the seeds of pittosporum (*Pittosporum*) are immersed.

seeds in **pulp**

pulverulent   Appearing dusty or powdery.

pulvinus   A swelling at the base of a leaf, leaflet or phyllode that functions as a leaf-moving organ.
see also **nyctinasty, phytochrome, thigmotropism**
pulvinate   Having a pulvinus.

pulvinus

puncticulate, punctulate   Minutely punctate.

ptyxis The configuration of a single leaf within the leaf bud before it opens. The leaf is typically folded, rolled or coiled.
see also **vernation**

Ptyxis

Leaf bent or folded   Leaf rolled   Leaf coiled

conduplicate   reclinate   plicate   convolute/contorted   supervolute   revolute   involute   circinate

punctum, *pl.* **punctae**
A small dot.
punctate With dot-like markings or depressions, often due to translucent or coloured glands, as the leaves of mints (*Mentha*).
punctiform In the form of a dot or point.

punctum

mint (*Mentha*)

pungent Tipped with a sharp rigid piercing point.
Having a strong acrid smell or taste.

leaf tip **pungent**

purebred Offspring having the same traits as their genetically similar parents.
These traits have been maintained and passed on unchanged through many generations.
cf. **hybrid**

pusticulate Minutely pustulate.

pustule A small raised swelling like a pimple or a blister, as the rust fungus (*Uromyces viciae-fabae*) infecting the leaves of faba beans (*Vicia faba*).
pustular, pustulate, pustulose Of a surface covered with pustules.

pustule

rust fungus

putamen The endocarp of some fruits, that encloses and protects the seed, as the hard stone of a plum, cherry or peach, or the shell of a walnut.
see also **pericarp**

putative Supposed. Commonly accepted as true without being proven, as applied to some hybrids.

pyramid A solid body with sloping triangular sides that meet at a point at the top. The base may be triangular, square, rectangular or polygonal.
pyramidal Resembling a pyramid in shape, as the canopy of liquidamber (*Liquidamber styractiflua*).

Pyramid

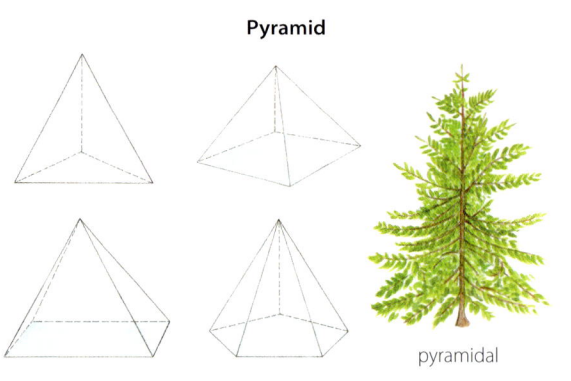

pyramidal

pyrene A seed covered by a hard endocarp, as the stone in the stone fruit genus *Prunus*.
A drupe-like fleshy fruit, derived from a multi-carpellate superior ovary, with two or more seeds, each surrounded by a stony endocarp, as holly (*Ilex*).

Pyrene

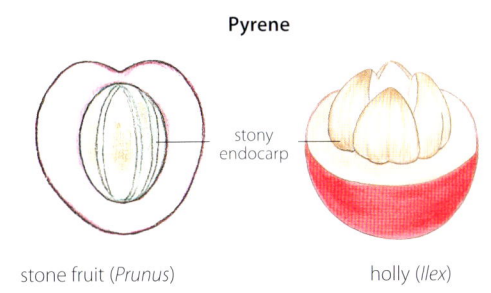

stony endocarp

stone fruit (*Prunus*)   holly (*Ilex*)

pyriform
Three-dimensional and
pear-shaped.

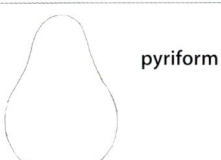

pyriform

pyrophile   A plant that is adapted to tolerate fire
or is stimulated by fire to germinate or regrow.
pyrophilous   Fire tolerant.

pyrophyte   A plant that is adapted to tolerate fire
or is stimulated by fire to germinate or regrow.

**pyxis, pyxide,
pyxidium,**
pl. **pyxides, pyxidia**
A capsule that
dehisces around the
circumference so
that the upper part
separates like a lid,
as plantain (*Plantago*).
= **circumcissile capsule**
pyxidate   Having a lid.

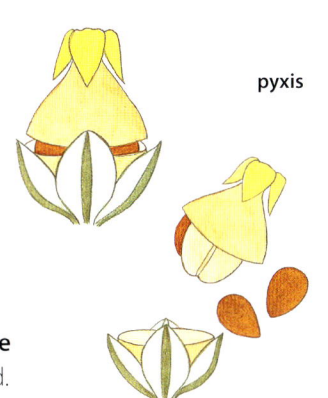

pyxis

quadri-   A prefix meaning four.

quadrijugate   Of a pinnate leaf having four pairs
of leaflets.
see **jugate**

quaternate
In sets of four.
Of leaves when four are
located at each node.

quaternate

quiescence   A period of inactivity, as the state of
delayed seed germination when environmental
conditions are not suitable.
quiescent   Inactive.
cf. **dormant, latent**

quillwort
The quillwort family
(Isoetaceae) comprises
one genus (*Isoetes*) of
mostly aquatic or semi-
aquatic vascular plants
that reproduce by spores
rather than seeds.
Plants have tufts of
spirally arranged single-
veined quill-like leaves.
Spores are of two kinds
(heterosporous).
see **fern allies**

quillwort

quillwort (*Isoetes*)

quincunx
Arrangement of an object with five parts, two are
outer, two are inner and the fifth has one margin
exterior and one margin interior.
A form of imbricate vernation.
quincunx aestivation   The arrangement of petals,
tepals or sepals in a bud.
quincunx vernation   The arrangement of young
leaves in an unopened leaf bud.
quincuncial   Arranged in a quincunx.

Quincunx

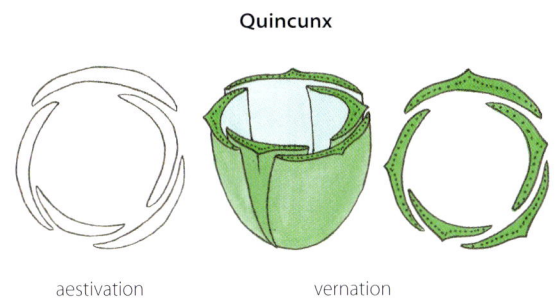

aestivation   vernation

quinque-   A prefix meaning five.

quinquefoliolate
Having five leaflets,
as a palmate leaf
with five leaflets
attached to the
tip of the petiole.

quinquefoliolate

quinquejugate   Of a pinnate leaf having five pairs
of leaflets.
see **jugate**

raceme   A racemose inflorescence in which the floral axis bears single flowers on stalks (pedicels) of about equal length.

The axis continues to grow indefinitely, with the youngest flower on the stem near the growing point at the apex and the oldest at the base.

A raceme can be simple, as lily of the valley (*Convallaria majalis*) or compound (then a panicle), as white hellebore (*Veratrum album*).

Of grasses (Poaceae), the arrangement of pedicellate spikelets, (rather than flowers), on the axis of the inflorescence.

An indeterminate or indefinite inflorescence.

see **acropetal, centripetal**

see also **racemose inflorescence**

racemiform   With the appearance, but not necessarily the structure, of a raceme.

racemose   Growing or arranged in a raceme. Bearing racemes.

cf. **cymose**

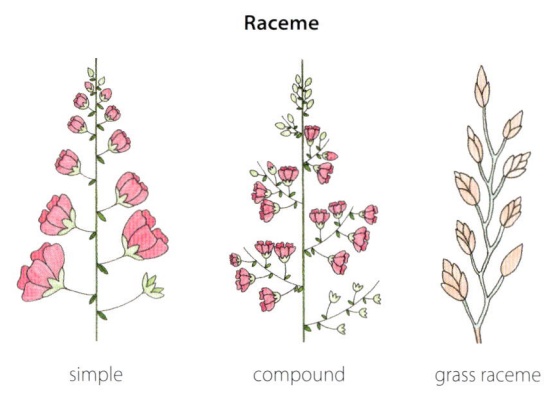

Raceme

simple   compound   grass raceme

racemose inflorescence   *see* page 247

rachilla, rhachilla, *pl.* **rachillae, rhachillae**

The secondary stem of a compound leaf, fern or inflorescence.

The stem of a grass or sedge spikelet above the glumes.

see **rachis**

Rachilla

rachilla   rachilla   rachilla

compound leaf   inflorescence   grass spikelet

rachis, rhachis, *pl.* **rachides, rhachides**

Of a pinnate leaf, the continuation of the petiole on which the leaflets (pinnae) are arranged.

Of an inflorescence, the continuation of the peduncle along which the flowers are arranged, as in a spike or a raceme.

Of ferns, the continuation of the stipe into the lamina of the frond to form the midrib.

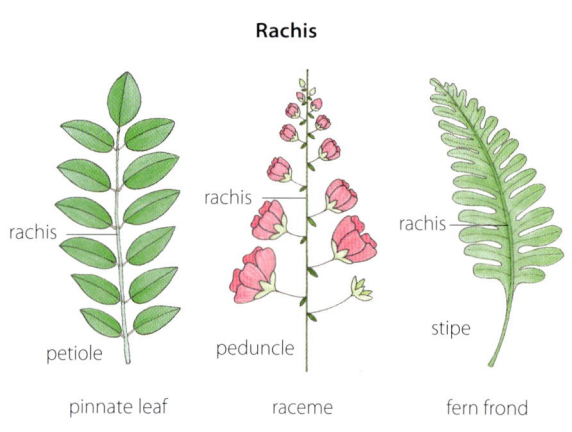

Rachis

rachis   rachis   rachis

petiole   peduncle   stipe

pinnate leaf   raceme   fern frond

radial vascular bundles   With xylem and phloem arranged separately on different radii.

Typical of roots of eudicots and monocotyledons.

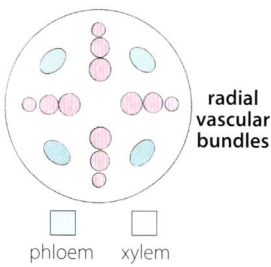

radial vascular bundles

phloem   xylem

radial symmetry

The quality of having a flattened circular surface that, looked at from above, has parts that divide equally anywhere by a line that passes through the centre.

cf. **bilateral symmetry, polysymmetric**

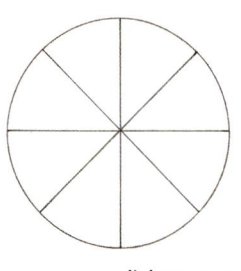

radial symmetry

radiate   Spreading from the centre like the spokes of wheel as the flower head of some daisies (Asteraceae).

radiate

racemose inflorescence   An inflorescence in which the main axis and lateral branches continue to grow indefinitely, with flowers arising along the axis and branches.

The first formed flower is at the base of the peduncle so that flowering begins at the base in ascending or acropetal succession.

The arrangement of flowers is centripetal, with the youngest flowers in the centre and the oldest flowers towards the outside.

The number of flowers is indefinite.

An indeterminate or indefinite inflorescence.

see **inflorescence, raceme**

cf. **cymose inflorescence**

Racemose inflorescence

Flowering ascending/acropetal   Flowering centripetal

spike   compound spike   raceme   compound raceme   panicle

corymb   compound corymb   umbel   compound umbel

head   compound head   ament   spadix

radical Of or relating to the root.
Of leaves arising directly from the rootstock.

radical

radicle The embryonic root in a seed embryo. Normally the first structure to emerge from the seed through the micropyle at germination.
In gymnosperms and eudicots it becomes a taproot, and in monocots it is replaced by adventitious roots.
cf. **plumule**

Radicle

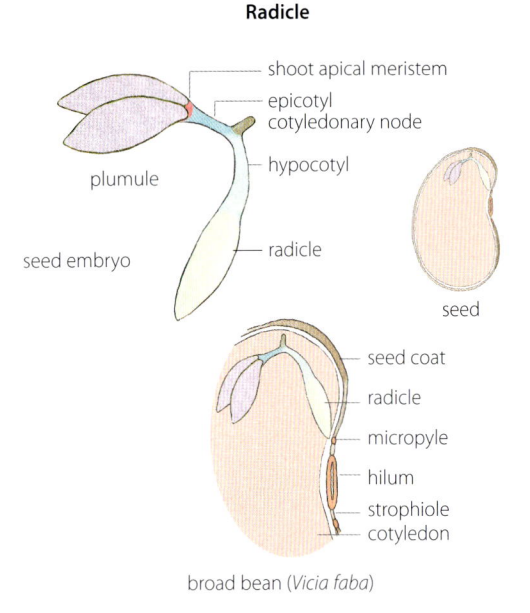

shoot apical meristem
epicotyl
cotyledonary node
hypocotyl
plumule
radicle
seed embryo
seed
seed coat
radicle
micropyle
hilum
strophiole
cotyledon
broad bean (*Vicia faba*)

radicular lobe The lobe sometimes visible on the seed coat under which the radicle is located.

Radicular lobe

radicular lobe
micropyle
hilum
strophiole
raphe
seed coat
radicular lobe
seed coat
radicle
hilum
bean seed (*Phaseolus*)
bean seed (*Vicia*)

rainforest Luxuriant forest characterised by high annual rainfall, extensive growth and poor soils.
It includes tropical rainforest found near the equator that has no distinct seasons and four layers of growth, (the emergent layer, the canopy layer, the understorey layer and the forest floor), monsoon rainforest that has a wet and a dry season, and cool temperate rainforests with a long wet winter and a shorter drier summer.
see also **storey**

rambler A plant with lax flexible stems that straggles over other vegetation, as rambling roses.

rambler

rame An inflorescence in grasses (Poaceae) consisting of paired sessile and pedicellate florets at each node, with each pair disarticulating at maturity, together with the internode above.
ramose Of a rame or series of rames.

Rame

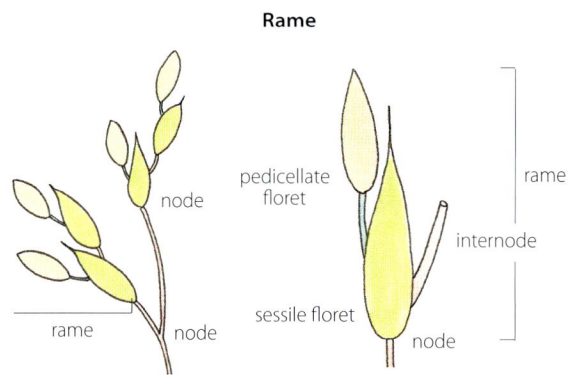

node
pedicellate floret
rame
internode
rame
node
sessile floret
node

ramentum, ramenta A thin chaffy scale.
ramentaceous With thin chaffy scales, as the stipes of many ferns.

ramentum

ramet A plant that is reproduced from a clone that is genetically identical to it.
A result of vegetative propagation.
A strawberry plant is a genet and the runners that can root and survive as separate plants are clones and individual member of a clone is a ramet.
cf. **genet**

ramification Of stems and roots, the process of branching and its general arrangement.

ramiflory The production of flowers on branches, as in the redbud (*Cercis canadensis*).
ramiflorous Exhibiting ramiflory.

ramulus, *pl.* **ramuli**
A small branch or twig.
ramulose, ramulous
With numerous small branches, as the twiggy daisy bush (*Olearia ramulosa*).

ramulus

ramus, *pl.* **rami** A branch or branching part.
ramal, rameal Of or belonging to a branch. Growing or originating on a branch, as some leaves.
ramiferous Having many branches.
ramose Bearing branches.
Full of branches.

ramus

rank A vertical row, as a row of leaves on a stem.

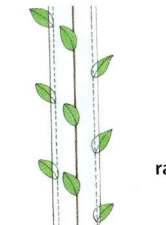
rank

rank A level in a taxonomic hierarchy.
In taxonomy, the successive levels from highest to lowest are: kingdom, division or phyum, class, order, family, genus and species.
The ranks of hybrid taxa (nothotaxa) are nothogenus and nothospecies.
Phylogenetic classification does not rank organisms but assigns a name to the position of a taxon within a tree diagram (cladogram).

raphe Of some ovules and seed, the part of the funicle that is adnate to the surface of the ovule and is visible as a line or ridge on the seed coat.

Raphe

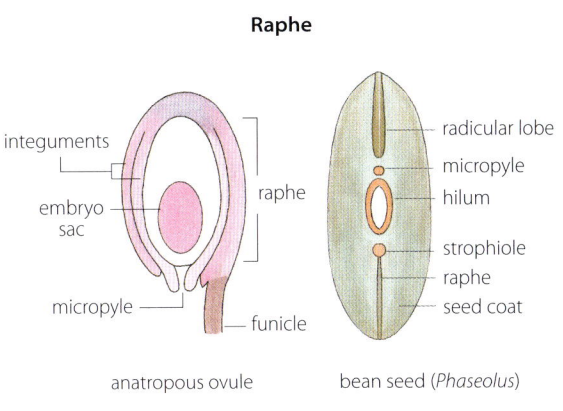

integuments

embryo sac

micropyle

raphe

funicle

anatropous ovule

radicular lobe
micropyle
hilum
strophiole
raphe
seed coat

bean seed (*Phaseolus*)

raphide One of the needle-shaped crystals of calcium oxalate found in an idioblast.

ratoon Regrowth from underground root buds after harvesting that produces a new crop, as sugar cane (*Saccharum officinarum*).

rattan A palm with very long stems and leaves not clustered in a crown. Leaves are pinnate and have long whip-like barbed tips by which the plant climbs to the top of the tree canopy in tropical rainforests.

rattan

ray The strap-shaped ligulate corolla of a ray floret in a daisy head (Asteraceae).
A branch of an umbel.
see **medullary ray, vascular ray**

Ray

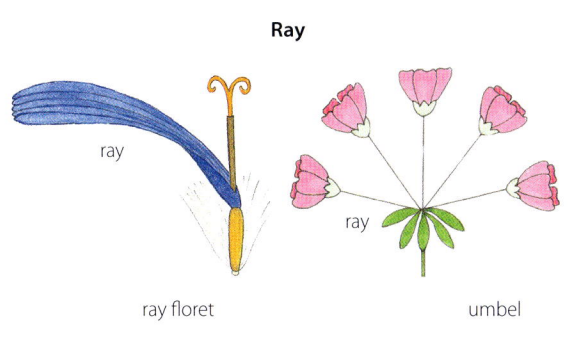

ray

ray floret

ray

umbel

ray floret In the inflorescence of a daisy (Asteraceae), a small tubular flower with the lobes united on one side into a strap-like blade.
= **ligulate floret**
see also **capitulum**
cf. **disc floret, tubular floret**

ray floret

ray initials One of two types of initial cells (the other being fusiform initials) in cambium, the meristematic tissue responsible for secondary growth in plants.
Ray initial cells are cuboidal in shape and give rise to horizontal growth in rays of wood that transport food and water horizontally through the secondary xylem and secondary phloem.

re- A prefix meaning again.

reaction wood Structurally abnormal wood formed in response to mechanical stress, as wind, on leaning stems and branches, that functions to bring the main stem or branch back to its normal position.
Called compression wood in softwoods like pine and tension wood in hardwoods like elm.

receptacle An expanded area at the top of a stem on which the sepals, petals, stamens and carpels of a single flower are inserted.
The expanded area at the top of a stem on which several flowers are inserted, as the genera *Dorstenia* and *Ficus* that are both in the mulberry family (Moraceae), or on which florets are inserted, as daisies (Asteraceae)
= **thalamus, torus**
cf. **hypanthium**
receptacular Borne on or relating to the receptacle.

Receptacle

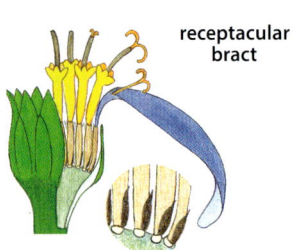

receptacle

single flower

receptacle

several flowers

receptacle

receptacle

receptacle

fig (*Ficus*) *Dorstenia* daisy (Asteraceae)

receptacular bract
One of the chaffy bracts or scales subtending a floret on the receptacle of a head in many daisies (Asteraceae).

receptacular bract

receptive Ready to receive, as a stigma that becomes receptive to pollen grains.

recessive Of a heterozygous individual, an allele of a gene that may be concealed by the expression of the other allele.
cf. **dominant**

reclinate Bent or turned downward towards so that the tip is lower than the base, as some leaves.
reclinate ptyxis Of a single leaf in bud that is folded inwards and lengthwise and has the apex bent downwards towards the base.

Reclinate

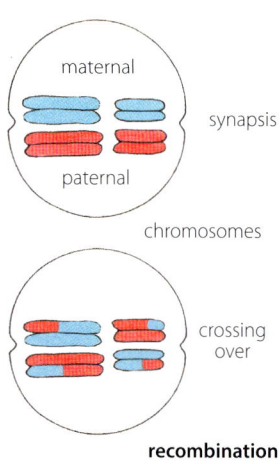

leaves ptyxis

recombination
In meiosis, the exchange of genetic material, during synapsis, between the maternal and paternal chromosomes so that the haploid chromosomes in the sperm and the egg will differ from the diploid parent chromosomes.
= **crossing over**

maternal

synapsis

paternal

chromosomes

crossing over

recombination

recti- A prefix meaning straight or upright.

recumbent Bent back until the apex is below the base, as leaves of common speedwell (*Veronica officinalis*).

recumbent

recurved   Curved or curled downward or backward, as some leaves on a stem.
Of leaf margins curved towards the abaxial side.
cf. **decurved, incurved, involute, revolute**

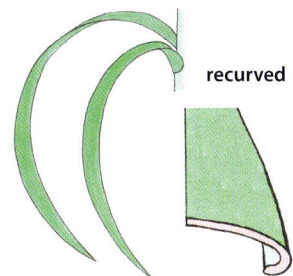

recurved

reduced   Decreased in size or number.

reduction
An atypical form, as the grass flower that lacks a calyx and corolla and is reduced to the stamens and pistil only.

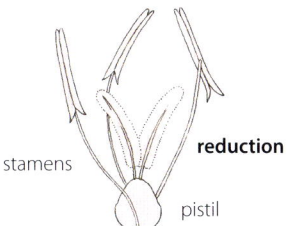

stamens

reduction

pistil

reduction division
The first division in meiosis.
The chromosome number in a diploid (2n) parent cell is halved to produce two haploid (n) daughter cells.
see **meiosis**
cf. **equational division**

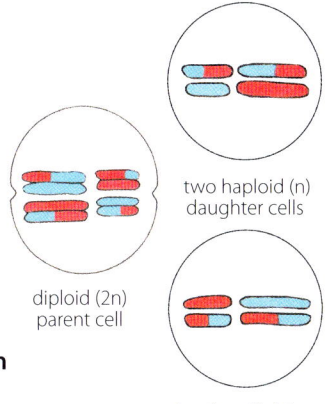

diploid (2n) parent cell

two haploid (n) daughter cells

reduction division

reduplicate   Of palm leaflets with margins bent backwards.
Upside-down V-shape in cross-section.
cf. **induplicate**

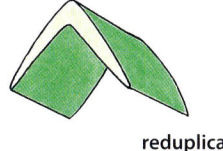

reduplicate

reed   A tall grass-like plant with slender jointed hollow stems that grows in water or on marshy ground.
see also **arundinaceous**

reflexed   Of leaves, bent sharply backwards or downwards towards the stem.

reflexed

regeneration   Renewal of growth in a disturbed area.

regma, *pl.* **regmata**   A dry schizocarpic fruit that splits into three one-seeded cocci.
The cocci in turn dry out, split open and eject the seed.
Derived from a three-carpelled, three-loculed syncarpous superior ovary, as the castor oil plant (*Ricinus communis*).

Regma

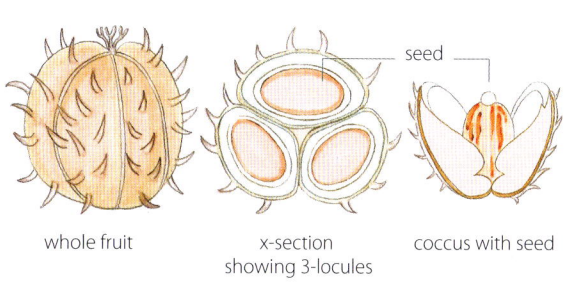

seed

whole fruit

x-section showing 3-locules

coccus with seed

regular   Of flowers with radial symmetry that divide through the centre into two or more like halves.
= **actinomorphic**
see **polysymmetric**
cf. **zygomorphic**

regular

regulator gene   A gene that controls the expression of another gene.

rein
Of palms (Arecaceae), a narrow strip along the margins of some pinnate leaves that peels away as the leaflets unfold. Usually shed or sometimes persists long after the leaf has opened.

pinnate palm leaf

rein

rejected name   A name rejected in favour of a conserved name.
= **nomen rejiciendum**

relict, relictual   The remains of a plant population that was once more widely spread at an earlier time.
Of conservation status, a small population of plants that has stabilised after declining or has been isolated from the core population.

remote   Separate or apart in space, at a distance.
cf. **adjacent**
Widely spaced, as leaves on a stem.
= **distant**

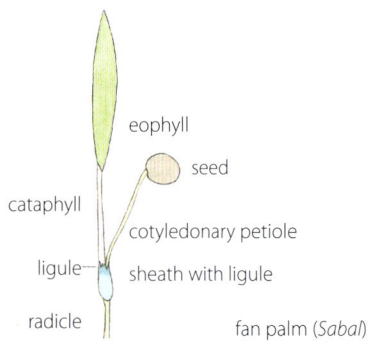

remote

remote germination   Of palms, having the seedling develop at some distance from the seed.
see **remote-ligular germination, remote-tubular germination**

remote-ligular germination   Of palms, one of three types of germination, as fan palms (*Sabal*). The cotyledon produces a petiole that grows downward into the soil at a distance from the seed. The plumular leaves emerge through the mouth of the sheath surrounding the cotyledonary petiole and the radicle emerges from its base.
The cotyledonary sheath has a ligule.
cf. **adjacent-ligular germination, remote-tubular germination**

Remote-ligular germination

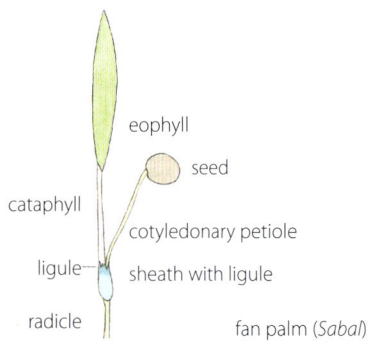

eophyll

seed

cataphyll

cotyledonary petiole

ligule

sheath with ligule

radicle

fan palm (*Sabal*)

remote-tubular germination   Of palms, one of three types of germination, as occurs in the date palm (*Phoenix dactylifera*).
The cotyledon produces a petiole that grows downward into the soil at a distance from the seed. Plumular leaves emerge through a slit in the sheath surrounding the cotyledonary petiole and the radicle emerges from its base.
The cotyledonary sheath lacks a ligule.
cf. **adjacent-ligular germination, remote-ligular germination**

Remote-tubular germination

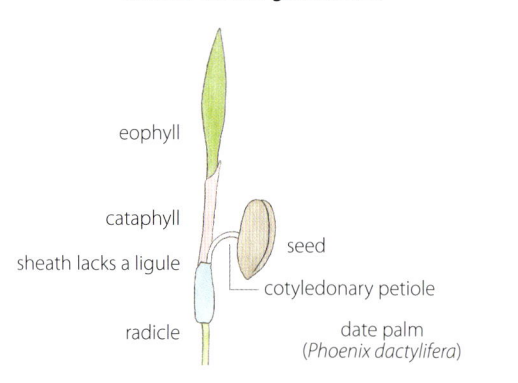

eophyll

cataphyll

sheath lacks a ligule

seed

cotyledonary petiole

radicle

date palm
(*Phoenix dactylifera*)

reniform
Kidney-shaped.

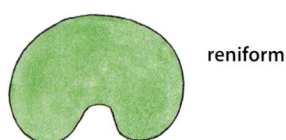

reniform

repand   With margins more or less flat in cross-section and curving slightly inward and outward like the movement of a snake, as the margins of some leaves.
Slightly sinuate.
= **sinuolate**
see also **sinuate**

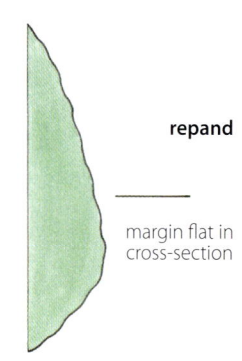

repand

margin flat in cross-section

repent, reptant
Prostrate and producing roots at the nodes, as creeping Charlie (*Glechoma hederacea*).
= **creeping**

repent

replacement tuber
Of orchids, the new tuber that will lie dormant until the following season at the end of a short root, a dropper, that grows down from the previous season's spent tuber.

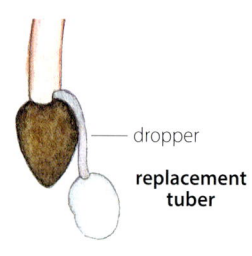

dropper

replacement tuber

leek orchid (*Prasophyllum*)

replication   In genetics, the process whereby chromosomes in a cell make an exact copy of themselves before cell division.
see **mitosis**

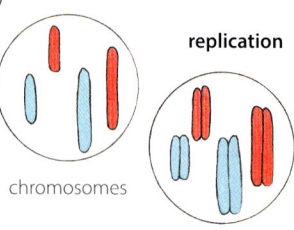

replication

chromosomes

replum The persistent frame-like placenta bearing seeds that remains after the valves fall away in some species of the mustard family (Brassicaceae). It usually has a membranous false septum.

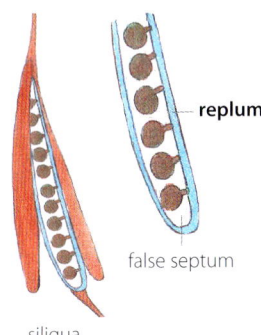

false septum

siliqua

reproduction The production of new individuals by sexual or asexual means.
see **asexual reproduction, sexual reproduction**

reproductive bud A flower bud that contains an embryonic flower. It is usually enclosed in the calyx.
see also **aestivation**
cf. **vegetative bud**

reproductive bud

reproductive cell An egg cell or sperm cell (gamete) of a plant that, as a result of meiosis, has half the number chromosome sets found in the non-reproductive (somatic) cells.
A set of these haploid (n) chromosomes is contributed to an organism by the the male parent and a set by the female parent.
cf. **somatic cell**

reproductive phase The phase in which plants transition from vegetative to reproductive growth. This involves transforming the meristem that produces vegetative structures, such as leaves, into meristem that produces reproductive structures, such as a flower or an inflorescence.
see **senescence, vegetative phase**
see also **alternation of generations**

resin A plant exudate, particularly from coniferous trees, that is insoluble in water but soluble in certain organic solvents.
cf. **gum**
resiniferous, resinous Containing or bearing resin. Resembling resin.

respiration The breakdown of glucose in the presence of oxygen to release energy.
Carbon dioxide and water are the by products.
see also **aerobic respiration, anaerobic respiration**
cf. **cellular respiration**

resting buds Buds that lie dormant until the onset of the next growing season, as those on the twigs of birch trees (*Betula*) in winter.
cf. **dormant buds**

resupinate Twisting through 180°, as the petiole of a leaf, resulting in the blade turning upside-down. Of the pedicel of a flower, resulting in the flower turning up-side down, as most orchids that have the dorsal sepal above the lateral sepals.
cf. **hyper-resupinate, non-resupinate**

Resupinate

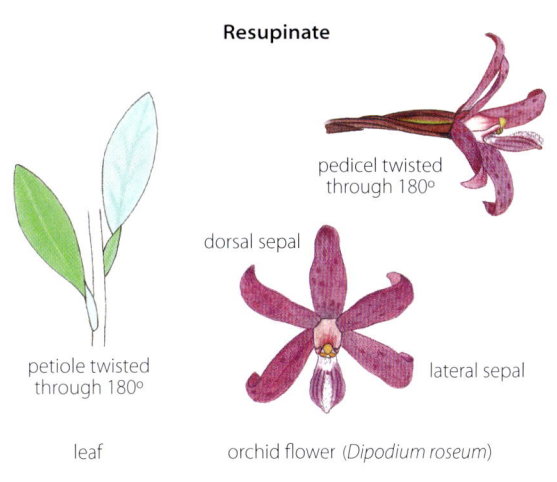

pedicel twisted through 180º

dorsal sepal

petiole twisted through 180º

lateral sepal

leaf

orchid flower (*Dipodium roseum*)

reticulate Forming a network of intersecting lines like those of a net. Of a pollen grain with a surface network of spaces (lumina) enclosed by ridges.
Of venation, forming a network of fine veins, as on a leaf or petal.
see also **reticulodromous**
cf. **areolate**

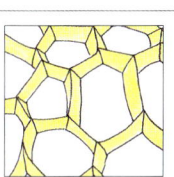

pollen

reticulate

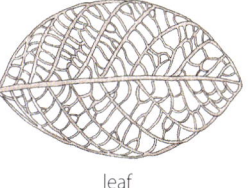

leaf

reticulation A network, as the net-like veins on a leaf.

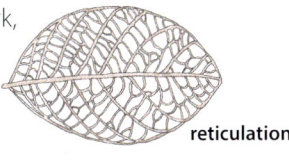

reticulation

reticulodromous Of leaves with secondary veins forming a network that becomes finer towards the margin.

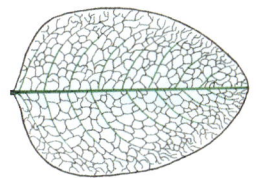

reticulodromous

reticulum, *pl.* **reticula**
A network-like pattern. Of pollen grains, a surface network of spaces (lumina) enclosed by ridges.

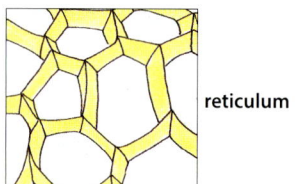
reticulum

retinaculum, *pl.* **retinacula**
In Asclepiadaceae, part of the translator arm that together with the caudicle connects a pollinium to the corpusculum.
= **translator**
Of orchids, another term for viscidium. Of the acanthus family (Acanthaceae), the hooked stalk on the seeds.

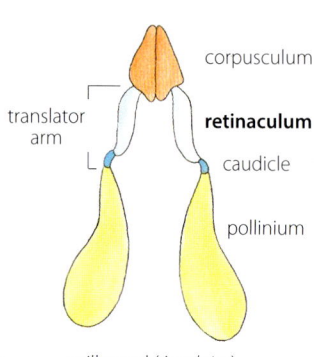
corpusculum
translator arm
retinaculum
caudicle
pollinium
milkweed (*Ascelpias*)

retrorse   Curved or bent downward towards the base, as hairs or spines.
cf. **antrorse**

retrorse

retuse   Having a rounded or obtuse apex with a central shallow notch.

retuse

revolute   Of a margin with the edges rolled under towards the lower surface, as the margins of some leaves.
cf. **involute**

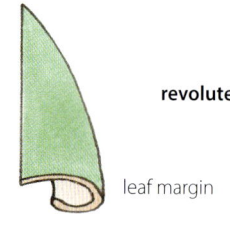
revolute
leaf margin

revolute aestivation
Of young petals, tepals or sepals in the unopened bud with margins rolled outwards towards the lower surface.

aestivation

revolute ptyxis
Of a single leaf in bud with margins rolled downward towards the lower outer surface.

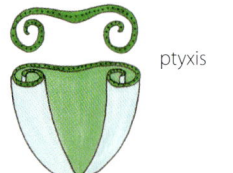
ptyxis

revolute vernation
Of young leaves in the unopened leaf bud with margins downward towards the lower outer surface and arranged in a circle.

vernation

rheophyte   Aquatic plants of flowing waters with narrow leaves that survive the currents or land plants along the edges of streams and rivers, like some palms, that are adapted to occasional fast-flowing currents, as flash floods.
cf. **limnophyte**

rhiphidium, *pl.* **rhiphidia, rhipidium,** *pl.* **rhipidia**
A cymose inflorescence that is flattened and fan-shaped, with successive branches alternating from one side to the other so that the axis is zigzagged.
see also **monochasium**
cf. **drepanium**

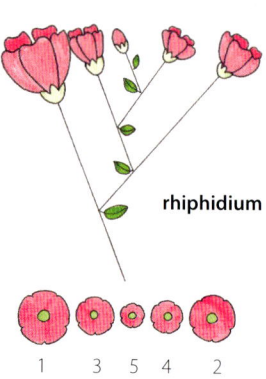
rhiphidium
1 3 5 4 2

rhizanthogene   An underground cleistogene. Spikelets are highly modified and are borne on highly specialised underground rhizomes.

rhizobium, *pl.* **rhizobia**   A bacteria, in roots and in nodules on the roots, of some plants that converts nitrogen into a form that is usable by plants.
see **nitrogen fixation**

rhizodermis   The epidermis of the root. It is of different origin to the epidermis of shoots but is continuous with it.
= **epiblem**

rhizoid   A thread-like outgrowth from a prothallus in some nonvascular plants, as ferns. It acts like a root and has similar functions of attachment to a substrate and absorption.

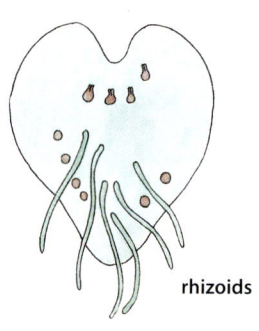
rhizoids
fern prothallus

rhizome   A specialised underground stem with nodes from which new plants are produced. Rhizomes can be horizontal or vertical.
see **caudex**
= **rootstock**
cf. **long-creeping, short-creeping, stolon**
rhizomatous   Bearing or producing rhizomes.

Rhizome

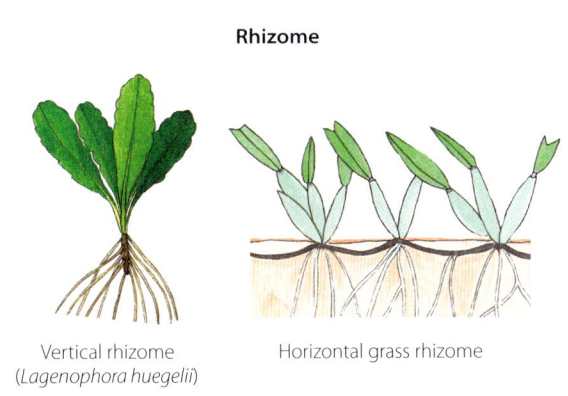

Vertical rhizome
(*Lagenophora huegelii*)

Horizontal grass rhizome

rhizoplane   The root epidermis to which soild particles, microorganisms and fungal hyphae adhere.
cf. **rhizosphere**

rhizosphere   The zone of soil, immediately surrounding a plant root, where complex interactions occur. It is rich in micro-organisms, root secretions and sloughed-off cells. Bacterial interactions are initiated here and mycorrhizal fungi and root nodules occur here.
see **nitrogen fixation**
cf. **rhizoplane**

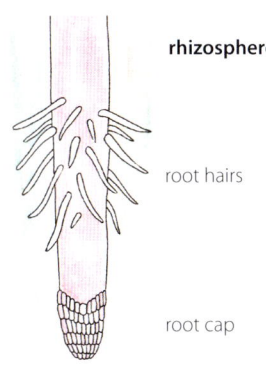

rhizosphere

root hairs

root cap

rhomboid   A four-sided figure with opposite sides parallel but adjacent sides of unequal length.
rhomboidal   Shaped like a rhomboid.

rhomboid

rhomboidal tetrad
A uniplanar tetrad arranged with the faces of two members in contact and the remaining two separated so that they cohere in a more or less diamond-shaped outline.
see **uniplanar, viscin thread**
see also **pollen tetrad**

rhomboidal pollen tetrad

rhombus   A four-sided figure with all sides equal.
rhombic   Diamond-shaped.

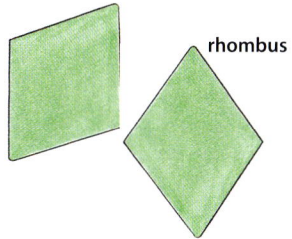

rhombus

rhytidome   Layers of dead phloem and periderm tissue that make up the exterior bark of a tree.

rib   The main vein, or other prominent vein, of a leaf.
ribbed   Of a leaf with prominent raised veins. Of a surface with marked, raised, roughly parallel bands, as the fruit of pennywort (*Hydrocotyl*) and the seeds of some oxalis (*Oxalis*).

Rib

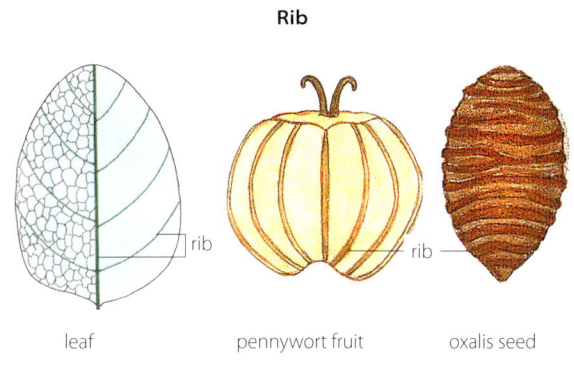

leaf

rib

pennywort fruit

rib

oxalis seed

ribose nucleic acid, RNA   The nucleic acid present in all living cells that plays an essential role in the synthesis of proteins.
see also **deoxyribose nucleic acid (DNA), ribosome, transcription, translation**

ribosome   An organelle that is the site of protein synthesis and is composed of protein and RNA (ribose nucleic acid).
It is either free in the cytoplasm or attached to the endoplasmic reticulum or outer nuclear membrane. Synthesised proteins may be stored in the endoplasmic reticulum or sent to the Golgi complex for modification.
see also **transcription, translation**

Ribosome

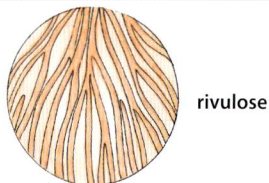

nuclear envelope
outer membrane
inner membrane
nuclear pore
perinuclear space
ribosome
nucleolus
nucleoplasm with chromatin (DNA)
nucleus
ribosomes
smooth endoplasmic reticulum
rough endoplasmic reticulum with ribosomes

rivulose   Marked with fine channels like rivulets.
cf. **striate, sulcate**

rivulose

rigid   Unable to bend or be forced out of shape, not flexible, not breaking.

ringent   Gaping.
Having a bilabiate corolla with lips separated and the throat open as the flower of thyme (*Thymus serpyllum*).
Lips may be lobed or not.
cf. **personate**

upper lip
throat
ringent
lower lip lobed
tube

riparian   Of or relating to the bank of a river or stream.

ripe   Fully developed, ready to harvest.

RNA   Ribose nucleic acid.
see also **DNA**

robbery   The removal of nectar by floral visitors without providing pollination, as bees that extract nectar from a hole pierced or bitten in the corolla.
cf. **thievery**

robust   Hardy and vigorous.

root   The usually underground part of a plant that absorbs water and nutrients and lacks nodes.
There are three types. A taproot is derived from the radicle in the seed and has different shapes as those of a turnip (napiform) or carrot (dauciform). Adventitious roots are roots growing on a short undergound stem, as those on a bulb.
Aerial roots grow in the air, as those of epiphytic orchids.

Roots

fusiform   dauciform   napiform
Taproots

Adventitious roots (bulb)   Aerial roots (epiphytic orchid)

root cap   The thimble-shaped mass of loose cells that covers and protects a root tip.

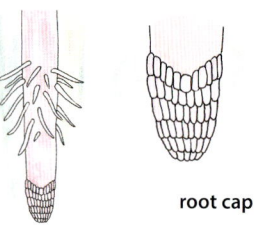

root cap

root hair   A hair-like outgrowth from an epidermal cell on a root that increases the total surface area available for the absorption of water and nutrients.
cf. **rootlet**

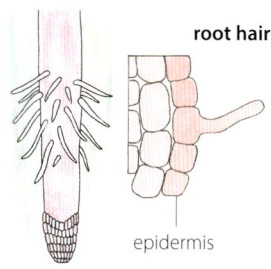

root hair

epidermis

root node In a phylogenetic tree, the most recent common ancestor of all of taxa represented on the tree.
see **node**

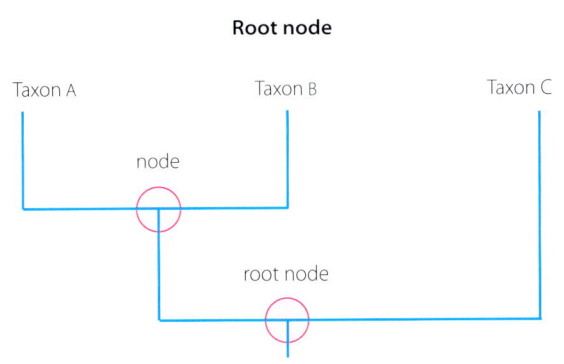

Root node

Taxon A   Taxon B   Taxon C

node

root node

root nodule A small swelling on roots of legumes, such as clover, and some other plants as a result of infection by nitrogen-fixing bacteria.
see also **nitrogen fixation, rhizobium**

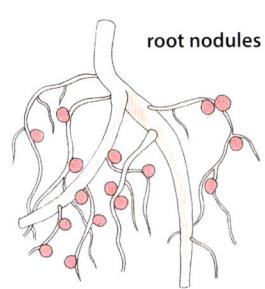

root nodules

root system One of two plant organ systems, the other being the shoot system, that consists of all roots and underground structures like tubers.

root tuber A swollen part of a root that functions as a storage organ and allows the plant to survive during dormancy. Unlike stem tubers, they do not have growing points called eyes. Root tubers propagate vegetatively by producing roots and shoots.
cf. **stem tuber, tuberoid**

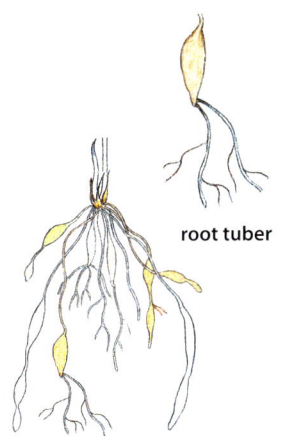

root tuber

rootlet A small root.
cf. **root hairs**

rootstock The underground part of a perennial plant from which roots arise, as a rhizome, corm or bulb.
In grafting, the rooted plant onto which a shoot or bud (scion) is grafted; that part below the graft gives rise to the lower main stem and root system of the graft. Also called the stock.

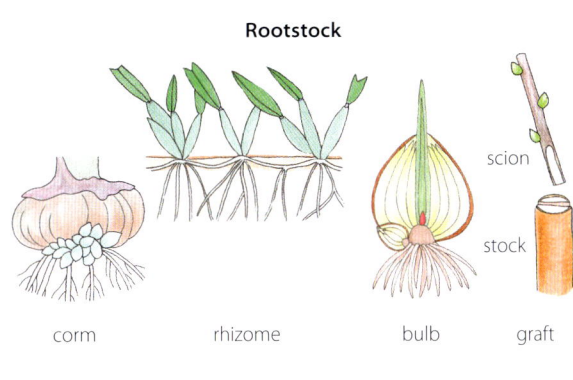

Rootstock

scion

stock

corm   rhizome   bulb   graft

roridous Dewy, appearing as if covered with fine dew drops.

roridulate, roridulous Finely roridous.

rosaceous Of, relating to or resembling the rose family (Rosaceae). Typically with a hypanthium bearing a four- or five-petalled radially symmetrical flower, many stamens and simple pistils, as roses, plums, blackberries and apples.

rosaceous

hypanthium

rosette Leaves borne on a contracted stem so that the internodes are close together and the leaves form a radiating cluster on or near the ground.
rosetted Arranged in a rosette.
= **rosulate**

rosette

rostellum, *pl.* **rostella rostellums** A small rostrum.
Of some orchids, a shelf-like or beak-like extension of the upper edge of the stigma that separates the pollinia from the fertile stigma. It forms the viscidium that secretes a viscous substance used in pollination.
see also **viscidium**
rostellate Having a rostellum.

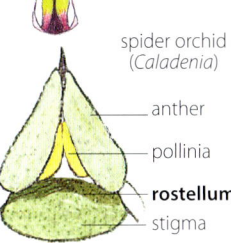

column

spider orchid
(*Caladenia*)

anther

pollinia

rostellum

stigma

rostrum   A beak-like projection, as the the carpophore of a crane's-bill (*Geranium*).
rostrate   Beak-like.

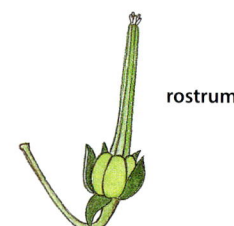

rosulate   Of leaves borne on a contracted stem so that the internodes are close together and the leaves form a radiating cluster on or near the ground.
= **rosetted**

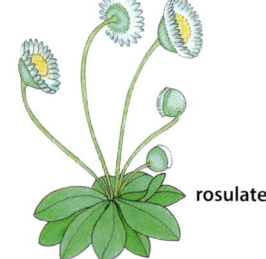

rotate   Wheel-shaped. Of a corolla with a short tube and a flattened, spreading circular limb, as some flowers of the nightshade genus (*Solanum*).

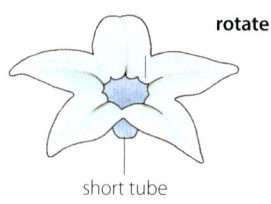

rough endoplasmic reticulum   Endoplasmic reticulum with ribosomes attached that give it a rough appearance.
The ribosomes synthesise proteins that may be stored in the rough endoplasmic reticulum or sent to the Golgi complex for modification.
cf. **smooth endoplasmic reticulum**

Rough endoplasmic reticulum

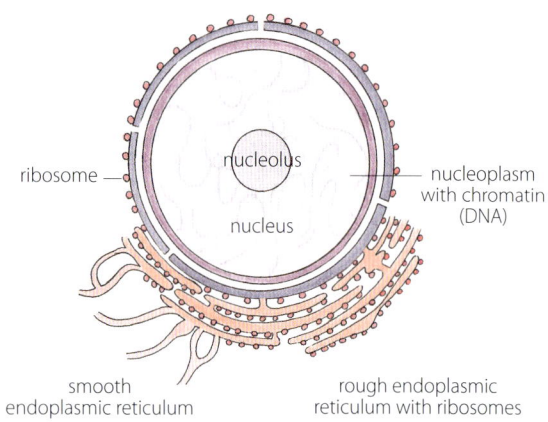

ribosome

nucleolus

nucleoplasm with chromatin (DNA)

nucleus

smooth endoplasmic reticulum

rough endoplasmic reticulum with ribosomes

rounded   In a smooth curve, as the base of some leaves.

leaf base **rounded**

rubescent, rubicund   Inclined to redness, turning rosy-red.

rubiginose, rubiginous   Rust-coloured, reddish-brown.

ruderal   Of plants growing on waste or disturbed lands. Commonly weedy or introduced plants, as nettles and rosebay willowherb (*Chamaenerion angustifolium*).

rudimentary   Imperfectly or incompletely developed, embryonic.
Being in the earliest stages of development, as the rudimentary plant in a seed comprised of a plumule, one or two cotyledons and a radicle.
cf. **abortive, vestigial**

rufescent   Tinged with red.

rufous   Dark red.

ruga, *pl.* **rugae**   A fold, crease or wrinkle.
rugate   Having rugae.

rugose   Deeply wrinkled, typical of leaves in the mint family (Lamiaceae).

rugose

rugula, *pl.* **rugulae**   A fine wrinkle or fold.
rugulate, rugulose   Finely wrinkled, as the pollen of elms (*Ulmus*).

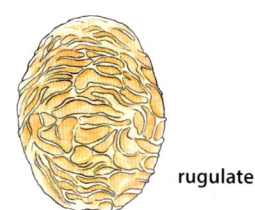

rugulate

ruminate   Appearing as if chewed. Praemorse.

ruminate

runcinate   Of a pinnately lobed leaf, usually oblanceolate in outline, with lobes pointing towards the base, as the leaves of dandelion (*Taraxacum*).

runcinate

runner   A slender prostrate stem with long internodes that forms new plants by putting down roots at the nodes or at the tips, as strawberry (*Fragaria*).
= **stolon**

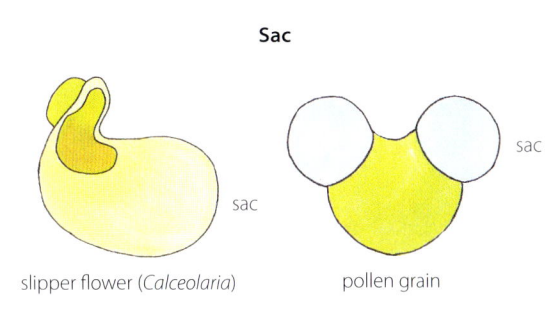

runner

rupestral, rupestrine, rupicolous   Growing on or among rocks, as many lichens and some ferns.
= **saxicolous**

rush   Common name for a plant in the Juncaceae family.
Tufted or rhizomatous, annual or perennial grass-like herbs, with solid or hollow, terete or laterally flattened stems that lack joints.
Flowers are inconspicuous, regular with six tepals and are wind-pollinated.
Fruit is a capsule.

sac, saccus *pl.* **saccii**   A little bladder, a pouch or cavity.
saccate   Pouched, inflated to form a sac.
Of the inflated corolla of slipper flowers (*Calceolaria*).
Of the pollen grains of some gymnosperms.

Sac

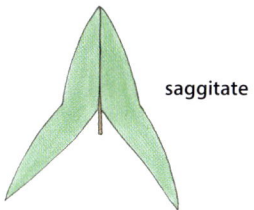

sac

sac

sac

slipper flower (*Calceolaria*)   pollen grain

saggitate   Shaped like an arrowhead, with two acute spreading lobes at the base, as some leaves.
cf. **hastate**

saggitate

saltmarsh   A low-growing plant community, typically dominated by halophytic vegetation and occupying the intertidal zone of coasts or fringing inland salt lakes.

salverform   Of a corolla with a long slim tube and an abruptly expanded limb spreading at 90° to the tube, as some primroses (*Primula*).
= **hypocrateriform**

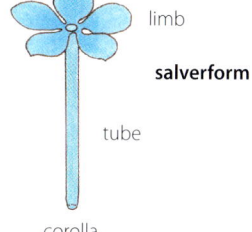

limb

salverform

tube

corolla

samara   A dry indehiscent fruit with one winged seed, as the fruit of elms (*Ulmus*). The wing is an extension of the fruit wall (pericarp). Derived from a two- or three-carpelled syncarpous superior ovary in which only one seed develops.
samaroid   Of or like a samara.

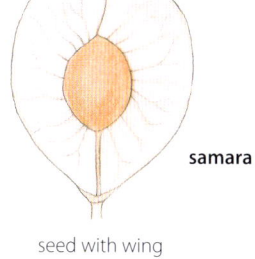

samara

seed with wing

samaracetum   An aggregate fruit composed of a cluster of samaras, as the tulip tree (*Liriodendron*).

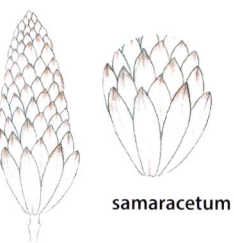

samaracetum

samphire   Hardy salt-tolerant grasses, succulents and low shrubs that form a boundary between the coastal shoreline and land vegetation. Dominated by the saltbush family (Chenopodiaceae).

sand   Small fragments or particles of rock ranging in size from 0.05 to 2 mm in diameter.

sanguine   A blood-red colour.

sap   The fluid that circulates through the vascular system of a plant carrying dissolved mineral salts, sugars and other nutrients to various tissues.

sapling   A young tree with a slender trunk. The stage between a seedling and a mature tree, usually defined as having a diameter between 5 and 10 cm and a height of 140 cm above ground level.

sapling

sapromyophile   A fly that visits dung, rotting flesh or plants that mimic their odours.
cf. **myophile**

sapromyophily   Dispersal of pollen and pollination by sapromyophile.
sapromyophilous   Pollinated by sapromyophiles.

saprophyte   A plant, fungus or microorganism that has no chlorophyll and lives on decaying organic matter, as the potato orchid (*Gastrodia sesamoides*).
cf. **epiphyte, parasite**
saprophytic   Of or relating to a saprophyte.

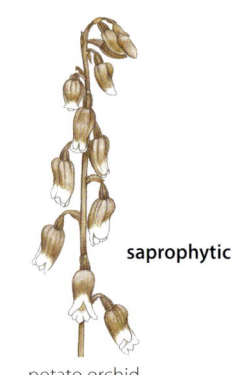

saprophytic

potato orchid
(*Gastrodia sesamoides*)

sapwood   The younger layers of wood between the heartwood and the bark. It contains the functioning vascular tissue in which the sap flows.
= **alburnum**

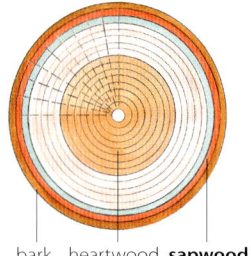

bark   heartwood   **sapwood**

sarcocarp   The fleshy mesocarp of a stone fruit, as a peach or a plum.
see **pericarp**

sarcotesta,
pl. **sarcotestae**   Of the unitegmic seed coat of gymnosperms, the outer fleshy parenchymatous layer, the other two layers being the middle sclerotesta and the innermost endotesta, as pine (*Pinus*).
Of the bitegmic seed coat of angiosperms, if all or part of the outer integument is fleshy it is then called a sarcotesta, as pomegranate (*Punica granatum*).
sarcotestal   Relating to or having a sarcotesta.

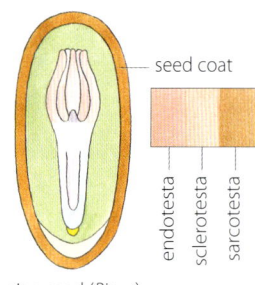

seed coat

endotesta
sclerotesta
sarcotesta

pine seed (*Pinus*)

sarcotesta

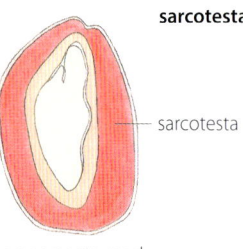

sarcotesta

pomegranate seed
(*Punica granatum*)

sarment   A long slender stolon or runner, as those of the strawberry (*Fragaria*).
A flexible new vine shoot, as those of the grapevine (*Vitis vinifera*).
sarmentiferous Bearing sarments.
sarmentose   Bearing or resembling sarments.

strawberry   **sarment**

grapevine

savanna, savannah   A plant community characterised by open grassland and very few trees due to a lack of water. Savannas are located between latitudes 30° north and 30° south of the equator. Temperatures are warm all year round. There is a characteristic long dry season in winter and a wet season in summer.
see **biome**

saxicolous   Growing on cliffs, rocks and screes.
= **rupestral, rupestrine, rupicolous**

scaberulent, scaberulose, scaberulous
Slighty scabrous, slightly rough to the touch.

scabrellate, scabrellous, scabridulous
Minutely scabrous, usually because of minute stiff bristles.

scabrid   Somewhat scabrous, slightly rough to the touch.

scabrosity   Roughness. An outgrowth that gives a surface a rough texture.
= **asperity**
scabrate, scabrous
Rough or harsh to the touch due to minute projections like tiny bristles or scales.

scabrosity

scalariform   Ladder-like, as a leaf with veins running parallel and appearing ladder-like.
see **percurrent venation**

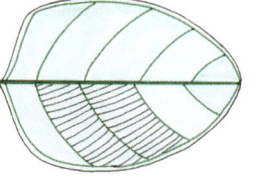

scalariform

scale A small, thin, usually flattened dry structure.

= **squama**

see also **perule, scurf**

A plant disease or infestation caused by scale insects.

scaly With a covering of scales, shedding scales or flakes, resembling scales.

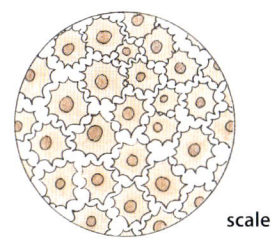

scale

lepidote scales

scale insect A parasitic sap-sucking insect that adheres to plants and robs them of essential nutrients.

scale leaf A small often membranous modified leaf, as on a rhizome or protecting a dormant bud. The fleshy modified leaf of a bulb. The leaves of some conifers, as juniper (*Juniperus*).

see also **tunic**

Scale leaf

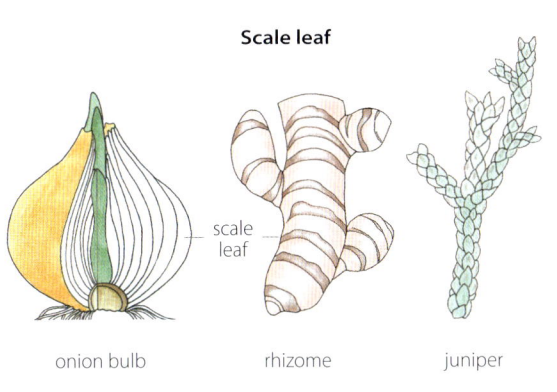

scale leaf

onion bulb   rhizome   juniper

scalloped Of a margin with rounded teeth.

= **crenate**

scalloped

scaly bulb A true bulb that consists of a compressed stem (basal plate), bearing roots and fleshy overlapping leaves that lack a tunic, as the lily genus (*Lilium*).

= **imbricate bulb, naked bulb**

see also **tunicate bulb**

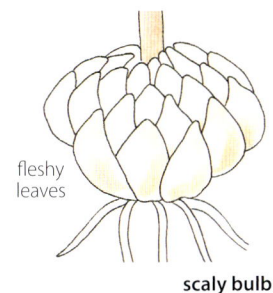

fleshy leaves

scaly bulb

scandent Of a plant that climbs as it grows, having a climbing habit.

scape A leafless stem bearing a solitary flower or an inflorescence that grows directly from the root or from a rosette of basal leaves.

scapigerous, scapose Having a scape.

scape

scapiflorous Having flowers borne on a scape. Of a flower or inflorescence stalk arising from ground level, with or without basal leaves.

cf. **cauliflorous**

scarious Thin, dry, membranous and somewhat stiff, as the bracts of golden everlasting (*Xerochrysum bracteatum*).

bract

scarious

scattered Having no regular order, as leaves around a stem.

scattered

schizocarp A fruit type intermediate between dehiscent and indehiscent fruits. The fruit splits into a number of indehiscent fruitlets that each contain a single seed, as a schizocarp of mericarps. Derived from a superior or inferior two- or more loculed ovary.

see also **carcerulus, cremocarp, lomentum, regma**

schizocarpic Of or bearing schizocarps.

Schizocarp

dehiscent fruit   indehiscent mericarp   seed

hemp bush (*Gynatrix pulchella*)

schizogenous   Of intercellular spaces formed by the breakdown of the common wall between adjacent glandular initial cells, resulting in a space lined by secretory epithelial cells.
cf. **lysigenous**

scientific name   The published and correct scientific name of a plant as opposed to the common name, as *Malus domestica* (scientific name) and apple (common name).
see **binomial nomenclature**

scion   A shoot, usually with two or three buds, or a single bud, that is implanted by grafting into the stock of another plant.
The aerial part of a graft.
see **budding**

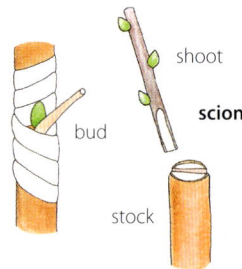

sciophile   A plant that tolerates or thrives in shade.
sciophilous   Shade-loving.
Of plants suited to diffuse light or low light intensity.

sciophyte   A plant that tolerates or thrives in shade, as forest understorey plants.
sciophytic   Tolerating or thriving in shade.

sclereid   One of variously shaped cells with thick often lignified walls.
Together with fibre cells they form sclerenchyma. Also occur singly as an idioblast or in groups, as in the flesh of a pear where they give a gritty texture.

sclerenchyma   Tissue, with hard tough cell walls that are impregnated with lignin, that functions as support and protects softer plant parts.
It is distributed throughout the plant body, both in primary and secondary growth, and is composed of sclereids and fibre cells.
Typically sclerenchyma is dead at maturity.
sclerenchymatous   With hard, tough cell walls that are impregnated with lignin.

sclerification   Thickening of cell walls, that may also involve lignification, to provide rigidity.
sclerified   Hardened.

scleroid   Having a hard or hardened texture, as the shell of a walnut.

sclerophyll   A type of vegetation characterised by hard leathery evergreen leaves that are adapted to reduce water loss and to survive hot dry summers and minimise the effects of fire, as the Proteaceae family.
Typical of Mediterranean-type ecosystems that include the garrigue, maquis, chaparral, matorral, fynbos and most Australian forests, heathlands and savannas.
sclerophyllous   Having tough leathery usually evergreen leaves adapted to reduce water loss.

sclerotesta,
pl. **sclerotestae**
Of the unitegmic seed coat of gymnosperms, the middle stony layer, the other two layers being the outer sarcotesta and the innermost endotesta, as pine (*Pinus*).
sclerotestal   Of, relating to or having a sclerotesta.

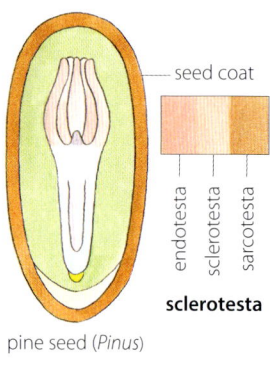

seed coat

endotesta sclerotesta sarcotesta

sclerotesta

pine seed (*Pinus*)

scobicular   In fine grains like sawdust, as most orchid seeds.
scobiform   Resembling sawdust.

scobicular

scorpioid   Curved or coiled to one side like a scorpion's tail.

scorpioid cyme
A spirally coiled cymose inflorescence with a single new stem developing from one axil only.
Branching continues to alternate from one axil to the other so that the axis is zigzagged.
= **cincinnus**
see **monochasium**
cf. **helicoid cyme**

scorpioid cyme

scrambler   A plant with long stems and a sprawling, climbing or creeping habit, as running postman (*Kennedia prostrata*).

scrambler

scrobiculate   Minutely pitted.
Of a surface with shallow depressions, grooves or pits, as the seeds of passion fruit (*Passiflora edulis*).

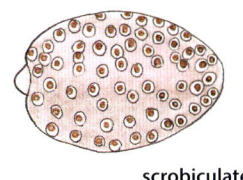

scrobiculate

sculpturing
Three-dimensional ornamentation on a surface, as on a pollen grain, seed or fruit.
see **ornamentation**

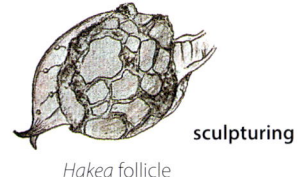

sculpturing

Hakea follicle

scurf   Small bran-like scales.
scurfy   Covered with minute loose flake-like scales.

scutellum, *pl.* **scutella**
Of a caryopsis, the more or less shield-shaped intermediate absorbing structure between the embryo and nutritive endosperm in the seed, as wheat and corn.
It is thought to be a modified cotyledon.

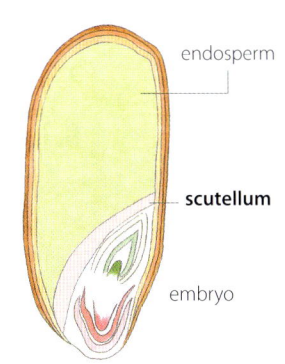

endosperm

scutellum

embryo

secondary cell wall
A rigid layer in some cells laid down between the primary cell wall and the cell membrane.
It is mainly composed of lignin.
see **middle lamella, primary cell wall**

adjacent cells

secondary cell walls

cell membrane
middle lamella

primary cell wall
secondary cell wall

secondary growth   Occurs in the second and subsequent years in woody eudicots and gymnosperms (it is absent in monocots).
It begins with the formation of lateral meristems (vascular cambium and cork cambium).
Vascular cambium of the stems and roots produces secondary xylem (wood) on the inside and secondary phloem (bark) on the outside to increase the thickness or girth of a plant.
Cork cambium produces periderm that replaces the epidermis in stems and roots.
cf. **primary growth**

secondary meristem   Two lateral meristem tissues, vascular cambium and cork cambium, found in woody plants that give rise to secondary growth.
Vascular cambium produces secondary xylem (wood) and secondary phloem (bark) that increase the girth of the plant. Cork cambium (phellogen) produces periderm (outer bark) that replaces the epidermis.
= **lateral meristem**
see also **intercalary meristem, primary meristem**

Secondary meristem

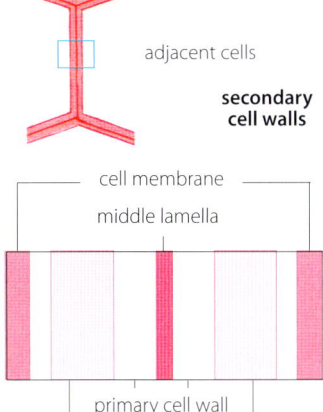

cork cambium
lataeral meristems
vascular cambium

periderm with cork cambium
cortex
primary phloem
secondary phloem
vascular cambium
secondary xylem
primary xylem
pith

Secondary growth

secondary phloem   A secondary tissue (the inner layer of bark) derived from vascular cambium.
It occurs in vascular tissue along the roots and stems of woody angiosperms (where it consists of sieve tube members and companion cells), and of gymnosperms (where it consists of sieve cells and albuminous cells).
It is composed of an axial and a radial cell system. The axial system is made up of conducting cells (sieve tube elements in angiosperms and sieve cells in gymnosperms), companion cells (in angiosperms), associated parenchyma cells and phloem fibres.
secondary phloem page 264 (cont.)

The radial system consists of phloem rays, composed of storage parenchyma, that are continuous with rays of the secondary xylem to make up vascular rays.

see **phloem, primary phloem**

Secondary phloem

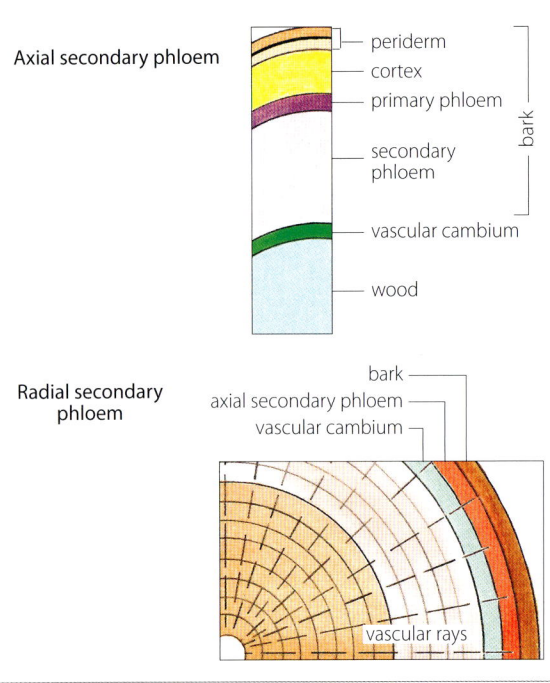

secondary tissue   Tissue derived from one of two secondary meristems (cork cambium or vascular cambium).

Cork cambium produces parenchymatous tissue and cork cells.

Vascular cambium produces secondary xylem and secondary phloem.

see **periderm**
cf. **primary tissue**

Secondary tissue

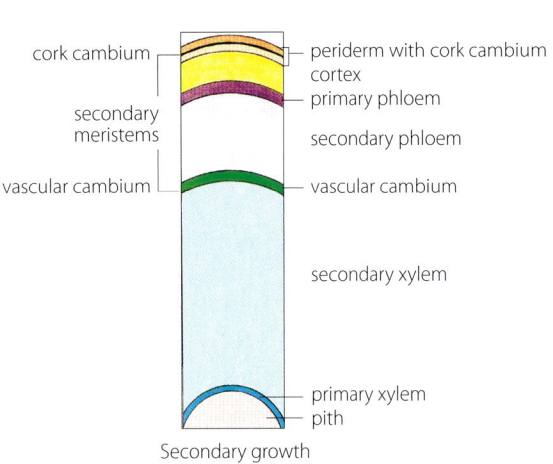

secondary vein

A vein arising from the main vein of a leaf.
A lateral vein.

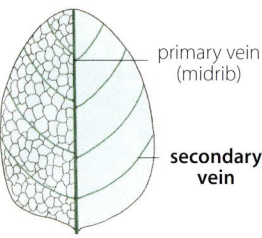

secondary xylem

Found in secondary growth, tissue in a vascular bundle that conducts water and mineral salts from the roots to other plant parts.

It is derived from vascular cambium, a lateral meristem, present in vascular tissue along the roots and stems of some flowering plants (woody eudicots) and gymnosperms.

The cells are thickened with deposits of lignin that provides support.

It is composed of an axial and a radial cell system. The axial system conducts water and is composed of fibres, tracheids, sclerenchyma and parenchyma, and vessel elements that are shorter and wider than those of primary xylem. It differentiates into sapwood and heartwood.

The radial system consists of xylem rays, composed of storage parenchyma, that are continuous with the rays of secondary phloem to make up vascular rays.

see **vascular bundle**
cf. **primary xylem**

Secondary xylem

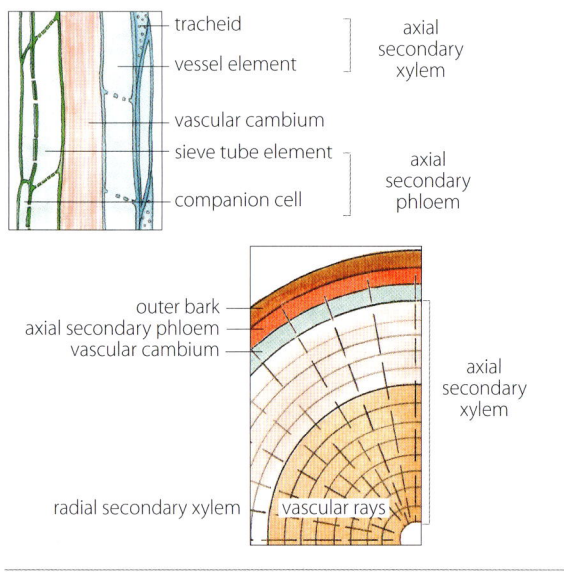

secretion   The process of manufacturing a substance inside a cell and releasing it from the interior of the cell to its exterior.
The substance released.
  secretory   Of or relating to secretion.
Producing a secretion, as glands, nectaries, laticifers and ducts.

-sect   A suffix meaning cut.

sectile   Of orchids, describes pollen when it is organised into discrete subunits called massulae.

section, sectio   Of taxonomic classification, the subdivisions of a large genus, below subgenus and above series.

secund   Arranged along one side only, as leaves or flowers along a stem.

secund

sedge   Common name for a plant in the Cyperaceae family.
Tufted or rhizomatous or sometimes stoloniferous or rarely tuber-producing, perennial grass-like herbs with solid, usually triangular stems that lack nodes.
Flowers are inconspicuous, generally lack a perianth and are wind-pollinated.
Fruit is an achene.

sedge

stem

seed   The propagation unit of seed plants (Spermatophyta) that include gymnosperms (conifers and related groups) and angiosperms.
A ripened ovule that is capable of germinating and developing into another plant.
It consists of a diploid embryo, food storage or nutritive tissue and a protective seed coat of one or more layers.
In angiosperms seeds are enclosed in the fruit (the ripened ovary).
In gymnosperms, seeds are borne naked on the surface of scales that form cones.
  see **albuminous ~, cotyledon, endospermic ~, exalbuminous ~, non-endospermic ~**

Seed

2 cotyledons   1 cotyledon   scutellum   2+ cotyledons

eudicot   monocotyledon   caryopsis   gymnosperm
Arabidopsis   *Allium*   *Triticum*   *Pinus*

seed coat
The outer covering of a seed that develops primarily from the integument(s) of the ovule after fertilisation.
It may have a mechanical layer that protects the embryo within or it may be variously embedded in a protective fruit wall (pericarp).
  see **bitegmic, unitegmic**

seed leaf   A cotyledon.

seed plants   Spermatophyta, the seed-producing angiosperms and gymnosperms.

seedling   The next growth phase after germination.
A plant resulting from seed germination rather than a cutting or other form of vegetative propagation.
Of trees, the stage before a sapling.

foliage leaves

eophylls

cotyledons

bean (*Phaseolus*)   **seedling**

roots

segment   A natural division or part, as a leaf that has three lobes or segments.

segment

segregate   In taxonomy, to separate or split off a part of a taxon.

self-compatible   Of a plant capable of producing viable seeds from its own pollen.
  = **self-fertile**

265

self-fertile   Of a plant capable of producing viable seeds from its own pollen.
= **self-compatible**

self-fertilisation   Fertilisation of an ovum of by a sperm cell from the pollen of the same flower or from another flower on the same plant.
= **selfing**

self-incompatible   Of plants that are unable to self-fertilise and produce viable seeds.
= **self-sterile**

self-pollination   The pollination of a flower by pollen from the same flower or from another flower on the same plant.
= **autogamy, selfing**
cf. **obligate outcrosser**

self-sterile   Of plants that are unable to self-fertilise and produce viable seeds.
= **self-incompatible**

selfing   Fertilisation of an ovum by a sperm cell from the pollen of the same flower or from another flower on the same plant.
= **autogamy, self-fertilisation**
cf. **obligate outcrosser**

semelparous   Of a plant that reproduces once in its lifetime then dies.
An annual is semelparous, as rice (*Oryza sativa*). Some semelparous plants are long-lived, as century plant (*Agave americana*), the talipot palm (*Corypha*) and some species of bamboo.
= **monocarpic**
cf. **hapaxanthic, iteroparous, pleonanthic, polycarpic**

semi-   A prefix meaning half or partly.

semi-craspedodromous
Of leaves with secondary veins forking just within the margin, one branch terminating at the margin, the other joining the vein above.
see **craspedodromous**

semi-craspedodromous

semi-inferior ovary
An ovary having the lower half embedded in the hypanthium and the upper part free.
= **half-inferior ovary**
cf. **inferior ovary, superior ovary**

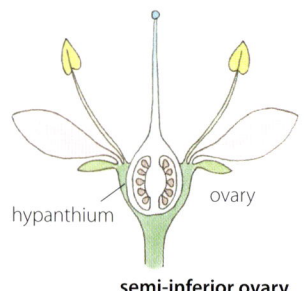
hypanthium   ovary
semi-inferior ovary

semi-parasite   Another term for hemiparasite.

semicarpous
Of a gynoecium with adjacent carpels partly fused and their styles and stigmas free.
cf. **synovarious**

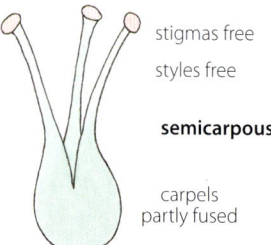
stigmas free
styles free
semicarpous
carpels partly fused

seminal   Relating to a seed.

seminal root system
Temporary roots, including the radicle, found in grasses (Poaceae). They arise from nodes on the root primordia in the embryo. Together they form the first or primary roots. These sustain the seedling until they are replaced by the adventitious roots that arise from the crown.

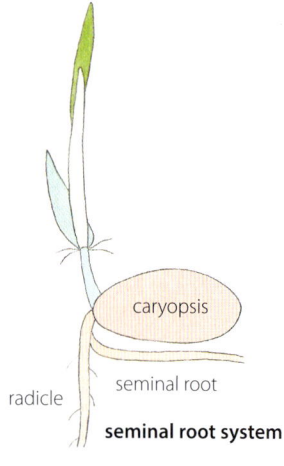
caryopsis
radicle   seminal root
seminal root system

senescence   Changes in the life cycle of a plant that lead to death of cells, tissues and eventually the whole organism.
senescent   Becoming old and losing the power of cell division and growth.

sens. *abbr.,* **sensu**   In the sense of.

sens. lat., s.l., *abbr.,* **sensu lato**   Used to refer to a taxon in a broad sense, as a taxon that may have smaller groups within it.

sens. str., s.str. *abbr.,* **sensu stricto**   Used to refer to a taxon in a narrow sense, as a reference to the type specimen.

sensu, *abbr.* **sens.**   In the sense of.

sensu lato, *abbr.* **sens. lat., s.l.**   Used to refer to a taxon in a broad sense, as a taxon that may have smaller groups within it.

sensu stricto, *abbr.* **sens. str., s.str.**   Used to refer to a taxon in a narrow sense, as a reference to the type specimen.

sepal   One of the segments of the calyx of a flower. Typically green and leaf-like, but petaloid in most monocotyledons.
sepaline   Relating to, attached to or resembling a sepal.
sepaloid   Having the form of a sepal.

sepalody   The abnormal development of floral parts into sepals or sepaloid organs.

separation layer
A layer of cells that disintegrates to facilitate the fall of a plant part, as a leaf or fruit.
It forms in the abscission zone of some plants.
= **abscission layer**

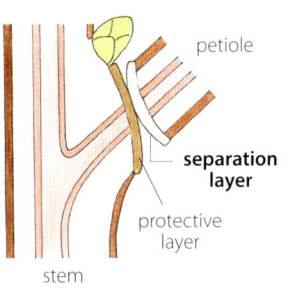

separation zone
The zone, as at the base of a leaf, in which shedding occurs. It includes the protection layer and the abscission layer.
= **abscission zone**

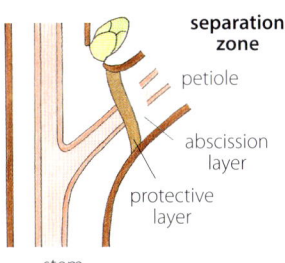

septenate
Having seven parts, as a palmate leaf with seven leaflets attached to the tip of the petiole.

septicidal capsule
A capsule that splits lengthwise through the fused sides (septa) of the carpels, as holywood (*Guaiacum sanctum*).

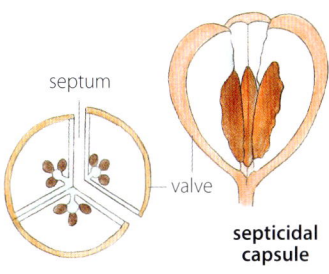

septifragal capsule
A capsule that splits so that the valves breakaway from the septa, as Argentine cedar (*Cedrela angustifolia*).
= **valvular capsule**

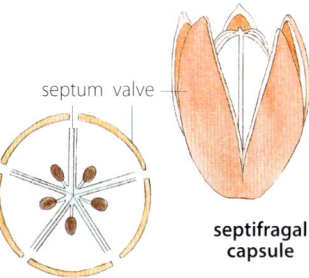

septum, *pl.* **septa**
A partition or wall separating two cavities. In an ovary or fruit, it is usually formed by the fusion of adjacent carpel walls.
= **dissepiment**
septate   Divided by one or more partitions, as an ovary divided into locules by internal walls (septa).

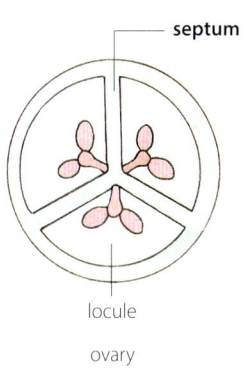

ser.   An abbreviation for series.

ser., series   In taxonomic classification, one of the subdivisions of a large genus.
The lowest rank below subgenus and section.

sere   One of the stages in the species structure of a plant community as it evolved over time.
see **ecological succession**
seral   Of or relating to a sere.

seriate   Arranged in one or more rows or whorls. Usually written with a prefix, as uniseriate (in one row or whorl), biseriate (in two rows or whorls), multiseriate (in many rows or whorls).
cf. **cyclic**

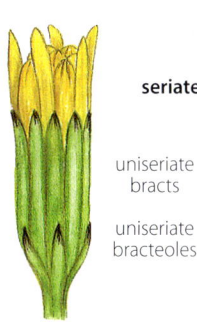

sericeous   Silky with fine, slender, soft, smooth and glossy appressed hairs, as the young leaf of *Neolitsea sericea*.
cf. **velutinous**

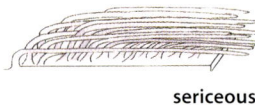
sericeous

series, *abbr.* **ser.**   In taxonomic classification, one of the subdivisions of a large genus.
The lowest rank below subgenus and section.

serotiny   Of seeds remaining in the fruit after maturity and the fruit staying on the plant, often for many years, until exposure to certain conditions, like fire, as the follicles of some banksias (*Banksia*).
serotinous   Relating to serotiny.

banksia (*Banksia*)
serotiny
follicle

serrate   Of a margin that is dentate but toothed like a saw with the teeth pointing forwards, as the margins of some leaves.

serrate

serrulate   Minutely serrate. Of a margin that is minutely toothed, like a saw with the teeth pointing forwards, as the margins of some leaves.

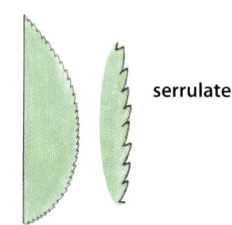
serrulate

sessile   Without a stalk and attached directly at the base, as a leaf without a petiole, a flower without a pedicel or a stigma without a style.

leaf **sessile**

seta, *pl.* **setae**   A bristle or stiff hair.
Of mosses and liverworts, the stalk supporting the capsule.
setiferous, setaceous Bearing bristles.
setiform   Bristle-like, bristle-shaped.
setose   Bristly.

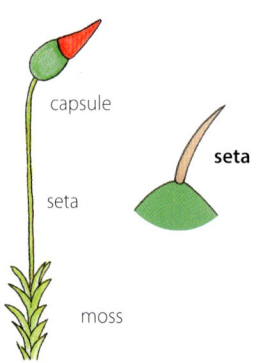
capsule
seta
seta
moss

setulose   Having small bristles or setae.

sexine   The outer, sculptured layer of the exine that lies above the nexine in the wall of a pollen grain.
see **pollen wall**

sexual phase   The production of gametes for reproduction.
The haploid phase in the life cycle of a plant represented by the male and female sex cells (gametes).
= **gametophyte generation**
see also **alternation of generations**

sexual reproduction   The fusion of the male sperm cell with the female egg cell, at fertilisation, to produce the beginnings of a new individual (a zygote).
cf. **asexual reproduction**

sheath   A tubular or rolled structure that surrounds, at least partly, another organ.
In grasses (Poaceae), the tubular or rolled part of the leaf that encloses the culm. Sheath margins may be separate, overlapping or closed to form a tube.
see **leaf sheath**
sheathing   Enclosing or closely enveloping, as the tubular or rolled part of the leaf that encloses the culm in grasses.
see also **vaginate**

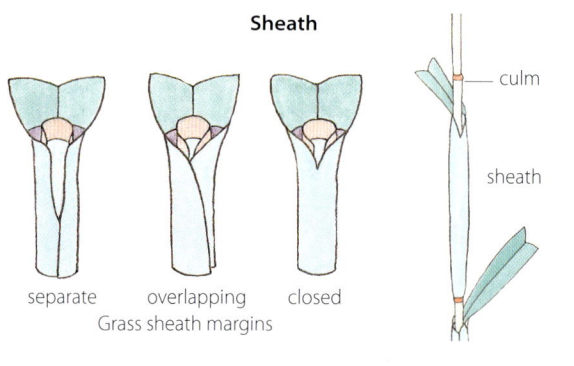
Sheath
culm
sheath
separate   overlapping   closed
Grass sheath margins

shoot   New growth, usually on the aerial part of a plant, as leaf or flower buds or new stems.
To sprout.
Of a seed, to germinate.

shoot

shoot system   One of two plant organ systems, the other being the root system.
The above-ground parts of the plant that include the vegetative parts, (the leaves and stems) and reproductive parts (the flowers and fruit).

short-creeping
Of ferns, having a rhizome that elongates slowly so that the fronds are clustered, as most lip ferns (*Cheilanthes*).
cf. **long-creeping**

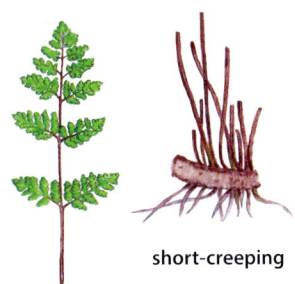

short-creeping

shrub   A low woody perennial plant with several stems and no distinct trunk.
= **bush**

shrub

sieve area   A region of a sieve cell wall in gymnosperms or a sieve element cell wall in angiosperms, with clusters of pores through which adjacent cells are interconnected for the transport of nutrients.

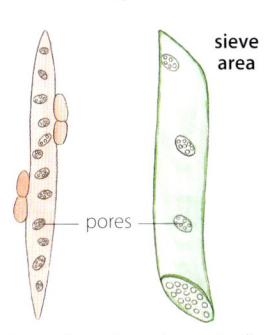

sieve area

pores

sieve cell   sieve element cell

sieve cell   An elongated food-conducting cell, associated with albuminous cells, in the phloem of gymnosperms and lower vascular plants.
They are arranged in longitudinal rows and are connected to each other by surface pores (sieve areas) through which nutrients flow.
Vertical rows of sieve cells perform the same function as sieve tubes in angiosperms.
cf. **sieve tube element**

Sieve cell

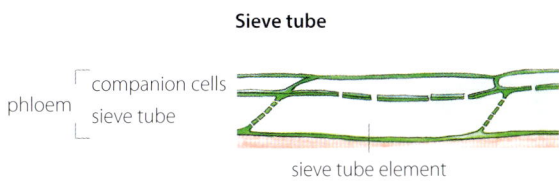

albuminous cells

sieve cell

sieve area

sieve element   The main conducting component of phloem, so called because the walls have pores (sieve areas).
There are two types of sieve elements: less specialised sieve cells, typically associated with

albuminous cells, in lower vascular plants and gymnosperms and more specialised sieve tube elements, typically associated with companion cells, in angiosperms.
= **sieve member**

sieve member   Another term for sieve element.

sieve plate
The perforated end wall of a sieve tube element in angiosperm phloem through which cytoplasm moves from one cell to another. Absent in the phloem of gymnosperms and lower vascular plants.
cf. **perforation plate**

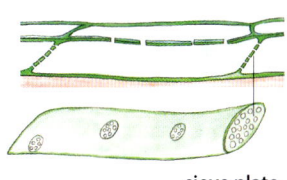

sieve plate

sieve tube   A longitudinal conducting tube in the phloem of angiosperms consisting of a vertical series of elongated sieve tube elements connected end to end by porous walls (sieve plates).
Its function is to translocate products of photosynthesis throughout the plant.

Sieve tube

phloem ⎡ companion cells
       ⎣ sieve tube

sieve tube element

sieve tube element   One of the elongated conducting cells, typically associated with a companion cell, in the phloem of angiosperms.
They are arranged in longitudinal rows to form a long sieve tube and are connected to each other by surface pores on the lateral surfaces and sieve plates on the adjoining ends of the cell walls.
= **sieve tube member**
see **sieve element**

Sieve tube element

phloem ⎡ companion cells
       ⎣ sieve tube

sieve tube element

lateral pores   sieve plate

269

sieve tube member Another term for sieve tube element.

sigmoid, sigmoidal
Shaped like the letter S, as the leaflets of some palms (Arecaceae).

sigmoid

silica, silica dioxide A colourless to white chemical compound.
In plants it is found in some hairs, in cell walls and in specialised silica cells in grasses.
siliceous Relating to or containing silica.

silicle, silicule, silicula
A stout dry fruit not more than twice as long as wide that splits into halves lengthwise usually leaving a persistent partition (false septum). Derived from a syncarpous two-carpelled superior ovary, as shepherd's purse (*Microlepidium pilosulum*).
see also **replum**
cf. **siliqua**

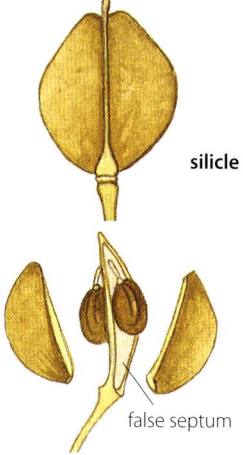

silicle

false septum

siliqua, silique,
pl. **siliquae**
A dry fruit that splits into halves lengthwise usually leaving a persistent partition (false septum), as bittercress (*Rorippa*). It is two or three times longer than it is broad. It is derived from a syncarpous two-carpelled superior ovary.
see also **replum**
cf. **silicle**

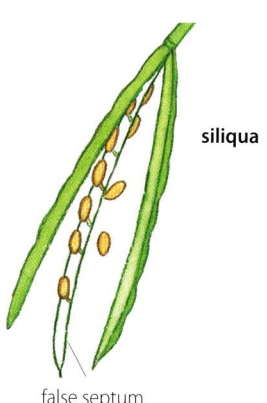

siliqua

false septum

silk The mass of long filiform styles at the top of female inflorescence (ear) of corn.
see **tassel**

silk

silky With fine soft hairs, as the bracts and seeds of quaking aspen (*Populus tremuloides*).

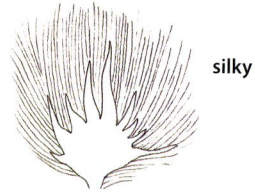

silky

silt A loose sedimentary soil composed of very fine particles ranging in size from 0.002 to 0.5 mm in diameter. Intermediate in size between sand and clay.

simple Undivided, unbranched, as a simple inflorescence.
Of a leaf, with the margin entire, lobed or toothed, but not divided into leaflets.
cf. **compound**

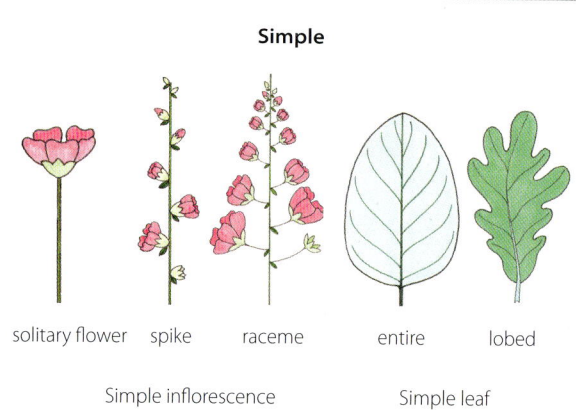

Simple

solitary flower spike raceme entire lobed

Simple inflorescence Simple leaf

simple craspedodromous
Of leaves with all of the secondary veins and their branches terminating at the margin.
= **marginal venation**
see **craspedodromous**

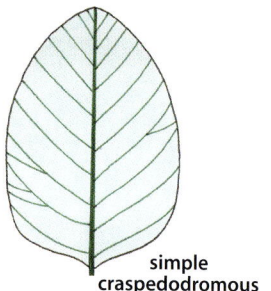

simple craspedodromous

simple fertilisation Two haploid sperm cells are produced by gymnosperms. When the nuclei of the two sperm cells meet the egg cell, one nucleus dies and the other unites with the egg nucleus to form a diploid zygote.
cf. **double fertilisation**

simple fruit   A fruit that develops from a single flower with one carpel.
It may be dry and dehiscent as a follicle, dry and indehiscent as an achene, or fleshy and indehiscent as a drupe.
cf. **accessory fruit, aggregate fruit, compound fruit, composite fruit, schizocarp**

Simple fruit

follicle   achene   drupe

simple gynoecium
A pistil with only one carpel, as a pea (*Pisum*).
cf. **compound gynoecium**

simple gynoecium

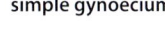
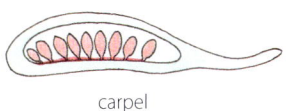

carpel

simple ovary
The ovary of a single free carpel.
A flower may have one simple ovary, as peas (*Pisum*), or more than one simple ovary, as buttercups (*Ranunculus*).
cf. **compound ovary**

simple ovary

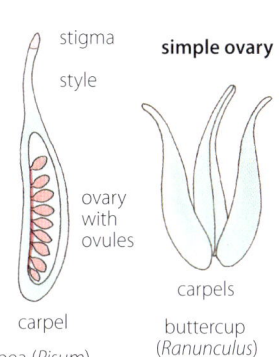

stigma
style
ovary with ovules
carpel
pea (*Pisum*)
carpels
buttercup (*Ranunculus*)

simple tissue   Tissue composed of one kind of cell, as ground tissue (parenchyma, collenchyma or sclerenchyma).
cf. **complex tissue**

sinistrorse   Twining from the base in a spiral from right to left, as seen from the side.
Twining in an anti-clockwise direction, as seen from above.
see also **twiner**
cf. **dextrorse**

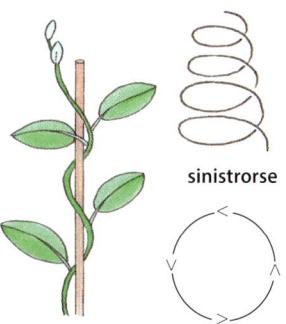

sinistrorse

sinker   A root that grows downward from a bulb or corm that bears a replacement bulb or corm.
Of orchids, the short root that bears the replacement tuber.
= **dropper**

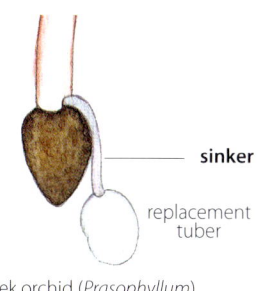

sinker
replacement tuber
leek orchid (*Prasophyllum*)

sinuate, sinuose, sinuous
With margins more or less flat in cross-section and curving strongly inward and outward like the movement of a snake, as the margins of some leaves.
see also **repand**
cf. **sinuolate, undulate,**

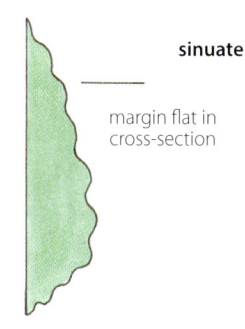

sinuate
margin flat in cross-section

sinuolate
With margins more or less flat in cross-section and curving slightly inward and outward like the movement of a snake, as the margins of some leaves.
Slightly sinuate.
= **repand**
cf. **sinuate**

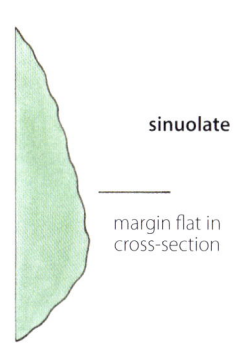

sinuolate
margin flat in cross-section

sinus   The gap between two lobes or segment.
It may be rounded to angular.
cf. **lobe**

sinus
sinus

siphonogamy   Process in which a non-motile male sperm is carried to the egg cell by a pollen tube, in all angiosperms from the stigma and in most gymnosperms from the pollen chamber.
siphonogamy page 272 (cont.)
see **pollination drop.**

271

Siphonogamy

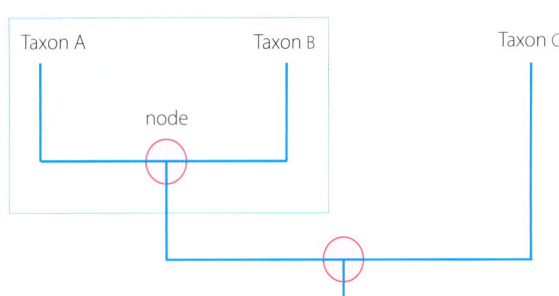

stigma

pollen tube

sperm cells

ovule
egg cell

angiosperm

pollination drop

pollen chamber

egg cell

ovule

gymnosperm

---

siphonostele   Stele having a central pith.
Surrounded by a cylinder of vascular tissue.
Common in ferns.

sister clade, sister group, sister taxa   In a
phylogenetic tree, two descendants that split from
the same node.
see also **speciation event**

Sister taxa

Taxon A   Taxon B   Taxon C

node

---

skeleton   Framework, as
veins are the framework
of a leaf.
skeletonised   To
reduce to a skeleton,
as occurs when some
insects attack a leaf.

skeleton

slip   Part of a stem, leaf
or root that is cut off and
grows roots and shoots
to produce a new plant.
= **cutting**
see also **grafting**

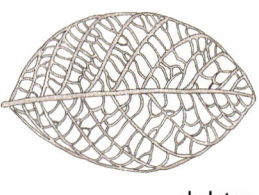

slip

smooth   Free from roughness or hairs, glabrous.

---

smooth endoplasmic reticulum
Endoplasmic reticulum that is 'smooth' because it
lacks ribosomes.
It is involved in the synthesis of lipids.
cf. **rough endoplasmic reticulum**

Smooth endoplasmic reticulum

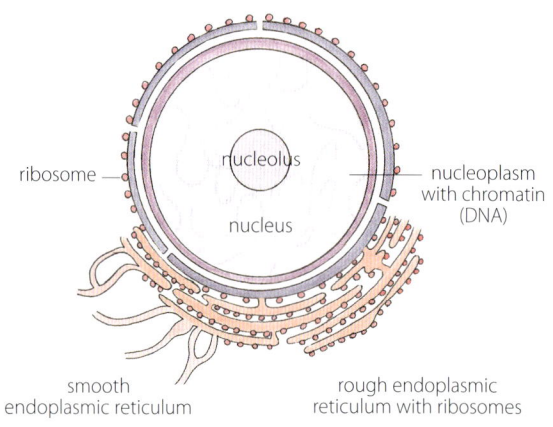

ribosome

nucleolus

nucleus

nucleoplasm
with chromatin
(DNA)

smooth
endoplasmic reticulum

rough endoplasmic
reticulum with ribosomes

---

sobol, sobole   A thin
rhizome or a runner
forming plants at short
distances.
soboliferous   Clump-
forming.
Bearing or producing
soboles, as some grasses.

sobole

sod   A section of grass-
covered surface soil
held together by matted
roots.
Turf.

sod

softwood   The wood of gymnosperms, though
the wood is not always soft.
Softwood trees include pines, redwoods and
larches.
cf. **hardwood**

solute   A substance dissolved in another substance
(a solvent) to create a solution, as salt (the solute)
dissolved in water (the solvent).

solution   A homogeneous mixture having one
substance dissolved in another, as salt (the solute)
is dissolved in water (the solvent), or carbon dioxide
gas (the solute) dissolved in water (the solvent).
The substances may be solids, liquids or gases.

solvent A substance in which another substance (solute) is dissolved to form a solution, as water (the solvent) in which salt (the solute) is dissolved.

somatic cell All the cells in a plant that are not reproductive cells, usually with two sets of chromosomes (2n).
cf. **reproductive cell**

sorosis, *pl.* **soroses**
Fruitlets on a common axis that are usually coalesced and derived from the unisexual flowers of a female inflorescence, as pineapple (*Ananas comosus*) and the fruit from the catkin of the white mulberry (*Morus alba*).
The fruit of the white mulberry consists of the succulent calyces of the inflorescence, with each calyx containing an achene.
see **composite fruit**

Sorosis

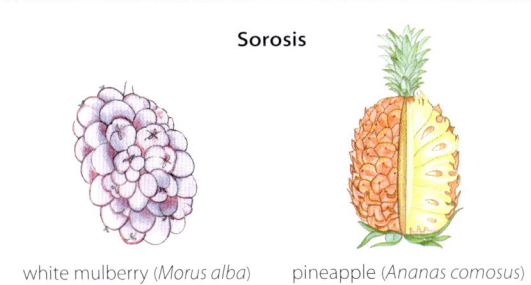

white mulberry (*Morus alba*)   pineapple (*Ananas comosus*)

sorus, *pl.* **sori** A cluster of sporangia variously arranged on the undersurface of a fern frond. It may or may not be covered by an indusium or false indusium.
soriferous Bearing sori.

Sorus

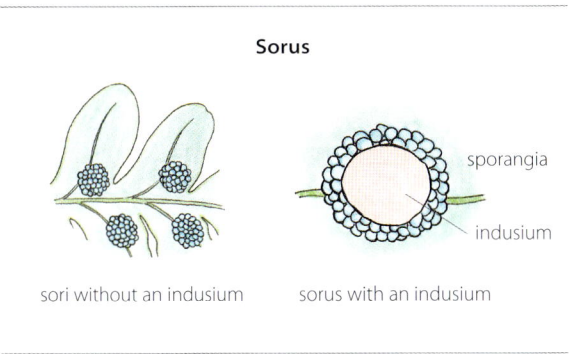

sori without an indusium   sorus with an indusium

sp., *pl.* **spp.** An abbreviation for species.

spadix, *pl.* **spadices**
A spike of minute flowers arranged on a fleshy receptacle and typically enclosed in a sheath-like spathe. Flowers are unisexual and commonly at the base of the elongated receptacle, as the arum lily genus (*Arum*). A racemose inflorescence.

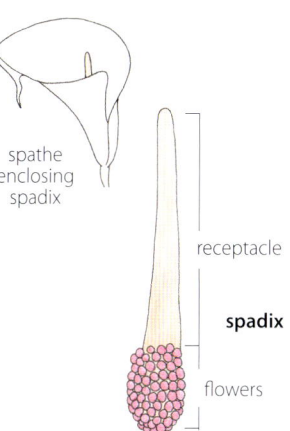

spathe enclosing spadix

receptacle

spadix

flowers

sparse Thinly scattered, not dense, as leaves on a stem.

sparse

spathe A large bract sheathing an inflorescence: one of the bracts that subtend the inflorescence of palms (Arecaceae).
Two or more bracts enclosing the flowers of the iris family (Iridaceae).
The large bract enclosing the flower spike (spadix) of the arum family (Araceae).
see **spadix**
spathaceous, spatheate Like or with a spathe.

Spathe

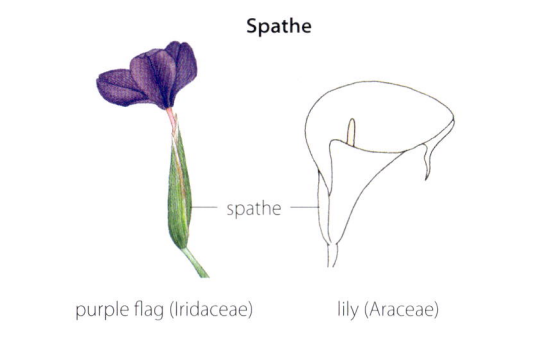

spathe

purple flag (Iridaceae)   lily (Araceae)

spathella
A membranous sac enclosing the immature flower in the riverweed family (Podostemaceae).

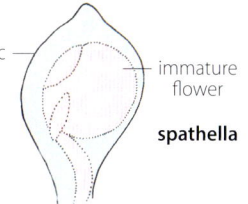

sac

immature flower

spathella

spatheole A small or secondary spathe.

spathulate, spatulate
Spatula-shaped.
With a broad, rounded
apex and narrowed at
the base.

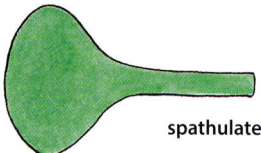

spathulate

speciation   The evolutionary formation of a new
species by the splitting of a single line of descent
into two or more genetically independent lineages.

speciation event   A lineage splitting event that
produces two or more new species.
The branching point that occurs at a node in a
cladogram represents a speciation event.

Speciation event

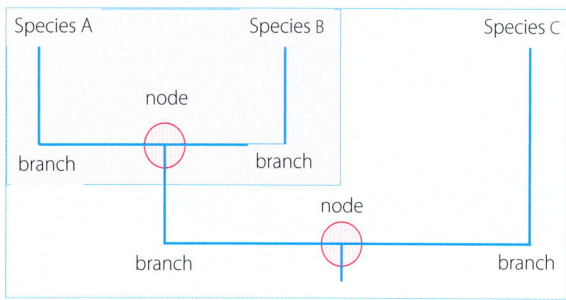

species, *abbr.* **sp.,** *pl.* **species,** *abbr.* **spp.**
The basic unit of taxonomic classification that is
a group of individuals, with common biological
characteristics, that are capable of interbreeding
and producing fertile offspring.
A taxonomic species has two names, the first is the
genus name, the second the species name (specific
epithet), as *Geranium robertianum.*
A phylogenetic species is the smallest set of
individuals that share a common evolutionary
ancestor.
see **taxonomic hierarchy**

species complex   A group of closely related
species, in which the limits between them are
unclear, that have been assigned to a single species.

specific epithet   Of a scientific name, the word
denoting species that follows the name of the
genus, as *Geranium robertianum.*
see **binomial nomenclature**

specimen   A plant, or part of a plant, collected and
preserved for scientific study, that is usually lodged
in a herbarium.

sperm cell, spermatogenous cell
In ferns, mosses, liverworts and hornworts, a male
gamete (sex cell) that unites with the female
gamete (egg cell) to form a zygote.
In angiosperms, one of two male microgametes,
derived from the generative cell in the pollen grain;
one unites with the female egg cell to form a zygote
and the other unites with the central cell to form
the seed's endosperm.
In gymnosperms, one gamete unites with the egg
cell and the second cell disintegrates.
see **microgametogenesis, pollen tube**
see also **double fertilisation, generative cell, spore**
cf. **egg cell, megagamete, ovum**

-sperma, -spermous   A suffix meaning seed.

Spermatophyta, spermatophytes
The seed plants that comprise the flowering plants
(angiosperms) and a group of non-flowering plants
(gymnosperms) that includes conifers and cycads.
see **euphylophytes**
= **phanerogam**
cf. **cryptogam**

spermatozoid, *pl.* **spermatozoa**
A motile male gamete, that moves by means of
whip-like hairs (flagellae).
In bryophytes and ferns it is produced in the
antheridia.
Also found in some gymnosperms, where it is
formed in the pollen tube prior to fertilisation.
= **antherozoid**
cf. **oosphere**

spermoderm   Seed coat.

sphere   A perfectly
round object but three-
dimensional as a ball.
A three-dimensional
surface with every point
on the surface equidistant
from the centre.
cf. **globose**
spherical   Relating to a
sphere.

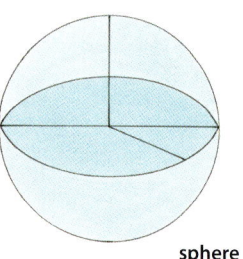

sphere

spheroidal

Of an object with the
polar axis and the
equatorial diameter
more or less equal.

= **isodiametric**

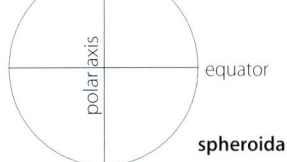

spheroidal

spicule   A spikelet.

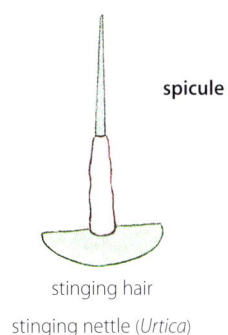

spicule

One of the minute silica
stinging hairs on the
leaves and stems of
stinging nettles (*Urtica*).

spiculate   Covered
with small spicules.

spicular   Resembling a
spicule. Bearing spicules.

stinging hair

stinging nettle (*Urtica*)

spike

A racemose inflorescence in which the floral axis
bears single sessile flowers.

The axis continues to grow indefinitely with the
youngest flower on the stem near the growing
point at the apex and the oldest at the base.

A spike can be simple or compound.

Of grasses (Poaceae) and some sedges
(Cyperaceae), the arrangement of sessile
spikelets (rather than flowers) on the axis of the
inflorescence.

An indeterminate or indefinite inflorescence.

see **acropetal, centripetal**

see also **cymose inflorescence**

cf. **digitate inflorescence, panicle, raceme**

spicate   Spike-like.

Of an inflorescence, arranged in a spike or spikes.

spiciform   Having the appearance, but not
necessarily the structure, of a spike.

Spike

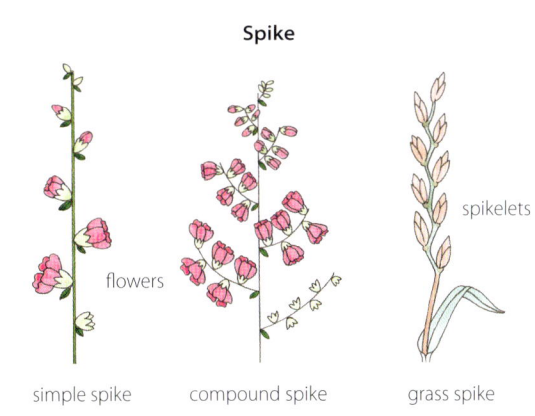

flowers

simple spike   compound spike   grass spike

spikelets

spikelet   A little spike.

A secondary spike in a compound spike.

In grasses (Poaceae), the basic unit of the
inflorescence, typically composed of two bracts
(glumes) at the base of an axis (rachilla), with one or
more florets arranged alternately in two ranks.

= **locusta**

Spikelet

florets

rachilla

spikelet

glume   glume

compound spike   grass spikelet

spike moss

The spike moss
family (Selaginellaceae)
comprises one genus
(*Selaginella*) of mossy or
fern-like vascular plants
with single-veined leaves.
Spores are of two kinds
(heterosporous).

see **fern allies**

Selaginella **spike moss**

spindle fibre   Any of

the filaments responsible
for pulling apart the
chromosomes during
nuclear division in
meiosis and mitosis.

spindle fibre

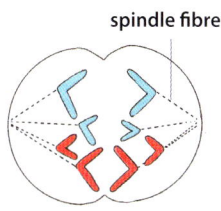

spine   A sharp hard outgrowth on a plant derived

from a modified plant organ, like the petiole or
stipule of a leaf, as those on barberry (*Berberis
vulgaris*).

The modified leaf of a cactus.

spine page 276 (cont.)

cf. **thorn**

spinescent   Ending in a spine, modified to form a
spine, tending to be spiny, as the margins of some
leaves.

spiniform   Shaped like a spine.

spinose, spinous   Bearing spines.

Spine

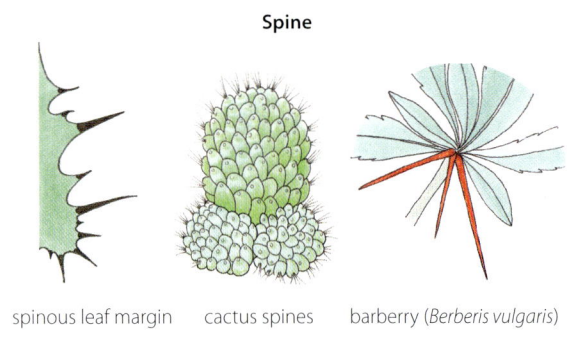

spinous leaf margin   cactus spines   barberry (*Berberis vulgaris*)

spinule   A small spine.
 spinulose   Bearing small spines.

spiracle   One of the mucilaginous threads released from hairs on the surface of the testa of some seed when they become wet, as gilia (*Giliastrum*).

spiral   A coil.
cf. **verticil, whorled**
 spiralled   The shape of or following the path of a coil, as leaves arranged around an axis.
see **phyllotaxy**
cf. **verticillate, whorled**

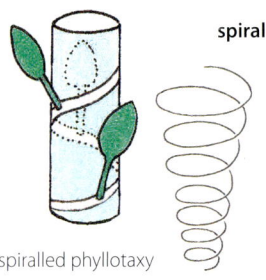

spiral

spiralled phyllotaxy

splitter   A taxonomist who uses subtle differences to subdivide species, resulting in a larger number of taxa.

spongiose   Porous in texture, compressible and absorbent.
 = **spongy**

spongy   Porous in texture, compressible and absorbent.
 = **spongiose**

spongy mesophyll   The layer of phyotosynthetic parenchyma cells internal to the lower epidermis of a eudicot leaf.
The irregularly shaped cells are separated by air spaces that allow for exchange of gases.
Spongy mesophyll cells communicate with guard cells, causing them to open or close the stomata depending on the concentration of gases in the air spaces.

Spongy mesophyll

waxy cuticle
epidermis

palisade mesophyll

spongy mesophyll

epidermis with guard cells

Eudicot leaf

spongy parenchyma   Another term for spongy mesophyll.

sporadic   Occurring at intervals that have no apparent pattern.

sporangiophore   The organ bearing sporangia on the cone of horsetails (*Equisetum*).

Sporangiophore

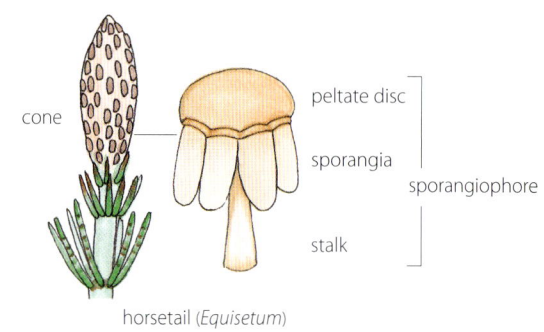

cone

peltate disc

sporangia

sporangiophore

stalk

horsetail (*Equisetum*)

sporangium, *pl.* **sporangia**   A structure in which spores are formed.
Seed plants (angiosperms and gymnosperms) have two kinds of sporangia: the male microsporangium and the female megasporangium.
In angiosperms, the microsporangia are the pollen sacs in the anther and in gymnosperms the microsporangia are the pollen sacs borne on the lower surface of the cone scales.
In angiosperms and gymnosperms, the megasporangium is the nucellus in the ovule.
In angiosperms, the ovule is enclosed in the ovary.
In gymnosperms, the ovule is exposed on a cone scale.
In ferns, the sporangia are capsules clustered in sori on the underside of the fronds.
In bryophytes (mosses, liverworts and hornworts), the sporangium structure is a capsule.
see **heterospory, homospory**

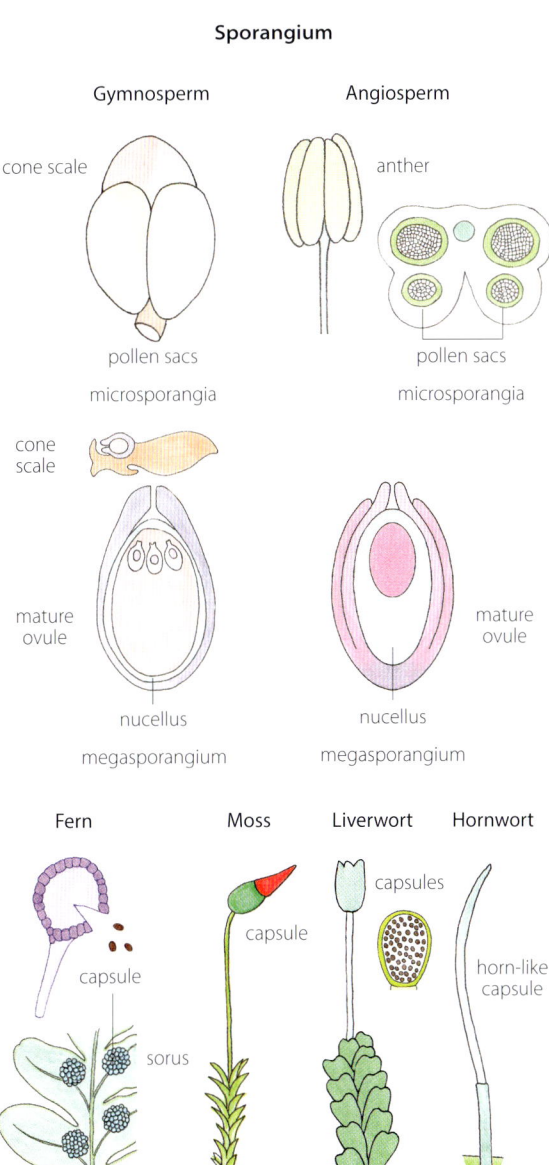

Sporangium

Gymnosperm

cone scale

pollen sacs
microsporangia

Angiosperm

anther

pollen sacs
microsporangia

cone scale

mature ovule

nucellus
megasporangium

mature ovule

nucellus
megasporangium

Fern

capsule

sorus

Moss

capsule

Liverwort

capsules

Hornwort

horn-like capsule

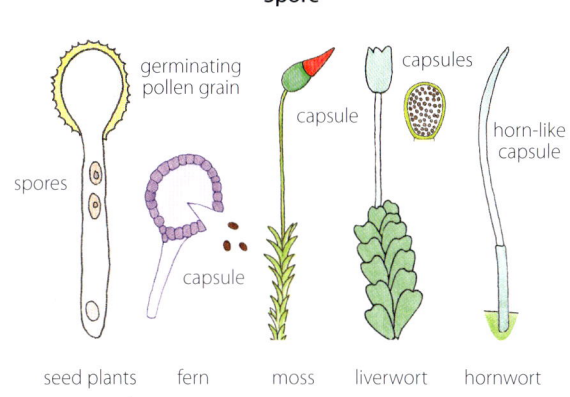

Spore

germinating pollen grain

spores

capsule

capsule

capsules

horn-like capsule

seed plants   fern   moss   liverwort   hornwort

Spore-bearing plants

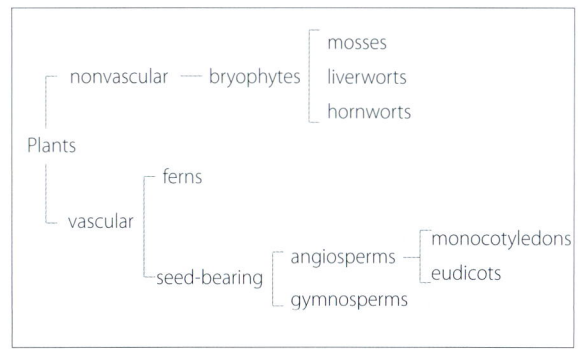

Plants
— nonvascular — bryophytes — mosses / liverworts / hornworts
— vascular — ferns
— seed-bearing — angiosperms — monocotyledons / eudicots
— gymnosperms

spore mother cell   A diploid cell in spore-bearing plants that gives rise, by meiosis, to four haploid spores.

= **sporocyte**

see also **megaspore mother cell, microspore mother cell, spore, sporogenesis**

sporiferous   Bearing spores.

sporocarp   A fruiting body that produces and releases spores in some heterosporous aquatic ferns like water clover (*Marsilea*).
Found only in the aquatic fern families Azollaceae, Marsileaceae and Salviniaceae.
The sporocarp includes one or more female megasporangia and/or several to many male microsporangia.

see **conceptacle, megasporocarp, microsporocarp**

Marsilea

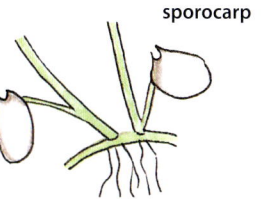

sporocarp

spore   A reproductive cell capable of developing into a new individual, either directly or after fusion with another spore.
In seed plants (angiosperms and gymnosperms), one of the two haploid cells in the pollen grain; one (the male gamete) unites with the female egg cell (the female gamete) to form a zygote.

= **sperm cell**

In nonvascular plants, like mosses, liverworts and hornworts, and seedless vascular plants, like ferns, clubmosses and whisk ferns, the haploid reproductive cell produced in the capsule (sporangium) that germinates to form the male and/or female haploid gametophyte.
Gametophytes produce male and female gametes.

see **megaspore, microspore**

see also **alternation of generations, heterosporous, homosporous**

cf. **gamete**

sporocyte   A diploid cell in spore-bearing plants that produces four haploid spores by meiosis.
= **spore mother cell**
see also **spore, sporogenesis**

sporoderm   The wall of a spore or pollen grain that encloses the cytoplasm.
A spore wall consists of perispore, exospore and endospore.
A pollen grain wall usually consists of exine and intine.
see also **pollen wall**

Sporoderm

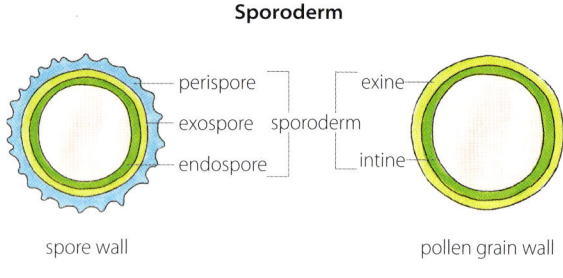

spore wall   pollen grain wall

sporogenesis   In plants, the process of forming haploid spores, by meiosis, from a diploid spore mother cell.

Sporogenesis

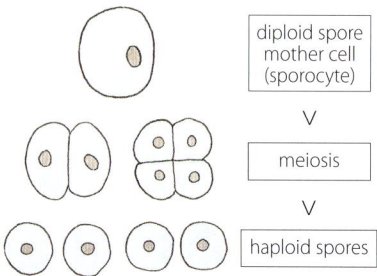

sporogenous   Producing or reproducing by means of spores.
Of tissue, spore producing, as that in an immature pollen sac.

sporophore   Of the adder's tongue family (Ophioglossaceae), the fertile spore-bearing part of a frond that shares a common stalk with the vegetative leaf-like part of the frond.
cf. **trophophore**

sporophore

trophophore

sporophyll   *see* page 279

sporophyte   *see* page 280

sporophyte generation   The usually diploid phase of a plant's life cycle that arises from the zygote.
see **alternation of generations, sporophyte**
cf. **gametophyte generation**

sporophytic apomixis   Of flowering plants (angiosperms), the production of an embryo directly from cells of the nucellus or integuments in the ovule.
A form of agamospermy.
= **adventitious embryony**
cf. **gametophytic apomixis**

sporopollenin   The main component of the tough outer wall (exine) of plant spores and pollen grains.
see **perispore, sporoderm**

sport   An individual or part of a plant, such as a shoot leaf or flower, that differs from the normal state of the rest of the species.
It is usually transient or it may be propagated vegetatively to form a new cultivar.

spreading   More or less at right angles.
Extending horizontally, as branches on a tree or leaves on a stem.

spreading

sprout   The shoot of a plant.
To send out new growth, as new growth on a cut down tree or eyes on a potato.
To shoot up suddenly.
Of a seed, to germinate.

Sprout

potato   seed

sporophyll   A modified leaf that bears spore-producing structures.

In angiosperms, ovules and the anthers are modified leaves that bear the spore-producing structure.

In gymnosperms, sporophylls are the cone scales that bear naked ovules or seeds on the female cone and pollen sacs on the male cone, as the genus of pines (*Pinus*).

In ferns, sporophylls are the fertile fronds that bear spore-producing sporangia in sori on the lower surface.

see **megasporphyll, microsporophyll, sporangium**
cf. **trophophyll**

Sporophyll

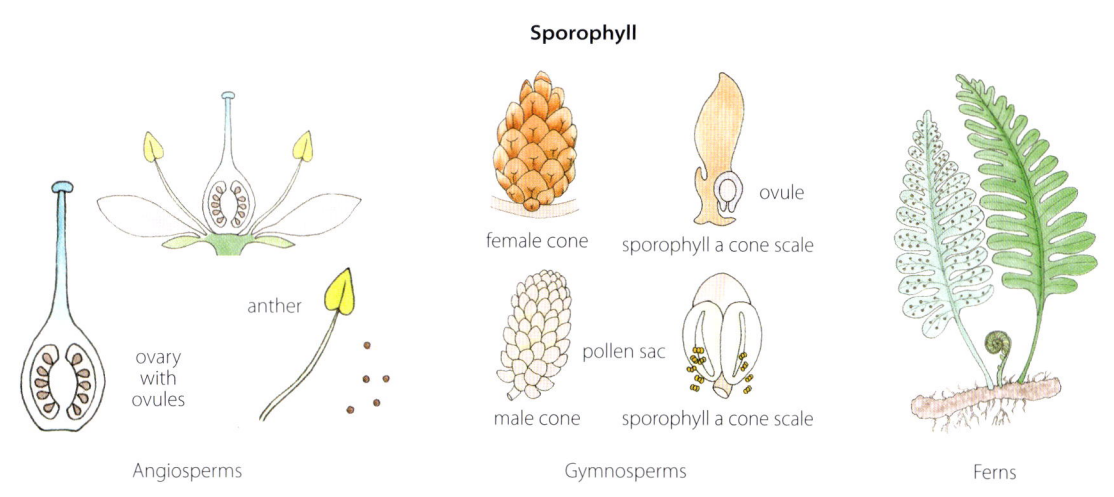

female cone

sporophyll a cone scale

ovule

anther

ovary with ovules

male cone

pollen sac

sporophyll a cone scale

Angiosperms

Gymnosperms

Ferns

spur   A slender tubular projection, especially of the calyx or corolla, that typically contains nectar, as the calyx of nasturtium (*Tropaeolum*).
spurred   Having a spur.
= **calcarate**

spur

squamula   One of usually two minute scales at the base of the pistil in a grass floret.
It swells to push the bracts (lemma and palea) apart during flowering.
= **lodicule**

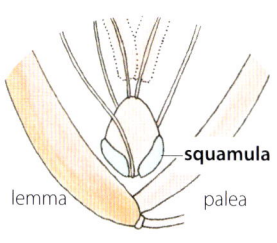

lemma

squamula

palea

squama, *pl.* **squamae**
Any kind of scale.
squamate, squamose, squamous   Covered with scales. Consisting of scales. Scale-like.
squamiform   Shaped like a scale.

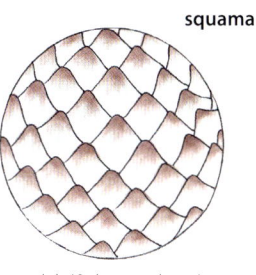

squama

salak (*Salacca zalacca*)

squarrose, squarrous
With a rough surface due to the reflexed tips of appendages like scales and bracts, as the involucral bracts of some daisies (Asteraceae).

squarrose

squamella, *pl.* **squamellae, squamula,**
pl. **squamulae, squamule**
A little scale.
Of grasses, a lodicule.
squamellate, squamulose   Covered with little scales.

squarrulose   Minutely or slightly squarrose.

stalk   A stem-like support.

stamen   The male reproductive organ of a flower. Usually a bilobed, pollen-bearing anther on a stalk-like filament.
stamen page 281 (cont.)
see also **androecium**
cf. **pistil, carpel**
staminal   Attached to or relating to the stamens.
staminoid   Resembling a stamen.

sporophyte   All plants have a life cycle alternating between a haploid gametophyte generation and a diploid sporophyte generation.

A sporophyte produces either spores or seeds.

In nonvascular plants, the tiny sporophyte grows on the larger gametophyte plant. It produces spores.

In vascular plants, the sporophyte generation is the larger familiar green plant. The gametophyte is microscopic and lives on or in the sporophyte. It produces seeds.

see **alternation of generations**

cf. **gametophyte**

Sporophyte

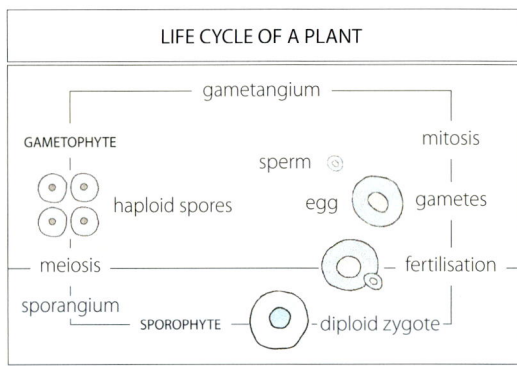

LIFE CYCLE OF A PLANT

Sporophyte of nonvascular plants that produce spores

Sporophyte of vascular plants that produce seeds

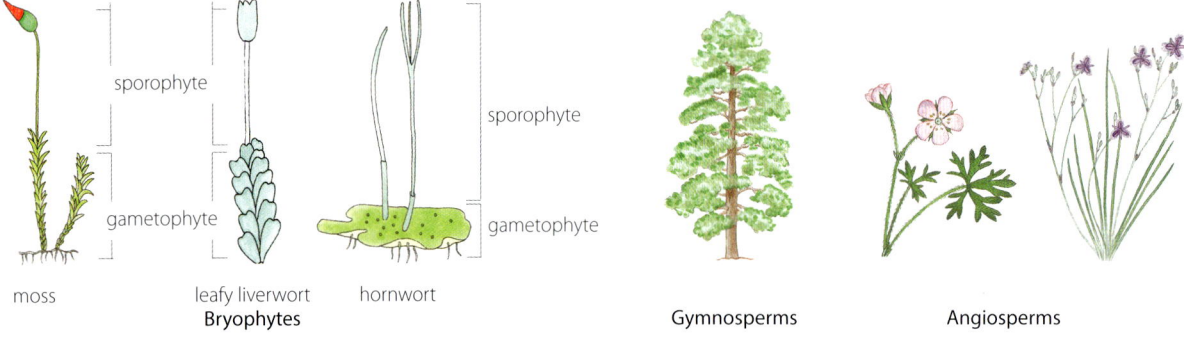

moss   leafy liverwort   hornwort
             Bryophytes

Gymnosperms   Angiosperms

Sporophyte of vascular plants that produce spores

Ferns

Pteridium   *Nephrolepis*   *Adiantum*   *Cyathea*   *Osmundia*   *Gleichenia*

Fern allies

Lygodium   *Equisetum*   *Psilotum*   *Isoetes*   *Selaginella*   *Lycopodium*

Stamen

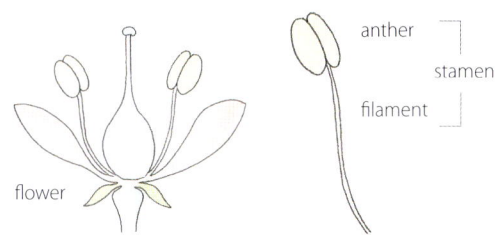

anther
stamen
filament

flower

Standard

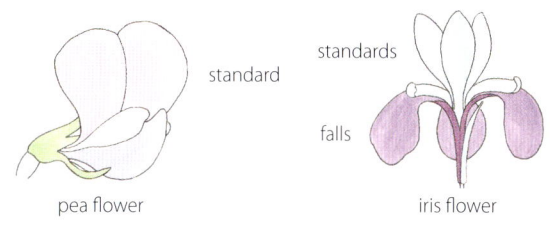

standard
standards
falls

pea flower
iris flower

staminal column

A tube formed by the fusion of the stamen filaments for much of their length.
Characteristic of, but not confined to, the mallow family (Malvaceae).

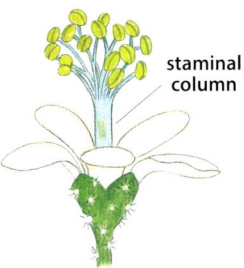

staminal column

staminate
Having stamens and no carpels, as the unisexual male flower of seaberry saltbush (*Rhagodia candolleana*).

cf. **pistillate**

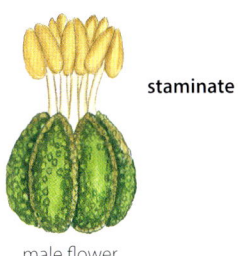

staminate

male flower

staminate cone
The male, pollen-producing cone of gymnosperms.

= **microstrobilus**
cf. **ovulate cone**

staminode, staminodium
pl. **staminodia**

A stamen that lacks an anther.
A sterile stamen.

staminodal   Relating to staminodes.

cf. **ovulode**

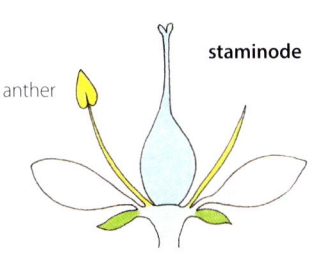

staminode

anther

staminody
The abnormal development of floral parts into stamens or staminoid organs.

standard
The large upper petal (banner or vexillum) of a pea flower (Faboideae).
One of the three inner more or less erect petals of an iris (*Iris*) flower as distinct from the three falls.

see **banner, vexillum**

starch
A complex carbohydrate produced by most green plants as a form of energy storage.
Found in seeds, fruits, tubers, roots etc.

= **amylum**

starch grain
A granule inside an amyloplast in which starch is stored.

starch grains

amyloplast

stat. nov. *abbr.*, status novus
A taxon that has been given a new rank in the taxonomic hierarchy, such as a subspecies that is raised to the rank of species or a species lowered to the rank of subspecies.

state
An identifying characteristic or quality, as having either a superior or inferior ovary.

status
In nomenclature, the standing of a taxon with regard to publication, legitimacy or correctness.
Rank of a taxon in the taxonomic hierarchy, that is, whether it is a species, genus etc.

status novus, *abbr.* stat. nov.
A taxon that has been given a new rank in the taxonomic hierarchy, such as a subspecies that is raised to the rank of species or a species lowered to the rank of subspecies.

stele
The central core of stems and roots comprising the vascular tissue and other tissues like pith.
The Theory of Steles describes the many different types of steles, including protostele, siphonostele, eustele and atactostele.
Different types of stele may be found in different regions of the same plant.

stele page 282 (cont.)

= **vascular cylinder**

Stele

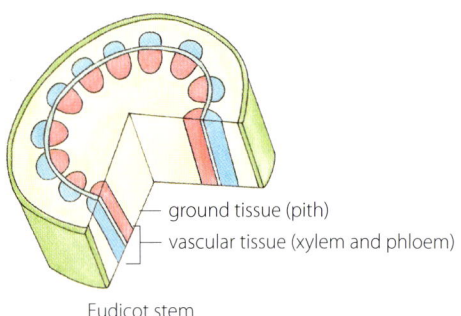

ground tissue (pith)

vascular tissue (xylem and phloem)

Eudicot stem

stellate Shaped like a star with points radiating from a common centre, as a sessile or stalked stellate hair.

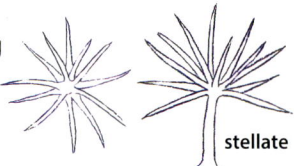

stellate

stem The main axis of a plant, typically above ground and bearing leaves and buds, but sometimes underground, as rhizomes. Develops from the plumule of the seed.
stem tissue *see* **tissue**

stem tuber
A thickening of an underground stem that functions as a storage organ. Stem tubers do not produce new tubers or offsets but have growing points called eyes from which the plant reproduces vegetatively, as potatoes (*Solanum tuberosum*).
cf. **tuberoid, tuberous root**

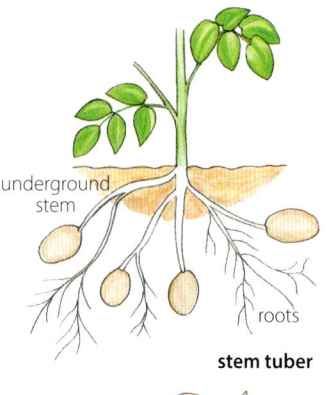

underground stem

roots

stem tuber

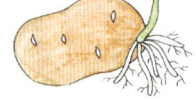

potato (*Solanum tuberosum*)

stenospermocarpy The development of fruit that is seedless from fruit that is pollinated and fertilised normally but the embryonic seed is aborted, as seedless grapes.
cf. **parthenocarpy**

stenothermal Able to tolerate growing in only a narrow range of temperature.

stephano- A prefix meaning situated on the equator.

stephanoaperturate Of a pollen grain with apertures situated at the equator.

steppe Treeless temperate semi-arid plains of low grasses that extend across southeastern Europe and Siberia.
see **biome, prairie**

stereome A strip of strengthening tissue in the bracts (phyllaries) of some daisies (Asteraceae), as lemon beauty heads (*Calocephalus citreus*).

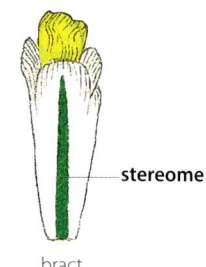

stereome

bract

stereomorphic Of a flower with basic radial symmetry that is three-dimensional, as the daffodil genus (*Narcissus*).

stereomorphic

sterigma, *pl.* **sterigmata** Of some conifers, the persistent, peg-like leaf base that remains on the twig after leaves fall.
see **brachyblast**

sterigma

sterile Unable to reproduce sexually. Of flowers lacking functional stamens and pistils, as the sterile flower of hydrangea (*Hydrangea*). Not producing viable seeds or fruit. Of shoots, branches, bracts etc. not producing flowers.
= **infertile**
cf. **fertile**

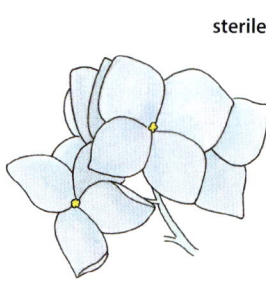

sterile

sterile flowers of *Hydrangea*

sterile bract
A bract that does not bear a flower.
cf. **fertile bract**

fertile bracts

sterile bract

stigma, *pl.* stigmata, stigmas

The pollen-receptive part of the pistil, commonly at the tip of the style, where pollen germinates.
It is variously shaped, including: capitate, clavate, crestate, decurrent, discoid, fimbriate, indusiate, lobed and plumose.
stigmatic   Of or relating to the stigma.

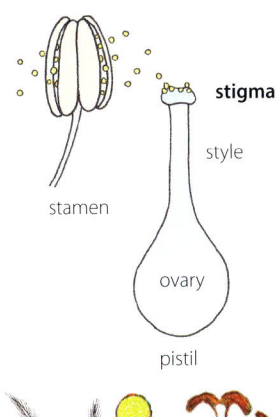

stigma
style
stamen
ovary
pistil

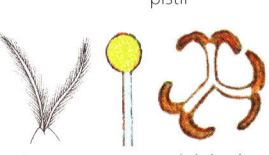

plumose   capitate   bilobed

stilt roots   Adventitious

roots that grow out from the lower trunk of a tree and into the soil to provide support.
Found in some mangroves as (*Rhizophora*) and the banyan (*Ficus benghalensis*).
= **prop roots**
cf. **buttress root**

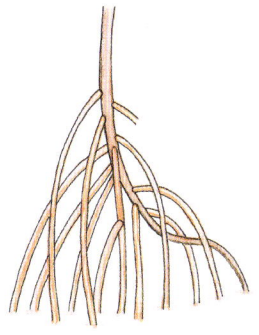

stilt roots

stipe   A small stalk that supports some other

structure.
Of flowers, the gynophore that bears the pistil above the other floral parts.
Of ferns, the petiole connecting the lamina of the frond to the rhizome.
The 'trunk' of a palm.
Of orchids, a strap or stalk, formed from columnar tissue, that connects the pollinia to the viscidium.
see also **caudicle, hamulus, tegula**
stipitate   Having a stipe.

Stipe

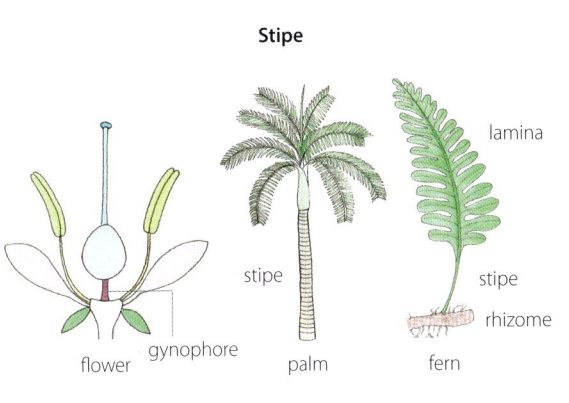

lamina
stipe
rhizome
flower   gynophore   palm   fern

stipel   A small

secondary stipule often found at the base of a leaflet or petiolule in a compound leaf.
stipellate   With stipels.
cf. **exstipellate**

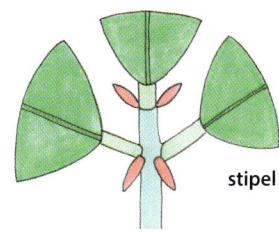

stipel

stipule   *see* page 284

stock   In grafting, a plant grown from seed that has

strong roots, vigour and a suitable habit, onto which a bud or scion from another plant is implanted, as tomato cultivars that are grafted onto a stronger species of the tomato genus (*Lycopersicum*).
A plant that provides slips or cuttings.
That part below the graft that gives rise to the lower main stem and root system of the new plant

stolon   A horizontal stem that grows at ground

level (usually above or slightly below) and produces roots and shoots at the nodes or at the tip.
= **runner**
cf. **rhizome**
stoloniferous   Bearing or producing stolons.
stoloniform   Resembling a stolon.
stolonoid   Spreading like a stolon.

Stolon

stolon

stolonoid   Of orchids,

a roots that grow sideways and terminates in reproductive daughter orchids that form colonies, as some greenhood orchids (*Pterostylis*).

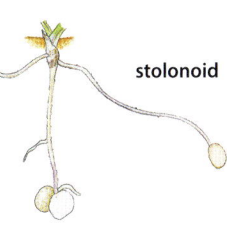

stolonoid

stipule An appendage at the base of the leaf-stalk (petiole), typically in pairs. Free or variously united to each other or to the petiole.
Shape and texture is variable, as lobed, spiny, membranous or tendril-like.
stipular Borne on or relating to stipules.
stipulate With stipules.
cf. **exstipulate**
stipuloid Resembling stipules.

Stipule

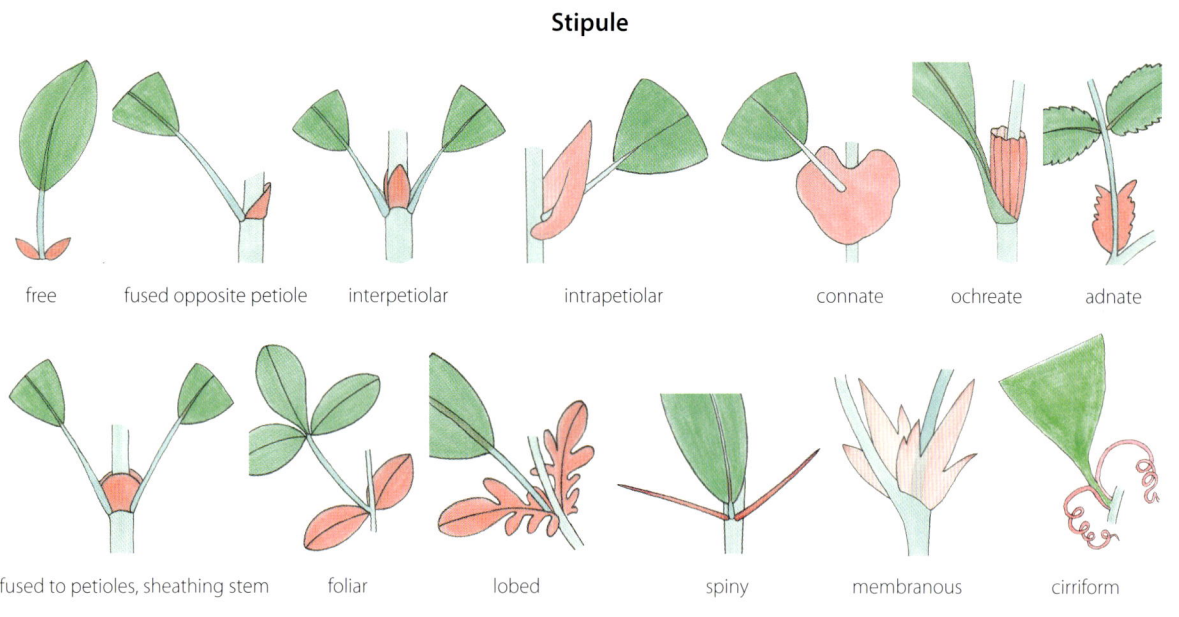

free   fused opposite petiole   interpetiolar   intrapetiolar   connate   ochreate   adnate

fused to petioles, sheathing stem   foliar   lobed   spiny   membranous   cirriform

stoma, *pl.* **stomata** A pore on the surface of a leaf that is surrounded by specialised guard cells. Guard cells regulate the opening and closing of a stoma and thus control the exchanges of gases and water vapour between the leaf and the atmosphere. Monocotyledons have stomata on both the upper and lower surfaces of their leaves and eudicots mostly have stomata only on the lower surface.

Stoma

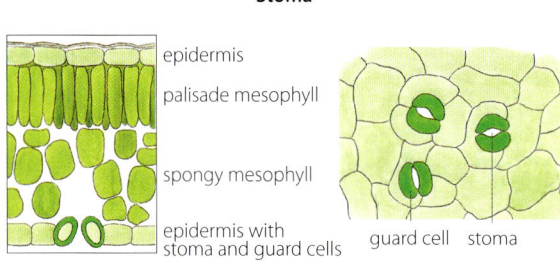

epidermis
palisade mesophyll
spongy mesophyll
epidermis with stoma and guard cells
guard cell   stoma

stomial groove In flowering plants, the region of dehiscence in a pollen sac through which spores are released from an anther.

Stomial groove

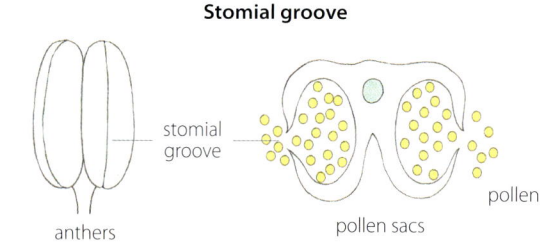

anthers   stomial groove   pollen sacs   pollen

stomium, *pl.* **stomia** The region of dehiscence of a sporangium or pollen sac.
In ferns, the opening in the lip cells of the annulus surrounding the sporangium through which spores are released.

Stomium

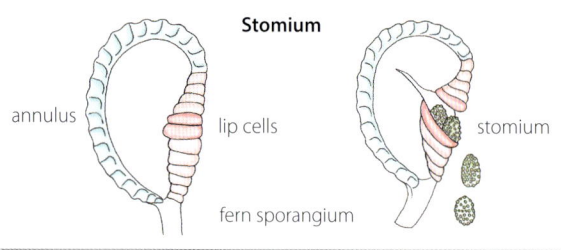

annulus   lip cells   stomium
fern sporangium

stone The hard endocarp and the enclosed seed in a cherry, plum or other fruit.
see also **pericarp**

stool A stump, woody base or root that annually produces shoots or suckers.
see also **layering**
stooling To produce shoots or suckers from a stool.

shoots
stool
sucker

stool layering A form of propagation whereby a plant is cut back to near ground level and covered with layers of soil as new shoots develop. Rooted shoots are later separated and grown as new plants.
= **mound layering**

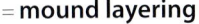
stool layering

storey One of the layers in a forest.
The canopy is more or less continuous and is made up of the crowns of trees.
The emergent layer consists of scattered taller trees that extend above the canopy.
The understorey receives less light and consists of shorter trees, shrubs and saplings.
The ground layer that is usually herbaceous but may also include prostrate woody plants.
Not all layers are represented in all forests.

Storey

emergent layer

canopy

understorey

shrub layer

ground layer

stramineous Like straw.
Straw-coloured, pale yellow.

strangler A plant that germinates in the canopy of a tree and sends aerial roots down into the ground. The host tree is starved of light and nutrients and eventually dies but the strangler continues to thrive independently, as strangler figs (*Ficus*).

Strangler

stria, *pl.* **striae** A fine groove or thread-like line.
striate Marked with striae.

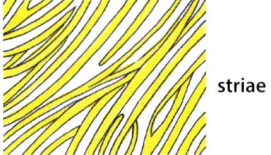
striae

striga, *pl.* **strigae** A sharp-pointed, rigid appressed bristle or hair.
strigose Covered with strigae.

strigae

strigulose Minutely or finely strigose.

strobile, strobilus, *pl.* **strobili** Of gymnosperms, the unisexual cone-like reproductive structure, typically with a central axis having spirally arranged scales.
The reproductive structure of some fern allies, as horsetails (*Equisetum*) and clubmosses (*Lycopodium*), with compressed modified spore-bearing leaves (sporophylls) at the tips of branches.
Strobili are often called cones.
strobiliferous Bearing a strobile or strobiles.
strobiliform, strobiloid Resembling a strobilus.
see **megastrobilus, microstrobilus**

Strobilus

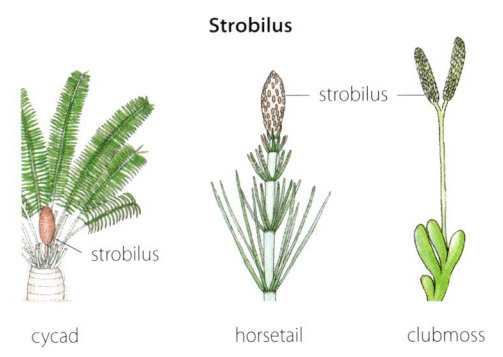
strobilus
strobilus
cycad
horsetail
clubmoss

stroma, *pl.* **stromata**
The matrix of a chloroplast in which the grana are embedded.

granum
stroma
chloroplast

strophiole Variously used as a synonym of caruncle and aril.

Specifically, an outgrowth on the testa of a seed in the region of the raphe, as greater celandine (*Chelidonium majus*).

One of the sites through which water is absorbed during seed germination, as beans (Fabaceae).

= **lens**

strophiolate Bearing or related to a strophiole.

Strophiole

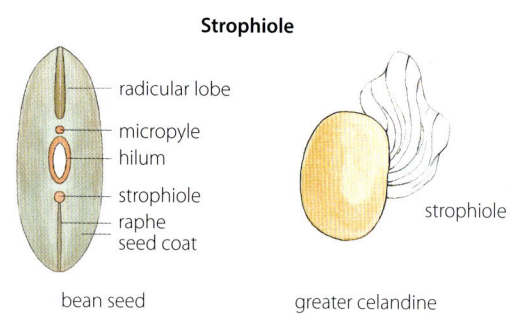

bean seed

greater celandine

struma, *pl.* **strumae**
A cushion-like swelling of or on an organ.

strumose Having a struma or strumae, as the filament of flax lilies (*Dianella*).

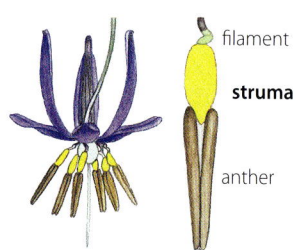

style The usually elongated part of a carpel that connects the stigma with the ovary. In orchids (Orchidaceae) it is an indiscernible part of the column.

stylar Of or relating to the style of a flower.

stylate Having a style.

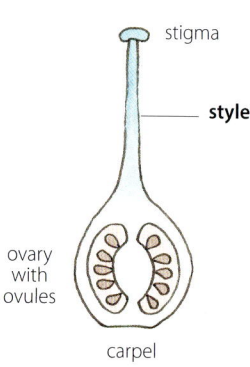

stylar canal Of flowering plants, the conducting tissue of the style through which the pollen tubes grow from the stigma to the ovary.

Stylar canal

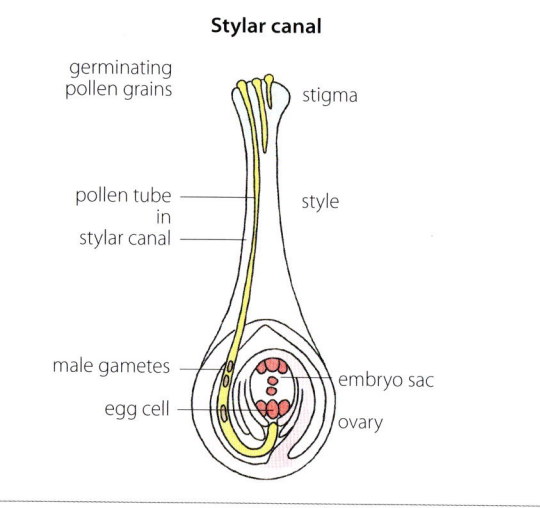

stylocarpellous
Of a carpel with a style and without a supporting stalk (stipe).
cf. **astylocarpellous, stylocarpepodic**

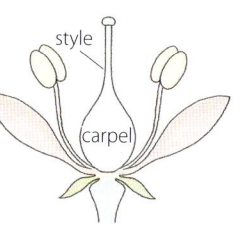

stylocarpellous

stylocarpepodic
Of a carpel with a style and a stipe.
cf. **astylocarpellous, stylocarpellous**

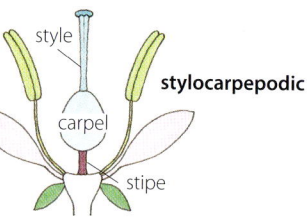

stylodious Of a flower having a single free carpel in the gynoecium.
= **monocarpellary, monocarpous, unicarpellate**

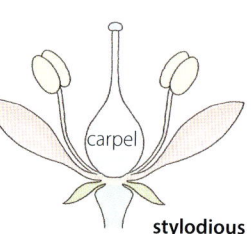

stylodium, stylodia
An elongated stigma that resembles a style, as the shrubs in the family Coriariaceae.

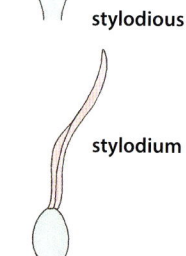

stylopodium The expanded nectar-secreting disc at the base of each of the two styles in the carrot family (Apiaceae).

suaveolent   Fragrant, sweet-smelling.

sub-   Under or less than.
Somewhat or almost, as suberect.
Under or inferior to, as subfamily that is a rank below family in taxonomic classification.

suberin   A waxy substance in the cell walls of some plant tissue, especially cork, that prevents water loss.
suberose, suberous   Relating to, resembling or consisting of cork, corky.
suberisation   The deposition of suberin in the cell walls of some plants.

submerged   Beneath the water, as the leaves of some aquatic plants.
= **submersed**
cf. **emergent, immersed**

submerged

submersed   Beneath the water, as the leaves of some aquatic plants.
= **submerged**
cf. **emergent, immersed**

suborbicular
A two-dimensional shape, with an almost circular outline.

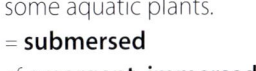
suborbicular

subradiate
Of daisies (Asteraceae), a heterogamous capitulum with the outer ray florets not exceeding the involucre.

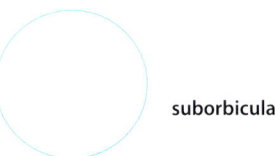
ray florets
subradiate
involucre

subshrub   A small shrub with a woody base and herbaceous new growth, as lavender (*Lavandula*) and thyme (*Thymus*).
= **suffrutex, undershrub**

subshrub

subsp., *pl.* **subspp.**   An abbreviation for subspecies.

subspecies, *abbr.* **subsp.,** *pl.* **subspecies,** *abbr.* **subssp.**   In taxonomic classification the rank below species and above variety, such as *Geranium robertianum* spp. *purpureum*.
see **taxonomic hierarchy**

subtend   To be inserted directly below a different organ or structure, as the bracts that subtend the calyx of the bindweed *Calystegia*.

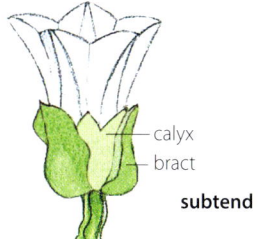
calyx
bract
subtend

subulate   Narrowly triangular and tapering gradually to a fine point.
= **awl-shaped**

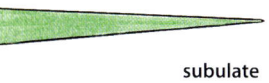
subulate

succession   In ecology, the process of change in species structure as a community establishes over time.
Primary succession occurs on previously uncolonised areas like lava flows.
Secondary succession occurs on disrupted or disturbed areas.
= **ecological succession**
see also **climax community**

Succession

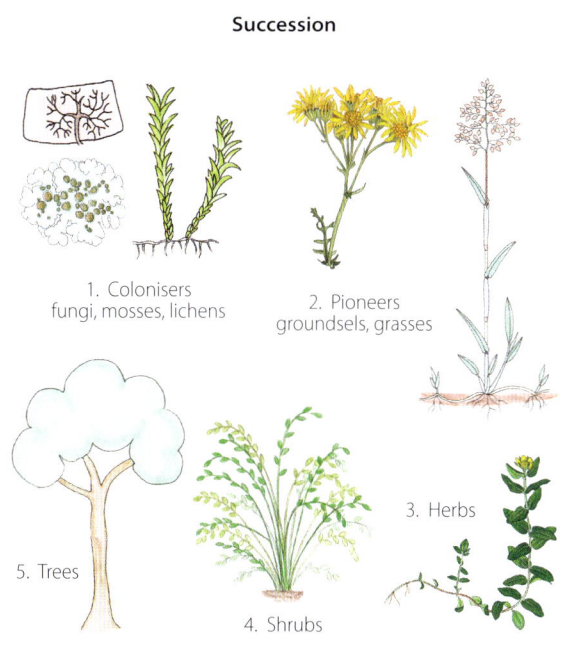

1. Colonisers fungi, mosses, lichens

2. Pioneers groundsels, grasses

3. Herbs

4. Shrubs

5. Trees

succubous   Of leafy liverworts, having leaves attached to the stem obliquely so that the upper margin of each leaf is overlapped by the base of the leaf above it. The new leaf begins above the older one.
cf. **incubous**

succubous

succulent   A plant characterised by fleshy water-storing tissues that allows it to survive arid conditions, as the pigface family (Aizoaceae). A common term for a member of the cactus family (*Cactaceae*). Having juicy fleshy tissue, as a peach.

succulent

sucker   A shoot that arises from an underground root or stem that gives rise to a new plant.
= **surculus**
suckering   Having or producing suckers.

shoots
stool
sucker

suffrutescent   Woody only at the base of the stem.
= **suffruticose**
see **suffrutex**

suffrutex   A small shrub with a woody base and herbaceous new growth, as lavender (*Lavandula*) and thyme (*Thymus*).
= **subshrub, undershrub**

suffrutex

suffruticose   Woody only at the base of the stem.
= **suffrutescent**
see **suffrutex**

sulculus, *pl.* **sulculi**
An elongated latitudinal aperture on a pollen grain that is not situated at a pole.
sulculate   Having a sulculus.
cf. **sulcus**

sulculus

sulcus, *pl.* **sulci**
A groove or furrow. The indentation along the side of the culm in the bamboo genus *Phyllostachys*. An elongated latitudinal aperture situated at the distal or proximal pole of a pollen grain.
sulcate   Having a sulcus.
cf. **colporus, colpus, pore, sulculus**

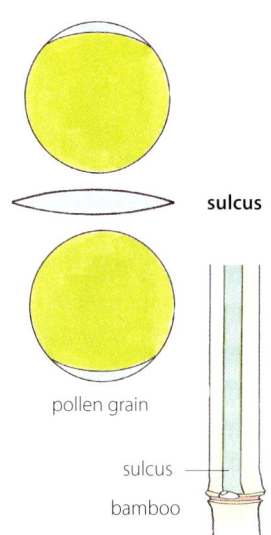
sulcus
pollen grain
sulcus
bamboo

super-   A prefix meaning above, as superorder that is a rank above order and below class in taxonomic classification.

superficial placentation   Having carpels fused but the internal walls (septa) lacking, creating a unilocular ovary, with ovules attached to septa-like placentas that project from the wall of the ovary, as water lilies (*Nymphea*).
= **lamellar placentation**
see **placentation**

Superficial placentation

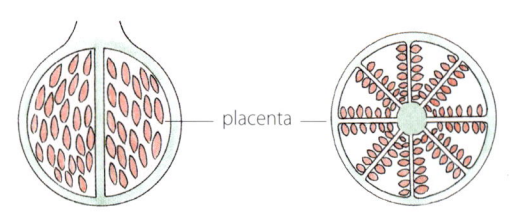
placenta

superior ovary   An ovary with the petals and sepals inserted below it on the receptacle.
cf. **inferior ovary, semi-inferior ovary**

superior ovary
ovary
receptacle

superposition

The placement of one part above another on the same radius, as stamens borne above the petals of a flower.

superposed Situated vertically on or above another part.

= **anteposition**

cf. **alternation**

superposition

supervolute

Having one margin rolled within the other.

supervolute ptyxis

Of a single leaf in bud with one margin rolled within the other.

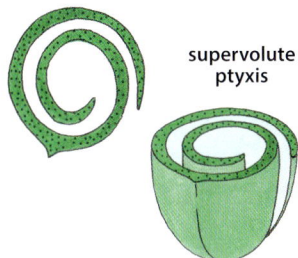

supervolute ptyxis

suppress To keep something from developing. To inhibit the growth or development of, as many eucalypts (*Eucalyptus*) suppress the growth of some species beneath their canopy.

supra- A prefix meaning above or over.

cf. **infra-**

supra-axillary Borne above the axil, as some flowers and fruit.

suprabasal

Above the base.

Of veins originating at some distance above the base of the leaf.

Of acrodromous leaf venation with the veins curving upwards from a single point above the base of the leaf.

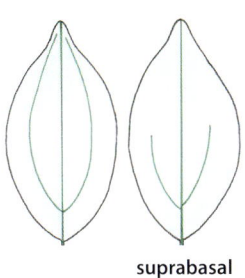

suprabasal

suprabasal-imperfect

Having veins originating at some distance above the base of the leaf and lateral veins that extend for less than two-thirds of the leaf surface.

see **acrodromous**

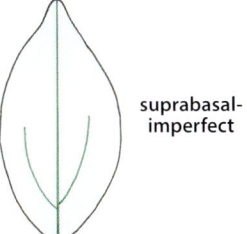

suprabasal-imperfect

suprabasal-perfect

Having veins originating at some distance above the base of the leaf and lateral veins that extend for at least two-thirds of the leaf surface.

see **acrodromous**

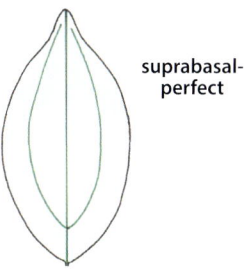

suprabasal-perfect

suprafoliar

Of an inflorescence that stands above the crown of leaves, as some palms (Arecaceae).

cf. **infrafoliar, interfoliar**

suprafoliar

surculus, *pl.* surculi

A shoot that arises from an underground root or stem that gives rise to a new plant.

= **sucker**

surculose Having or producing suckers.

shoots

stool

surculus

suspensor Of eudicot embryogenesis, a stalk-like region that develops from cell division of the zygote. It connects the nourishing endosperm to the embryo.

The suspensor degenerates when the endosperm is absorbed by the embryo.

see **embryogenesis**

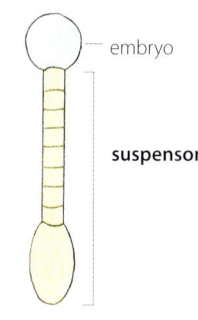

embryo

suspensor

suture The line of junction between two fused parts, as the fused valves of a capsule.

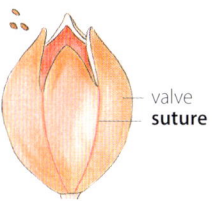

valve

suture

swamp A permanently saturated freshwater or saltwater wetland usually fringed by trees. They form around lakes or river outlets along coastlines.

cf. **bog, mangrove, marsh**

switch plant   A plant, such as broom, with leaves soon falling or absent, and photosynthesis occurring in the stems.

switch plant

sword fern
Sword ferns (*Nephrolepis*) are one of about twenty genera of mostly terrestrial ferns in the wood fern family (Dryopteridaceae). They have simple or divided tufted fronds and long-creeping to erect scaly rhizomes.
see **fern**

sword fern

sword fern (*Nephrolepis*)

syconium, *pl.* **syconia**   The fruit derived from a fleshy, hollow, inverted receptacle, bearing an inflorescence of numerous sessile flowers that develop into fruitlets.
Characteristic of the fig genus (*Ficus*).
see **composite fruit, hypanthodium**

Syconium

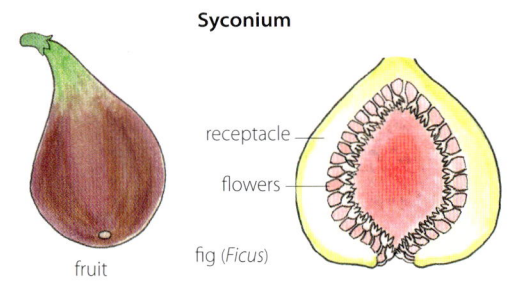

receptacle

flowers

fruit

fig (*Ficus*)

symbiont, symbiote   An organism living in symbiosis with another of a different species, as the coexistence of algae and fungi in lichens.

symbiosis   A close and mutually beneficial relationship between organisms of different species, as mycorrhiza associated with some trees.
The mycorrhizal fungus extracts nutrients from the soil for the tree and the fungus receives sugars from plant photosynthesis.
see **amensalism, commensalism, mutualism, parasitism**
symbiotic   Of or relating to symbiosis.

symmetry   The quality of being divisible into equal halves, either facing each other or around an axis.
symmetric, symmetrical
With any plane through the centre producing like halves.
cf. **asymmetric**

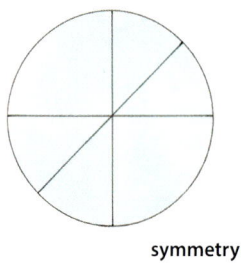

symmetry

sympatric   Of distribution, occurring in the same geographic area and having more or less overlapping ranges.
cf. **allopatric, parapatric**

sympetalous   Of a flower with petals fused, at least at the base.
= **gamopetalous**
cf. **apopetalous**

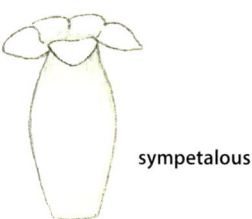

sympetalous

symplast   All of the protoplasm of all the cells in a plant that is interconnected by plasmodesmata in the cell walls.
It can be considered as one continuous mass. The symplast together with the apoplast make up the whole plant.
cf. **apoplast**

symplastic pathway   Diffusion of water and solutes through the cellulose cell wall then across the cell membrane through the cytoplasm, and from cell to cell by way of the plasmodesma.
One of the pathways of movement of water and solutes radially from the root epidermis through the cortex and endodermis to the vascular cylinder where it will be transported vertically in the xylem.
see also **apoplastic pathway, Casparian strip, symplast**

symplesiomorph, symplesiomorphy
In cladistics, an ancestral or primitive character shared by two or more taxa.
A shared plesiomorphy, as hair that is an ancestral character for all primates.
symplesiomorphic   Of or relating to a symplesiomorph.

sympodium   A pattern of growth in which the apex of the main stem ceases to grow due to the abortion of the apical bud or the development of a flower or another structure, as a tendril.
Growth continues below the apex from a succession of axillary branches with a similar growth pattern.
A rhizome that grows horizontally for a while then turns upward to form a new shoot.
The growth pattern of a non-invasive clumping bamboo.
see **leptomorph**
cf. **amphipodium, pachymorph, sympodium**
sympodial   Lacking a persistent terminal growing point and having growth occur in successive lateral branches, as elms (*Ulmus*).
Of determinate or definite growth, as a cymose inflorescence.

Sympodium

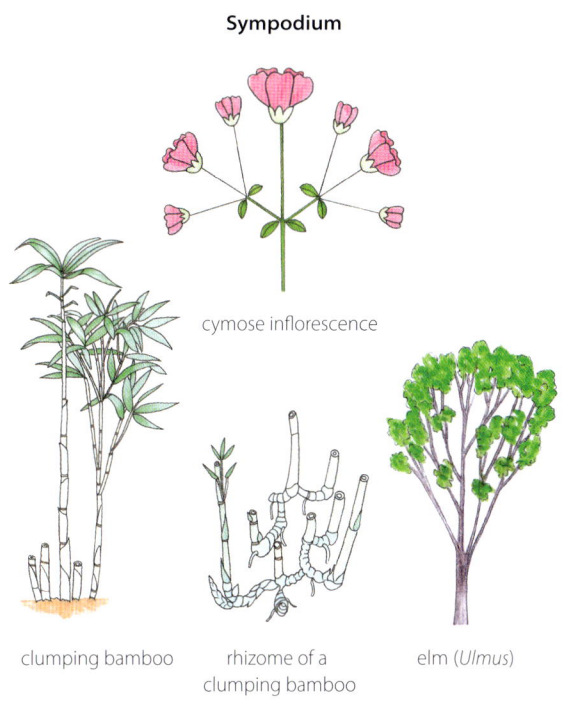

cymose inflorescence

clumping bamboo   rhizome of a   elm (*Ulmus*)
                        clumping bamboo

syn-   A prefix meaning with or together.

synandrium   The structure formed by stamens that are partially or completely fused along both their filaments and anthers, as the Aroideae genus *Taccarum*.
cf. **syngenesious**
synandry   Fusion of the stamens into a tube-like structure.
synandrous   Of stamens partially or completely fused along both their filaments and anthers.

Synandrium

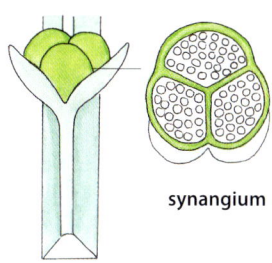

stamens   anthers

filaments

synangium,
pl. **synangia**
A structure formed by the fusion of two or more sporangia, as in whisk ferns (*Psilotum*) and the staghorn fern genus (*Platycerium*).

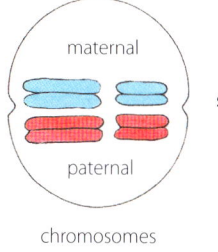

synangium

synanthesis   Having the anthers and stigmas of a flower mature at the same time.
cf. **synanthous**

synanthous   Having flowers and leaves appear at the same time.
cf. **hysteranthous, precocious, synanthesis**

synanthy   Abnormal fusion of two or more flowers.

synapomorph, synapomorphy   In cladistics, a character, derived from a common ancestor, that is shared between two or more taxa.
An apomorphy that is found in two or more species.
synapomorphic   Of or relating to a synapomorph.

synapsis
The time early, in meiosis, when homologous chromosomes that are usually separate lie side by side in pairs.
see **bivalent, univalent**

maternal

synapsis

paternal

chromosomes

syncarp A fruit formed from two or more fused carpels of a single flower, as flax (*Linum*), or the fused carpels of all the flowers of an inflorescence, as mulberry (*Morus*).
see **multiple fruit, syncarpous**
cf. **aggregate fruit, apocarp**

Syncarp

From a single flower
flax (*Linum*).

From an inflorescence
mulberry (*Morus*)

syncarpous
Of a flower having a compound gynoecium of two or more carpels with the ovaries styles and stigmas fused together to form one unit.
Of a fruit formed from the fused carpels of a single flower or of an inflorescence.
see **syncarp**
cf. **apocarpous, paracarpous, polygynous, semicarpous, synovarious**
syncarpy The condition of being syncarpous.

Syncarpous

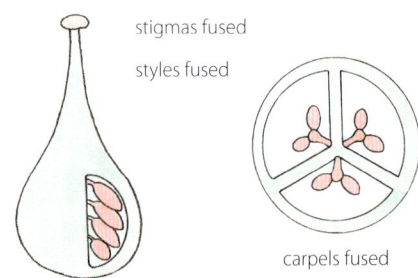

stigmas fused
styles fused
carpels fused

syndrome A set of conditions that form an identifiable pattern, as a system of pollen transfer.
see **pollination syndrome**

synergid One of two cells on either side of the egg cell at the micropylar end of an embryo sac.

Synergid

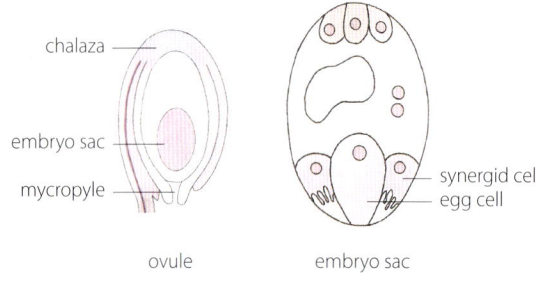

chalaza

embryo sac

mycropyle

synergid cell
egg cell

ovule

embryo sac

synflorescence
A compound inflorescence, composed of a terminal inflorescence and one or more lateral inflorescences.
see also **coflorescence, paraclade**

synflorescence

syngenesious
Of stamens with anthers united in a tube and filaments free, as the daisy family (Asteraceae).
cf. **synandrous**

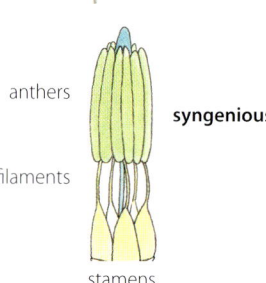

anthers

syngenious

filaments

stamens

synonym In nomenclature, one or more names for the same taxon.
A name that has been superseded and is no longer valid.

synovarious
Of a gynoecium with ovaries of adjacent carpels fused and their styles and stigmas free.
cf. **semicarpous**

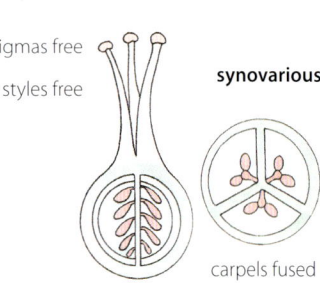

stigmas free

styles free

synovarious

carpels fused

synsepalous
With sepals fused, at least at the base.
= **gamosepalous**
cf. **aposepalous**

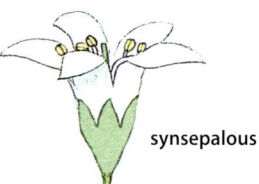

synsepalous

synsepalum

A structure formed by the fusion of two or more sepals, as the two fused lateral sepals of greenhood orchids (*Pterostylis*).

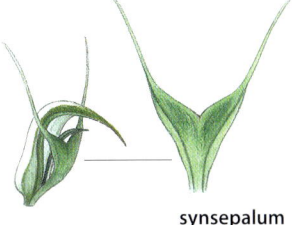

synsepalum

synstylovarious

Of a gynoecium with ovaries and styles of adjacent carpels fused and their stigmas free.
cf. **semicarpous**

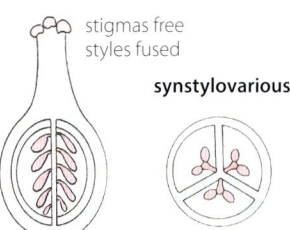

stigmas free
styles fused

synstylovarious

carpels fused

syntepalous

With tepals fused, at least at the base, as the flowers of lily of the valley (*Convallaria majalis*).
= **gamotepalous**
cf. **apopetalous**

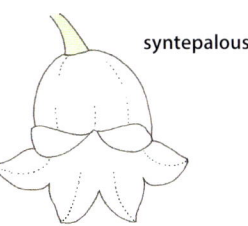

syntepalous

syntype

Any of two or more specimens used in the original description of a taxon when a holotype was not designated.

system

A group of organs that work together to carry out a particular task.
Plants have two systems with related organs.
The shoot system includes stems, leaves and flowers, and the root system includes all roots and underground structures, such as tubers.
Cells are organised into tissues, tissues are organised into organs and organs function together in systems.

systematics

The branch of botany that deals with the identification, classification, naming (nomenclature) and evolutionary relationships (phylogenetics) of plants.
see **phylogenetic systematics, taxonomy**

T-shaped tetrad

A uniplanar tetrad arranged with two members of the tetrad perpendicular to the other two so that they cohere in the shape of the letter T.
see **uniplanar, viscin thread**
see also **pollen tetrad**

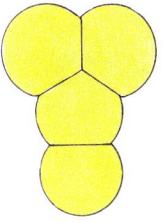

T-shaped pollen tetrad

taiga

The most northerly biome of the temperate zone that forms a nearly continuous belt across North America and Eurasia.
It is characterised by coniferous forests, long harsh winters and short summers.
see **boreal**

tannin

A brownish bitter-tasting substance made of tannic acids, present in wood-bark, leaves, roots and fruit of some plants.

tapetum, *pl.* tapeta

Of angiosperms and gymnosperms, the innermost wall of an immature pollen sac (microsporangium) that nourishes the microspore mother cells.
see **microsporangial wall**
Of ferns, the nutritive layer surrounding the developing spores in the sporangium.
tapetal Of or relating to a tapetum.

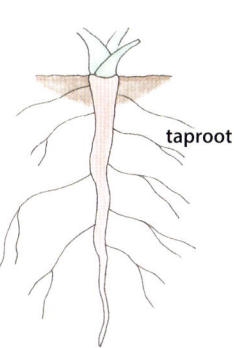

sporangium wall

tapetum

spores

fern sporangium

taproot

In eudicots and gymnosperms, an enlarged primary root together with the lateral roots that emerge from it. It develops from the radicle.
see **conical taproot, fusiform ~, napiform ~**
cf. **fibrous roots**

taproot

tartareous   Having a rough crumbling surface. Crumbling spontaneously.

tassel   The male inflorescence at the tip of a corn plant.
see **silk**

tassel

tawny   Brownish-yellow, the colour of tanned leather.

taxon, *pl.* **taxa**   In traditional taxonomy, a general term for an entity with its members having characteristics in common, as a family, genus, species or variety.
In phylogenetics, the members of a named entity with a common evolutionary history of descent, rather than shared characteristics, that may or may not be currently alive.

taxonomic hierarchy   *see* page 295

taxonomy   The branch of botany that deals with the identification, description, classification and naming (nomenclature) of plants by comparing selected characters associated with their structure, function or other attributes.
see **phylogenetics, systematics**
taxonomic   Of or relating to taxonomy.
Of the description, classification and naming of organisms.

tectum   A layer of the wall of a pollen grain that forms a roof over the columellae and other elements.
see **pollen wall**

tegmen, *pl.* **tegmina**
Of the angiosperm ovule, the inner of two integuments surrounding the nucellus.
Of the bitegmic seed coat of angiosperms, the derivatives of the inner integument of the ovule form the tegmen.
cf. **testa**
see **tegmic seed, testal seed**

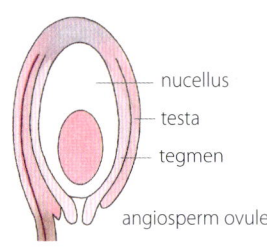

nucellus
testa
tegmen
angiosperm ovule

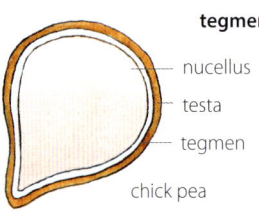

tegmen
nucellus
testa
tegmen
chick pea

tegmic seed   Of angiosperms, one having the mechanical layer in the exotegmen, the mesotegmen or the endotegmen.
cf. **testal seed**
see **endotegmic seed, exotegmic seed, mesotegmic seed**

tegula   Of orchids, a stipe formed mostly from the epidermis of the rostellum.
cf. **hamulus**

temperate   The biome that lies mainly between latitudes 30° and 60° in both hemispheres. Generally receives plentiful rainfall and has four seasons that are neither extremely hot nor extremely cold.
It includes deciduous and evergreen forests and savannas.
see **biome**

tendril   A slender coiling structure modified from a plant part (as a petiole, shoot or stem) that twines around a support.

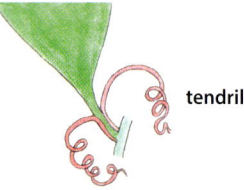

tendril

tension wood   Structurally abnormal wood formed as support in response to stress.
Found on the upper side of a lean, the side under tension, of stems in angiosperms.
see also **reaction wood, compression wood**

tenui-   A prefix meaning thin.

tenuinucellate   Of an ovule with one or few layers of cells in the nucellus.
Having a thin nucellus.
cf. **crassinucellate**

tepal   A segment of the perianth of a flower when the petals and sepals are similar in appearance, as the day lily (*Hemerocallis*).
cf. **petal, sepal**

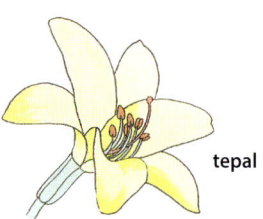

tepal

taxonomic hierarchy   A series of terms classifying plants in levels or ranks from the highest and most complex to the lowest and least complex, as kingdom, division, class, order, family, genus and species.
see **PhyloCode**

Taxonomic hierarchy of plants

Taxonomic hierarchy of plants	
Kingdom	
Subkingom	
Infrakingdom	
Superdivision	
Division	
Subdivision	
Class	
Subclass	
Superorder	
Order	
Suborder	
Family	
Subfamily	
Genus	
Subgenus	
Species	
Subspecies	
Variety	
Form	

KINGDOM
Plantae

algae   bryophytes   ferns   gymnosperms   angiosperms

DIVISION
Tracheophyta
(vascular plants)

bryophytes   ferns   fern allies   gymnosperms   angiosperms

CLASS
Magnoliopsida
(flowering plants)
eudicots

Geraniaceae   Fabaceae   Vitaceae   Ranunculaceae

FAMILY
Fabaceae

Cullen   *Acacia*   *Lotus*   *Gompholobium*   *Phaseolus*

GENUS
Acacia

Acacia melanoxylon   *Acacia acinacea*   *Acacia paradoxa*

SPECIES
Acacia acinacea

Acacia acinacea

terete   Cylindrical or slightly tapering, with a circular transverse section.

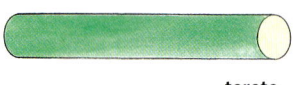

terete

tergeminate
Of a compound leaf having a pair of leaflets at the base then forking, with a pair of leaflets on each branch, as the powderpuff plant (*Calliandra tergemina*).
see **geminate**

tergeminate

terminal   At the apex. Borne at the tip of a stem, as leaves, an inflorescence or other structure.
cf. **axillary, basal, cauline, intercalary inflorescence**

terminal

ternate   Arranged in groups of three. A trifoliolate compound leaf, it can be pinnately ternate or palmately ternate.
Leaflets can be petiolulate or sessile.
see **trifoliolate**

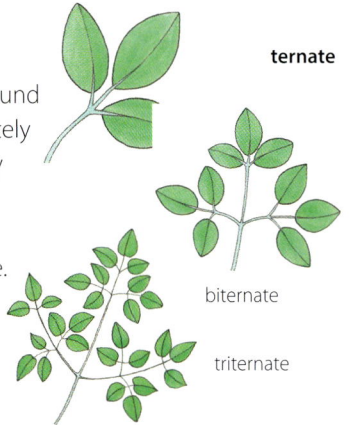

ternate

biternate

triternate

terrestrial   Of or relating to land, as opposed to an aquatic habitat.
cf. **amphibious, aquatic**

tertiary vein   A small vein.
The ultimate visible division of a vein on a leaf.
= **veinlet, venule**

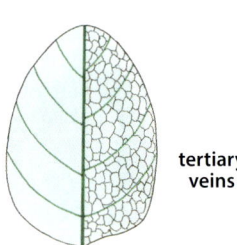

tertiary veins

tesselate, tesselated
With markings in small squares, chequered, as the tesselated venation of a leaf.

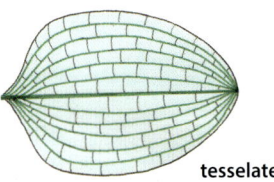

tesselate

testa, *pl.* **testae, testas**   A term generally used as a synonym of seed coat.
Of the angiosperm ovule, the outer of two integuments surrounding the nucellus.
Of the bitegmic seed coat of angiosperms, the derivatives of the outer integument of the ovule form the testa.
If whole or part of the outer integument is fleshy it is then called a sarcotesta, as pomegranate (*Punica granatum*).
cf. **tegmen**
see **tegmic seed, testal seed**

Testa

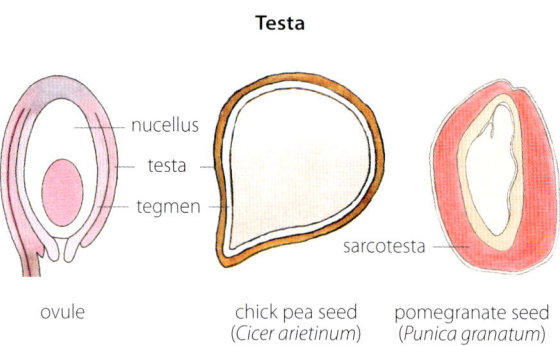

nucellus

testa

tegmen

sarcotesta

ovule

chick pea seed (*Cicer arietinum*)

pomegranate seed (*Punica granatum*)

testal seed   Of angiosperms, one having the mechanical layer in the exotesta, the mesotesta or the endotesta.
cf. **tegmic seed**
see **endotestal seed, exotestal seed, mesotestal seed**

tetra-   A prefix meaning four.

tetracyclic   Having four whorls, as a flower with sepals, petals, stamens and carpels.
see also **cyclic**

tetracyclic

stamens

carpels

petals

sepals

tetrad   A group of four. Four flowers grouped together in an inflorescence, as those of some mistletoes. A unit of four variously cohering pollen grains or spores.
see also **diad, monad, pollen dispersal, pollen tetrad, polyad, triad**

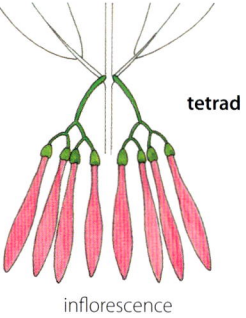

tetrad

inflorescence

tetrad mark   A single groove (monolete) or three-rayed, Y-shaped groove (trilete) on the face of a spore.
Also sometimes, one branch of a trilete laesura.
It marks the way in which the four spores of a tetrad were in contact with each other after meiosis.
It is the area of weakness in the wall through which a spore germinates.
= **laesura**

Tetrad mark

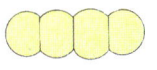

pollen tetrad linear

pollen grain
laesura monolete

pollen tetrad tetrahedral

pollen grain
laesura trilete

tetradynamous, tetradidynamous
Having six stamens, four long and two short, as the flowers of some members of the mustard family (Brassicaceae).
cf. **didynamous, tridynamous**

tetradynamous

tetragon   A flat shape with four sides and four angles, as a square or rhombus.
tetragonal   Of or in the form of a tetragon.

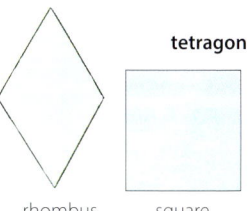

tetragon

rhombus   square

tetragonal tetrad
A uniplanar tetrad arranged with all four cohering members in contact at the centre and the orientation of the walls forming a cross.
see also **pollen tetrad**
see **uniplanar, viscin thread**

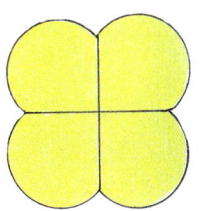

tetragonal pollen tetrad

tetrahedral tetrad
A multiplanar tetrad with the four cohering members arranged in a pyramid, with each member of the tetrad in contact with the other three.
see **pollen tetrad**

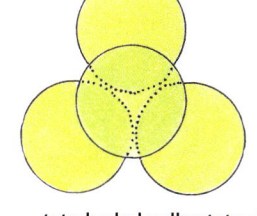

tetrahedral pollen tetrad

tetrahedron   A solid figure enclosed by four triangles, as some pyramids.
tetrahedral   Of a solid figure having four triangular faces.

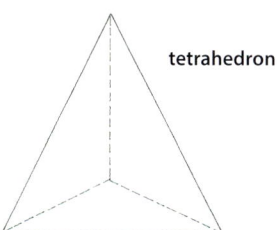

tetrahedron

tetramerous   Having flower parts, such as petals, sepals and stamens, in whorls of four or multiples of four. 4-merous.
see **-merous**

tetrandrous   Having four stamens, as the flowers of bedstraw (*Galium*).
cf. **diandrous, monandrous, pentandrous, polyandrous, triandrous**

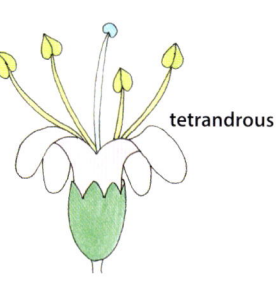

tetrandrous

tetraploid   Having four complete sets (4x) of chromosomes in each somatic cell.
see **ploidy**

tetrasporangiate   Of an anther with four pollen sacs, two in each anther lobe (theca). The pairs of pollen sacs coalesce before dehiscence.
cf. **monosporangiate, unisporangiate**

Tetrasporangiate

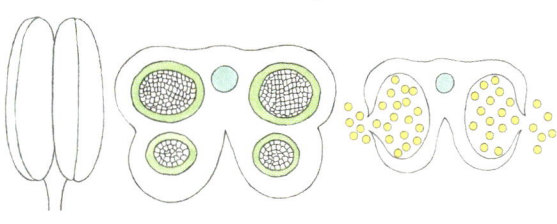

two-lobed   four pollen sacs   pollens sacs coalesce
anther

texture   The feel, appearance or consistency of a surface or substance, as of a leaf or soil.

thalamus   An expanded area at the apex of a stem bearing the organs of a single flower (sepals, petals, stamens and carpels).
= **receptacle, torus**
cf. **hypanthium**

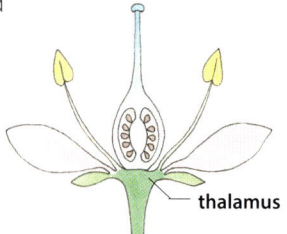

thallus, *pl.* **thalli**   A flattened, usually photosynthetic vegetative body with no vascular system, root-like rhizoids and no differentiation into a stem or leaf-like structures.
Of bryophytes, in hornworts and liverworts, the gametophyte phase of the life cycle that supports the dependent sporophyte phase.
Moss gametophytes lack a thallus and have an upright stem-like structure with leaf-like blades and root-like rhizoids.
thalloid, thallose   Relating to, resembling or consisting of a thallus.
see **bryophyte, prothallus**

Thallus

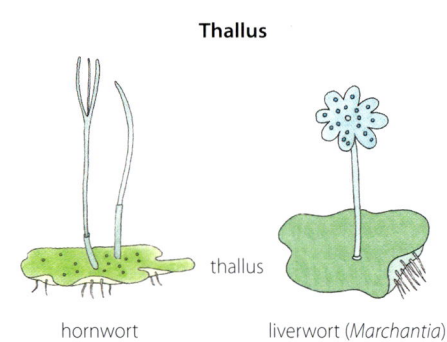

hornwort   liverwort (*Marchantia*)

thatch   An intertwined layer of dead roots, stems and blades that builds up on the soil surface at the base of living grass plants.

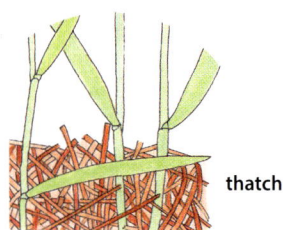

theca, *pl.* **thecae**   One half of an anther that contains two pollen sacs.
The urn-shaped, spore-bearing part of a moss capsule.

Theca

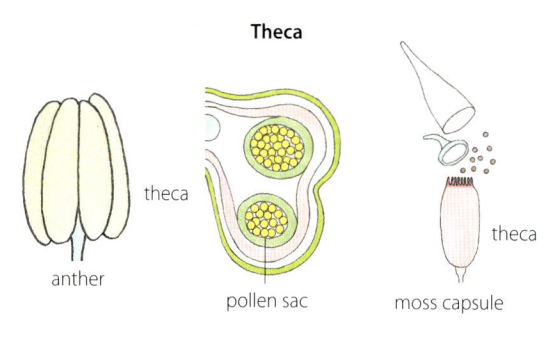

anther   pollen sac   moss capsule

therophyte   A plant that completes its life cycle in one season (an annual) and survives as seeds.

thicket   A dense growth of shrubs, bushes or small trees; a coppice.

thievery   The removal of pollen by an insect that, because of its shape, cannot contact the stigma and pollinate the flower.
cf. **robbery**

thigmotropism   Growth of a plant towards or away from touch, as a tendril that curves in the direction of support and coils around it.
see **tropism**
thigmotropic   Of or relating to thigmotropism.

thorn   A sharp hard outgrowth on a plant derived from woody tissue, as hawthorn (*Crataegus*).
cf. **prickle, spine**

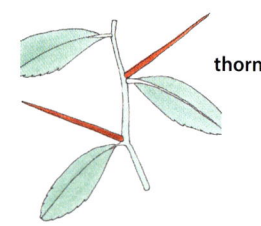

threatened   According to the IUCN, a conservation status covering species that are at risk of becoming extinct.
Three subdivisions are commonly recognised: critically endangered, endangered and vulnerable.

three-ranked
Of leaves arranged in three vertical rows, with any fourth leaf above the one below it.
= **tristichous**
see also **orthostichy**

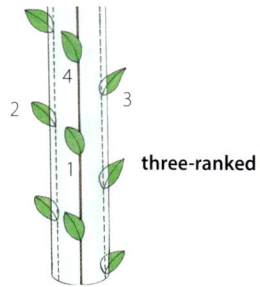

throat　The opening of the tubular part of a calyx, corolla or perianth where it joins the limb.
cf. **palate**

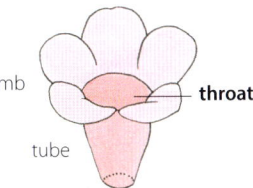

limb

throat

tube

throwback　The reappearance of a character of a distant ancestor after several generations.
An atavism.

thrum, thrum-eyed
Presentation of the anthers above the level of the stigma so that floral visitors contact the anthers first, as the cowslip (*Primula vulgaris*).
see **herkogamy**
cf. **pin**

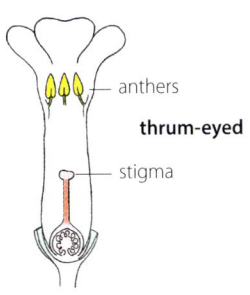

anthers

thrum-eyed

stigma

thylakoid
A disc-shaped membranous sac in a chloroplast that functions as the site of photosynthesis.
Chlorophyll is situated in the thylakoid membranes.
A stack of thylakoids is a granum.

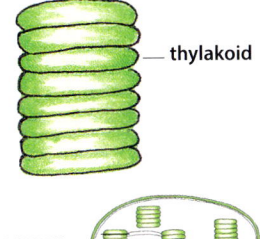

thylakoid

granum

chloroplast

thyrse, thyrsus　A mixed inflorescence, with the main axis racemose (indeterminate) and the lateral branches cymose (determinate), as lilac (*Syringa*).
see also **determinate, indeterminate**
cf. **panicle**
thyrsiform　With the appearance, but not necessarily the structure, of a thyrse.
thyrsoid　Like a thyrse.

Thyrse

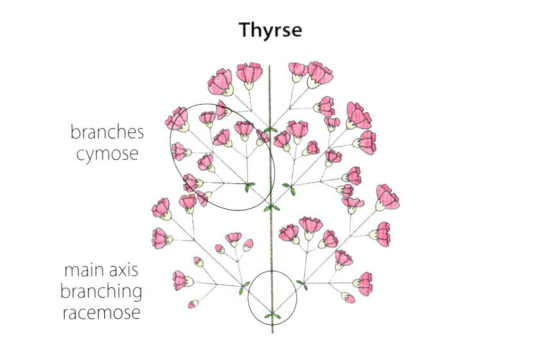

branches cymose

main axis branching racemose

tigellum　The plumule, epicotyl, hypocotyl and radicle that together form the embryo axis.
it is attached to the cotyledon(s) at the cotyledonary node.
It represents the axis of the future plant.
= **embryo axis**

Tigellum

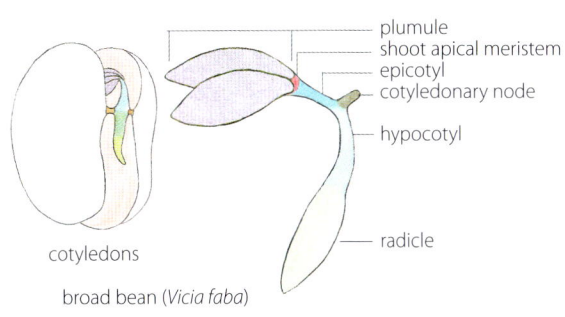

plumule
shoot apical meristem
epicotyl
cotyledonary node

hypocotyl

radicle

cotyledons

broad bean (*Vicia faba*)

tiller　A shoot that arises at or near the base of the stem of a grass or sedge.
Tillers are separated from the parent plant for vegetative propagation or can produce an inflorescence and seeds.

tiller

tissue　A group of cells, with a similar structure, that act together to perform a specific function.
A level of organisation between cells and organs.
Plants have four tissue types: meristematic tissue for differentiating into new tissues, vascular tissue for transport, dermal tissue for protection and ground tissue that makes up the remainder of the plant.
see **complex ~, primary ~, secondary ~, simple ~**

Stem tissues

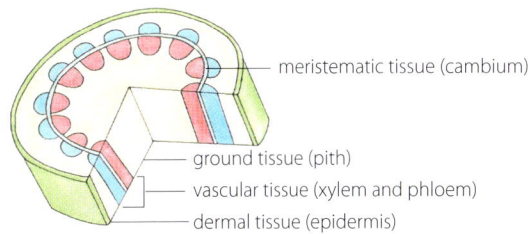

meristematic tissue (cambium)

ground tissue (pith)
vascular tissue (xylem and phloem)
dermal tissue (epidermis)

tissue system　Plants have three tissue systems:
the dermal tissue system (epidermis and periderm), the vascular tissue system (xylem and phloem) and the ground tissue system (parenchyma, collenchyma and sclerenchyma).
Each tissue originates from a particular meristem.
tissue system page 300 (cont.)

299

Tissue system

Tissue System	Tissue	Meristem Origin
dermal	epidermis periderm	protoderm cork cambium
ground	parenchyma collenchyma sclerenchyma	ground meristem
vascular	primary xylem primary phloem secondary xylem secondary phloem	procambium vascular cambium

tomentulose   Slightly or minutely tomentose.

tomentum   A covering of dense matted short soft cottony or woolly hairs, as on cudweed (*Gnaphalium*).
tomentose   With a tomentum.

tomentum

tonoplast
The membrane surrounding a plant cell vacuole that maintains water pressure through osmosis.
see **cell sap**
see also **turgour**

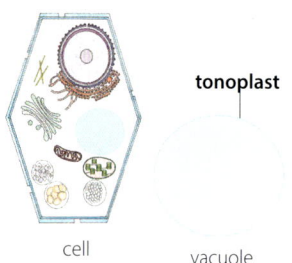

tonoplast

cell   vacuole

toothed   Having teeth. With shallow tooth-like projections that are like an equilateral triangle, at right angles to the margin.
= **dentate**
cf. **crenate, edentate, serrate**

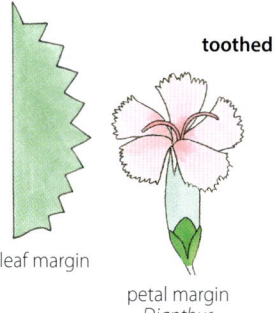

toothed

leaf margin

petal margin
Dianthus

torose, torous
Cylindrical with alternate swellings and contractions but less markedly so than moniliform, as the stems of glassworts (*Salicornia* and *Sarcocornia*).

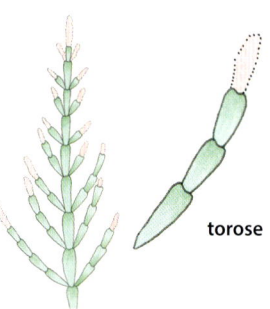

torose

tortuous   Twisted in different directions, as some leaves.

tortuous

torulose   Minutely torose, somewhat torose.

torus   An expanded area at the apex of a stem bearing the organs of a single flower (sepals, petals, stamens and carpels).
= **receptacle, thalamus**
cf. **hypanthium**

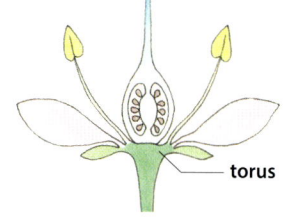

torus

toxin   A poisonous substance produced within living cells or organisms.

trabecula,
pl. **trabeculae**
A transverse partition dividing or partly dividing a cavity, as in the sporangium of a quillwort (*Isoetes*).
A transverse line or ridge, as on some seeds.
trabeculate   Of or having a trabecula.

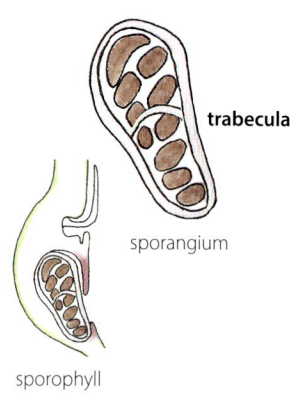

trabecula

sporangium

sporophyll

trace element   Nutrients essential to plant growth and health that are only needed in very small quantities, including iron, manganese, boron, zinc, molybdenum, chlorine and copper.
= **micronutrient**
cf. **macronutrient**

trachea   In the secondary xylem of angiosperms, a column of dead cells (vessel elements) with the joining ends perforated or totally degraded to form a tube.
The woody cell walls are pitted so that water and minerals can flow sideways as well as upwards from one cell to another.
Gymnosperms generally do not have vessels.
= **vessel**

Trachea

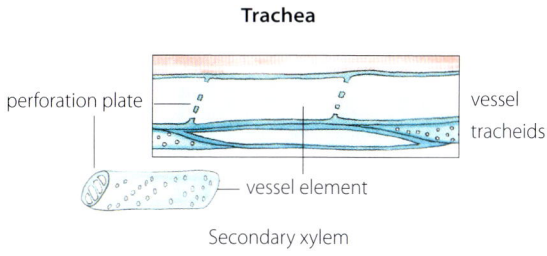

Secondary xylem

tracheary element   Either of two types of specialised elongated cells in xylem, tracheids and vessels, for transporting water and solutes up the plant.
Tracheids are found in gymnosperms, ferns and angiosperms, and vessels are found only in angiosperms.

Tracheary element

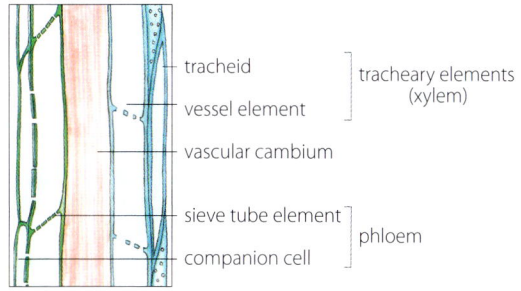

tracheophytes   Vascular plants that have xylem and phloem for conducting water and nutrients.
includes flowering plants (angiosperms), conifers and other gymnosperms, ferns, clubmosses and horsetails.
cf. **nonvascular plants**

trailing   Of stems, hanging loosely or spreading over the ground, as trailing beach morning glory (*Ipomoea pes-caprae*).

trailing

trait   A genetically determined characteristic or condition.

transcription   The copying of one kind of nucleic acid, DNA, to form another kind, mRNA (messenger ribose nucleic acid), that is the code used for making proteins.
see also **translation**

translation   The conversion of the code in mRNA (messenger RNA) into the of amino acids that make up a protein.
see **transcription**

translator   One of the thread-like arms that attaches a pollinium to the corpusculum in the milkweed family (Apocynaceae).
= **retinaculum**
Equivalent to the caudicle in the orchid family (Orchidaceae).

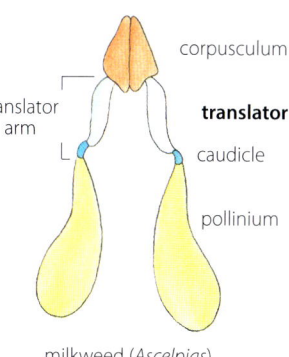

milkweed (*Ascelpias*)

translocation   The movement of soluble nutrients from one part of a plant to another, as sap containing nutrients in phloem tubes.
In genetics, the attachment of a broken part of a chromosome to a different chromosome or to a different part of the same one.

translucent   Allowing light to pass through but not transparent.

transparent   Sheer. Of a substance, clear and transmitting light so that it can be seen through.
So fine in texture that it can be seen through.
cf. **translucent**

transpiration   The movement of water into, through and out of a plant.
Water enters through the roots, is transported upwards through the xylem and evaporates mainly from the stomata on the leaves.
see also **evaporation**

transverse   Situated or extending across something at right angles to the axis.
Of anthers borne ar right angles to the tip of the filament.
see **anther attachment**
cf. **explanate**

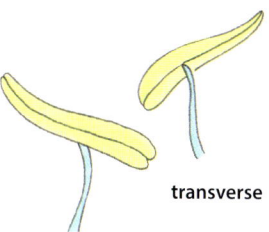

transverse

transverse dehiscence
Of anthers, opening across the anther lobe to release pollen.
see also **anther dehiscence**

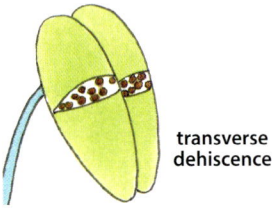

transverse dehiscence

trapezium   A four-sided figure with two sides parallel of unequal length.
trapeziform   Shaped like a trapezium.

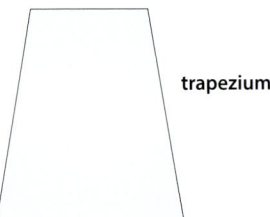

trapezium

tree   A tall perennial woody plant with a single main stem (trunk) from which branches extend to form a characteristic canopy or crown of leaves.

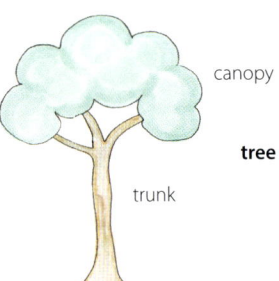

canopy

tree

trunk

tree fern   Tree ferns primarily belong to the tree fern family (Dicksoniaceae) and the scaly tree fern family (Cyatheaceae).
Some species, as *Cyathea,* have trunk-like stems composed of rhizomes that grow vertically. The tip of the stem produces a cluster of large divided fronds.
see **fern**

tree fern

tree fern (*Cyathea*)

treeline   The altitude above which trees are unable to grow. It varies according to latitude and other effects like exposure and soil.

tri-   A prefix meaning three.

triad   A group of three, as three flowers grouped together in an inflorescence, as those of some mistletoes.
cf. **diad, monad, polyad, tetrad**

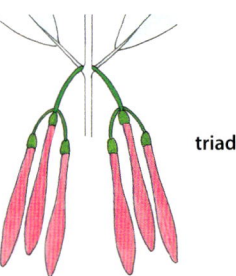

triad

triadelphous   Of stamens united by their filaments into three bundles, as some members of the gourd family (Curcurbitaceae).
see **adelphous**

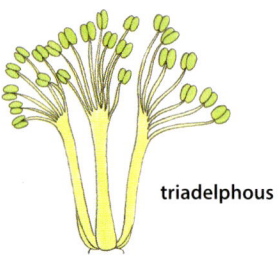

triadelphous

triandrous   Having three stamens, as the flowers of the iris family (Iridaceae).
cf. **diandrous, monandrous, pentandrous, polyandrous, tetrandrous**

triandrous

triangular
With three straight sides and three angles.

triangular

tribe, tribus   In taxonomic classification, a subdivision of a large family or, if present, subfamily. Names of tribes end in *-eae*.

trichome   An outgrowth from the epidermis, as a prickle, glochid or various types of hairs.
They may be glandular or non-glandular, unicellular or multicellular, branched or unbranched.
cf. **spine, thorn**

trichotomosulcate
Of a pollen grain with a three-armed sulcus.

trichotomosulcate

trichotomous
Branching once or regularly into three parts.
cf. **dichotomous**

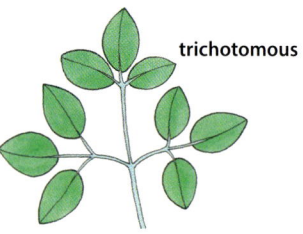

trichotomous

tricyclic   Having three whorls, as a flower with a whorl or tepals, a whorl of stamens and a whorl of carpels.
see also **cyclic**

tridentate   Three-toothed, as the leaflet tips of three-toothed cinquefoil (*Sibbaldiopsis tridentata*).

tridynamous   Having six stamens, three long and three short, as the pale grass lily (*Caesia parviflora* var. *parviflora*).
cf. **didynamous, tetradynamous**

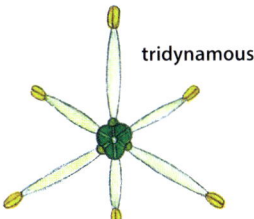
tridynamous

trifid   Split by deep clefts into three lobes, as a trifid leaf.

trifid

trifoliate   With three leaves.
see **foliate**
cf. **bifoliate, unifoliate**

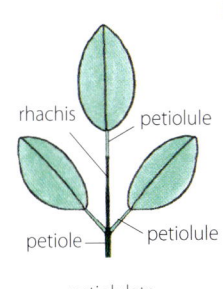
trifoliate

trifoliolate   Of a compound leaf with three leaflets.
It can be pinnate or palmate and leaflets can be petiolulate or sessile.
see also **ternate**
cf. **trifoliate**

Trifoliolate

Pinnate

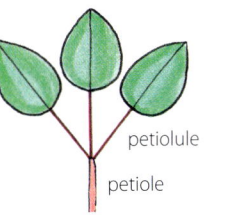

rhachis   petiolule

petiole   petiolule

petiolulate

rhachis

petiole

sessile

Palmate

petiolule

petiole

petiolulate

petiole

sessile

trifurcate   Divided into three more or less equal branches or prongs. Forked.
see **furcate**

trifurcate

trigger   A pollinating mechanism in trigger plants (Stylidiaceae). The sensitive column arches back on one side of the flower and springs forward to deposit pollen on, or to collect pollen from, an insect visitor.

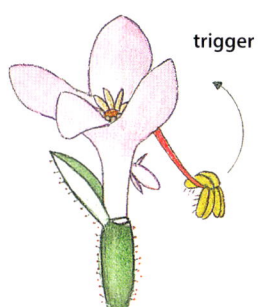
trigger

trigonous   Triangular in cross-section with angles obtuse.

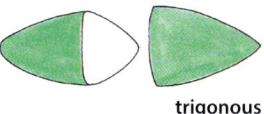
trigonous

trijugate   Of a pinnate leaf having three pairs of leaflets.
see **jugate**

trijugate

trilete   Of a spore with a three-branched, Y-shaped laesura that is a result of the way the four spores of the multiplanar tetrad were in contact with each other after meiosis.
It is the area of weakness in the wall through which the spore germinates.
cf. **alete, trilete**

trilete

Y-shaped laesura

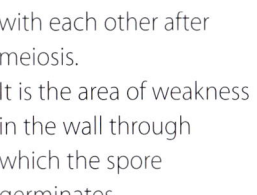

multiplanar spore tetrad

trilobate, trilobed   Having three lobes, as some leaves.

trilobate

trilocular   Of an ovary, anther or fruit, having three locules or cavities for ovules, pollen or seeds.
cf. **plurilocular, unilocular**

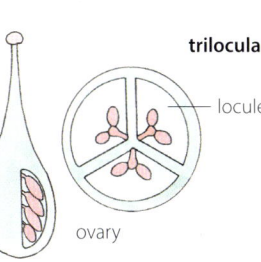
trilocular

locule

ovary

trimerous   Having flower parts, such as petals, sepals and stamens, in whorls of three or multiples of three. 3-merous.
see **-merous**

trimorphic   Having three distinct forms, as daffodils (*Narcissus*) that can have three different style lengths and different stamen lengths.
cf. **dimorphic, monomorphic, polymorphic**

Trimorphic

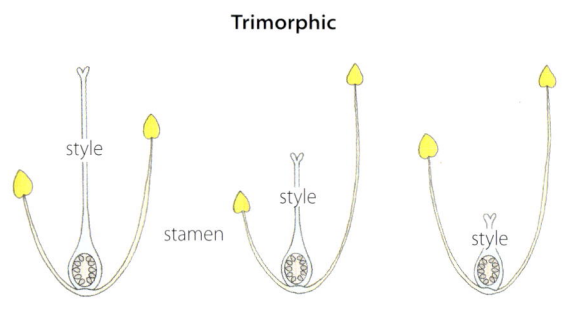

trioecious   Of a species having plants with male flowers, plants with female flowers and plants with bisexual flowers.
see **androgynomonoecious**
cf. **diclinous, dioecious, monoecious**

tripartite   Divided almost to the base into three lobes, as a tripartite leaf.

tripartite

tripinnate   Of a pinnate leaf with the primary divisions (pinnae) themselves divided into leaflets (pinnules) and the pinnules again divided into pinnules. A three times pinnately divided compound leaf.

tripinnate

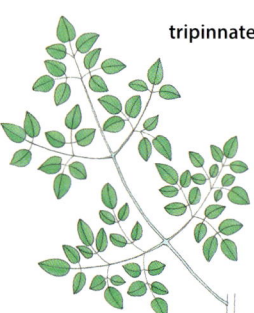

tripinnatifid   Thrice pinnatifid.

tripinnatipartite   Thrice pinnatipartite. Of a pinnately lobed leaf with the primary lobes twice pinnatipartite.

tripinnatisect   Thrice pinnatisect. Of a pinnately lobed leaf with the primary lobes twice pinnatisect.

tripinnatisect

triple-nerved, triple-ribbed, triplinerved   Having two prominent lateral veins emerge from the midrib a little above its base.

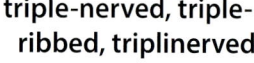

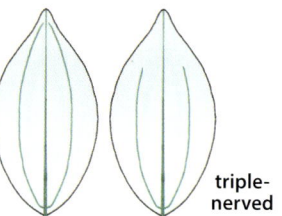

triple-nerved

triploid   Having three or more complete sets (3x) of chromosomes in each somatic cell.
see **ploidy**

tripterous   Of fruit or seed, having three wing-like expansions.

triquetrous   Triangular in cross-section with angles acute, as some leaves.
cf. **trigonous**

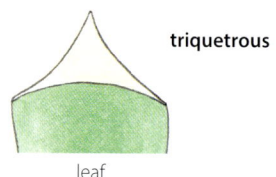

triquetrous

leaf

triquetrous vernation   Of young leaves in the unopened leaf bud having the bud triangular in section and leaves equitant at each angle.

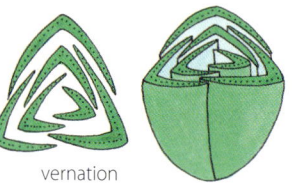

vernation

tristichous   Arranged on a stem in three vertical rows, as some leaves, with any fourth leaf above the one below it.
= **three-ranked**
see **orthostichy**

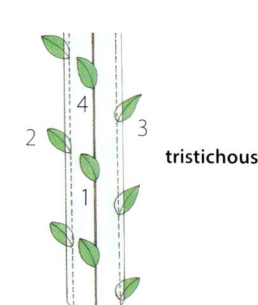

tristichous

tristyly   Having styles of three different lengths in flowers on the same plant, as purple loosestrife (*Lythrum salicaria*).
cf. **distyly, heterostyly**

tristylous   Exhibiting tristyly.

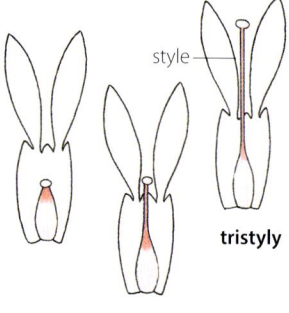

style

tristyly

triternate   Consisting of three parts, with each part divided into three and each of these three again divided into three, as a triternate leaf.

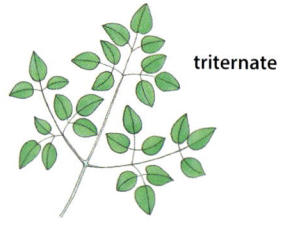

triternate

trophic   Of or relating to the nutrition of an organism in a food chain.
see **autotrophic, chemoautotrophic, dystrophic, ectotrophic, endotrophic, eutrophic, heteromycotrophic, heterotrophic, holomycotroph, hypereutrophic, mesotrophic, mycotrophic, oligotrophic, photoautotrophic**

trophophore   Of the adder's tongue family (Ophioglossaceae), the sterile photosynthetic vegetative leaf-like part of a frond that shares a common stalk with the fertile spore-bearing part of the frond.
cf. **sporophore**

trophophyll   Of ferns, a vegetative frond that is usually similar in appearance to the fertile fronds (sporophylls) that bear sori, as hard ferns (*Blechnum*), but often different, as flowering ferns (*Osmundia*).
cf. **sporophyll**

Trophophyll

hard fern (*Blechnum*)   flowering fern (*Osmundia*)

trophopod,
   pl. **trophopodia**   The enlarged and modified persistent base of a frond filled with starch storage tissue, as the lady fern (*Athyrium filix-femina*).

tropic response   Growth of a plant in a direction towards or away from a stimulus, such as light, gravity or moisture.
= **tropism**
see also **apogeotropism, geotropism, gravitropism, heliotropism, hydrotropism, orthotropism, phototropism, thigmotropism**
cf. **nastism**

tropical rainforest   Luxuriant forest characterised by high annual rainfall, extensive growth and poor soils.
Tropical rainforest is found near the equator, has no distinct seasons and four layers of growth (the emergent layer, the canopy layer, the understorey layer and the forest floor).
see **biome**
see also **storey**

tropics   The region between the Tropic of Cancer at latitude 23½° north of the equator and the Tropic of Capricorn at latitude 23½° south of the equator. Characterised by relatively constant temperatures, a wet and a dry season in the monsoon regions and tropical savanna, and a year-long wet season in tropical rainforest regions.
Also called the torrid zone.
tropic, tropical   Of or occurring in the tropics.
see **biome**

tropism   Growth of a plant in a direction towards or away from a stimulus, such as light, gravity or moisture.
= **tropic response**
see also **apogeotropism, geotropism, gravitropism, heliotropism, hydrotropism, orthotropism, phototropism, thigmotropism**
cf. **nastism**

tropophyte   A plant adapted to grow in a climate with alternating rainy and dry seasons, as yams (*Dioscorea alata*).
cf. **mesophyte, xerophyte**

true fruit   Fruit derived only from the the tissue of the ovary, as a drupe.
cf. **false fruit**

trullate   Shaped like a bricklayer's trowel.
cf. **obtrullate**

truncate   Appearing cut off at the base or the apex. Terminating suddenly rather than gradually.
= **abrupt**

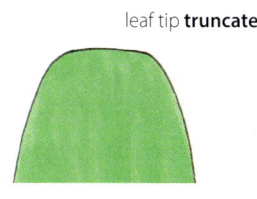

trunk   The main stem of a tree, usually thick and upright, that bears branches at some distance above the ground.
cf. **crown, bole**

trunk

tryma   A nut surrounded by a fused involucre that dehisces at maturity, as the pecan (*Carya*).
see **accessory fruit**
cf. **pseudodrupe**

tryma

nut

involucre

tryphine   A sticky material coating the pollen grains of the mustard family (Brassicaceae).
see **pollen kit**

tube   Of a flower, the fused, usually more or less cylindrical portion of a calyx or corolla.
cf. **limb**
tubular   Cylindrical and hollow.

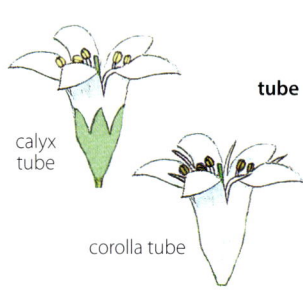

tube

calyx tube

corolla tube

tube cell   One of the two cells, lacking a cell wall, in a pollen grain.
After pollen germination, it develops into the pollen tube.
see also **generative cell**

Tube cell

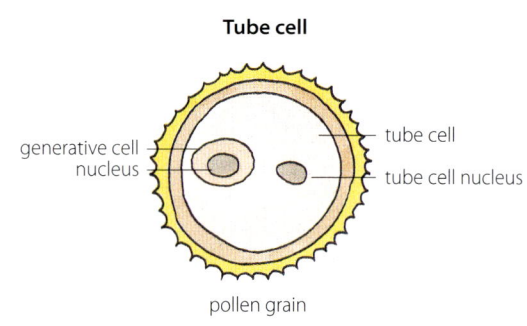

generative cell nucleus

tube cell

tube cell nucleus

pollen grain

tuber   The part of an underground root or stem, rarely an aerial stem, as Madeira vine (*Anredera cordifolia*), that is swollen with food reserves and survives from season to season.
see **root tuber, stem tuber, tuberoid**
tuberous   Bearing tubers or tuberoids.

tubercle   A small rounded wart-like protuberance.
A nodular growth on the roots of legumes.
see **nitrogen fixation**
tubercular   Like a tubercle.
Having tubercles, as the seeds of some chickweeds (*Stellaria*).
tuberculate, tuberculose   Covered with small blunt warty projections.

tubercle

chickweed seed tuberculate

tuberoid   A fleshy storage organ on a root, usually on plants that die back over winter.
It has an apical bud and can form roots at the base to produce a new plant.
Daughter tuberoids, that develop on the ends of additional roots, can produce a colony of new plants.
Often only two tuberoids are present, with one being a replacement for the previous season's tuberoid.
Common in terrestrial orchids.
see also **dropper, tuber, tuberous root**

Tuberoid

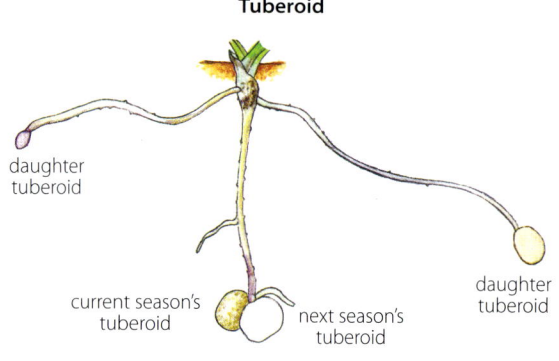

daughter tuberoid

current season's tuberoid

next season's tuberoid

daughter tuberoid

tuberous root   Roots with swollen parts that function as a storage organ and allows the plant to survive during dormancy.
see **root tuber**

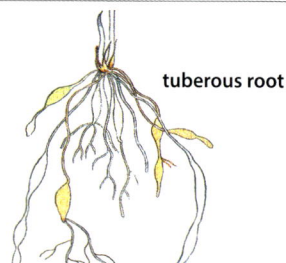

tuberous root

tubular floret   A small tubular flower usually with lobes.
= **disc floret**
cf. **ligulate floret, ray floret**

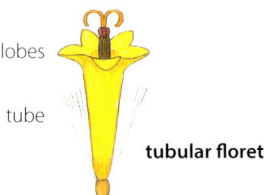

lobes

tube

tubular floret

tuft   A dense bunch or cluster, as of leaves or hairs, that are attached to or close together at the base. Of grasses (Poaceae), tillers joined together at the base by very short stems or apparently stemless.
cf. **mat grass**
A small clump of trees or bushes.
tufted   In a dense bunch, cluster or clump.
= **caespitose**

Tuft

willow-herb seed
(*Epilobium*)

leaves

grass

tumid   Swollen and distended, especially due to high fluid content.
= **turgid**
tumescent   Becoming tumid or nearly tumid.
tumidity   The state of being tumid.

tundra   A vast mostly flat treeless arctic biome in Europe, Asia and America, with a patchy low vegetation of mosses, lichens, herbs and small shrubs.
Surface soil freezes and thaws seasonally and subsoil remains permanently frozen.
see **biome, permafrost**

tunic   The outer dry papery covering of bulbs and corms.
tunicate   Having or covered with a tunic, as the fleshy cylinder-like leaves of an onion bulb.

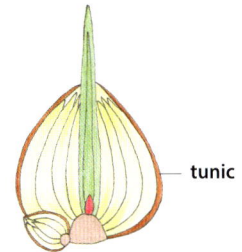

tunic

tunicate bulb   A true bulb that consists of a compressed stem with nodes (basal plate) bearing cylinder-like leaves arranged in concentric circles, that surround the leaf shoot and flower bud for the following season. The fleshy leaves are covered by an outer sheath (tunic) of dry membranous scale leaves.
see also **imbricate bulb**

tunicate bulb

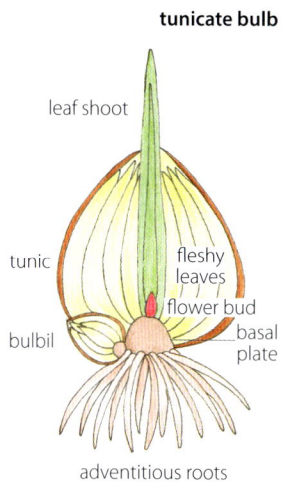

leaf shoot

tunic

bulbil

fleshy leaves

flower bud

basal plate

adventitious roots

turbinate   Cone-shaped with the narrow end at the base. Shaped like a top.

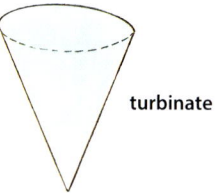

turbinate

turf   A layer of earth covered with grass. Sod.

turgidity   The state of being turgid.
see **turgor**
turgid   Of a plant cell having high turgor pressure.
It occurs when the vacuole increases its water content and swells causing outward pressure against the cell wall.
cf. **flaccid, tumid**

turgidity

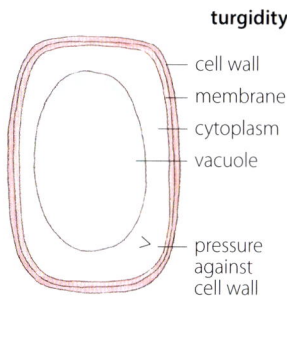

cell wall

membrane

cytoplasm

vacuole

pressure against cell wall

turgor, turgour   Of a plant cell, the normal state of rigidity caused by pressure of the contents of a cell against the hard cellulose cell wall.

tugor

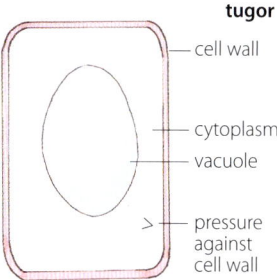

cell wall

cytoplasm

vacuole

pressure against cell wall

307

turgor pressure   The pressure on the cell wall caused by the degree of swelling of the vacuole. Pressure from excessive water in the vacuole causes the cell walls to become distended.
cf. **flaccidity, plasmolysis**

Turgor pressure

cell wall

cytoplasm

vacuole

pressure against cell wall

normal cell turgor pressure   excessive cell turgor pressure

turion   In some aquatic species, an over-wintering perennating bud that separates from the parent plant and either sinks to the bottom, shooting when conditions are favourable, or disperses and colonises a new habitat, as European frog-bit (*Hydrocharis morsus-ranae*).
A scaly shoot that develops from an underground bud, as asparagus (*Asparagus officinalis*).
cf. **gemma**

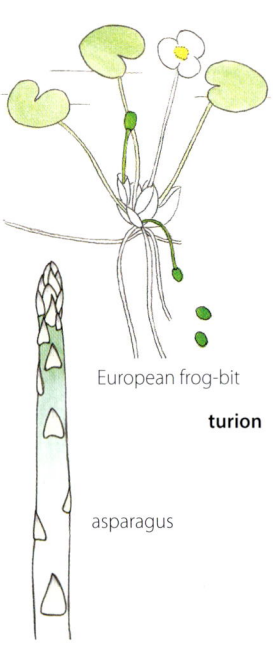

European frog-bit

turion

asparagus

turpentine   A resinous exudate from the turpentine tree (*Pistacia terebinthus*) and some conifers.
see **oleoresin**

tussock   A plant growing in a dense tuft, usually separately from similar tufts, as serrated tussock (*Nassella trichotoma*).
see also **hummock grass**

tussock

twig   The lowest order of branching in a woody plant, usually of the current season's growth and concentrated in the outer extremities.

twiner   A plant that winds itself spirally around a support as it grows, as honeysuckle (*Lonicera caprifolium*) and twining glycine (*Glycine clandestina*).
cf. **climber**

twiner

two-ranked   Of leaves arranged alternately on opposite sides of the stem, in two vertical rows, with any third leaf above the one below it.
= **distichous**
see also **orthostichy**

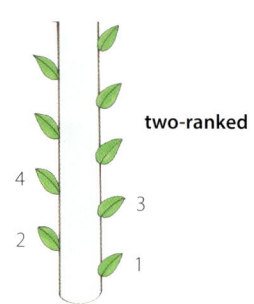

two-ranked

4

2

3

1

type   A preserved specimen, or sometimes an illustration, that serves as the reference point when naming or describing a new species, genus or family.
= **type specimen**
see **holotype, isolectotype, isosyntype, isotype, lectotype, neotype, nomenclatural type, paratype, syntype**

type specimen   A preserved specimen, or sometimes an illustration, that serves as the reference point when naming or describing a new species, genus or family.
= **type**

ulculus, *pl.* **ulculi**
A more or less circular aperture not situated at a pole on a pollen grain.
cf. **ulcus**
ulculate   Of a pollen grain with an ulculus.

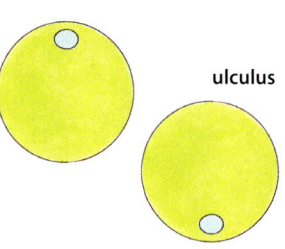

ulculus

ulcus, *pl.* **ulci**   A more or less circular aperture situated at either pole of a pollen grain.
cf. **ulculus**
ulcerate   Of a pollen grain with an ulcus.

ulcus

umbel   A flat-topped to almost spherical racemose inflorescence, with the flower stalks (pedicels) arising from the same point at the top of the main stem.
Typical inflorescence of the carrot family (Apiaceae).
An umbel may be simple or compound.
An indeterminate or indefinite inflorescence.
cf. **corymb**
umbellate   Bearing umbels.
umbelliferous   Bearing or producing an umbel or umbels.
umbelliform   Having the shape, but not necessarily the structure, of an umbel.

Umbel

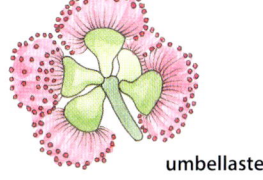

simple umbel   compound umbel

umbellaster
The unique umbel-like inflorescence of some eucalypts.

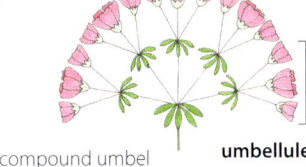

umbellaster

umbellet, umbellule
A secondary umbel in a compound umbel.
A small or partial umbel.

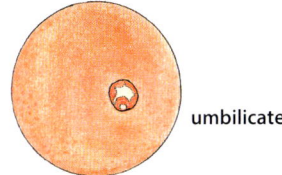

compound umbel   **umbellule**

umbilicate   Having a central depression resembling a navel, as on a navel orange.

umbilicate

umbo   A projection arising from a surface.
The raised part of the apophysis on the scale of a pine cone.
umbonate   Bearing or like an umbo.

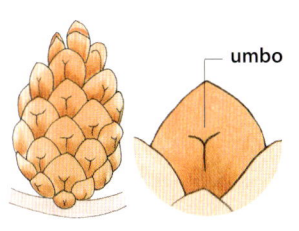

umbo

pine cone

umbraculate, umbraculiferous
Having the shape of an open umbrella, as the elaborate style of the pitcher plant genus *Sarracenia*.

umbraculate

style of *Sarracenia*

un-   A prefix meaning negation or reversal.

unarmed   Lacking thorns, spines or prickles.
= **inermous**
see **armature**

uncinate   With a hook at the apex, as some leaves.

uncinate

unctuous   Smooth and greasy in texture or appearance, as camphor, the oil extracted from the camphor laurel (*Cinnamomum camphora*).

undershrub   A small shrub with a woody base and herbaceous new growth, as lavender (*Lavandula*) and thyme (*Thymus*).
= **subshrub, suffrutex**

undershrub

understorey, understory
The layer of vegetation between the shrub layer and the canopy in a forest.
It is made up of shade-tolerant trees and the saplings of canopy and emergent species.

emergent layer

canopy

understorey

shrub layer

ground layer

undescribed   A taxon, such as a new species, that has not yet been formally described and named.

undifferentiated   With no distinctive characteristics, as the unspecialised cells of a meristem or the sepals and petals of some lilies.

undulate   Wavy. With margins curving up and down in cross-section, as the margins of some leaves. Not flat.
cf. **sinuate**

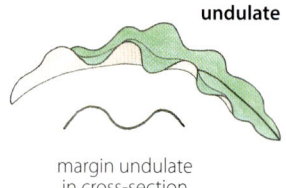

margin undulate
in cross-section

unequal   Asymmetrical, as a leaf or leaflet larger on one side of the midrib than on the other.
see also **oblique**

unequal

unguis   A claw.
unguiculate   Narrowed at the base into a claw, as some petals and bracts.

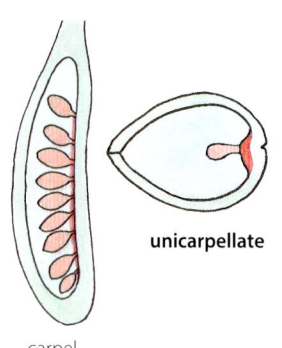

bract   petal

uni-   A prefix meaning having only one.

unicarpellate, unicarpellous
Of a flower having a carpel with one locule, as peas.
= **monocarpellary, monocarpellate, monocarpous, stylodious**
cf. **apocarpous, syncarpous**

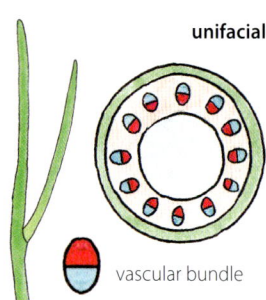

carpel

unifacial   Having one usually cylindrical surface, as the leaf of an onion (*Allium cepa*) with no distinct upper and lower surfaces.
= **centric**
cf. **bifacial, equifacial**

unifacial

vascular bundle

uniflorescence
One of the inflorescence units in a conflorescence, including any pedicels and bracts when present, as the paired flowers of grevilleas (*Grevillea*) and banksias (*Banksia*).

uniflorescence

unifoliate   With one leaf, as the orchid pink fairies (*Caladenia latifolia*).
see **foliate**
cf. **bifoliolate, trifoliate**

unifoliate

unifoliolate   Of a compound leaf having a single leaflet, with the leaflet on a petiolule attached to the top of the petiole, as some bossiaea (*Bossiaea*) or lemons (*Citrus*).
cf. **unifoliate**
= **monofoliolate**

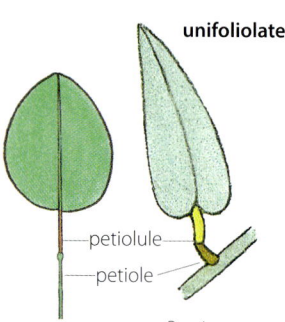

unifoliolate

petiolule
petiole

Bossiaea

unijugate   Paired. Of a pinnate leaf having one pair of leaflets.
see **jugate**

unijugate

unilateral   Arranged on one side only.
cf. **bilateral**

unilocular   Of an ovary, anther or fruit, having a single locule or cavity for ovules, pollen or seeds.
cf. **bilocular, plurilocular**

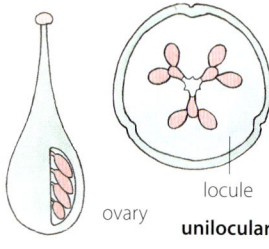

ovary

locule

unilocular

uniparous   Of a cymose inflorescence forming a single stem at each branching point, as a helicoid or scorpioid cyme.
cf. **biparous, multiparous**

helicoid cyme   scorpioid cyme

uniparous

uniplanar   Lying on an imaginary two-dimensional flat or level surface, as pollen grains in tetrads that are tetragonal, T-shaped, rhomboidal or linear.
see **plane, pollen tetrad**
cf. **multiplanar**

uniseriate   Arranged in one row or whorl.
= **monoseriate**
see also **seriate**

unisexual   Of a flower with either stamens or a pistil or pistils fertile, but not both.
Of an inflorescence or plant with fertile flowers of one sex only.
cf. **bisexual, neuter**

unisporangiate   Of an anther with only one pollen sac, as the dwarf mistletoes (*Arceuthobium*).
= **monosporangiate**
cf. **bisporangiate, tetrasporangiate**

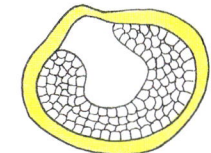
pollen sac   **unisporangiate**

united   Of parts fused or joined together.
cf. **adherent, coherent**

unitegmic   Of an ovule, with one integument surrounding the nucellus, as gymnosperms.
Of a seed coat, having a single integument comprising three layers: an outer fleshy sarcotesta, a middle stony sclerotesta and an innermost parenchymatous endotesta that generally collapses to form a membranous layer, as gymnosperms.
cf. **ategmic, bitegmic**

Unitegmic

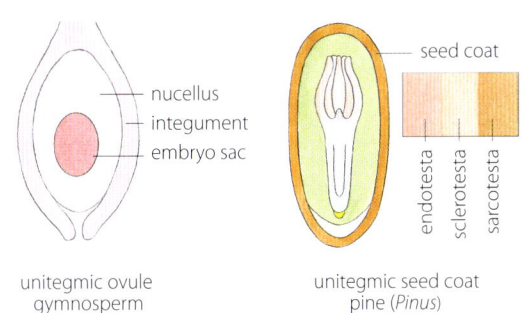
nucellus
integument
embryo sac
seed coat
endotesta
sclerotesta
sarcotesta
unitegmic ovule gymnosperm
unitegmic seed coat pine (*Pinus*)

univalent   In meiosis, of a chromosome that is not paired with its homologous chromosome during synapsis.

urceolate   Urn-shaped.
of a globose to sub-cylindrical corolla with a constriction at or below the short limb, as manzanita (*Arctostaphylos*).

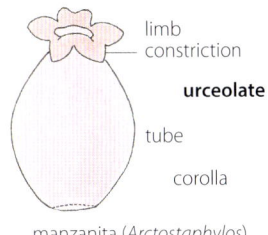
limb
constriction
urceolate
tube
corolla
manzanita (*Arctostaphylos*)

urent   Stinging.
Causing irritation when touched, as hairs of the stinging nettle (*Urtica dioica*).

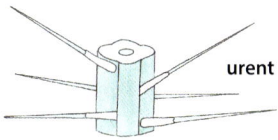
urent

urticate   Causing a stinging sensation, as hairs of the stinging nettle (*Urtica dioica*).

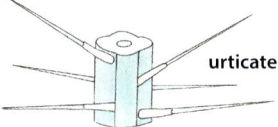
urticate

utricle   A bladder-like structure.
A membranous indehiscent single-seeded fruit, as bluebush (*Maireana*).
A bladder-like sac, as the traps of bladderworts (*Utricularia*).
The perigynium that encloses the female flower in the sedge genus *Carex*.

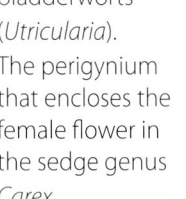
seed
bluebush (*Maireana*)

utricle

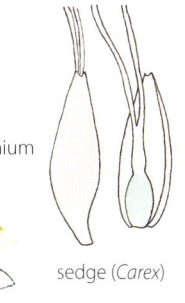
perigynium
sedge (*Carex*)

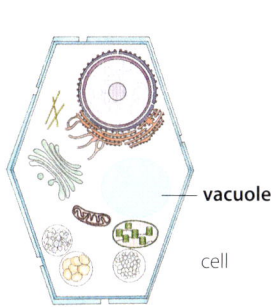
bladderwort (*Utricularia*)

vacuole   A fluid-filled membranous sac in the cytoplasm of a plant cell. Its functions include digestion, storage of wastes and toxins, and maintenance of water pressure.
see **tonoplast, cell sap**

vacuole
cell

vaginate   Enclosed in a sheath, as the leaf base that surrounds the stem of most grasses.
see also **sheathing**

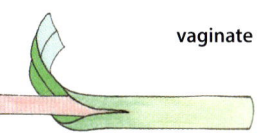
vaginate

valid name   A name published in accordance with the International Code of Nomenclature.

validly published   Published in accordance with the International Code of Nomenclature.
= **effectively published**

valleculla, *pl.* **valleculae**
A groove or furrow.
A furrow between the ribs on the fruit of the carrot family (Apiaceae).
valleculate   Having valleculae.
vallecular   Relating to valleculae.

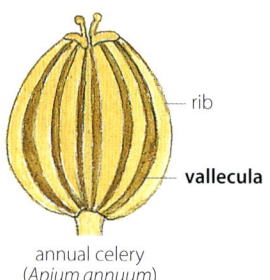

rib

vallecula

annual celery
(*Apium annuum*)

valvate   With similar parts touching at the edges but not overlapping.
Relating to valves.
Opening by valves, as some capsules.

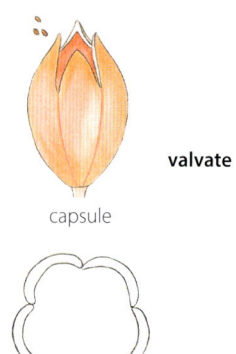

valvate

capsule

valvate aestivation
The arrangement of petals, tepals or sepals in a bud with the margins meeting but not overlapping.

aestivation

valvate vernation
The arrangement of young leaves in an unopened leaf bud with the margins meeting but not overlapping.

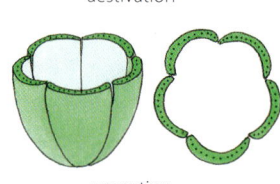

vernation

valvate capsule
A capsule that splits at the tip along the margins of the valves, as some of the carnation family (Caryophyllaceae).

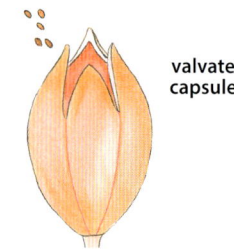

valvate
capsule

valvate dehiscence, valvular dehiscence
Of anthers, with flap-like valves that open upwards to release pollen, as the laurel family (Lauraceae).
see also **anther dehiscence**

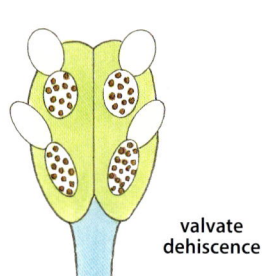

valvate
dehiscence

valve   A segment of a dry fruit, as a capsule or siliqua, or an anther, that splits open at maturity.
It may remain attached or fall off.
Derived from the fruit wall.
valvular   Of or relating to a valve.
see **valvate**

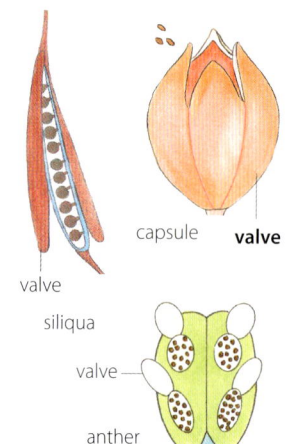

capsule   **valve**

valve

siliqua

valve

anther

valvular capsule
A capsule that splits so that the valves breakaway from the septa, as Argentine cedar (*Cedrela angustifolia*).
= **septifragal capsule**

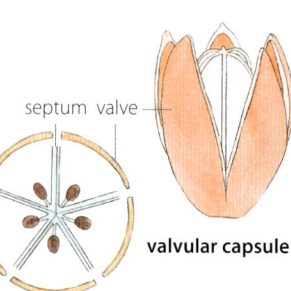

septum valve

valvular capsule

VAM   Vesicular arbuscular mycorrhiza.

vaporisation   The change of a substance from a liquid state to a gaseous state.
There are two types of vaporisation, evaporation and boiling.
cf. **condensation**

var.   An abbreviation for variety.

variation   The naturally occurring genetic and/or morphological differences among individuals of the same species.

variegation   Having differently coloured patches, spots or streaks in plant parts, as leaves and petals.
variegated   Exhibiting variegation, as the leaves of some pelargoniums (*Pelargonium*).
cf. **concolorous, discolorous**

variegation

pelargonium
leaves

variety, varietas, *abbr.* **var.** In taxonomic classification, a subdivision of species below subspecies and above forma, as *Geranium sanguineum* var. *strictum*.
see **taxonomic hierarchy**

vascular Having a system of vessels (xylem and phloem), for conducting water and nutrients.
It includes flowering plants (angiosperms), conifers and other gymnosperms, ferns, clubmosses and horsetails.
see also **cryptogams, nonvascular plants**

vascular bundle A strand of vascular tissue (xylem and phloem) and associated tissues (as cambium in secondary growth) that provides support and supplies nutrients to plant parts.
It originates in the procambium of the apical meristem.
In monocotyledon stems the bundles appear scattered in cross-section and in eudicot stems they are arranged in a ring inside the epidermis.
Open vascular bundles in secondary growth have a layer of cambium, as eudicots. Closed vascular bundles lack a cambium layer and secondary growth, as monocotyledons.
Xylem and phloem are usually bundled together (conjoint vascular bundles) but are typically arranged separately (radial vascular bundles) in the roots of monocots and eudicots.
In the stem and root the vascular bundles are continuous throughout the length of the axis.
= **vascular strand**
see also **stele**

Vascular bundle

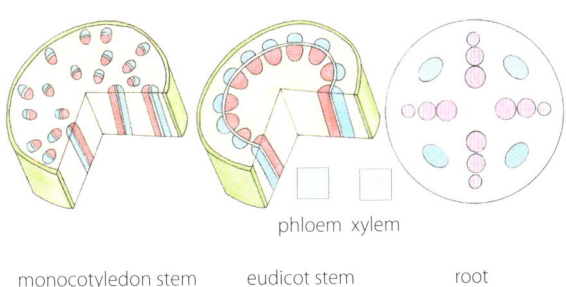

phloem xylem

monocotyledon stem   eudicot stem   root

vascular cambium A layer of meristematic cells, between the xylem and phloem in the vascular tissue of eudicots and gymnosperms, that produces secondary xylem (wood) and secondary phloem (inner bark).
In primary growth it is derived from procambium, as well as the parenchyma between the vascular bundles.
It is found in secondary growth and eventually forms a continuous ring.
see **vascular bundle**
see also **cambium, fusiform initials, interfascicular cambium, lateral meristem, ray initials**

Vascular cambium

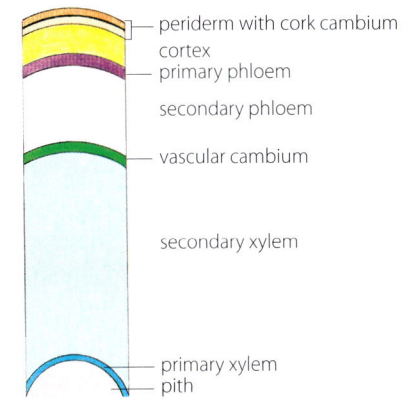

periderm with cork cambium
cortex
primary phloem
secondary phloem
vascular cambium
secondary xylem
primary xylem
pith

vascular cylinder The central core of eudicot and gymnosperm stems and roots comprising the vascular tissue and other tissues like pith.
= **stele**

Vascular cylinder

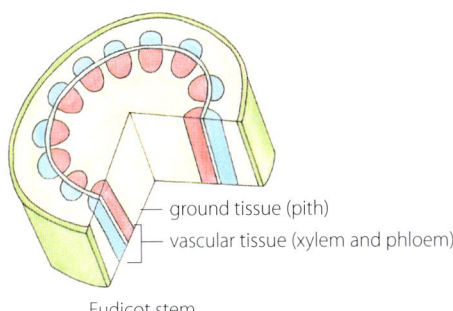

ground tissue (pith)
vascular tissue (xylem and phloem)

Eudicot stem

vascular plants Plants with a specialised conducting system (xylem and phloem) for transporting water and nutrients.
The three kinds of vascular plants are: flowering plants (angiosperms), conifers and other gymnosperms, and seedless plants like ferns, clubmosses and horsetails.
vascular plants page 314 (cont.)
= **tracheophyte**
cf. **nonvascular plants**

Vascular plants

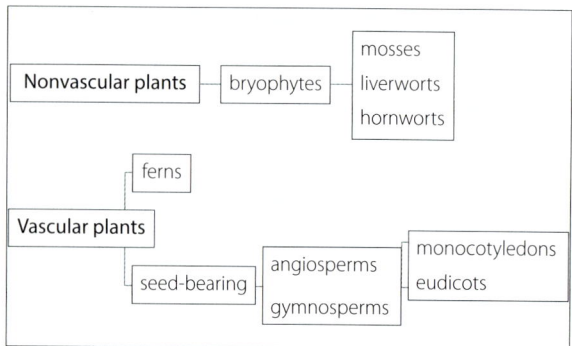

vascular ray   Of secondary growth in eudicots and gymnosperms, one of the bands of mostly parenchyma tissue produced by the vascular cambium that extends through the secondary xylem (growth rings) and the secondary phloem in woody plants.
Vascular rays store nutrients and transport them radially. Vascular bundles transport nutrients vertically.
see **medullary ray**
see also **ray initials**

Vascular ray

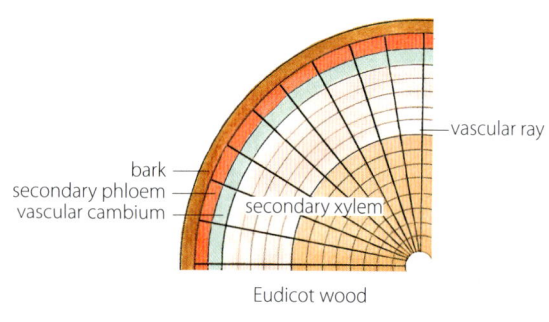

Eudicot wood

vascular strand   A vascular bundle.

vascular system   Specialised tissues, xylem and phloem and associated tissues, that provide support and conduct water and nutrients throughout a plant.

vascular tissue   Tissue arranged in long strands (vascular bundles) that provide support and transport water and nutrients throughout a plant.
see **complex tissue, primary tissue**

vase-shaped   Of a tree having a tall central trunk with a canopy of spreading branches that is widest at the top, as American elm (*Ulmus americana*).

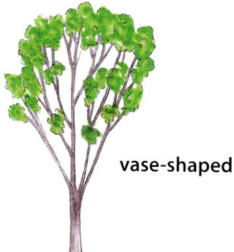

vase-shaped

Vavilovian mimicry   A form of imitation where a weed comes to share one or more characteristics with a domesticated plant, especially of a food crop, as the grass *Lolium remotum* is a nearly obligate weed of flax (*Linum usitatissimum*).
cf. **Bakerian mimicry, Dodsonian mimicry, Pouyannian mimicry**

vector   A DNA molecule that carries and inserts a foreign gene into another cell.
see also **genetic engineering**

vegetation   The general form or appearance of the plant life of a particular community.
cf. **flora**

vegetative   Relating to the non-floral parts of a plant, as stems, leaves and roots.

vegetative apomixis
Vegetative buds or bulbils produced in the place of flowers.
May be the only means of propagation for the plant, as tree onion (*Allium cepa* var. *proliferum*).

bulbils

vegetative apomixis

vegetative bud   A leaf bud, a bud composed of embryonic leaves.
see also **ptyxis, vernation**
cf. **reproductive bud**

vegetative phase   The phase after germination in which plant shoots increase in size and photosynthetic capacity.
The diploid generation of a plant that begins with the zygote and includes its roots, stems and leaves, and in angiosperms the flowers and fruit, and in gymnosperms the cone.
The gamete-bearing phase in the life cycle of a plant.
= **sporophyte**
see **reproductive phase, senescence**
see also **alternation of generations**

vegetative reproduction   A form of asexual reproduction where a new plant grows from parts of a parent plant and is identical to it (a clone). It is either natural, as from stems (stolons and rhizomes), roots (tubers) and bulbils, or artificial, as from cuttings, grafting or layering.
see also **apomixis**

vein   A vascular bundle in a leaf or other plant part that circulates water, minerals and other substances. It typically divides or branches and provides support and strength.
= **nerve**

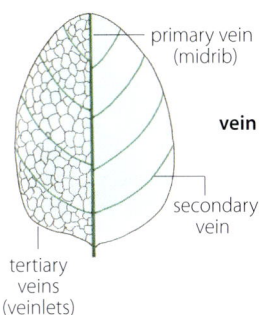

primary vein (midrib)

vein

secondary vein

tertiary veins (veinlets)

veinlet, venule
A small vein. The ultimate visible division of a vein on a leaf.
= **tertiary vein**

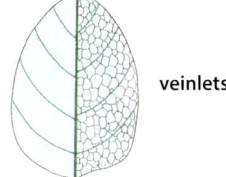

veinlets

velamen   A water-retaining outer layer on the aerial roots of some epiphytes, especially orchids.

velum, *pl.* **vela**   The membranous flap covering the sporangium in quillworts (*Isoetes*).

Velum

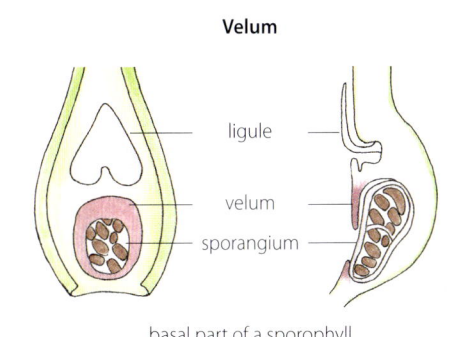

ligule

velum

sporangium

basal part of a sporophyll

velutinous   Velvety. Covered with silky, short, fine, erect hairs of an even length.
cf. **sericeous**

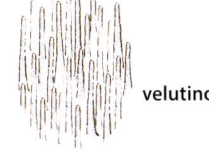

velutinous

velvety   Covered with silky, short, fine, erect hairs of an even length.
= **velutinous**
cf. **sericeous**

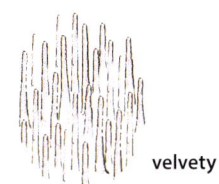

velvety

venation   The arrangement of veins on a leaf. It can be open, with free-ending veins and margins from toothed to lobed or compound (mostly eudicots), or closed, with veins fused into loops and margins entire with smooth edges (mostly monocotyledons).
Venation can be pinnate, palmate or parallel. Venation can be net-like (reticulate), branching (dendritic) and with veins of clearly different diameters (hierarchical), as midrib, vein and veinlets.
= **nervation**

Venation

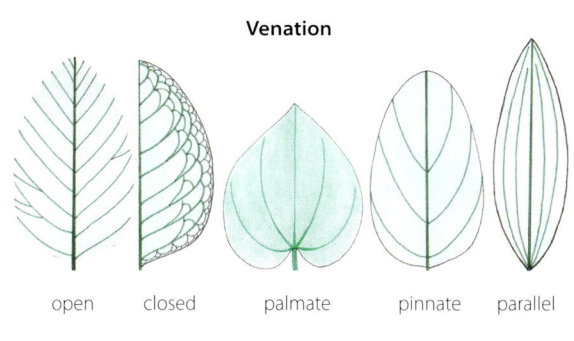

open   closed   palmate   pinnate   parallel

venose   Having numerous or conspicuous veins.

venter   The expanded basal part of an archegonium in which the oosphere is formed.

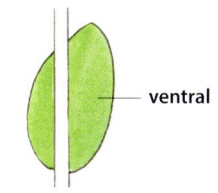

oosphere   **venter**

archegonium

ventral   The front. The inner side facing towards the axis, as a leaf on a stem.
cf. **dorsal**

ventral

ventricose   Swollen on one side as the corolla tube of some flowers in the figwort family (Scrophulariaceae) and the mint family (Lamiaceae). Swelling out in the middle, as the stem of the palm *Iriartea ventricosa*.
cf. **gibbous**

palm

ventricose

corolla tube

ventriculose   Slightly ventricose.

ventrifixed Attached on or by the front, as a stamen filament attached to the connective somewhere along the front of an anther.
cf. **dorsifixed, medifixed**
see **anther attachment**

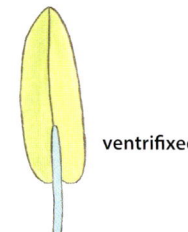

ventrifixed

vernacular name The common name for a plant as opposed to the scientific name, as apple (common name) and *Malus domestica* (scientific name).

vernal Of or appearing in spring.
cf. **aestival, autumnal, hibernal**

vernation *see* page 317

vernicose
Having a shiny surface as though varnished, as the phyllodes of the varnish wattle (*Acacia verniciflua*).

vernicose

verruca, *pl.* **verrucae**
A wart-like projection.
verrucate, verrucose, verrucous Warty. Bearing small wart-like projections, as the seeds of some species of skullcap (*Scutellaria*).
verruciform Wart-like.

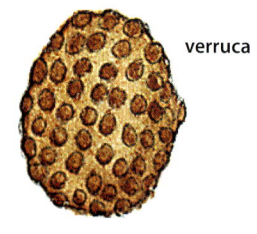

verruca

skullcap seed verrucose

verrucula, *pl.* **verruculae** A small wart.
verruculate, verruculose With small warts. Slightly verrucose, finely verrucose.

versatile Of anthers, swinging freely about the point of attachment to the filament.
see **anther attachment**

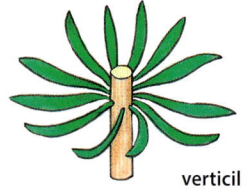

versatile

verticil A circular arrangement of parts around an axis, as leaves or petals around a stem.
= **whorl**
see **verticillaster**
cf. **spiral**

verticil

verticillate Arranged in one or more verticils.
= **whorled**

verticillaster A cymose inflorescence at a node on a stem, resembling a whorl but composed of two opposite, usually sessile, axillary cymes, as the genus *Salvia*.
see **fascicled cyme, glomerule**

verticillaster

vesicle A small bladder-like sac containing fluid. Of a mycorrhiza, a hyphal swelling that acts as a lipid-filled storage structure. It is found inside or outside cells.
vesicular Of or relating to vesicles.

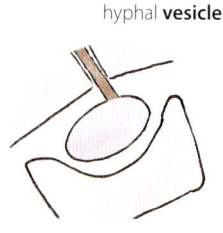

hyphal **vesicle**

vesicular arbuscular mycorrhiza
A mutually beneficial symbiosis formed between Glomerocyta fungi and the roots of many vascular plants. Vesicles are present as well as hyphae that penetrate the cells of the root cortex and form branching structures called arbuscules. One of the endomycorrhizas.
see **mycorrhiza**

Vesicular arbuscular mycorrhiza

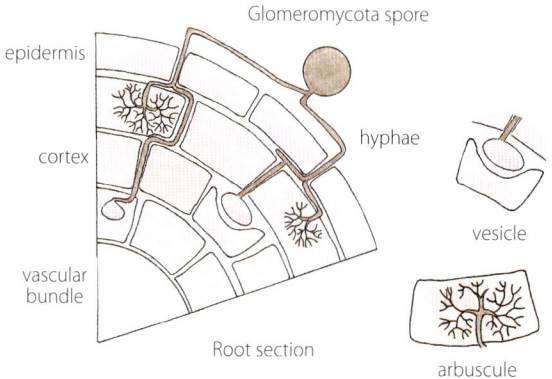

Glomeromycota spore

epidermis

cortex

hyphae

vascular bundle

vesicle

Root section

arbuscule

vespertine
Relating to, occurring or active at dusk, as flowers that open or emit fragrance at this time.
cf. **crepuscular, diurnal, matutinal, nocturnal**

vernation   The arrangement of young leaves in the unopened leaf bud.

Typically, they may be flat or slightly convex, rolled or folded.

= **prefoliation**

see also **ptyxis**

cf. **aestivation**

Vernation

Vernation
Arrangement of young leaves
in the unopened leaf bud.

Vernation equitant.
Ptyxis conduplicate.

Ptyxis
Arrangement of a young leaf
in the unopened leaf bud.

Ptyxis conduplicate.

Leaves flat or slightly convex

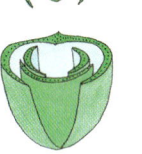

valvate   imbricate   contorted   quincunx   cochlear   opposite   alternate

Leaves rolled

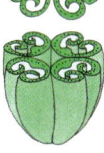

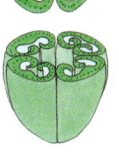

convolute   revolute   involute

Leaves folded

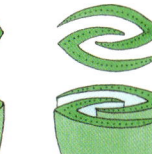

induplicate   equitant   obvolute/half-equitant   triquetrous

---

vessel

In the secondary xylem of angiosperms, the main water-conducting structure composed of a column of dead cells (vessel elements), with the joining ends perforated or totally degraded to form a tube. The woody cell walls are pitted so that water and minerals can also flow sideways from one cell to another.

Gymnosperms generally do not have vessels.

= **trachea**

see **perforation plate**

Vessel

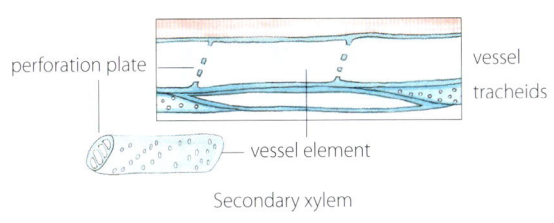

perforation plate   vessel
                           tracheids

vessel element

Secondary xylem

vessel element   A water-conducting cell with lignified walls in the xylem of some angiosperms. The end walls between two cells are perforated or totally degraded to form a perforation plate. A column of dead vessel elements that forms a tube is a vessel.

= **vessel member**

Vessel element

perforation plate — | — vessel tracheids

— vessel element

Secondary xylem

vessel member   Another term for vessel element.

vestigial   A mere trace in one organism of that which is fully developed in another organism, as the unisexual flowers of rice flowers (*Pimelea*) that have vestigial organs of the opposite sex.
Part of an organism that is reduced from the fully developed ancestral condition and is no longer functional.

cf. **abortive, rudimentary**

Vestigial

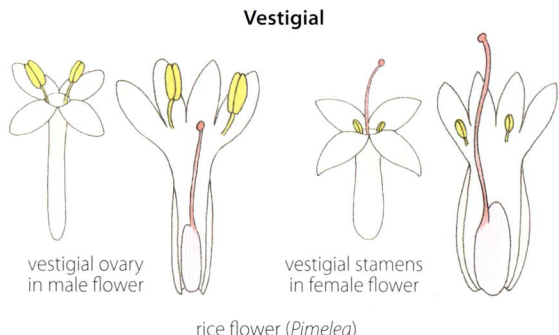

vestigial ovary in male flower

vestigial stamens in female flower

rice flower (*Pimelea*)

vesture, vestiture   A covering on or arising from the surface of a plant or plant part, as hairs, scales, thorns and tubercles.

see also **indumentum**

Some vesture types

matted hairs   warty   scales   glandular hairs

vexillary aestivation
Having five petals with the largest (the standard or vexillum) overlapping the two lateral petals (the wings) which in turn overlap two petals (the keel). Typical of peas (Faboideae).

= **papilionaceous aestivation**

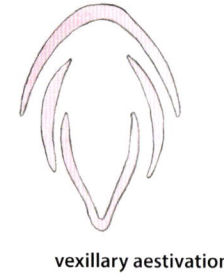

vexillary aestivation

vexillum   The large upper petal of a pea flower (Faboideae).

= **banner, standard**

vexillary   Of or pertaining to the vexillum.

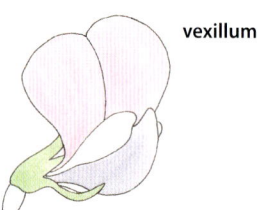

vexillum

villose, villous   Shaggy, with dense long and soft hairs.

cf. **pilose**

villous

vine   A thin-stemmed climber or scrambler, as a grapevine (*Vitis vinifera*), that uses tendrils, hooks or twining shoots for support, usually on another plant.
It may be herbaceous or woody, annual or perennial.

cf. **bine, strangler**

vine

virescence   The appearance of green pigmentation in plant tissues that are not ordinarily green.

virgate   Elongated straight and slender, as the stems of pretty heath (*Epacris virgata*).

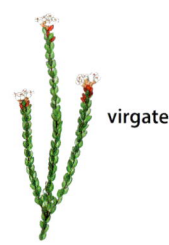

virgate

viscid, viscous
Covered with a
sticky substance.
Thick and sticky in
consistency, as the
secretion from the
glands on the leaves
of sundews (*Drosera*).

viscid

viscidium *pl.* **viscidia**   Present in some orchids, a
sticky pad formed by the rostellum that is removed,
together with the pollinia, when it attaches to a
pollinator.
= **retinaculum**

Viscidium

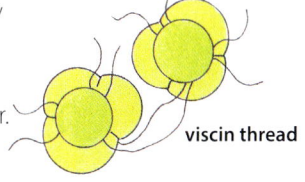

viscidium

stigma

Removal of the pollinia by a pollinator.
sun orchid (*Thelymitra*)

viscin thread   A sticky
thread arising from the
surface of pollen grains
that links them together.

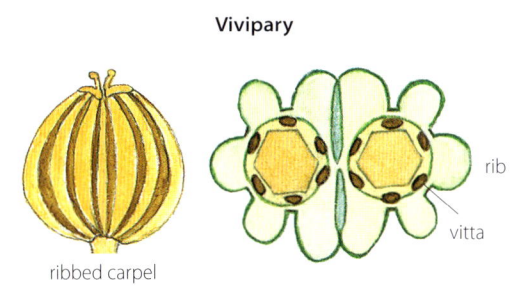

viscin thread

vitreous   Transparent like glass.
cf. **hyaline**

vitta, *pl.* **vittae**   Longitudinal ducts, that secrete
aromatic oils, situated under or between the ribs on
the carpels of the carrot family (Apiaceae).
vittate   Bearing oil tubes (vittae).
Striped longitudinally.

Vivipary

rib

vitta

ribbed carpel

**viviparous
germination**
Producing seeds that
germinate before
becoming detached
from the parent plant,
as the red mangrove
(*Rhizophora mangle*).
Having a single seed
that germinates
within the fruit
before it is shed.
cf. **epigeal germination,
hypogeal germination**

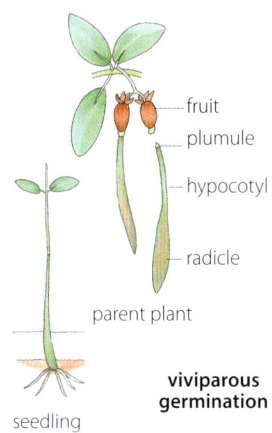

fruit

plumule

hypocotyl

radicle

parent plant

viviparous
germination

seedling

vivipary   The germination of seed before it is shed
from the parent plant, as red mangrove (*Rhizophora
mangle*).
The asexual production of buds that start to
grow while still attached to the parent plant, into
plantlets, as on the leaves of the piggyback plant
(*Tolmiea menziesii*) and on the fronds of some ferns.
viviparous   Reproducing by vivipary.

Vivipary

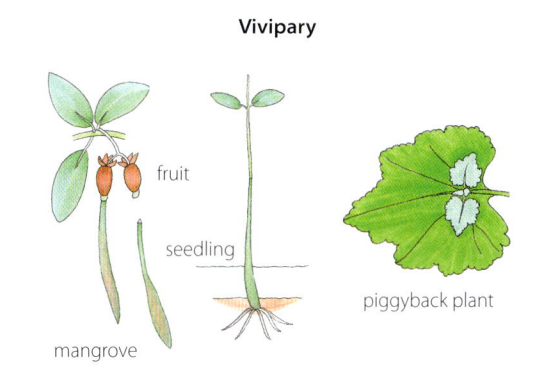

fruit

seedling

piggyback plant

mangrove

volute   Rolled up.
see **involute, revolute**

voucher specimen   A specimen of a plant
preserved and stored in a herbarium for reference
and study.

vulnerable   According to IUCN, a conservation
status covering species that are considered to be
facing a high risk of extinction in the medium term.

wart   A hard or firm
outgrowth.
warty   Having warts
or as though covered
with warts, as the seeds
of some species of
skullcap (*Scutellaria*).

wart

skullcap
seed

wax   An oily water-resistant substance that is solid at room temperature, as beeswax.

waxy   Resembling wax in texture or appearance.

weather   The condition of the atmosphere, including temperature, precipitation, humidity and wind, at a particular place over a short period of time.
cf. **climate**

weed   Any plant, usually one that grows profusely, where it is not wanted, as many thistles.

weeping   Drooping gracefully downward, as the branches and leaves of weeping willow (*Salix babylonica*).

weeping

wetland   A coastal or inland area with water covering the soil, or at or near the surface of the soil, for all or part of the year.
A saline or freshwater transition zone between aquatic and terrestrial ecosystems, as a marsh, swamp or bog, or on the edges of lakes and streams and in river estuaries.

whip   A general term for cirrus and flagellum.
Of palms (Arecaceae), the barbed climbing organ of rattans.

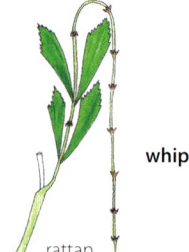

whip

rattan

whisk fern   The whisk fern family (Psilotaceae) comprises one genus (*Psilotum*) of epiphytic or terrestrial vascular plants that reproduce by spores rather than seeds. Stems are undivided or branched dichotomously with small or scale-like leaves.
Spores are of only one kind (homosporous).
see **fern allies**

sporangium

whisk fern

Psilotum

whorl   A circular arrangement of parts around an axis, as leaves or petals around a stem.
= **verticil**
cf. **spiral**
whorled   Arranged in one or more whorls.
= **verticillate**

whorl

widespread   Distributed over a large area or occurring in many different places.

wild   Growing in a natural state, not cultivated.

wing   A thin, often membranous extension, as that on the seeds of elms (*Ulmus*).
One of two clawed lateral petals of a pea flower (Faboideae).
= **ala**

Wing

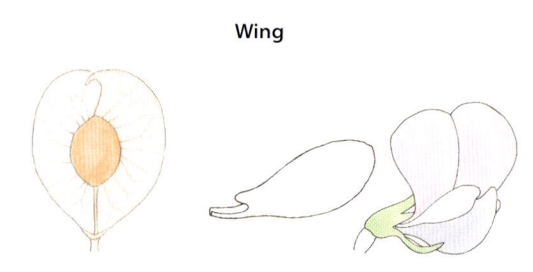

winged seed of elm (*Ulmus*)   lateral petal of pea flower

wood   The secondary xylem that accumulates as a cylinder beneath the bark in the stems, branches and roots of woody plants.
A dense strengthening and nutrient conducting tissue composed of cellulose and lignin.
see **heartwood, sapwood**
see also **woodland**

woodland   A plant community with large trees widely spaced so that their crowns form an open canopy that limits shade and moisture.
The understorey includes grasses, shrubs and herbs.

woolly   Densely covered with long tangled fine soft curly or wavy hair, as the white-woolly bracts of blanket leaf (*Bedfordia arborescens*).
= **lanate**

woolly

xanthophyll   A yellow pigment in the chromoplasts of plant cells that protects them from excessive solar radiation during photosynthesis.
see **carotenoid**

xenogamy   Pollination between flowers on different plants of the same species.
see **allogamy, geitonogamy**
cf. **autogamy**

xeric   Of, relating to or adapted to an environment that receives only a small amount of water, as the Gobi Desert and the Sahara Desert.
cf. **hydric, mesic**

xero-   A prefix meaning dry.

xerochasy   Hygroscopic movement caused by the loss of water in plant parts that are mostly dead, resulting in the opening of follicles, pods and some capsules.
cf. **hydrochasy**
xerochastic   Of or related to xerochasy.

xeromorph   A plant that has the morphological characteristics of a xerophyte but may not be able to resist drought.

xerophyte   A plant adapted to withstand extremely dry conditions, as occurring in deserts and on sand dunes.
Adaptations include surviving drought as seeds and growing only during short rainy periods.
Succulents can store large amounts of water and deep-rooted plants can tap water occurring at great depths.
cf. **mesophyte, tropophyte, xeromorph**

xerosere   An ecological succession that starts on dry, bare land.

xylem   Tissue in a vascular bundle that provides support and conducts water and dissolved nutrients from the roots to other parts of a plant.
There are two types of xylem. Primary xylem is associated with vertical primary growth and forms from procambium. Secondary xylem is associated with lateral secondary growth and is formed from vascular cambium.
Both types consist of less specialised tracheids and more specialised vessel elements that form tubes.

Secondary xylem occurs in woody plants and differentiates into sapwood and heartwood.
see **metaxylem, protoxylem, vascular bundle**
xylar   Relating to xylem.

Xylem

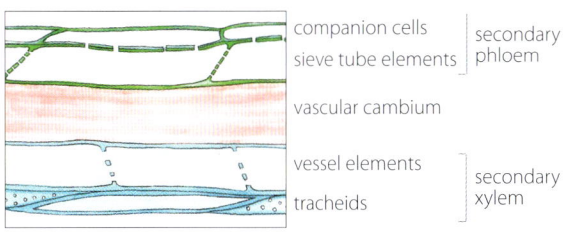

---

zona-   A prefix meaning ring-like.

zoni-, zono-   A prefix meaning located equatorially.

zoochory   Dispersal of pollen, spores, seeds or fruit by animals.
zoochorous   Of or relating to zoochory.

zoophily   Dispersal of pollen and pollination by animals.
zoophilous   Pollinated by animals.

zygomorphic
Divisible through the centre, on one plane only, into exactly similar halves, as flowers in the pea family (Faboideae). Bilaterally symmetrical.
= **monosymmetric**
cf. **actinomorphic**

zygomorphic

pea flower

zygote   The diploid cell that results, at fertilisation, from the fusion of the nuclei of the haploid male gamete (sperm cell) and the haploid female gamete (egg cell).
Once a zygote begins to undergo cellular division it becomes an embryo.

Zygomycota   A phylum of fungi known as pin mould fungi, the 'pin' being the thick-walled resting spore.
Arbuscular mycorrhizas, the most prevalent type of mycorrhiza, develop between plant roots and species of Zygomycota.
see **fungus, mycorrhiza**